Modern Selective Fungicides

Modern Selective Fungicides

– Properties, Applications, Mechanisms of Action –

Editor

Professor Dr. H. Lyr

In cooperation with

Prof. Dr. H. Buchenauer (FRG), Dr. L. C. Davidse (NL), Prof. Dr. J. Dekker (NL), Dr. C. J. Delp (USA), Dr. M. A. De Waard (NL), Dr. W. Edlich (GDR), Dr. M. Gasztonyi (Hungaria), Prof. Dr. S. G. Georgopoulos (Greece), Dr. D. W. Hollomon (UK), Prof. Dr. F. Jacob (GDR), Dr. A. Kerkenaar (NL), Dr. K. H. Kuck (FRG), Dr. M. Kulka (Canada), Dr. S. Lambros (UK), Dr. G. Lorenz (FRG), Dr. A. F. Marchington (UK), Dr. St. Neumann (GDR), Dr. E.-H. Pommer (FRG), Dr. N. Ragsdale (USA), Prof. Dr. H. D. Sisler (USA), Prof. Dr. T. Schewe (GDR), Prof. Dr. H. Scheinpflug (FRG), B. von Schmeling (USA), Dr. H. H. Schmidt (GDR), Dr. B. Schreiber (FRG), Prof. Dr. F. Schwinn (Switzerland), Dr. T. Staub (Switzerland)

Longman Scientific & Technical
Copublished in the United States with
John Wiley & Sons, Inc., New York

Longman Scientific & Technical,
Longman Group UK Limited,
Longman House, Burnt Mill, Harlow,
Essex CM20 2JE, England
and Associated Companies throughout the world.

Copublished in the United States with
John Wiley & Sons, Inc., 605 Third Avenue, New York, NY 10158

First published 1987

British Library Cataloguing in Publication Data
Modern selective fungicides : actions,
applications and mechanisms of action.
1. Fungicides
I. Lyr, Horst
668'.652 SB951.3
ISBN 0-582-00461-6

Library of Congress Cataloguing in Publication Data
Modern selective fungicides.
Bibliography: p. 383
Includes index.
1. Fungicides. I. Lyr, H. (Horst) II. Title:
Selective fungicides.
SB951.3.M63 1987 632'.952 86-27442
ISBN 0-470-20799-X

Set in Modern Extended
Printed in the German Democratic Republic
by Magnus Poser, Jena

List of Contributors

Prof. Dr. Heinrich Buchenauer
Institut für Pflanzenkrankheiten und Pflanzenschutz der Universität Hannover
3000 Hannover 21, Herrenhäuserstr. 2
Federal Republic of Germany (FRG)

Dr. Leen C. Davidse
Department of Phytopathology, Agricultural University Wageningen
6709 PD Wageningen, Binnenhaven 9
The Netherlands

Prof. Dr. Johan Dekker
Head of the Department of Phytopathology of the Agricultural University Wageningen
6709 PD Wageningen, Binnenhaven 9
The Netherlands

Dr. Marten A. De Waard
Department of Phytopathology, Agricultural University Wageningen
6709 PD Wageningen, Binnenhaven 9
The Netherlands

Dr. Charles J. Delp
Washington D.C. 20003, 145 Kentucky Ave. SE
USA

Dr. Wilfried Edlich
Institute for Plant Protection Research Kleinmachnow of the Academy
of Agricultural Sciences of the GDR
DDR - 1532 Kleinmachnow, Stahnsdorfer Damm 81
German Democratic Republic (GDR)

Dr. Maya Gasztonyi
Plant Protection Institute, Hungarian Academy of Sciences
H - 1525 Budapest, Hermann Otto ut 15
Hungaria

Prof. Dr. S. G. Georgopoulos
Head of the Laboratory of Plant Pathology, Athens College of Agricultural Sciences,
Votanikos Gr.
11855 Athens
Greece

Dr. Derek Hollomon
Long Ashton Research Station, University of Bristol, Department of Agricultural Sciences
Bristol BS18 9AF
United Kingdom (England)

Prof. Dr. Friedrich Jacob
Department of Biosciences, Martin-Luther-University Halle–Wittenberg
4020 Halle/Saale, Am Kirchtor 1
German Democratic Republic (GDR)

Dr. Antonius Kerkenaar
TNO Institute of Applied Chemistry, Division of Technology for Society, Group leader of Microbiology
3700 AC Zeist
The Netherlands

Dr. K. H. Kuck
Institute for Plant Diseases, Pflanzenschutzzentrum Monheim Bayer AG
D - 5090 Leverkusen, Bayerwerke
Federal Republic of Germany (FRG)

Dr. Marshall Kulka
Uniroyal Ltd., Research Laboratories
Guelph/Ontario N1H 6N3, 120 Huron Street
Canada

Dr. Sandra Lambros
Imperial Chemical Industries PLC (ICI)
Plant Protection Division, Jealotts Hill Research Station
Bracknell, Berkshire RG12 6EY
United Kingdom (England)

Dr. Gisela Lorenz
Research Department of Microbiology BASF AG, Agricultural Research Station
D - 6703 Limburgerhof
Federal Republic of Germany (FRG)

Prof. Dr. Horst Lyr
Head of the Department of Pesticide Research, Institute for Plant Protection Research Kleinmachnow of the Academy of Agricultural Sciences of the GDR
DDR - 1532 Kleinmachnow, Stahnsdorfer Damm 81
German Democratic Republic (GDR)

Dr. A. F. Marchington
Imperial Chemical Industries PLC (ICI), Plant Protection Division Fenhurst
Haslemere Surrey GU27 3JE
United Kingdom (England)

Dr. Stefanie Neumann
Department of Biosciences, Martin-Luther-University Halle–Wittenberg
4020 Halle/Saale, Am Kirchtor 1
German Democratic Republic (GDR)

Dr. Ernst-Heinrich Pommer
Head of the Research Department of Microbiology, BASF AG,
Agricultural Research Station
D - 6703 Limburgerhof
Federal Republic of Germany (FRG)

Dr. Nancy N. Ragsdale
Pesticide Coordinator, Cooperative State Research Service, USAD
Washington DC 20251
USA

Prof. Dr. Hans Scheinpflug
Head of the Institute for Plant Diseases, Pflanzenschutzzentrum Monheim, Bayer AG
D - 5090 Leverkusen, Bayerwerke
Federal Republic of Germany (FRG)

Prof. Dr. Tankred Schewe
Institute of Biochemistry of the Humboldt-University of Berlin
DDR - 1040 Berlin, Hessische Str. 3—4
German Democratic Republic (GDR)

Boe von Schmeling
Vice President Crop Protection Technology, Chemical Group Uniroyal Ldt.
World Headquaters, Middlebury CT 06749
USA

Dr. Hans Hermann Schmidt
Institute for Plant Protection Research Kleinmachnow of the
Academy of Agricultural Sciences of the GDR
DDR - 1532 Kleinmachnow, Stahnsdorfer Damm 81
German Democratic Republic (GDR)

Dr. Bernhard Schreiber
Department Agricultural Development, Hoechst AG
D - 6234 Hattersheim am Main, Hessendamm 1—3
Federal Republic of Germany (FRG)

Prof. Dr. Franz J. Schwinn
Agricultural Division Ciba-Geigy SA and Institute of Botany of the
University of Basle
CH - 4002 Basle
Switzerland

Prof. Dr. Hugh D. Sisler
Department of Botany, University of Maryland, College Park
Maryland 20742
USA

Dr. Theo Staub, Phytopathologist
Ciba-Geigy AG
CH - 4002 Basle
Switzerland

Contents

Pages

Foreword 10
D. Spaar/GDR

Preface 11
H. Lyr/GDR

1. Principles of uptake and systemic transport of fungicides within the plant . . . . 13—30
F. Jacob and St. Neumann/GDR

2. Selectivity in modern fungicides and its basis. . . . . 31—38
H. Lyr/GDR

3. Development of resistance to modern fungicides and strategies for its avoidance. . . . . 39—52
J. Dekker/NL

4. The genetics of fungicide resistance . . . . 53—62
S. G. Georgopoulos/Greece

5. Aromatic hydrocarbon fungicides . . . . 63—74
H. Lyr/GDR

6. Mechanism of action of aromatic hydrocarbon fungicides. . . . . 75—90
H. Lyr/GDR

7. Dicarboximide fungicides . . . . 91—106
E.-H. Pommer and G. Lorenz/FRG

8. Mechanism of action of dicarboximide fungicides . . . . 107—118
W. Edlich and H. Lyr/GDR

9. Carboxin fungicides and related compounds . . . . 119—132
M. Kulka and B. v. Schmeling/Canada — USA

10. Mechanism of action of carboxin fungicides and related compounds 133—142
T. Schewe and H. Lyr/GDR

11. Morpholine fungicides . . . . 143—158
E.-H. Pommer/FRG

12. Mechanism of action of morpholine fungicides . . . . 159—172
A. Kerkenaar/NL

13. Sterol biosynthesis inhibiting piperazine, pyridine, pyrimidine and azole fungicides. . . . . 173—204
H. Scheinpflug and K. H. Kuck/FRG

14. Mechanism of action of triazolyl fungicides and related compounds 205—232
H. Buchenauer/FRG

15. Benzimidazole and related fungicides . . . . 233—244
C. J. Delp/USA

Foreword

Even today, crop losses in agriculture that are due to biotic pests are estimated at about 35 per cent worldwide. About 12 per cent of these losses are caused by pathogens. Phytopathogenic fungi continue to be the most important group of pathogens. Efficient control of these pathogens is an urgent necessity if food is to be provided for the growing world population. The control of pathogenic fungi in agricultural crops will be increasingly fitted into integrated control strategies that are based on comprehensive use of soil and plant hygiene techniques, cultivation of resistant and tolerant varieties, combination of biological and chemical control on the basis of precise monitoring and forecasting, and objective decisions on control action on the basis of ecological and economical threshold values. In such a plant protection strategy, fungicides will keep well in the future their prominent position among the active, direct and purpose-oriented measurements of control. They will be used even more widely in modern high-intensity agriculture. The concept of integrated pest management will contribute to a higher effectiveness of fungicide use and, at the same time, be more demanding in terms of the efficiency and selectivity of these preparations and their application. The research and use of fungicides is undergoing a dynamic development. Intensive research work is being carried out to prevent breakdowns in effectiveness due to resistance development. Biochemical and plant physiological research is extending our understanding of the adhesion, penetration, mobility and stability of fungicides, and of their side-effects in plant, soil and biocoenosis, and hence offers starting points for new developments. By a more detailed investigation of the differential biochemistry of phytopathogenic organisms it will be possible also to develop preparations against problematic pathogen groups that are still difficult to control. And last but not least, the elucidation of biochemical causal processes of structure and action will help accelerate the systematic development of new efficient fungicides. In view of the rapid development of that field of science it meets with general approval that the editor of this book together with his competent co-authors from nine countries, has provided a survey of the activity, application and mechanism of action of modern fungicides, conveying most recent findings in that specific area of knowledge. May this useful book find good acceptance and wide distribution in professional circles.

D. Spaar
1. Vice President
of the Academy of Agricultural Sciences
of the German Democratic Republic

Preface

Although several excellent books on fungicides have been published during the past two decades, the editor and all contributors of this book felt the need to issue a comprehensive, up to date survey on modern, selective fungicides which deals with practical applications, the mechanism of action and problems of acquired resistance.

During the past few years major progress has been made in all these fields which is of importance from a practical as well as from a theoretical point of view. Modern fungicides are powerful tools for increasing agricultural productivity and making more food available for an increasing world population. Of course we are aware that it is difficult to say which fungicides are modern and which are not. Some fungicides have been in use for many years, whereas newer ones have disappeared from the market for various reasons and older ones have come back to fill the gap in the market. The build up of resistance in fungal populations is a major driving force for the development of new compounds as well as for using older compounds with a low risk for resistance problems. Therefore, we have also considered some of the more classic compounds in this book, either because of their practical importance or because of the theoretical value regarding their interesting mechanism of action.

Although there are many excellent compounds on the market, a need still exists for new fungicides that give a more efficient control of diseases. For example phloem mobility, low risk for resistance development and high disease control activity combined with low mammalian toxicity are some improvements that can be made.

By using selective fungicides as a tool for biochemical investigations we learned a lot about fungal biochemistry and specific differences between various genera and species. At present, we are still far from a complete understanding of the differential biochemistry of fungi, and therefore the directed design of new antifungal compounds is still very limited, but this approach is increasing in importance.

We include also discussions of fungal control by antifungal compounds acting indirectly as well as problems of synergism and antagonism which are gaining practical interest.

I wish to thank all contributors to this book, who are internationally known experts, for their cooperation and contributions as well as for their stimulating ideas for the design of the book, which we hope will fulfil the main demands of the readers.

H. Lyr

Chapter 1

Principles of uptake and systemic transport of fungicides within the plant

F. JACOB and St. NEUMANN
Department of Biosciences, Martin-Luther-University, Halle-Wittenberg, GDR

Introduction

Although several protective fungicides, which do not penetrate into the plant have a stable place on the fungicide market, most of the modern, selective fungicides have systemic properties or at least are locally mobile. A necessary prerequisite is a sufficient selectivity which allows control of the fungal pathogen without damaging the host organism. Therefore, the main principles of penetration and transport of xenobiotics, especially fungicides, are summarized in this chapter.

The specific behaviour of the various fungicides is described in chapters 5—21.

Reviews and general descriptions

In the past fifteen years results of studies on uptake and translocation of xenobiotic substances in higher plants were repeatedly summarized (CRAFTS and CRISP 1971; CRISP 1972; JACOB *et al.* 1973; CROWDY 1973; SHEPHARD 1973; PRICE 1977, 1979; CHRIST 1979; HARTLEY and GRAHAM-BRYCE 1980; JACOB and NEUMANN 1983; NEUMANN *et al.* 1985). A special interest was taken in herbicides, because of their systemic distribution is performed not only in the xylem system but partly also in the phloem system. Some reports dealt with the systemic behaviour and pattern of distribution of fungicides: EVANS (1971), VAN DE KERK (1971), CROWDY (1972), ERWIN (1973), GROSSMANN (1974), FEHRMANN (1976), PETERSON and EDGINGTON (1977), EDGINGTON (1981), and HASSALL (1982).

Terminology of transport processes

Transport in tissues

1. apoplastic — transport in the apoplast, the coherent network of free space, cell walls and non-living cells

1.1. euapoplastic — apoplastic movement without any passage through protoplasts

1.2. pseudoapoplastic — apoplastic movement with occasional passage through or retention in protoplasts

2. symplastic — transport in the symplast, the coherent network of protoplasts connected by plasmodesmata

Long-distance transport

1. xylem-mobile — apoplastic transport in vessels and tracheids of xylem by means of the transpiration stream (xylem-systemic)*
2. phloem-mobile — symplastic transport in the sieve tubes of the phloem by means of the mass flow from source to sink (phloem-systemic)
3. ambimobile — transport in xylem and phloem (ambisystemic)
4. locally mobile — transport within the organ of application (locosystemic)
5. amobile — no long-distance transport from site of application (non-systemic)

General aspects

The chemical control of phytopathogenic fungi has to be adapted to the biology of the pathogen and the sites of its occurrence. Establishment within the plant tissue requires systemic fungicides, taken up by leaves, roots, seeds or fruits and translocated a short distance in the parenchym (locosystemic) or a long distance in the flow of solutions through the xylem (xylem-systemic) or the phloem (phloem-systemic) or both (ambisystemic). The two ways of long-distance transport differ in their anatomic properties and their physiological mechanisms. The applied substances are dependent on the engaged transport system and are subjected to characteristic patterns of distribution which are important for practical use. To determine comparatively the mode of mobility a test with *Sinapis alba* seedlings was developed (Jacob and Neumann 1983; Neumann *et al.* 1985). The obtained quotient of translocation Q_{tr} allows the relationship between phloem and xylem mobility to be characterized by means of a numerical value. In this way it was possible to prove the existence of all transitional stages from phloem mobility through ambimobility to xylem mobility, using an artificial classification of systemic chemicals suitable for practical use.

The equal distribution of substances on the whole leaf surface is difficult to arrange. Therefore, systemic fungicides by their possibilities for redistribution of fungicides may ensure a better disease control. But for several reasons (costs, resistance problems etc.), both systemic and non-systemic antifungal compounds will be applied in future. The use of systemic fungicides obviously has advantages as well as disadvantages, as evidenced by increased rates of decomposition and phytotoxic or growth-regulating side effects. The development of resistance (cf. chapter 3) in fungi for instance to benomyl (Dekker 1977) and thiabendazole (Georgopoulos 1977) can lead to ineffectivity of some compounds for a practical disease control (monosite inhibitors).

Locosystemic fungicides show a limited spread from the site of application into the plant and the mobility behaviour is intermediate between systemic and non-systemic. Whether the penetration takes place apoplastically or symplastically (or both) is very often unclear. The systemic fungicides move *via* uptake, translocation and distribution inside the plant to reach the phytopathogenic fungus. A successful application, therefore, requires certain properties of mobility as well as stronger fungicidal activity. All outer cell wall layers, cells, and intercellular spaces which have to be

*) The term "xylem-systemic" characterizes the ability of an absorbed pesticide to furnish the whole plant with the pesticidal activity by means of the xylem transport; "xylem-mobile" indicates the transport in the xylem of any substance.

crossed by a compound may influence the strength of the fungicidal effect by chemical or physical interactions with the active substances. A diminished effect can be the consequence of adsorption, chemical binding or metabolic decomposition, an improved result may be effected by the translocation flow in the plant, an accumulation in the area of targets or metabolic release of activated substances.

Till now the absorption of fungicides by pathogenic fungi in host-parasite combinations has not been investigated intensively in a quantitative respect. Undoubtedly non-polar substances and bases as well as acids in non-ionic forms are favoured in penetrating the lipo-proteinic plasmalemma. Xenobiotic compounds are taken up more easely by germ hyphae than by spores (HASSALL 1982). Pesticides as xenobiotics penetrate mainly passively through the cell membrane and — with few exceptions — not by a carrier-mediated transport. Differences in chemical structures result in different physico-chemical properties such as lipo-hydrophilicity, solubility, mol volume, steric parameter, ionization and distribution of charge density. Such features influence the behaviour with regard to uptake and transport in plants. Metabolic changes as well as non-metabolic changes of the compounds modify these properties and by this, often the mobility. There is a close connection between mobility, stability and strength of effect. The relevant concentration of a chemical in the tissue of higher plants and in the cells of parasitic fungi is determined by the equilibrium between rate of absorption and rate of destruction.

While some principles of the relation between structure of molecules and mobility in higher plants are well known and can be explained, we are still far away from being able to conclude from certain antifungal properties of a chemical compound on structure requirements for a favourable mobility. By suitable formulations the absorption into the plant undoubtedly can be increased thereby possibly the translocation of the compound. But previous experiences show that significant changes in mobility behaviour cannot be expected.

Translocation in the xylem

The pathway in the apoplast

Xylem translocation takes place in tracheids and vessels forming a continous network of dead elongated cells with more or less lignified secondary cell walls. Pattern and extent of movement in the xylem are determined by the gradient in the water potential between soil and air. Thus, within the plant, transport is usually directed from the root to transpiring areas, especially leaves. Factors controlling the intensity of transpiration (relative humidity, temperature, light, and phytohormones (especially ABA) also have some influence on the translocation rate, including the distribution of xenobiotics dissolved in the xylem sap.

The distribution of xylem-mobile xenobiotics is characterized by the following peculiarities (PETERSON and EDGINGTON 1975a).

1. Substances are accumulated at the sites of high transpiration, e.g. at the tips and margins of leaves.
2. Translocation occurs only to a limited extent into plant organs with negligible transpiration, e.g. fruits and young leaves.
3. Xylem-mobile substances do not undergo any downward movement from expanded leaves. If they are applied to the base of a leaf, they are mainly transported to the tip. Transport in the opposite direction is extremly rare.

Xylem- mobile xenobiotics are absorbed along with water, mainly in the root hair zone, this means 5—50 mm from the root tip. In some cases the absorption of xenobiotics seems to be more intensive than that of water (SHONE and WOOD 1974). In contrast to the leaves the rhizodermis is not covered by a well developed cuticula and is therefore not expected to be a barrier for the passage of xenobiotics to the root (ESAU 1965). Radial movement through the cortex zone occurs either symplastically or apoplastically (Fig. 1.1). Symplastically moving substances cross the plasmalemma and are transported via protoplasts and plasmodesmata of cortex cells and than through the endodermis cells to the vessels of the xylem. Transfer in the apoplast is supposed to be accomplished in the free space of the cell walls. Since this route is blocked by lipophilic incrustations of the Casparian strips at the endodermis level, xenobiotics should enter the symplast at this site to be transferred to the xylem. Entrance into the symplast requires lipophilic properties, therefore extremely hydrophilic compounds like the fluorescent dye tri-sodium-3-hydroxy-5,8,10-pyrene-

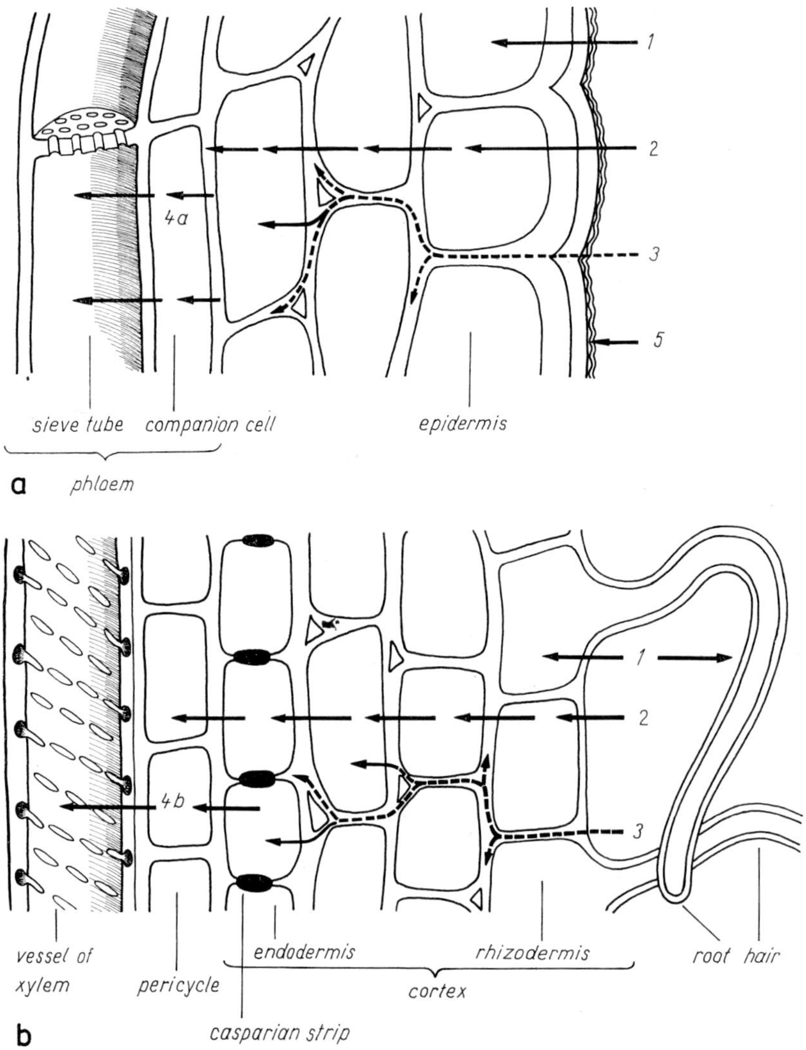

Fig. 1.1 The pathway of xenobiotic substances in plant tissue by penetrating (a) the leaf or (b) the root surface. 1 uptake in outer cells, 2 symplastic transport, 3 apoplastic transport, 4a absorption in sieve tubes, 4b transfer to vessels, 5 absorption in cuticular layers.

trisulfonate (PTS) are retained in the free space of the cortex. Thus with one exception they are not translocated in the long distance stream of the xylem to the leaves: As far as the endodermis continuity is disturbed by emerging lateral branches of the root (BONNET 1969; DUMBROFF and PEIRSON 1971; KARAS and MCCULLY 1973) temporary apoplastic connections seem to be responsible for the appearance of highly hydrophilic substances in the xylem in very poor concentrations (PETERSON and EDGINGTON 1975b; PETERSON *et al.* 1981).

Uptake of xenobiotics by roots

Time courses of uptake of fungicides (CROWDY and RUDD-JONES 1956a; LEROUX and GREDT 1983; DOBE *et al.* 1986) and herbicides (MOODY *et al.* 1970; SHONE *et al.* 1974; DONALDSON *et al.* 1973) were initially found to be rapid, followed by a phase of declining rate of absorption. This initial part represents the uptake into the apoplast, the second is correlated to translocation of the absorbed substances from the root to the leaves depending on the extent of transpiration (MINSHALL 1954; CROWDY and RUDD-JONES 1956b; SHONE and WOOD 1972; WALKER and FEATHERSTONE 1973). Linear concentration response of absorption, low temperature coefficient and insensibility to metabolic inhibitors suggest uptake by diffusion (SHONE and WOOD 1974; DONALDSON *et al.* 1973; LEROUX and GREDT 1983; DOBE *et al.* 1986). In hydroponic culture the concentration of xenobiotics that is actually available for absorption by the plants entirely corresponds to the concentration applied. In contrast various physical and chemical interactions between xenobiotic compounds and soil matter affect the quantities available for uptake following root application (BRIGGS 1973; HANCE 1983). Due to the pH-value of the soil solution the cationic moieties of dissociating fungicides like carbendazim are adsorbed to a certain extent by negatively charged soil matter resulting in a smaller amount of molecules being accessible for uptake (AHARONSON and KAFKAFI 1975; AUSTIN and BRIGGS 1976). An analogous effect can be attributed to hydrophobic interaction between non-ionised lipophilic molecules and soil components. The adsorption of non-ionised compounds was shown to be positively correlated to the hydrophilic-lipophilic balance of substances characterized by the log P-values (octanol-water partition coefficient) (BRIGGS 1973; BRIGGS *et al.* 1977). The derivatives of the fungicide isoprothiolane are adsorbed better, the longer the alkyl side chain is, the move is the concentration in the soil solution is accordingly decreased (UCHIDA and KASAI 1980; UCHIDA and SUZUKI 1982). From studies with oxymecarbamate and phenylureas, a log P of 0.5 was found to be optimal for root uptake following soil application. In hydroponic culture an optimum uptake can be expected for compounds with log P 1.7 (BRIGGS *et al.* 1977). The uptake of xenobiotics by the root can be quantitatively characterized by the root concentration factor (RCF) defined by SHONE and WOOD (1974) as

$$\text{RCF} = \frac{\text{concentration in the root}}{\text{concentration in the external solution}}$$

The RCF is usually independent of the concentration in the external solution, supporting the view that entry is accomplished by diffusion. Furthermore log P and RCF are positively correlated: Increasing lipophilicity is linked with increasing absorption (SHONE and WOOD 1974; BRIGGS *et al.* 1977). This is in accordance with findings by LEROUX and GREDT (1983) who demonstrated the pH- dependent uptake of benzimidazoles and thiophanates. The same refers to the uptake of ethirimol investigated by SHONE and WOOD (1974). RCF-values greater than unity indicate accumulation and are due to association of lipophilic uncharged xenobiotics with lipophilic cell components. Cationic fungicides like carbendazim ($pK_a = 4.1$), thiabendazole ($pK_a = 4.7$) and ethirimol ($pK_a = 4.8$) which became protonated in physiological pH ranges are, independently of lipophilic interactions, also accumulated by adsorption to negative charges of the cell wall. Uptake of these compounds is obviously pH dependent and was inhibited by addition of bivalent cations (Ca^{2+} and Mg^{2+}) (Leroux and Gredt 1983; Shone *et al.* 1974). Briggs *et al.* (1982) suggested that there exists a general relationship between RCF and P expressed by the function $\log(RCF - 0.82) = 0.77 \log P - 1.52$, which is valid for young roots without lignified cell walls. Subsequently the lipophilic nature of root constituents involved in adsorption of xenobiotics, seems to be similar among different plant species, at least in young stages of root development.

Properties of xylem-systemic substances

According to their physico-chemical properties, xenobiotics attain a characteristic equilibrium of adsorbed moieties, and those which are freely mobile in the root apoplast and symplast. This equilibrium is responsible for the extent of translocation from the root to the shoot. Since xylem translocation is preferably determined by transpiration, its efficiency can be described by the so-called "transpiration stream concentration factor" (TSCF) (Shone and Wood 1974)

$$TSCF = \frac{\text{concentration in transpiration stream}}{\text{concentration in external solution}}$$

TSCF values are independent of the concentration in the external solution. They are always less than unity and those of non-ionised compounds are correlated to log P.

From investigations by Crowdy *et al.* (1959) on mobility of griseofulvin derivatives after root application in *Vicia faba*, it was concluded that substances with a hexan water partition coefficient of 0.3 are favourably transported and 0.8 are sufficiently transported in xylem, whereas P values lower than 6 indicate amobility and retention in the root. Transport of oxymecarbamates and phenylurea derivatives in *Hordeum* (Briggs *et al.* 1977) as well as dialkyl-1,3-dithiolan-2-ylidenemalonates (derivatives of the fungicide isoprothiolane) in rice (Uchida 1980) show a similar correlation between partition coefficient and transport in xylem. Briggs *et al.* (1977, 1982 and 1983) demonstrated further, that rates of translocation, expressed as TSCF-values, against partition coefficient yield a Gaussian curve: A sharp increase of TSCF was found, when log P exceeds —0.5, a decline was noted above 3 and the optimum lipophilicity-hydrophilicity balance is around 1.7. Uchida (1980) also showed a log P dependent increase of transport in xylem up to 2.9, substances with log P larger than 4 are not translocated. Furthermore the group of analogous investigations and results includes studies on transport of atrazine and linuron in lettuce, parsnip, carrots and turnip (Walker and Featherstone 1973). Summarizing this, transport in the xylem is correlated with partition coefficients. Retention of lipophilic xenobiotics occurs by adsorption to cell constituents in the root and additionally in the lower and middle parts of the stem. Substances of strong lipophilicity never appear in transpiring leaves. The question which arises is, how are lipophilic compounds adsorbed in the root and along the pathway of the water stream? Lignin is generally accepted to play a decisive role in adsorption causing retention. Barak *et al.* (1983a, b) found a linear correlation between log P and adsorption of the fungicides carbendazim, triarimol, triadimefon, nuarimol, fenarimol (cf. chapter 13, 15) and of the herbicide fluometuron to "ground stems" as well as to purified lignin of *Phaseolus vulgaris, Capsicum annuum* and *Gossypium hirsutum*. Association to cellulose, polygalacturonic acids and proteins seems to be negligible. Nevertheless methylation of "ground stems" resulted in a two- to threefold increase of adsorption. Both hydrophylic interaction of undissociated carbendazim molecules with lignin and the association of their protonated moieties with negative charges of the cell wall are pH-dependent.

Ca^{2+} inhibited adsorption to *Pinus* lignin. The latter phenomenon also became obvious from experiments with cationic dyes (VAN BEL 1978), cationic xenobiotics (EDGINGTON and DIMOND 1964; WILHELM and KNÖSEL 1976) and positively charged amino acids (HILL-COTTINGHAM and LLOYD-JONES 1968, 1973). Translocation in the xylem, limited by adsorptive binding to cell wall charges, was promoted by the addition of Ca^{2+} or Mg^{2+}. Due to a proceeding lignification, adsorption of xenobiotics on lignin increases with the age of stems, which could be the reason for differences in transport of ethirimol and diethirimol in herbaceous and woody plants (SHEPHARD 1973).

Practical consequences

Absorption of fungicides or other xenobiotics by the root, and subsequent transport in the xylem, result from different consecutive stages of partition between hydrophilic and lipophilic compartments. From the particular knowledge of these processes information becomes available about optimum partition coefficients, pH values in the medium, desirable responses and sites of action. Considering the distribution of xenobiotics in soil and plants, BRIGGS *et al.* (1977) concluded that xylem-systemic fungicides should on the one hand exhibit log P in the range between —0.5 to 3, if responses are expected in leaves. On the other hand more lipophilic compounds (log P 3—5) seem to be favourable for acting in roots because of confined movement. Generally speaking, hydrophilic compounds are transferred through the soil and rapidly absorbed by the root whereas lipophilic chemicals undergo a retarded movement in soil and long-lasting uptake into the plant. Nevertheless, two important points have not get been considered, the transport limitation of lipophilic substances in xylem, which is dependent on the age of the plants, and the proportion of the xenobiotic which is effective.

Translocation in the phloem

The pathway through the sieve tubes

Phloem transport occurs in the continuous network of sieve tubes, consisting of elongated living cells (sieve elements) with characteristic cell wall perforations, the so-called sieve areas or sieve plates. Both the nucleus and tonoplast have disappeared in completely differentiated sieve elements. The pH of the sieve tube protoplasm amounts to 7.2—8.7, which is markedly higher than in surrounding parenchyma cells. This pH is maintained by the activity of a plasmalemma-bound ATPase pumping protons into the apoplast. Thus the apoplast pH is about 5 (GIAQUINTA 1983). Sieve elements build with companion cells a physiological unit: the sieve element-companion cell complex (se-cc complex). Both types of cells seem to be involved in the loading of assimilates, whereas the longitudinal movement, i.e. the long distance stream in the phloem, takes place in the sieve tubes exclusively. Phloem transport is accomplished by the physiological gradient between sites of assimilate production and export. Usually fully expanded green leaves are sources and growing or storing organs with demand for photosynthates are sinks. The distribution patterns of phloem and ambimobile xenobiotics are characterized by the following peculiarities (CRAFTS and CRISP 1971; MÜLLER 1976):

1. According to the developmental stage and its source-sink relationship, xenobiotics are translocated in phloem only together with photosynthates to sites of assimilate demand.

2. As soon as assimilate transport ceases the movement of xenobiotics also stops. Furthermore movement into fully expanded exporting leaves is absent.

3. Following root application only small amounts of phloemmobile compounds are transported to upper parts, whereas ambimobile compounds show the characteristic distribution patterns of xylem-systemic xenobiotics.

In contrast to xylem-mobile substances, which do not exhibit common structural features, xenobiotics mobile in the phloem (phloem- and ambimobile) are usually acids. In many cases the acidity depends on the presence of carboxylic groups (phenoxy acids) and also phenolic hydroxy groups (maleic hydrazide) or on the arrangement of $—SO_2—NH—$ in so-called nitrogen acids (mefluidid, asulam) (Jacob *et al.* 1973). It was shown that phloem mobility disappears with elimination of COOH groups. Conversely the introduction of carboxylic groups into xylem-mobile chemicals resulted in ambimobile derivatives (Mitchell *et al.* 1954; Jacob *et al.* 1973; Crisp and Look 1979; Jacob and Neumann 1983; Neumann *et al.* 1985).

Xenobiotica uptake into the leaf

Unlike the rhizodermis of the root, the epidermis as surface layer of aerial plant parts si not adapted for absorption of dissolved substances. The leaf epidermis with stomata and cuticula enables the regulation of the gaseous exchange avoiding an appreciable loss of water. The cuticula covering leaves as a 0.5—14 μm thick layer consists of cutin, intracuticular lipids, polysaccharides, polypeptides and phenols (Martin and Juniper 1970; Norris 1974; Schönherr and Huber 1977; Hunt and Baker 1980; Wattendorf and Holloway 1980; Holloway 1982). Epicuticular waxes impede the adhesion of compounds in aqueous solution on the surface of leaves. Adhesion

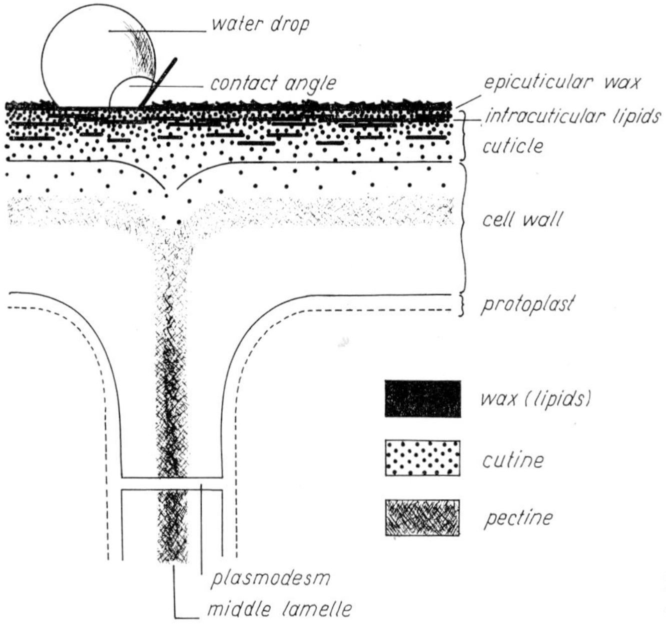

Fig. 1.2 Structure of a leaf epidermis.

can be quantified by the so-called contact angle (HOLLOWAY 1970; RENTSCHLER 1971; NETTING and WETTSTEIN-KNOWLES 1973) (Fig. 1.2).

Nevertheless the main barrier for the penetration of polar substances through the cuticula are the intracuticular lipids, because of this, penetration rates vary greatly according to the proportion of intracuticular lipids in the cuticula of different plant species. Additionally intracuticular lipids are expected to retain lipophilic xenobiotics and consequently to have some depot effect (RIEDERER and SCHÖNHERR 1984, 1985). Consequently undissociated molecules of acidic compounds penetrate preferably. Entrance via the abaxial epidermis was found to be more efficient than through the adaxial one, facilitated by a lower content of intracuticular lipids and a higher number of stomata (NORRIS and BUKOVAC 1968). The promoting impact of stomata seems to be due to the absence of epicuticular waxes in the stomatal chamber, in which there is a high relative humidity (DYBING and CURRIER 1961; PRASAD *et al.* 1967; MANSFIELD *et al.* 1983). Generally, high humidity and other environmental factors which cause opening of stomata enhance absorption (SHARMA and VAN DEN BORN 1973; GREENE and BUKOVAC 1974). Within the leaf, transfer to the conducting bundles can be carried out either symplastically or apoplastically. Xenobiotics are probably taken up into the sieve tubes of minor veins from the apoplast in the same way as photosynthates (GIAQUINTA 1983).

Uptake of the xenobiotics by the phloem

In contrast to the loading of sucrose by a carrier mediated proton-cotransport (GIAQUINTA 1983), the uptake of xenobiotics into the se-cc-complex is accomplished by a diffusion (GRIMM *et al.* 1983). Both phloem- and xylem-mobile substances enter the sieve tube, but xylem-mobile ones are leaked out very rapidly. The velocity of uptake was found to be proportional to log P. Phloem- and ambimobile xenobiotics can usually be accumulated by an ion trapping mechanism e.g. MCPA absorption by isolated conducting tissue of *Cyclamen persicum* results in a three- to fourfold accumulation. Following leaf application in *Sinapis alba* 98.5 % of the absorbed xylem-mobile defenuron is retained in the treated cotyledon. By introduction of a —COOH group defenuron is converted to the ambimobile N-(p-carboxyphenyl)-N′-methylurea (CPMU). Both compounds were taken up by isolated conducting tissue of *Cyclamen persicum* in linear proportion to the external concentration without any accumulation. However in efflux experiments defenuron, as well as other xylem-mobile xenobiotics (SHONE and WOOD 1974; SHONE *et al.* 1974; BOULWARE and CAMPER 1973), was leaked out very quickly in contrast to the moderate escape of CPMU. The total concentration in the tissue after leaching for 2 h still amounted to 32 % of the ambimobile compound but only 2.4 % of the xylemmobile one. 2,4-D, maleic hydrazide and amitrole were also found to be leaked out moderately from *Cyclamen* conducting tissue (GRIMM *et al.* 1985). Similar experiments with potato tuber disks revealed a markedly slower escape of phloem-mobile substances compared with xylem-mobile ones (EDGINGTON and PETERSON 1977; MARTIN and EDGINGTON 1981). The moderate efflux seems to be a characteristic feature of phloem- and ambimobile xenobiotics. Neither the uptake into, nor the accumulation in the se-cc complex should be expected to be the only important requirement for subsequent translocation of any compound in the phloem. The really decisive step, however, is the prolonged retention of absorbed substances in the phloem sap preventing any efflux along the pathway. Acidic phloem-mobile and ambimobile xenobiotics undergo retention by an ion trapping mechanism (JACOB and NEUMANN 1983; NEUMANN 1985; NEUMANN *et al.* 1985). According to the pH regime in the phloem and pk_a-values, the acidic substances dissociate to different extents in the apoplast (pH = 5), and in the symplast (pH =

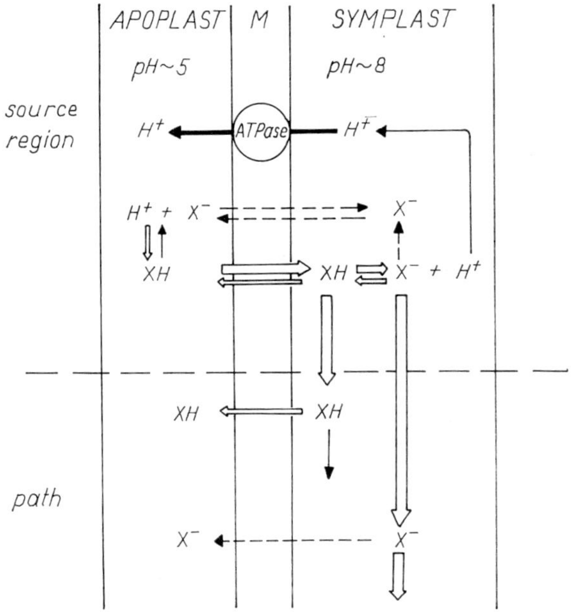

Fig. 1.3 Diagram of phloem transport of xenobiotics. Uptake of phloem- and ambimobile, as well as xylem-mobile xenobiotics into se-cc complex by diffusion of the undissociated molecules (XH). Retention of the relatively hydrophilic anions (X^-) by ion trap mechanism. Along the pathway the relatively lipophilic, undissociated molecules XH are leaked out very quickly from the sieve tube, while the relatively hydrophilic anions X^- escaping to a lesser extent are translocated together with the assimilates to sink regions.

8). The proportion of dissociated moieties is higher in the alkaline medium (sieve tube) than in the acidic one (apoplast). Since the more hydrophilic anions penetrate membrane barriers at lower velocities and undissociated molecules are usually lipophilic enough to cross the plasmalemma by diffusion with sufficient velocity, the latter represent the principal form in which xenobiotics enter the sieve tube. Due to their pk_a values which are usually smaller than 7, acidic phloem- and ambimobile xenobiotics undergo a nearly complete dissociation in the sieve tube. Along the path the relatively lipophilic, undissociated molecules are leaked out very quickly from the sieve tube, while the relatively hydrophilic anions which are escaping to a lesser extent are translocated together with the assimilates to sink regions (Fig. 1.3). Among the ambimobile xenobiotics there are also some without acidic properties, e.g. the nematicide oxamyl (Peterson *et al.* 1978), the herbicide amitrole (Lichtner 1983) and the fungicide pyroxychlor (Small *et al.* 1970). These substances are characterized by a high degree of water solubility (oxamyl 60 g/l; amitrole 280 g/l; pyroxychlor 26 g/l), and seem to be retained by the so-called "intermediate-diffusion-mechanism" (Tyree *et al.* 1979). Of the two combined steps in phloem transport of xenobiotics, transfer through the plasmalemma of the se-cc complex, and longitudinal translocation to sinks, the latter is large enough to prevent substantial efflux along the transportation route. The permeability coefficients of the above mentioned substances are in the optimum range to submit to this mode of translocation.

Xenobiotics without any structural prerequisites to be submitted, either by ion trapping or by retention by optimum permeability in the phloem sap never usually appear in the sink organs. Despite the rapid leakage of such xylem-mobile compounds along the phloem path, a limited translocation in sieve tube can be observed, while the distances that they are carried away from the sites of application are determined by the hydrophilicity-lipophilicity-balance (expressed as log P), and the concentration of the compounds applied. An increase in the concentration applied to the source leaves, results in an elongated path (SOLEL *et al.* 1973; PRICE 1979). Thus highly effective xylem-mobile pesticides can be expected to show some response in basipetal parts after leaf application.

Metabolism and mobility

The fungitoxic quality of any substance is dependent on its chemical structure, which normally does not offer favourable conditions for absorption into higher plants and translocation. Metabolic changes and decomposition of active substances in plants mainly result in detoxification and loss of effectiveness. Slightly water soluble substances often become more strongly polar, a process changing the properties of mobility too. The formation of conjugates may lead to inactivity and changed mobility, furthermore to delayed decomposition with the consequence of a depot effect. In these particular cases a species specific conversion may be advantageous for the control by pesticides (WAIN and CARTER 1967; GASZTONYI and JOSEPOVITS 1979). Finally modifications of active ingredients desirable in respect of mobility, or metabolites with new effectiveness and favourable conditions for uptake and translocation may be formed by metabolism. Thus the herbicides chlorfenprop-methyl (BOLDT and PUTNAM 1980) and dichlofop-methyl (SHIMABUKURO *et al.* 1979) have been applied as esters, transformed metabolically within the cells to acids and translocated in the phloem according to its acidity.

Among fungicides benomyl is well known by its rapid conversion after absorption into cells, by removing the butylisocyanat side chain and producing the methyl ester of benzimidazole carbamic acid (CLEMONS and SISLER 1969). That is also known as carbendazim, which is the real fungitoxic compound. It is less soluble in lipophilic solvents compared with benomyl, but has equally low water solubility. Both xylem-mobile compounds are absorbed in equal amounts by pumpkin leaves according to YOUNG and HAMMET (1973). N-butylisocyanat has been identified as an inhibitor of cutinase, and may presumably cause an additional protective effect against fungi which penetrate the cuticula in infection (KÖLLER *et al.* 1982) (cf. chapter 23).

The piperazin derivative triforine (chapter 13, Fig. 13.1) has only a small water solubility and is distinctly xylem-mobile. However, slight transport from the leaves to the roots could be ascertained when the plants were illuminated and showed photosynthetic activity (MARSH 1977; GREEN *et al.* 1977). It is unresolved whether these findings can be based on the formation of phloem-mobile metabolites. The decomposition to the main metabolite piperazin, which is only recognizable after some days should not be considered (FUCHS *et al.* 1976; HASSALL 1982).

After application of triadimefon (chapter 13, 14), two diastereomere forms of the likewise antifungal triadimenol appear (BUCHENAUER 1979; BUCHENAUER and RÖHRER 1982), of which the triadimenol I is fungitoxic to a higher degree. According to PERKOW (1985) triadimenol is nearly four times more water soluble than the initial substances. The experiments up to now do not permit the conclusion that pseudo-apolastic translocation is changed.

Carboxin (Fig. 9.1.a), which is taken up very well in plants will be oxidized quickly to the non fungitoxic sulphoxide (BRIGGS *et al.* 1974). The carboxin sulphon, known as the fungicide oxycarboxin (Fig. 9.1.b), has a higher persistence, but is absorbed in a smaller amount, as it was proved by SNEL and EDGINGTON (1970), by means of bean roots. It can be supposed, that the reason for the higher uptake is the distinct stronger lipophilicity of carboxin, but a general advantage of that compound can not be found after leaf application, presumably because of its fast decomposition.

In summary one can state that metabolic processes give rise to changed behaviour in mobility of absorbed substances, which can be advantageous for transport and distribution in plants, as far as pesticidal properties are preserved.

This process hardly appears to play a role in the fungicides used up to now, however our knowledge of the activities and mobility of metabolites, or of the metabolism itself, is still insufficient. It remains necessary to take such correlations more into consideration in the future and to make them useable by the application of suitable precursors of active substances.

Practical conclusions drawn from the mobility behaviour of fungicides

Since the introduction of carboxins for the control of the loose smut pathogen *Ustilago nuda* by SCHMELING and KULKA (1966) (chapter 9), systemic fungicides have very quickly gained in importance (EVANS 1971; GROSSMANN 1974; HASSALL 1982). However, the molecular configurations of most of the present available systemic fungicides, allow only a distribution by the xylem stream, i.e. they are xylem-systemic, and only very few of them are phloem-systemic to a limited extent. Therefore, as a rule the protection can be expected in the whole plant only after fungicide application to the soil, a process which is economically acceptable and ecologically harmless only in the case of seed treatment, or the use of granules or incrustations. From this consideration results the demand for new phloem-mobile or ambimobile fungicides, since by the normally practicized leaf application they reach many parts of the plant including the roots, and can also migrate into the new grown parts if the persistance is sufficiently high. On account of the thinning factor, a high effectiveness of such substances is a decisive requirement.

Besides the disadvantages of xylem-systemic pesticides in application to the roots via soil, some advantages can be seen, for instance in substances such as ethirimol and diethirimol (chapter 19) which were adsorbed at the soil matter and are released continously to the roots. In this case the soil exerts a depot effect as long as the rate of decomposition of the substance is not very high (GRAHAM-BRYCE and COUTTS 1971). Moreover the absorption of systemic substances by roots takes place easier than by leaves with their outer wall barriers. If there is a sufficient high concentration of active substance in the xylem, it can reach the cortex by lateral translocation as was demonstrated with carbendazim by PETERSON and EDGINGTON (1970).

Any application of a systemic antifungal compound leads to its close contact with the living plant cells, which on the one side increases the danger of a fast decomposition and on the other side raises the chance of formation of useful antifungal metabolites. The aptitude of fungicides for systemic use presupposes the realization of some physiological conditions. One of these conditions is a largely undisturbed passage through living protoplasts — at least the endodermis in the case of xylem-systemic compounds — by preservation of membrane integrity. Furthermore, the transport in the xylem requires an uninfluenced transpiration, that means that opening of stomata must not be disturbed by accumulated active ingredients in the leaf. A phloem-systemic export from the leaves can only be performed when they have finished their growth when the "import leaves" become "export leaves".

The importance of formulation of fungicides in order to increase mobility shall not be dealt with in detail here. According to our present knowledge it influences exclusively the uptake and by this has an indirect effect on far translocation. Under no circumstances does a specific change of the mode of far transport take place. Many results concerning the modified mobility of active substances in combinations, can presumably be reduced to a formulation-analogous effect of one of the active substances.

Conclusions

1. Systemic fungicides differ from non-systemic ones by their ability to be distributed in the plant. The mode of the involved far transport system of xylem, phloem or of both, however, is essential for distribution and the chemotherapeutic effect. The highly effective systemic fungicides discovered till now are mainly xylem-systemic ones.
2. The absorption of fungicides in roots, leaves, seeds or fruits occurs passively by diffusion. The chemical structure or its metabolic changes decide whether the tissue transport takes place in an apoplastic or symplastic manner, and whether the substances are subjected to a far transport or not. The outer cell wall layers of the organ of application play an important role in the process of absorption.
3. The relationship of hydro- and lipophilicity of the applied compound is essential for the mobility behaviour, in the case of dissociating molecules depending on the degree of dissociation and also on the pH value of the solution. Like hydrophilic substances, extreme lipophilic substances are not or badly translocated.
4. Accumulation of substances in cells and tissues can be the consequence of water loss by transpiration (e.g. in leaves), adsorption, metabolic change of mobility (e.g. formation of conjugates) or of an effect of ion trap. The latter has been recognized as the essential mechanism for the accumulation of substances in sieve tubes of the phloem.
5. The interrelations between the chemical and physical properties of substances and their mobility explored hitherto, should also be used in the control of fungal diseases and for the development of new agents with better characteristics of application.

References

Aharonson, N., and Kafkafi, U.: Adsorption, mobility and persistence of thiabendazole and methyl-2-benzimidazolecarbamate in soils. J. Agricult. Food Chem. **23** (1975): 720—724.

Austin, D. J., and Briggs, G. G.: A new extraction method for benomyl residues in soil and its application in movement and persistence studies. Pesticide Sci. **7** (1976): 201—210.

Barak, E., Dinoor, A., and Jacoby, B.: Adsorption of systemic fungicides and a herbicide by some components of plant tissues, in relation to some physico-chemical properties of the pesticides. Pesticide Sci. **14** (1983): 213—219.

— Jacoby, B., and Dinoor, A.: Adsorption of systemic pesticides on ground stems and in the apoplastic pathway of stems, as related to lignification and lipophilicity of the pesticides. Pesticide Biochem. Physiol. **20** (1983): 194—202.

Boldt, P. F., and Putnam, A. R.: Selectivity mechanisms for foliar applications of diclofop-methyl. I. Retention, absorption, translocation and volatility. Weed Sci. **28** (1980): 474—477.

Bonnett, H. T.: Cortical cell death during lateral root formation. J. Cell Biol. **40** (1969): 144—159.

Boulware, M. A., and Camper, N. D.: Sorption of some ^{14}C-herbicides by isolated plant cells and protoplasts. Weed Sci. **21** (1973): 145—149.

Briggs, D. E., Waring, R. W., and Hackett, A. M.: The metabolism of carboxin in growing barley. Pesticide Sci. **5** (1974): 599—607.

Briggs, G. G.: A simple relationship between soil adsorption of organic chemicals and their octanol/water partition coefficients. Proc. 7th Brit. Insecticide Fungicide Conf. (1973): 83—86.

— Bromilow, R. H., Edmundson, R., and Johnston, M.: Distribution coefficients and systemic activity. In: McFarlane, N. R. (Ed.): Herbicides and Fungicides – Factors Affecting Their Activity. Chem. Soc., London 1977, pp. 129—134.

— — and Evans, A. A.: Relationship between lipophilicity and root uptake and translocation of non-ionised chemicals by barley. Pesticide Sci. **13** (1982): 495—504.

— — — and Williams, M.: Relationship between lipophilicity and the distribution of non-ionised chemicals in the barley shoot following uptake by roots. Pesticide Sci. **14** (1983): 492—500.

Buchenauer, H.: Conversion of triadimefon into two diastereomers, triadimenol-I and triadimenol-II by fungi and plants. IX. Congr. Plant Prot. Washington (1979): Abstract 939.

— and Röhrer, E.: Aufnahme, Translokation und Transformation von Triadimefon in Kulturpflanzen. Z. Pflanzenkrankh. Pflanzenschutz **89** (1982): 385—398.

Christ, R. A.: Physiological and physicochemical requisites for the transport of xenobiotics in plants. In: Geissbühler, H. (Ed.): Adv. Pesticide Sci. Vol. **3**. Pergamon Press, Oxford 1979, pp. 420–429.

Clemons, G. P., and Sisler, H. D.: Formation of a phytotoxic derivative from Benlate. Phytopathology **59** (1969): 705—706.

Crafts, A. S., and Crisp, C. E.: Movement of exogenous substances. In: Phloem Transport in Plants. Freeman and Comp., San Francisco 1971, pp. 168—264.

Crisp, C. E.: The molecular design of systemic insecticides and organic functional groups in translocation. In: Tahori, A. S. (Ed.): Proc. 2nd Congr. Pesticide Chemistry **1** (1972): 211–264.

– and Look, M.: Phloem loading and transport of weak acids. In: Geissbühler, H. (Ed.): Adv. Pesticide Sci. Vol. **3**. Pergamon Press, Oxford 1979, pp. 430—437.

Crowdy, S. H.: Translocation. In: Marsh, R. W. (Ed.): Systemic Fungicides. Longmans, London 1972, pp. 92—115.

— Patterns and processes of movement of chemicals in higher plants. Proc. 7th Brit. Insecticide Fungicide Conf. (1973): 831—835.

— Grove, J. F., and McCloskey, P.: The translocaction of antibiotics in plants. **4**. Systemic fungicidal activity and chemical structure in griseofulvin relatives. Biochem. J. **72** (1959): 241—249.

— and Rudd-Jones, D.: The translocation of sulphonamides in higher plants. I. Uptake and translocation in broad beans. J. exp. Bot. **7** (1956a): 335—346.

— — Partition of sulphonamides in plant roots: a factor of their translocation. Nature **178** (1956b): 1165—1167.

Dekker, J.: Resistance. In: Marsh, R. W. (Ed.): Systemic Fungicides. 2nd edit. Longmans, London 1977, pp. 176—197.

Dobe, H., Sieber, K., Neumann, St., and Jacob, F.: Untersuchungen zur Aufnahme und Translokation von Aldimorph in ausgewählten Pflanzenspecies. Arch. Phytopathol. Pflanzensch. **21** (1986): 415—425.

Donaldson, T. W., Bayer, D. E., and Leonard, O. A.: Absorption of 2,4-dichlorophenoxyacetic acid and 3-(p-chlorophenyl)-1,1-dimethylurea (monuron) by barley roots. Plant Physiol. **52** (1973): 638—645.

Dumbroff, E. B., and Peirson, D. R.: Probable sites for passive movement of ions across the endodermis. Can. J. Bot. **49** (1971): 35—38.

Dybing, C. D., and Currier, H. B.: Foliar penetration by chemicals. Plant Physiol. **36** (1961): 169—174.

Edgington, L. V.: Structural requirements of systemic fungicides. Ann. Rev. Phytopathol. **19** (1981): 107—124.

— and Dimond, A. E.: The effect of adsorption of organic cations to plant tissue on their use as systemic fungicides. Phytopathology **54** (1964): 1193—1197.

— and Peterson, C. A.: Systemic fungicides: Theory, uptake, and translocation. In: Siegel, M. R., and Sisler, H. D. (Eds.): Antifungal Compounds. Vol. 2. Dekker, New York 1977, pp. 51—89.

Erwin, C. E.: Systemic fungicides: Disease, control, translocation and mode of action. Ann. Rev. Phytopath. **11** (1973): 389–422.

Esau, K.: Plant Anatomy. 2nd edit. John Wiley and Sons. New York-London-Sydney 1965.

Evans, E.: Problems and progress in the use of systemic fungicides. Proc. 6th Brit. Insecticide Fungicide Conf. (1971): 758—764.

Fehrmann, H.: Systemische Fungizide — ein Überblick. Phytopath. Z. **86** (1976): 67—89.

Fuchs, A., De Vries, F. W., and Aalbers, A. M. J.: Uptake, distribution and metabolic fate of ^{3}H-triforine in plants. I. Short-term experiments. Pesticide Sci. **7** (1976): 115—126.

Gastonyi, M., and Josepovits, G.: The activation of triadimefon and its role in the selectivity of fungicide action. Pesticide Sci. **10** (1979): 57—65.

Georgopoulos, S. G.: Development of fungal resistance to fungicides. In: Siegel, M. R., and Sisler, H. D. (Eds.): Antifungal Compounds. Vol. 2, Dekker, New York 1977, pp. 409—495.

Giaquinta, R. T.: Phloem loading of sucrose. Ann. Rev. Plant Physiol. **34** (1983): 347—387.

Graham-Bryce, I. J., and Coutts, J.: Interaction of pyrimidine fungicides with soil and their influence on uptake by plants. Proc. 6th Brit. Insecticide Fungicide Conf. (1971): 419.

Green, M. B., Hartley, G. S., and West, T. F.: Chemicals for Crop Protection and Pest Control. Pergamon Press, Oxford 1977.

Greene, D. W., and Bukovac, M. J.: Stomatal penetration, effect of surfactants and role in foliar absorption. Amer. J. Bot. **61** (1974): 100—106.

Grimm, E., Neumann, St., and Jacob, F.: Aufnahme von Saccharose und Xenobiotika in isoliertes Leitgewebe von *Cyclamen*. Biochem. Physiol. Pflanzen **178** (1983): 29—42.

— — — Transport of xenobiotics in higher plants. II. Absorption of defenuron, carboxyphenylmethylurea and maleic hydrazide by isolated conducting tissue of *Cyclamen*. Biochem. Physiol. Pflanzen **180** (1985): 383—392.

Grossmann, F.: Möglichkeiten und Grenzen des Einsatzes systemischer Fungizide. Z. Pflanzenkrankh. Pflanzensch. **81** (1974): 670—677.

Hance, R. J.: Processes in soil which control the availability of pesticides. Proc. 10th Int. Congr. Plant Prot. Brighton (1983): 537—544.

Hartley, G. S., and Graham-Bryce, I. J.: Physical Principles of Pesticide Behaviour. Vol. 2, Academic Press, London-New York-Toronto-Sydney-San Francisco 1980.

Hassall, K. A.: The Chemistry of Pesticides. Their Metabolism, Mode of Action and Uses in Crop Protection. Verlag Chemie, Weinheim-Deerfield-Beach-Florida-Basel 1982.

Hill-Cottingham, D. G., and Lloyd-Jones, C. P.: Relative mobility of some nitrogenous compounds in the xylem of apple shoots. Nature **220** (1968): 389—390.

— — A technique studying the adsorption and metabolism of amino acids in apple stem tissue. Physiol. Plant. **28** (1973): 443—446.

Holloway, P. J.: Surface factors affecting the wetting of leaves. Pesticide Sci. **1** (1970): 156—163.

— The chemical constitution of plant cutin. In: Cutler, D. F., Alvin, K. L., and Price, C. E. (Eds.): The Plant Cuticle. Academic Press, New York-London 1982, pp. 45—85.

Hunt, G. M., and Baker, E. A.: Phenolic constituents of tomato fruit cuticles. Phytochemistry **19** (1980): 1415—1419.

Jacob, F., Neumann, St.: Quantitative determination of mobility of xenobiotics and criteria of their phloem and xylem mobility. In: Miyamoto, J., and Kearney, P. C. (Eds.): IUPAC Pesticide Chemistry — Human Welfare and the Environment. Vol. 1. Pergamon Press, Oxford-New York-Toronto-Sydney-Paris-Frankfurt 1983, pp. 357—362.

— — and Strobel, U.: Studies on mobility of exogen applied substances in plants. In: Transactions of the 3rd Symp. on Accumulation of Nutrients and Regulators in Plant Organism. Warsaw. Proc. Res. Inst. Pomology Skierniewice Poland, Series E (1973): 315—330.

Karas, I., and McCully, M. E.: Further studies on the histology of lateral root development in *Zea mays*. Protoplasma **77** (1973): 243—269.

Köller, W., Allan, C. R., and Kolattukudy, P. E.: Inhibition of cutinase and prevention of fungal penetration into plant by benomyl — a possible protective mode of action. Pesticide Biochem. Physiol. **18** (1982): 15—25.

Leroux, P., and Gredt, M.: Uptake of systemic fungicides by maize roots. Netherl. J. Plant Path. **83**, Suppl. 1 (1977): 51—61.

Lichtner, F. T.: Amitrole absorption by bean (*Phaseolus vulgaris* L. c.v. Red Kidney) roots. Plant Physiol. **71** (1983): 64—72.

Mansfield, T. A., Pemasada, M. A., and Snaith, P. J.: New possibilities for controlling foliar absorption via stomata. Pesticide Sci. **14** (1983): 294—298.

Marsh, R. W. (Ed.): Systemic Fungicides. 2nd edit. Longmans, London 1977.

Martin, J. T., and Juniper, B. E.: The Cuticles of Plants. Arnold, London 1970.

Martin, R. A., and Edgington, L. V.: Comparative systemic translocation of several xenobiotics and sucrose. Pesticide Biochem. Physiol. **16** (1981): 87—96.

Minshall, W. H.: Translocation path and place of action of 3-(4-chlorophenyl)-1,1-dimethylurea in bean and tomato. Can. J. Bot. **32** (1954): 795—798.

Mitchell, J. W., Marth, P. C., and Preston, W. H.: Structural modification that increases translocability of some plant-regulating carbamates. Science **120** (1954): 263—265.

Moody, K., Kust, C. A., and Buchholz, K. P.: Uptake of herbicides by soybean roots in culture solution. Weed Sci. **18** (1970): 642—647.

Müller, F.: Translokation von ^{14}C-markiertem MCPA in verschiedenen Entwicklungsstadien einiger mehrjähriger Unkräuter. Acta Phytomedica (Suppl. J. Phytopath.) **4** (1976): 5—160.

Netting, A. G., and Wettstein-Knowles, P.: The physico-chemical basis of leaf wettability in wheat. Planta **114** (1973): 289—309.

Neumann, St.: Die Ionenfalle als Akkumulations- und Retentionsmechanismus bei der Aufnahme von xenobiotischen Substanzen in das Phloem. Colloquia Pflanzenphysiologie Humboldt-Univ. Berlin **8** (1985): 117—130.

— Grimm, E., and Jacob, F.: Transport of xenobiotics in higher plants. I. Structural prerequisites for translocation in the phloem. Biochem. Physiol. Pflanzen **180** (1985): 257—268.

— Petzold, U., and Jacob, F.: Mobilität von Xenobiotika in höheren Pflanzen — eine tabellarische Übersicht. I. Allgemeine Prinzipien der Aufnahme und des Transports. II. Fungizide. Wiss. Z. Univ. Halle **XXXIV** '85M (1985): 56—70.

Norris, R. F.: Penetration of 2,4-D in relation to cuticle thickness. Amer. J. Bot. **61** (1974): 74—79.

— and Bukovac, M. J.: Structure of pear leaf cuticle with special reference to cuticular penetration. Amer. J. Bot. **55** (1968): 975—983.

Perkow, W.: Wirksubstanzen der Pflanzenschutz- und Schädlingsbekämpfungsmittel. 2. Aufl. Parey, Berlin-Hamburg 1983.

Peterson, C. A., De Wildt, P. Q., and Edgington, L. V.: A rationale for the ambimobile translocation of the nematicide oxamyl in plants. Pesticide Biochem. Physiol. **8** (1978): 1—9.

— and Edgington, L. V.: Transport of the systemic fungicide, benomyl, in bean plants. Phytopathology **60** (1970): 475—478.

— — Factors influencing apoplastic transport in plants. In: Lyr, H., and Polter, C. (Eds.): Systemic Fungicides (Int. Symp. Reinhardsbrunn 1974). Akademie-Verlag, Berlin 1975, pp. 287—299.

— — Uptake of the systemic fungicide methyl-2-benzimidazolecarbamate and the fluorescent dye PTS by the onion roots. Phytopathology **65** (1975): 1254—1259.

— Emanuel, M. E., and Humphreys, G. B.: Pathway of movement of apoplastic fluorescent dye tracers through the endodermis at the site of secondary root formation in corn *(Zea mays)* and broad bean *(Vicia faba)*. Can. J. Bot. **59** (1981): 618—625.

Prasad, R., Foy, C. L., and Crafts, A. S.: Effects of relative humidity on absorption and translocation of foliarly applied dalapon. Weed Sci. **15** (1967): 149—156.

Price, C. E.: Penetration and translocation of herbicides and fungicides in plants. In: McFarlane, N. R. (Ed.): Herbicides and Fungicides — Factors Affecting Their Activity. Chem. Soc., London 1977, pp. 42—66.

— Movement of xenobiotics in plants — perspectives. In: Geissbühler, H. (Ed.): Adv. Pesticide Sci. Vol. **3**, Pergamon Press, Oxford 1979, pp. 401—409.

Rentschler, I.: Die Wasserbenetzbarkeit von Blattoberflächen und ihre submikroskopische Wachsstruktur. Planta **96** (1971): 119—135.

Riederer, M., and Schönherr, J.: Accumulation and transport of (2,4-dichlorophenoxy)acetic acid in plant cuticles. I. Sorption in the cuticular membrane and its components. Ecotoxicol. Environ. Safety **8** (1984): 236—247.

— — Accumulation and transport of (2,4-dichlorophenoxy)acetic acid in plant cuticles. II. Permeability of the cuticular membrane. Ecotoxicol. Environ. Safety **9** (1985): 196—208.

Schmeling, B. v., and Kulka, M.: Systemic fungicidal activity of 1,4 oxathiin derivatives. Science **152** (1966): 659—660.

SCHÖNHERR, J., and HUBER, R.: Plant cuticles are polyelectrolytes with isoelectric points around three. Plant Physiol. **59** (1977): 145—150.

SHARMA, M. P., and VAN DEN BORN, W. H.: Uptake, cellular distribution and metabolism of ^{14}C-picloram by excised plant tissues. Physiol. Plant **29** (1973): 10—16.

SHEPHARD, M. C.: Barriers to the uptake and translocation of chemicals in herbaceous and woody plants. Proc. 7th Brit. Insecticide Fungicide Conf. (1973): 841—850.

— Factors which influence the biological performance of pesticides. Proc. Brit. Crop Protect. Conf. (1981): 711—721.

SHIMABUKURO, R. H., WALSH, W. C., and HOERAUF, R. A.: Metabolism and selectivity of diclofop-methyl in wild oat and wheat. J. Agricult. Food Chem. **27** (1979): 615—623.

SHONE, M. G. T., BARTLETT, B. O., and WOOD, A. V.: A comparison of the uptake and translocation of some organic herbicides and a systemic fungicide by barley. II. Relationship between uptake by roots and translocation to shoots. J. exp. Bot. **25** (1974): 401—409.

— and WOOD, A. V.: Factors affecting absorption and translocation of simazine by barley. J. exp. Bot. **23** (1972): 141—151.

— — A comparison of the uptake and translocation of some organic herbicides and a systemic fungicide by barley. I. Absorption in relation to physico-chemical properties. J. exp. Bot. **25** (1974): 390—400.

SMALL, L. W., MARTIN, R. A., and EDGINGTON, L. V.: A comparison of the translocation within plants of pyroxychlor and a 6-amino analogue. Netherl. J. Plant Path. **83**, Suppl. 1 (1977): 63—70.

SNEL, M., and EDGINGTON, L. V.: Uptake, translocation and decomposition of systemic oxathiin fungicides in beans. Phytopathology **60** (1970): 1708—1716.

SOLEL, Z., SCHOOLEY, J. M., and EDGINGTON, L. V.: Uptake and translocation of benomyl and carbendazim (methylbenzimidazole-2-ylcarbamate) in the symplast. Pesticide Sci. **4** (1973): 713—718.

TYREE, M. T., PETERSON, C. A., and EDGINGTON, L. V.: A simple theory regarding ambimobility of xenobiotics with special reference to the nematicide oxamyl. Plant Physiol. **63** (1979): 367—374.

UCHIDA, M.: Affinity and mobility of fungicidal dialkyl dithioanylidenemalonates in rice plants. Pesticide Biochem. Physiol. **14** (1980): 249—255.

— and KASAI, T.: Adsorption and mobility of fungicidal dialkyl dithioanylidenemalonates and their analogs in soil. J. Pesticide Sci. **5** (1980): 553—559.

— and SUZUKI, T.: Affinity and mobility of fungicides in soils and rice plants. 5th Int. Congr. Pesticide Chem. (IUPAC) Kyoto (1982): Abstract IIe—3.

VAN BEL, A. J. E.: Behaviour of differently charged amino acids towards tomato wood powder. Z. Pflanzenphysiol. **89** (1978): 313—320.

VAN DER KERK, G. J. M.: Systemic fungicides new solutions and new problems. Proc. 6th Brit. Insecticide Fungicide Conf. (1971): 791—802.

— Evolution of the chemical control of plant diseases; an evaluation. Netherl. J. Plant. Path. **83**, Suppl. 1 (1977): 3—13.

WAIN, R. L., and CARTER, G. A.: Uptake, translocation and transformation by higher plants. In: TORGESON, D. C. (Ed.): Fungicides. Vol. 1. Academic Press, New York-London 1967, pp. 561—611.

WALKER, A., and FEATHERSTONE, R. M.: Absorption and translocation of atrazine and linuron by plants with implications concerning linuron selectivity. J. exp. Bot. **24** (1973): 450—458.

WATTENDORF, J., and HOLLOWAY, P. J.: Studies on the ultrastructure and histochemistry of plant cuticles: The cuticular membrane of *Agave americana* L. *in situ*. Ann. Bot. **46** (1980): 13—28.

WILHELM, H., and KNÖSEL, H.: Penetration und Translokation von ^{3}H-Tetracyclin-hydrochlorid in pflanzlichem Gewebe. Z. Pflanzenkrankh. Pflanzenschutz **83** (1976): 241—251.

YOUNG, H., and HAMMET, K. R. W.: Comparative uptake of benomyl-2-^{14}C and methylbenzimidazole-2-ylcarbamate-2-^{14}C by cucumber leaves. New Zealand J. Sci. **16** (1973): 535—541.

LYR, H. (Ed.): Modern, Selective Fungicides — Properties, Applications, Mechanisms of Action. Longman Group UK Ltd., London, and VEB Gustav Fischer Verlag, Jena, 1987.

Chapter 2

Selectivity in modern fungicides and its basis

H. LYR

Institute for Plant Protection Research of the Academy of Agricultural Sciences of the GDR, Kleinmachnow

Introduction

Although the word "fungicide" might suggest that such a compound unselectively kills all fungi, this is not the case even with classic, multisite inhibitors such as the dithiocarbamates, phthalimides, quinones (Tab. 2.1) and similar compounds. Within this group of fungicides remarkable differences in activity against various fungi can be observed. Therefore, even compounds with an unspecific mechanism of action (TORGESON 1969) cannot be used in practice to control all fungal diseases. Even within a species, remarkable variance in the tolerance to such fungicides has been described. Only strong biocidal agents like mecurial compounds, phenols and other disinfectants act very unselectively.

(Problems of acquired resistance are not dealt with in this chapter, although some common biochemical properties may be involved in the phenomena of selectivity and resistance.)

Facts and problems

With the introduction of modern, highly efficient fungicides, practice and science have been more urgently confronted with the problem of fungicide selectivity. This phenomenon may have positive as well as negative aspects. A positive effect can be a more selective interference with the fungal microflora in the soil or in the phyllosphaera. As a hypothetical example, selective control of *Pythium* in the soil may favour its antagonists, such as *Trichoderma* species, which can support the controlling

Table 2.1 Spectrum of activity of some fungicides with an unspecific mechanism of action (ED_{50} values for mycelial radial growth on malt medium, in mg/l)

Species	Thiram	Maneb	Captan	Dichlone
Phytophthora cactorum	20	10	30	5
Pythium ultimum	20	8	20	5
Mucor mucedo	30	100	100	100
Botrytis cinerea	3	40	20	3
Fusarium oxysporum	20	100	40	100
Gloeosp. fructigenum	20	40	—	50
Mycosph. pinodes	4	30	—	30
Trametes versicolor	20	30	30	100
Verticillium albo-atrum	15	10	—	10
Rhizoctonia solani	10	35	30	4

effect of the chemical (Beth *et al.* 1984). On the other hand, selective control of *Pythium* can lead to increased severity of diseases caused by *Phytophthora*, *Fusarium* or *Rhizoctonia* species. A similar situation occurs after application of soil disinfectants or partial soil sterilisation, where quickly colonizing fungi like *Pythium*, *Phytophthora* or *Trichoderma* may develop as a consequence of decreasing the microbial competitors. A practical example has been described where the control of *Pseudocercosporella herpotrichoides* by benomyl favours a *Rhizoctonia* attack. Several other cases of this type are known. Similar processes probably occur on leaf surfaces although our knowledge in this area is very limited. Application of fungicides can cause a drastic change in the phyllosphaera, which may normally exhibit an antibiotic or at least an antitrophic effect against leaf pathogens. A selective control of mildew may stimulate rusts and other leaf pathogens which normally are of minor importance. *Alternaria*, *Epicoccum* and *Trichoderma* significantly antagonize the infection of lettuce plants by *Sclerotinia* (Mercier and Reefelder 1984). Other antagonistic effects are described by Hindorf and Ketterer 1984; Leinhos and Buchenauer 1984; Stenzel 1984. The basis of such effects can be different. Advantages or disadvantages of selectivity are still in discussion even with insecticides where much more investigations have been performed regarding the utility of parasites, hyperparasites and predators in the control of pests. The opinions are still quite divergent, not only from a commercial point of view but also from biological considerations.

Spectra of activity

Modern fungicides have indeed distinct spectra of activity. Some examples for fungicides with different mechanisms of action are demonstrated in Tab. 2.2. Similarities or differences are not simply related to the mechanism of action. This can be seen within the group of sterol biosynthesis inhibiting fungicides or within the group of Aromatic Hydrocarbon Fungicides, where not only quantitatively but also qualitatively different spectra of activity can occur. This even is true for very specific fungicides with a narrow spectrum of activity for example against Oomycetes. Even within the genus *Phytophthora*, considerable differences in the sensitivity to a fungicide can occur among the various species. *Pythium* species, although closely related to *Phytophthora* species, may behave very differently from the latter in respect to fungicide sensitivity (Tab. 2.3).

It cannot be ignored that a certain variance in the response to a fungicide exists also within a species. This is demonstrated in Tab. 2.4 for *Phytophthora infestans* and various pathotypes from an area in the German Democratic Republic where the population at that time never previously had contact with the fungicide metalaxyl (Zollfrank and Lyr 1987).

The data, therefore, reflect the natural variance of a species in a population with the exclusion of resistance problems. The same type of variation in sensitivity to tolclophos-methyl can be seen for strains of *Rhizoctonia solani* (Tab. 2.4).

Of course the variance within species may make the values indicated in Tab. 2.2 somewhat inaccurate, but in spite of this problem, knowledge of the spectrum of activity of a fungicide is of great importance for its application.

Because similarity or dissimilarity of antifungal spectra are hard to describe we attempted to express this phenomenon quantitatively in terms of an analogy factor comparable to a correlation coefficient. Tab. 2.5 indicates a close similarity between the selectivity spectrum of the fungicide chloroneb and that of dichlofluanid, vinclozolin or pentachloronitrobenzene, but a great dissimilarity between the selectivity spectrum of chloroneb and that of triarimol (Lyr and Casperson 1982).

Table 2.2 Comparison of the spectra of activity of some modern fungicides (malt-agar-growth test) ED_{50} values in mg/l (* barley leaf-test)

	Tridemorph	Triforine	Triarimol	Triadimefon	Etridiazole	Chloroneb	Dicloran	Tolclophos-methyl	Vinclozolin	Benomyl	Thiophanate-methyl	Carboxin	Dichlofluanid
Phytophthora cactorum	300	20	100	200	0.5	2	200	100	200	50	100	100	4
Pythium ultimum	60	15	100	200	0.5	10	200	100	50	100	100	100	2
Mucor mucedo	300	20	10	200	8	2	0.5	—	10	100	100	50	40
*Erysiphe graminis**	0.02	1	0.02	0.006	20	20	—	—	3	7	100	20	3
Penicillium chrysogenum	150	200	10	5	90	500	0.5	1.6	2	0.2	50	50	2
Botrytis cinerea	0.4	20	4	40	20	5	1	1.9	0.5	0.4	3	10	0.5
Fusarium oxysporum	300	20	3	50	70	500	200	100	40	2	70	200	50
Gloeosp. fructigenum	10	50	100	200	80	500	200	—	200	0.2	0.2	300	6
Verticillium alb.-atrum	0.3	80	0.2	3	40	10	8	100	40	0.2	3	5	9
Rhizoctonia solani	40	60	10	100	70	15	200	0.1	20	5	200	5	3
Trametes versicolor	6	30	0.3	0.5	30	2	200	—	100	100	200	5	2

Table 2.3 Selectivity of action of some fungicides with relative specificity for the genus *Phytophthora* or Oomycetes (inhibition of radial mycelial growth as % of control on agar medium, concentrations in mg/l) (Andoprim is a new pyrimidin derived fungicide)

Species	Meta-laxyl	Bena-laxyl	Al-phos-ethyl	Cymo-xanil	Chlo-roneb	Etri-diazole	Hyme-xazole	Andro-prim
Conc.	(0.1)	(1)	(100)	(10)	(10)	(1)	(100)	(3)
P. cactorum	100	100	57	3	51	67	30	57
P. cinnamomi	28	31	74	11	52	66	17	90
P. citricola	17	10	65	0	33	47	18	100
P. citrophthora	15	11	65	7	39	73	28	50
P. cryptogaea	68	42	34	0	54	80	32	40
P. infestans	85	83	22	100	47	56	98	48
P. megasperma	75	50	48	76	73	89	75	58
P. nicotianae	38	45	71	30	64	77	42	32
P. palmivora	77	37	12	34	50	73	19	88
P. parasitica	72	35	65	52	65	66	17	97
Pyth. ultimum	56	50	—	24	0	100	100	8
Botrytis cinerea	0	0	—	5	50	2	—	0
Fusarium oxyspor.	0	0	—	0	0	0	—	3
Rhizoctonia solani	0	0	—	3	30	0	—	0

Table 2.4 Variability in sensitivity of fungal strains to the fungicides metalaxyl and tolclophos-methyl. A: Sensitivity of strains of *Phytophthora infestans* (various pathotypes collected in the GDR) to metalaxyl. B: Sensitivity of strains of *Rhizoctonia solani* to tolclophosmethyl (results mainly from information of Sumitomo Chemical Co.). The inhibitory concentrations are expressed as ED_{50} values in mg/l (AG means Ogoshi's hyphal anastomosis groups)

A: *Phytophthora infestans*		B: *Rhizoctonia solani*	
Pathotypes:	ED_{50}	strains:	ED_{50}
1.2.3.4.	0.004	coll. IPF	0.6
1.5.	0.007	AG-1	0.15
1.3.4.	0.012	AG-2	0.10
1.3.4.5.7.10.11.	0.015	AG-2	0.12
1.2.3.4.5.7.10.	0.056	AG-3	0.43
4.1.	0.022	AG-4	0.01
4.5.	0.008	AG-5	0.39

Table 2.5 Correlation coefficients regarding the similarity of the antifungal activity spectrum of several fungicides to that of chloroneb

Fungicide	Correlation coefficient	Number of fungal species
PCNB	0.88	47
Vinclozolin	0.91	35
Etridiazole	0.51	42
Triarimol	0.58	29
Dichlofluanid	0.98	7

Of course several other combinations can be evaluated by this method. Such evaluations show, that there is no strict correlation between the spectra of activity and the mechanism of action of fungicides (chloroneb and dichlofluanid differ in their mechanism of action and fungi do not show cross resistance to these compounds, whereas chloroneb and etridiazole have much more in common in respect to their mode of action). As shown in Tab. 2.2, a correlation between activity of a fungicide and taxonomic position of a fungus exists only for very few fungicides. Although the carboxins act mainly on Basidiomycetes (EDGINGTON and BARRON 1967) (chapter 9) and carbendazim or benomyl are known to be inactive towards fungi from the *Porosporaceae* within the Ascomycetes (EDGINGTON *et al.* 1971) (chapter 15), very often large differences in response to a certain fungicide occur within a fungal family. This is shown in Tab. 2.3 for the *Phytophthoraceae*.

This situation points to the fact that fungi are not only very different in their phylogeny, but that they can also be very different in their biochemistry, even within closely related groups. Unfortunately very little is known of the differential biochemistry in fungi (LYR 1977).

Basis of selectivity

The reasons for selectivity of modern fungicides are very poorly understood. From a theoretical point of view and from experimental data, it can be deduced that one or more of the following possibilities account for the differences in sensitivity of various fungi to a fungicide.

1. Differences in the accumulation of a fungicide in the cell
2. Different structures of the receptor or target systems
3. Differences in ability to toxify (activate) a compound
4. Differences in ability to detoxify a compound
5. Different degrees of importance of a receptor or target system for survival of the fungus.

Differences in the accumulation of fungicides within the fungal cell

Numerous investigations have revealed that there are few cases of selective action based on differences in the accumulation of fungicides in the fungal cell. The following are some examples that have been reported: cycloheximide (WESTCOTT and SISLER 1964), pentachloronitrobenzene (NAKANISHI and OKU 1969) and few others mentioned by RICHMOND (1977).

A new mechanism for regulating fungicide accumulation was detected by the work of DE WAARD and VAN NISTELROY (1982 and earlier papers) (chapter 24) who demonstrated that some fungi have the ability actively to extrude fungicides such as fenarimol from the cell by a still unknown process, and thereby lower the level of the toxicant within the cell. Whether this mechanism is widely distributed among fungi and more affects a greater number of fungicides is still an open question.

Different structure of the receptor or target systems

A prerequisite for the involvement of this principle as a basis of selectivity is the interaction of a fungicide with a specific receptor. A well known example is the selective action of carboxin and related compounds against Basidiomycetes which results from their binding to a specific receptor structure at site II within the mitochondria of sensitive species (chapter 10). The introduction of a phenyl group at the ortho position of the carboxin molecule (F 427) (Fig. 9.1) causes a dramatic change of selectivity towards Ascomycetes and even Phycomycetes (EDGINGTON and BARRON 1967).

Receptor affinity is also involved in the selectivity of benzimidazole fungicides of the carbendazim type. It is highly remarkable that certain phenylcarbamates act exclusively against benomyl insensitive fungi (and resistant strains) which seem to have a changed receptor structure (SUMITOMO patent EP 51 871ff.; KATO *et al.* 1984) (chapter 15, 24).

An interaction with a specific receptor can be expected for the azolylfungicides (chapter 14). Here cytochrome P-450 in combination with sterolbinding proteins seems to represent the receptor structure, but the specific reasons of the selective effects of these compounds are not very clear.

An interaction with receptors can be assumed too for morpholine-, dicarboximide- and aromatic-hydrocarbon-fungicides, but the situation regarding their selectivity has not been elucidated.

A target system is not as specific as a receptor structure but in some cases necessary for a fungicidal action of a compound. The action of polyenic antibiotics like nystatin may serve as an example. The complex formation with normal sterols (ergosterol) of fungal membranes is the basis for the strong and broad activity of these antibiotics. Oomycetes or bacteria without exogenous supply of sterols are insensitive against these antibiotics. The activity of polyoxins can be ascribed to the target system "chitin biosynthesis" because only fungi capable of chitin or chitosan synthesis are sensitive to these and related antibiotics. Of course the enzyme chitin synthetase may be regarded also as receptor because it is the target of these antibiotics.

Differences in uptake processes and degradation can also be involved in the selectivity of these compounds.

Differences in the ability to toxify (activate) a compound

Many well-known fungicides seem to be only prefungicides which means that they must be converted to the biochemically active toxicant. This was described by DE WAARD (1974) for pyrazophos (chapter 20), a rather selective fungicide which must be oxidatively converted to the active compound "PPO". It was surprising that triadimefon, a highly active compound needs enzymatic reduction to triadimenol, which is the actual fungicide (chapter 13, 14).

This principle seems to be more widely distributed than it is generally assumed. Questionable candidates for activation may be dichlofluanid, etridiazole, biphenyl, triforine, cymoxanil and others. If the conversion products are not very stable, the proof of activation is difficult. However, the possibility that this phenomenon may occur with a compound should not be overlooked.

Differences in the ability to detoxify a compound

Metabolic detoxication can be the reason for the relative selectivity of some compounds The enzymes involved in detoxication are of very different types and surely not all are known at this time (LYR 1984). Distribution of such enzymes and their level of activity in various species can be responsible for selective actions of several fungicides. Because many combinations are possible, and even important fungi are very poorly defined biochemically, the situation is very complicated and no generalizations can be made at present.

Enzymes involved in these processes are probably oxido-reductases (flavin enzymes), esterases, phenoloxidases, cytochrome P-450-monoxygenase, glutathione-S-aryl-transferase, glutathione-S-acyltransferase, thiolreductase and others, which in many cases are still of unknown substrate specifity. Detoxication processes may occur in fungi with fungicides such as PCNB, dicloran, tolclophosmethyl and other organo-phosphorus fungicides, carboxins, polyoxins, cymoxanil and several other compounds, but the biochemistry of these processes remains largly unknown.

Different degrees of importance of a receptor or target system

Very little is known at present of the natural function of some receptor or target systems and their importance for survival or fitness of a fungus. A variation among species regarding the essential role of such systems or possible mechanisms of circumvention of a block in the metabolism can affect the selectivity of fungicides. However, more information is needed in this field.

This survey should give the impression that the phenomenon of fungicide selectivity is very complex in nature and that the knowledge of the biochemical processes involved, is still scanty. A better understanding of this phenomenon is not only of theoretical value but of great importance from a practical point of view e.g. for possibilities to enlarge the spectrum of activity of given antifungal substances. Advances in the chemical control of fungal diseases will depend to a high degree on a better understanding of the factors which control the selective action of a compound.

References

Anonymus: Sumitomo Patent EP 51 871.

Beth, H., Denzer, H., and Schlösser, H.: Antagonistische Wirkungen von *Trichoderma harzianum* auf einige phytopathogene Pilze. Mitt. Biol. Bundesanstalt f. Land- u. Forstwirtsch. **223** (1984): 255.

De Waard, M. A.: Mechanism of action of the organophosphorus fungicide pyrazophos. Ph. D. Dissertation. Wageningen 1974.

— and van Nistelroy, J. G. M.: An energy dependent efflux mechanism for fenarimol in a wild-type strain and fenarimol resistant mutants of *Aspergillus nidulans*. Pesticide Biochem. Physiol. **13** (1980): 255—266.

Edgington, L. V., and Barron, G. L.: Fungitoxic spectrum of oxathiin compounds. Phytopathology **57** (1967): 1256—1257.

— Khew, K. L., and Barron, G. L.: Fungitoxic spectrum of benzimidazole compounds. Phytopathology **61** (1971): 42—44.

Hindorf, H., and Ketterer, N.: Einfluß von Stoffwechselprodukten pilzlicher Antagonisten auf *Sclerotium rolfsii* und *Verticillium fungicola*. Mitt. Biol. Bundesanstalt f. Land- und Forstwirtsch. **223** (1984): 258—259.

Kato, T., Suzuki, K., Takanashi, J., and Ramoshita, K.: Negatively correlated cross resistance between benzimidazole fungicides and methyl-N-(3,5-dichlorophenyl)carbamate. J. Pesticide Sci. **9** (1984): 489—495.

Leinhos, G., and Buchenauer, H.: Untersuchungen zu Antagonismus und Hyperparasitismus einiger ausgewählter Pilze gegenüber Getreiderostpilzen. Mitt. Biol. Bundesanstalt f. Land- u. Forstwirtsch. **223** (1984): 255.

Lyr, H.: Effect of fungicides on energy production and intermediary metabolism. In: Siegel M. R., and Sisler, H. D. (Eds.): Antifungal Compounds. Vol. 2. Marcel Dekker Inc., New York 1977, pp. 301—332.

Lyr, H.: Biochemical mechanisms of fungi for toxification and detoxification of fungicides. In: Tagungsbericht Akad. Landwirtschaftswiss. DDR **222**, 1984.

— and Casperson, G.: On the mechanism of action and the phenomenon of cross resistance of aromatic hydrocarbon fungicides and dicarboximide fungicides. Acta Phytopath. Acad. Sci. Hungary **17** (1982): 317—326.

Mercier, J., and Reefelder, R.: A method for screening phylloplane antagonists to *Sclerotinia sclerotiorum* on lettuce. Phytopathology **74** (1984): 804 and personal communication 1984.

Nakanishi, T., and Oku, H.: Metabolism and accumulation of pentachloronitrobenzene by phytopathogenic fungi in relation to selective toxicity. Phytopathology **59** (1969): 1761—1762.

Richmond, D. V.: Permeation and migration of fungicides in fungal cells. In: Siegel, M. R., and Sisler, H. D. (Eds.): Antifungal Compounds. Vol. 2. Marcel Dekker Incorp., New York 1977, pp. 251—276.

Stenzel, K.: Untersuchungen zur verminderten Ausbreitung Echter Mehltaupilze auf induziert resistenten Pflanzen. Mitt. Biol. Bundesanstalt f. Land- u. Fortwirtsch. **223** (1984): 238.

Torgeson, D. C.: Fungicides, an advanced treatise. Vol. II, Academic Press Inc., New York-London 1969.

Westcott, E. W., and Sisler, H. D.: Uptake of cycloheximide by a sensitive and a resistant yeast. Phytopathology **54** (1964): 1261—1264.

Zollfrank, G., and Lyr, H.: Die Variabilität der Metalaxyl-Empfindlichkeit bei verschiedenen Pathotypen von *Phytophthora infestans*. Arch. Phytopath. u. Pflanzenschutz (1987) (im Druck).

Lyr, H. (Ed.): Modern, Selective Fungicides-Properties, Applications, Mechanisms of Action. Longman Group UK Ltd., London, and VEB Gustav Fischer Verlag, Jena, 1987.

Chapter 3

Development of resistance to modern fungicides and strategies for its avoidance

J. DEKKER
Agricultural University, Wageningen, The Netherlands

Introduction

Like many other living organisms, fungi may become resistant to toxicants. In a population of a fungal plant pathogen, sensitive to a particular fungicide, cells may emerge by mutation or otherwise which are significantly less sensitive. The frequency of mutation towards resistance, usually between 10^{-4}—10^{-10}, may increase considerably by exposure of the fungus to mutagenic agents. Administration of the fungicide concerned will favour the resistant cells by elimination of competition from the sensitive cells. The fungicide itself does not induce resistance, but acts solely as a selecting agent. If a particular fungicide would appear to be also mutagenic, it should not be used on agricultural crops which are grown for consumption. When under selection pressure by the fungicide build-up of a resistant pathogen population occurs, it may result in failure of disease control. This has happened many times, especially after the introduction in the nineteen sixties of specific-site fungicides, among which were many systemic fungicides. In a review, compiled in 1979 (OGAWA *et al.* 1983), more than 100 plant pathogen species are listed, which have become resistant to fungicides under field conditions, a number which has grown considerably in recent years. Failure of disease control due to acquired resistance has been reported, among others, for acylalanines, benzimidazole fungicides, thiophanates, carboxanilides, hydroxypyrimidines, organophosphates (see chapters 17, 15, 19, 20) and for antibiotics such as kasugamycin, used against *Pyricularia oryzae* on rice (MIURA *et al.* 1975) and polyoxin B, used against *Alternaria kikuchiana* on Japanese pear (NISHIMURA *et al.* 1973). In other cases a moderate decrease in sensitivity has been observed without a rapid loss of disease control, e.g. with some inhibitors of sterol biosynthesis (triazoles, pyrimidines, imidazoles) and with dicarboximides, which will be discussed below (see also chapter 7, 13).

Development of fungicide-resistance is now one of the major problems in plant disease control. It causes unexpected crop losses for the grower, and may put him in a difficult position if no adequate substitutes are available. It may reduce the profits of the manufacturer, which has developed the fungicide at high cost, and may even lead to costly lawsuits. Should this occur repeatedly, the agrochemical industry might become hesitant to invest in the costly development of modern fungicides, which in the long run would reduce the arsenal of available disease control agents, again to the disadvantage of the farmer. The development of resistance may further have consequences for the extension officers, the regulatory authorities, and in some cases even for the consumer and the national economy (SCHWINN 1982).

The problems with fungicide-resistance have given impetus to genetic, biochemical and epidemiological studies, in order to clarify the origin of resistance and to understand the factors which govern the build-up of a resistant pathogen population in the field. In this chapter the principles underlying the fungicide-resistance phenomenon will be treated, illustrated by examples. The genetic aspects are discussed in chapter 4 by GEORGOPOULOS.

Attention will further be paid to strategies to delay or avoid the occurrence of fungicide-resistance. No attempt has been made to prepare an exhaustive review of resistance to all fungicides concerned. For additional information various reviews are available, among others by GEORGOPOULOS (1977), OGAWA *et al.* (1983), DEKKER (1985) and a book on "Fungicide-resistance in crop protection" by various authors, edited by DEKKER and GEORGOPOULOS (1982).

Definitions

Fungicide-resistance may be defined as the stable, inheritable adjustment by a fungus to a fungicide, resulting in a less than normal sensitivity to that fungicide. The term is used for strains of a sensitive fungal species, which have become, usually by mutation, significantly less sensitive to the fungicide.

Resistance to fungicides with specific-site action is often easily obtained on agar, containing a fungicide concentration, which is lethal to the wild-type fungus. However, the appearance of so-called "laboratory-resistance" does not necessarily imply that resistance problems will arise in the field. This will depend on the level and frequency of resistance. Further, it is obvious that failure of disease control should not be easily attributed to resistance before appropriate tests have shown that it is indeed caused by the presence of resistant strains.

Quite often the term tolerance is used instead of or interchangeably with the term resistance, and sometimes also in addition to resistance to indiciate a less severe type of resistance. The latter view has no grammatical basis. Moreover, the concept tolerance is already used in another way, namely as the maximum amount of a pesticide residue that may lawfully remain on or in food. In view of this the FAO panel of experts on pest resistance to pesticides, meeting in 1978, "agreed that resistance should continue to apply to hereditable resistance in fungi and bacteria, and recommended that the word tolerance should not be used, since it may be ambiguous" (Anonymous 1979). Some plant breeders prefer the term insensitivity instead of resistance, in order to avoid confusion with resistance of the plant to diseases. A disadvantage of the word insensitivity, however, is that it suggests a complete loss of sensitivity, which will seldom occur.

The level of resistance, or resistance factor, may be expressed as the ratio: ED_{50} or EC_{50} resistant strain/ED_{50} or EC_{50} wild-type fungus. This term should not be confused with frequency of resistance, which is the percentage of resistant isolates in the pathogen population.

The term (positive) cross-resistance designates resistance to two or more toxicants, mediated by the same genetic factor. When such a factor mediates resistance to one fungicide and at the same time increased sensitivity to a second fungicide, the term negative cross-resistance is used. Cross-resistance should not be confused with multiple resistance, which means resistance to two or more toxicants, mediated by different genetic factors.

Mechanisms of resistance

General

Nowadays several fungicides are known which interfere at specific sites with biosynthetic processes in fungi (e.g. synthesis of nucleic acids, proteins, ergosterol, chitins), with respiration, membrane structure or nuclear function. Development of resistance to such specific-site inhibitors is often due to single gene mutation, resulting in a slightly changed target site, with reduced affinity to the fungicide. It is obvious that resistance to multi-site inhibitors, comprising most conventional fungicides cannot be achieved in this way, as it would require simultaneous changes at many sites. This does not imply, however, that reducttion of sensitivity to these fungicides will not occur. It may happen when a change in the fungal cell occurs, which prevents the fungicide from reaching the site of action. This may happen by binding at other places, conversion into non fungicidal compounds, or reduced uptake. Experience during a century of fungicide use in agriculture has learned, however, that this does not occur easily. In fact, with conventional fungicides only rarely resistance problems arose.

The mechanism of resistance is not only of scientific interest, but also of practical significance, as it may influence the fitness of resistant strains. This will be discussed on page 43. Resistance to specific-site fungicides is mostly caused by a change at the site of action or by reduced uptake, whereas with insecticides conversion into an inactive compound is the predominant mechanism.

Change at site of action

There is an increasing number of fungicides, where resistance can be attributed to a change at the site of action, which reduces its affinity to the fungicide. This may be illustrated by a few examples. For more extensive information the reader is referred to a recent review (Dekker 1985) and other chapters in this book.

Carbendazim, the toxic principle of benomyl and thiophanate-methyl, binds to tubulin, the major constituent of microtubules, which constitute the spindle. The assembly of microtubules is prevented, and as a consequence mitosis and other cellular processes in which microtubules are involved, are inhibited. Resistance to carbendazim is caused by single gene mutation, resulting in slightly changed tubulin with a reduced affinity to carbendazim (Davidse 1982).

The antibiotic kasugamycin, used for control of rice blast, inhibits protein synthesis in some bacteria and fungi by attaching to one of the subunits of the ribosome. After several years of application in Japan, failure of disease control occurred, due to development of resistance. This appeared to be caused by single mutation in each of three different loci for resistance in the causal organism, *Pyricularia oryzae* (perfect stage *Magnaporthe grisea*) (Taga *et al.* 1978). Studies with cell free systems learned that resistance was due to a change in the ribosome, resulting in reduced affinity to the antibiotic (Misato and Ko 1975).

Also metalaxyl, an acylalanine active against Oomycetes, has met serious resistance problems in the field. In studies with *Phytophthora megasperma* f. sp. *medicaginis* Davidse *et al.* (1983) showed that metalaxyl inhibits RNA synthesis by specific interference with template bound RNA polymerase activity. Endogenous polymerase activity of nuclei isolated from a metalaxyl resistant mutant was not inhibited by the fungicide, which indicates a modification of the target site.

Changes in the respiratory chain may be responsible for resistance to fungicides, which specifically act at certain steps in this chain. Carboxin, a carboxamide and mainly active against Basidiomycetes, inhibits fungal respiration by specific binding to the succinate-ubiquinone reductase complex (complex II), thus blocking the electron flow (Mowery *et al.* 1977). In laboratory experiments carboxin resistant mutants of *Ustilago* spp. emerged readily. Evidence has been obtained that at least some of these mutants have a slightly changed complex II, resulting in a decreased afficity for carboxin (Georgopoulos 1982b).

Reduced uptake, detoxification

The fungal cell may become less sensitive to a fungicide by changes, which keep the fungicide from reaching the site of action in sufficient quantity. These changes may hamper entrance by the fungicide through the membrane, or they may lead to an increased efflux immediately after entrance, preventing accumulation. Further there may be changes which increase the capacity of the cell to detoxify the fungicide. This may happen by conversion of the fungicide into non fungitoxic compounds or by binding to other cell constituents before the sites of action have been reached. A few examples may illustrate this.

Polyoxin antibiotics, which interfere with chitin synthesis in fungi, are used for control of black spot in Japanese pear. Strains of *Alternaria kikuchiana*, resistant to polyoxin B, appeared in orchards treated with the antibiotic, resulting in failure of disease control. Resistance was not due to a change at the target site, chitin synthetase as this enzyme was equally inhibited in cell free systems of resistant and sensitive strains, but it appeared to be caused by a change in the fungal membrane, resulting in reduced uptake (MISATO *et al.* 1977). Lack of accumulation may also be due to the presence of a constitutive, energy dependent efflux mechanism, as has been shown in studies with fenarimol and *Aspergillus nidulans* (DE WAARD and VAN NISTELROOY 1980a) (cf. chapter 24).

Fungicide-resistance due to conversion into non fungitoxicants seems to occur only rarely. Kitazin-P (5-benzyl 0,0-diisopropyl phosphorothiolate), which among others is used for control of rice blast, interferes with the biosynthesis of phosphatidylcholine. Strains of the causal organism, *Pyricularia oryzae*, were obtained with moderate, others with a high level of resistance to this fungicide. The resistance mechanism of the latter has not yet been elucidated, but that of the former appears to be due to cleavage of the S—C bond of the molecule by the pathogen, which gives non-fungitoxic derivatives (UESUGI and SISLER 1978). It is interesting that resistance can also be based on a lack of conversion of an itself non-fungitoxic compound into a fungicide. Resistance to pyrazophos in *P. oryzae* appeared due to the inability of resistant strains to convert pyrazophos into the actual toxic principle: 2-hydroxy-5-methyl-6-ethoxycarbonylpyrazolo(1,5 a) pyrimidine (DE WAARD and VAN NISTELROOY 1980b).

Other resistance mechanisms

In principle, resistance could be caused also by metabolic changes in the fungal cell which result in a compensation for the inhibiting effect, for example by an increase in the production of an inhibited enzyme, or by changes which result in circumvention of the blocked site by an alternate pathway. No clear cut examples of such a mechanism seem available.

Build-up of a resistant pathogen population

General

The emergence of fungicide-resistant strains on agar medium in laboratory experiments does not necessarily mean that failure of disease control has to be expected after application of this compound in the field. In the first place the frequency of resistant strains may stay low, because their properties and the environmental conditions are not conducive to the build-up of a resistant pathogen population, and, secondly, the level of resistance may stay relatively moderate, so that the fungicide continues to provide control. Failure of disease control will not arise before a considerable proportion of the pathogen population has become resistant. In some cases this happened shortly after introduction, e.g. with benomyl and metalaxyl, but in other cases it took many years, e.g. Kitazin-P resistance in *P. oryzae* (UESUGI 1982) and dodine resistance in *Venturia inaequalis* (GILPATRICK 1982). Factors which influence the speed of build-up of a resistant pathogen population are: the fitness of resistant strains in the presence or absence of the fungicide, the nature of the pathogen and the disease, the selection pressure by the fungicide and others.

Fitness of resistant strains

Fitness is a comparative concept: a strain may be more or less fit than another one in a particular situation. It is further a rather complex property with many parameters involved, including infection chance, speed of colonization of the host tissue and degree of sporulation. Resistance may be linked to reduced fitness when the change in the fungal cell, responsible for resistance, has a disadvantage under normal conditions, i.e. in absence of the fungicide. This may be illustrated by studies with the antibiotic pimaricin, an experimental fungicide active against bulb rot in narcissus, caused by *Fusarium oxysporum* f. sp. *narcissi* (DEKKER and GIELINK 1979a). This fungicide complexes with ergosterol in the fungal membrane, thus causing leakage and cell death. In mutants of yeasts with resistance to pimaricin or the closely related nystatin, ergosterol appeared to be replaced by one of its precursors. This resulted in two effects: resistance, because of lower affinity of this precursor to pimaricin, and at the same time reduced fitness, because the more primitive precursor functions less well in the membrane stabilization than ergosterol.

There are indications that a link between resistance and reduced fitness might exist also for certain commercial fungicides, such as fenarimol (DE WAARD and VAN NISTELROOY 1982). Reduced fitness of fenarimol resistant strains can possibly be explained by the assumption that the constant presence of an energy dependent efflux mechanism is an extra burden for the cell, which is disadvantageous in absence of the fungicide. Reduced fitness of resistant strains has also been reported for other sterol biosynthesis inhibitors, such as triforine (FUCHS *et al.* 1977). Another example, may be, is dicarboximide resistance in *Botrytis cinerea*. From vineyeards, attacked by grey mold, some strains of *Botrytis cinerea* were isolated with a low level and others with a higher level of resistance. The latter strains appeared to possess a higher osmotic sensitivity than the wild-type pathogen. This might be the cause of reduced virulence of such strains in view of the high sugar content of the grapes (BEEVER and BYRDE 1982) (but compare chapter 7). Decreased fitness has further been reported for ethirimol resistant strains of *Erysiphe graminis* (SHEPHARD *et al.* 1975) and pyrazophos resistant strains of *Sphaerotheca fuliginea* (DEKKER and GIELINK 1979b). It is obvious that reduced fitness will hamper the build-up of a resistant pathogen population, because the less fit strains will lose ground to the sensitive pathogen population. Moreover, when resistance is inversely related to fitness, the level of resistance may be limited. This may also have consequences for disease control.

There are, however, also fungicides where resistance is not linked to reduced fitness. This means that also resistant strains may emerge with a fitness which does not significantly differ from that of sensitive strains. Examples are benzimidazoles and acylalanine fungicides. In such cases there are, apparently, changes possible at the sites of action (namely the tubulin monomers of the microtubules and the RNA polymerase-template complex, respectively) which are not disadvantageous to the cell. The emergence of such strains will favour the shift towards a resistant pathogen population.

It may be concluded that the fitness of fungicide-resistant strains seems related to the mechanism by which the pathogen becomes resistant to that fungicide. Some metabolic changes, conferring resistance, are linked to lower fitness in absence of the fungicide, others are not. On this basis some fungicides seem more risky with respect to development of resistance than others. However, the possibility may exist that strains, in which originally resistance seems linked to lower fitness in the absence of the fungicide, may increase their fitness in the course of time. It should further be stressed that development of resistance in the field will depend also on other factors.

Nature of pathogen and disease

Resistance will build up more rapidly in an abundantly sporulating pathogen on aerial plant parts than in a pathogen which sporulates scarcely and spreads slowly, such as with certain soil borne root or foot diseases. For example, resistance to metalaxyl developed rapidly in *Phytophthora infestans* in potatoes (Davidse *et al.* 1981) so that control of this disease by the single compound even had to be abandoned, but problems with control of various *Phytophthora* root diseases in other crops, seem to occur less readily. However, also here problems may arise. Indications were obtained that reduced control of *Phytophthora cinnamomi* in avocado by metalaxyl was due to development of fungicide-resistance (Kotzé 1983).

The influence which the fungal life cycle may have was shown in apple orchards, treated with benomyl for control of scab and rust. After some time failure of scab control occurred but there were no problems with control of rust. The explanation may be that *Venturia inaequalis* does have a repetitive summer cycle on apple, which favours build-up of resistance, but *Gymnosporangium juniperi viriginianae* does not (Gilpatrick 1982).

Thus in calculating the risk for development of resistance, also the type of disease should be taken into consideration.

Selection pressure by fungicide

The degree of selection pressure will mainly be determined by the doses applied, the frequency of application, the persistence of the fungicide on the crop or in the soil, and by the thoroughness of the treatment. Also the method of application may play a role, as, for example, treatment of the seed or the soil by a systemic compound may make a long lasting uptake of the chemical possible. Further, the size of the area treated with a particular fungicide may contribute to selection pressure, since this influences the influx of sensitive strains from outside. Insight in these factors is important for the design of counter measures to avoid resistance (Dekker 1982).

Environmental and other factors

In addition to the above-mentioned factors, there may be other phenomena, which influence the build-up of a resistant pathogen population. Samoucha and Cohen (1984) made the interesting observation that synergism occurs between metalaxyl-sensitive and -resistant strains of *Pseudoperonospora cubensis*. They found that the sensitive strains stimulated the release of zoospores and the infection by the resistant strains. Wild and Eckert (1982) found synergism between a benzimidazole-sensitive and benzimidazole-resistant isolate of *Penicillium digitatum*. The former, although unable to infect citrus fruit in the presence of benomyl, germinated and produced pectolytic enzymes. This caused breakdown of cell wall material, followed by release of nutrients, which increased infection by resistant isolates.

Further, environmental conditions, among others the weather, may play a role. For example, failure of disease control due to resistance will sooner occur under conditions which favour the outbreak of an epidemic.

Level of resistance in relation to disease control

In addition to the proportion of resistant strains in the pathogen population, also the level of resistance is important. If it is only low or moderate, fungicide applications may continue to provide satisfactory disease control, even when the majority of the pathogen population has become less sensitive to the fungicide. It is obvious that the level of resistance will be limited in those cases where an inverse relation exists between resistance and fitness as has been suggested for resistance to dicarboximides. Although resistant strains of *Botrytis cinerea* have been found in many areas, where dicarboximides had been frequently used for several years, loss of field control has not become a major field problem (POMMER and LORENZ 1982). They attributed this to a loss of vigour in dicarboximide resistant strains. LORENZ and EICHHORN (1982) found that the proportion of resistant strains increases during periods of high selection pressure, but decreases again in between two growing seasons. Lower fitness of resistant strains, however, is not a guarantee that no problems will arise in the future. Under continuous selection pressure by the fungicide there may be an evolution towards higher fitness among resistant strains, by acquiring the ability to compensate for the reduced fitness in some way or another. Recently indeed there have already been reports about reduced disease control, due to decreased dicarboximide sensitivity (LEROUX and BESSELAT 1984). Further there may be exceptions to the rule that high resistance is connected with lower fitness, as shown by GRINDLE and TEMPLE (1985) for one out of a number of vinclozolin-resistant isolates of the non-pathogen *Neurospora crassa*. However, comparable variants of fungal pathogens have, so far, not been recovered from field populations. If they would occur, they might compete with the normal wild type pathogen and become a problem in practice.

Reduced fitness of resistant strains has also been reported for sterol biosynthesis inhibitors as fenarimol, triforine and triadimefon. This may explain why the use of triadimefon against barley powdery mildew in the U. K. has not resulted in clear cut failure of disease control, in spite of a considerable increase in the frequency of less sensitive strains (WOLFE *et al.* 1984). Several surveys are being carried out to investigate whether the level of resistance to sterol biosynthesis inhibitors increases in the course of the years. An example is the one carried out from 1982—1984 in commercial cucumber greenhouses in The Netherlands, where several biosynthesis inhibitors have been in use for powdery mildew control during a number of years (SCHEPERS 1985). Although a significant shift towards fungicide resistance was observed with fenarimol, imazalil and triforine, complete failure of disease control did not yet occur. It still has to be awaited whether the level of resistance will continue to rise in the years to come.

Tactics to avoid development of resistance

General

On the basis of the knowledge obtained, measures may be taken to counteract the development of resistance. This should be done before a build-up of resistance has taken place. Once the pathogen population has become resistant, the only possibility is to change to other fungicides with a different mechanism of action, or to non-chemical control measures, if available.

Firstly, it is desirable to obtain some information about the resistance risks of a new fungicide. When a fungicide has to be used which is resistance prone, it is imperative to take stringent measures to avoid or at least delay the development of resistance. In cases where resistance may reach a high level beyond control by the fungicide, a reduction of selection pressure by the fungicide concerned should be considered, when necessary and possible in combination or alternation with other fungicides. On the other hand, when only a moderate level of resistance is likely to occur, an increase of the selection pressure, for as far technically and economically feasible, may be considered in order to kill also the moderate-level resistant strains. However, monitoring is desirable to keep track of the level of resistance, which may increase in the course of years, and the frequency of resistance.

For more information on tactics to avoid resistance the reader is referred to Dekker (1982), Delp (1980), Staub and Sozzi (1984).

Prediction

Experiments on an artificial medium, with or without mutagenic agents, may inform us whether emergence of resistant cells is possible, by mutation or otherwise. When no resistant mutants emerge in this way, in the pathogen concerned or in related species, it is unlikely that they will appear in the field. But if they do, or when they are already present at low frequency in a wild-type population, their fitness should be tested on the plant. Although this information may provide an indication of the resistance risk of the fungicide concerned, it will be difficult to predict exactly what is going to happen in practice. Conditions in the greenhouse are never exactly the same as in the field, and even field experiments may not yield the results that can be obtained in large-scale application in practice, as also the size of the area may play a role. Nevertheless, experiments may indicate some of the risks involved, which is useful when devising counter measures.

Reduction of selection pressure

It should be clear that a continuous selection pressure by one particular fungicide or by fungicides which show cross-resistance, will increase the chance for build-up of resistance, if not on short notice, than anyway in the long run. It should be avoided to carry out unnecessary applications and to apply larger quantities of the chemical than needed, not only to avoid resistance, but also for economical and environmental reasons. It is not advisable to use the same type of chemical for treatment of seed or plant material, spraying of the crop and post harvest treatment, since development of resistance in the earlier treatments may jeopardize the effect of the later applications. Also the treatment of a crop in a whole region or country with the same type of chemical should be avoided, as it increases the selection pressure. It should further be realized that also a very thorough treatment of the crop will increase the selection pressure and therefore favour the build-up of resistance. It will be obvious, however, that the possibilities to reduce the selection pressure, without use of other fungicides, are limited.

Combination and alternation of fungicides

A prerequisite for the use of two or more fungicides to avoid resistance, is that these fungicides do now show cross-resistance. In a mixture the companion should preferably be a conventional and anyway a low risk fungicide, as a combination of two risky fungicides may lead to multiple resistance. Considering the usually extremely large populations of fungal cells, this may happen rather easily (Wild 1983).

With respect to the effect of a mixture on development of resistance, there is much difference of opinion in the literature. Some authors claim that the use of a mixture counteracts resistance, but others do not observe a delay in the build-up of resistance. Resistance to the risky chemical in such a mixture may build up in that part of the pathogen population which has not been killed by the companion compound. The speed with which this occurs depends on various factors, among which, according to a model by Kable and Jeffery (1980), the "escape" is the single most important factor. When there are only few fungal cells which escape from being hit by the mixture and when selection pressure is continuous, the speed of build-up of resistance of the risky compound, expressed in percentage of the pathogen population, will be the same for the mixture as when the risky compound alone is applied. In that case resistant strains face hardly any competition from sensitive strains during spray intervals. Such a situation will not likely occur in the field, and in most cases a mixture will at least delay the build-up of resistance. When the duration of released selection pressure between treatments, and reduced fitness of resistant strains allow a sufficient drop in the proportion of resistant cells, resistance problems might even be avoided indefinitely.

With alternating use of a risky and a non risky fungicide, there will only be selection pressure towards resistance during the periods that the risky chemical is present in the plants. During the periods that only the non risky chemical is present, the proportion of resistant cells may remain the same, or it may decrease, depending on the fitness of resistant strains in absence of the fungicide at risk. If it remains the same, and increases only in presence of the risky compound, there will be a build-up of resistance in steps, which eventually may result in failure of disease control. If the proportion of resistant cells decreases during periods that the risky compound is absent, the overall step-wise increase of resistance will be slower than in the former case, or not happen at all. It is not possible to make a general statement, whether combination or alternation is a better tactic to avoid resistance. The mixture may be more effective under certain conditions and the alternations under other conditions, depending on the values of the parameters involved.

In addition to combinations or simple alternations more complex application sequences may be followed. A mixture has the disadvantage that the compound at risk is continuously applied, and with alternation the application of the non risky compound is interrupted for no good reason. Better results might therefore be obtained by a sequential scheme in which mixtures and rotation are combined in such a way that the non risky compound is constantly present and that only the application of the risky compound is interrupted. Calculations, based on the model by Kable and Jeffery, support this assumption (Tab. 3.1).

Table 3.1 The effect of combined or alternating use of two fungicides on build-up of a resistant pathogen population. Simulation model, assuming specific values for spray coverage[1], efficacies[2] and initial resistance level[3] (DEKKER 1982, modified after KABLE and JEFFERY 1980)

Fungicide	Percentage of population resistant after 5—40 sprayings				
	5	10	20	30	40
S—S—S—S	0.0	82.6	100.0	100.0	100.0
(S+C)—(S + C)—(S + C)—(S + C)	0.0	0.0	99.6	100.0	100.0
S—C—S—C	0.0	0.0	82.6	100.0	100.0
(S + C)—C—(S + C)—C	0.0	0.0	0.0	26.1	99.6

[1] E (escape) = proportion of population escaping fungicide contact, set at 5 %.

[2] S (systemic fungicide) with efficacy against sensitive and resistant subpopulations set at 90 % and 10 %, respectively.
C (conventional fungicide) with efficacy against both S-sensitive and S-resistant subpopulations set at 80 %.

[3] Initial resistance frequency set at 10^{-9}.

In addition to the valuable but rather limited model of KABLE and JEFFERY, other authors have presented models with more parameters (SKYLAKAKIS 1982; LEVY *et al.* 1983). Although these models cannot yet be used for application in practice, they certainly help to increase our knowledge about the influence which particular parameters have on the build-up of resistance.

Integrated control

For avoidance of resistance a combination of chemical with non-chemical methods may be considered, e.g. natural host resistance, cultural measures, biological control. WOLFE (1984) advocates an approach in which different fungicides are used in combination with several varieties, where resistance is based on different resistance genes. Exerting selection pressure on the pathogen in repeatedly changing directions should "confuse" the pathogen.

Detection and monitoring

In order to detect build-up of resistance in an early stage, so that counter measures might still be possible, it is often advised to monitor for resistance in the field. However, if development of resistance occurs very fast, the information may come too late for adequate counter measures. Timely detection of resistance might offer perspectives in those cases, where build-up of resistance occurs only slowly, and requires more than one growing season.

Another difficulty is that detection of resistance in an early stage, e.g. at the 1 % level, would require a fairly large scale and costly sampling procedure. It is easier and less costly to monitor for performance of the fungicide, since in that case only few samples have to be analyzed from spots where failure of disease control is suspected to be caused by resistance. Although it will than be too late for counter measures in the field concerned, the information will provide a warning to take measures in other fields.

Long term strategies

Availability of a broad fungicide arsenal

To be able to avoid the development of fungicide resistance in the future it is important that a broad arsenal of fungicides is available, so that flexibility exists in the design of counter measures. Therefore the industry should continue to search for and develop new fungicides, especially fungicides with yet unknown mechanisms of action. Special attention should be given to site-specific inhibitors that show a low risk for development of resistance. In view of this the search for new disease control agents should include chemicals which are not fungitoxic in itself, but which interfere in other ways with the relation between plant and parasite. Such compounds may induce or increase the natural resistance of the host, or may decrease the capability of the pathogen to attack the plant. The compounds which have shown activity along these lines have been reviewed recently (Dekker 1983; Wade 1984). Some of these might not or might less readily encounter resistance. In addition to strengthening our research efforts to discover new chemicals, we must ensure that the number of conventional fungicides is not needlessly decreased by regulatory agencies. They may be needed to fall back upon in case resistance to specific-site fungicides occurs, and they may be used in combinations or rotations to avoid resistance.

Negative cross-resistance

A quite interesting and potentially promising phenomenon is negatively correlated cross-resistance between two fungicides. When one fungicide is especially active against strains which have become resistant to a second fungicide, and when this second fungicide is more than normally active against strains resistant to the first fungicide, the development of resistance will in principle be precluded. However, this will only be succesful when negative cross-resistance holds for all strains, and when no other resistance mechanisms become apparent. Negative cross-resistance has been observed for benomyl and thiabendazole (Van Tuyl *et al.* 1974), for some carboxamide fungicides (Georgopoulos 1982a) and for phosphoramidate and phosphorothiolate fungicides (Uesugi 1982). The problem with these cases of negative cross-resistance was that it did not hold for all resistant strains. Remarkable is the negative cross-resistance, shown by *Botrytis cinerea* and three other plant pathogens for benomyl and methyl N-(3,5-dichlorophenol) carbamate (MDPC) which seems to hold for all resistant isolates (Kato *et al.* 1984). In experiments with *Venturia nashicola*, however, Ishii *et al.* (1984) found negative cross-resistance to these compounds only for the highly carbendazim resistant strains, but not for the strains with intermediate or low carbendazim resistance. Rosenberger and Meyer (1985) observed that benomyl-resistant strains of *Penicillium expansum* were more sensitive to diphenylamine than the wild type pathogen, and that this effect was temperature dependent.

Negative cross-resistance has further been reported for two sterol biosynthesis inhibitors. Fenarimol-resistant mutants of *Penicillium italicum*, which cause post harvest rot in citrus, appeared more sensitive to fenpropimorph than the wild type strain (De Waard and Van Nistelrooy 1982), but negative cross-resistance with respect to these compounds was not observed for *Aspergillus nidulans*. The use of negative cross-resistance as a tool to avoid resistance should further be explored (De Waard 1984).

Synergism

Several reports mention synergistic action between two fungicides. This phenomenon might appear to have also value for avoidance of resistance if the second fungicide interferes with the resistance mechanism towards the first fungicide. As an example the synergism between fenarimol and captan might be mentioned. As has been mentioned above, resistance to fenarimol depends on an energy-dependent efflux. This efflux is inhibited by addition of captan or other compounds, which exert an inhibitory effect on respiration (De Waard 1984). It is desirable that such possibilities of the phenomenon of synergism are further investigated (see chapter 24).

Implementation

Any long- term strategy should include the creation of possibilities for the implementation of tactics to prevent or manage fungicide resistance. For this reason it is necessary to establish and maintain an efficient communication system among growers, extension officers, teachers, research workers, manufacturers, salesmen, the press, regulatory agencies and the government.

Concluding remarks

Fungicide-resistance has become one of the major problems for control of fungal plant diseases. Tactics to prolong the life of badly needed fungicides and long term strategies to ensure efficient disease control are therefore much necessary. For the design of such tactics and strategies knowledge about the underlying genetical and biochemical principles of the fungicide-resistance phenomenon is needed, which can only be obtained by research. Equally important is insight in the behaviour of resistant strains in the field, to which the use of epidemiological models may contribute.

A prerequisite for the design of tactics to avoid resistance is the availability of a varied arsenal of fungicides, conventional as well as systemic fungicides, which act at different sites in the fungal cell, and against which different mechanisms of resistance operate. This requires continued research efforts and a prudent and conservative policy by regulatory agencies.

References

Anonymus. Pest resistance to pesticides and crop loss assessment. Report on the 2nd session of the FAO panel of experts, held in Rome, 28 August—1 September 1978. FAO Plant Product. and Protect. Paper **6**, 2 (1979): 1—41.

Beever, R. E., and Byrde, R. J. W.: Resistance to the dicarboximide fungicides. In: Dekker, J., and Georgopoulos, S. G. (Eds.): Fungicide Resistance in Crop Protection. Pudoc, Wageningen 1982, pp. 101—117.

Davidse, L. C.: Benzimidazole compounds: selectivity and resistance. In: Dekker, J., and Georgopoulos, S. G. (Eds.): Fungicide Resistance in Crop Protection. Pudoc, Wageningen 1982, pp. 60—70.

— Hofman, A. E., and Velthuis, G. C. M.: Specific interference of metalaxyl with endogenous RNA polymerase activity in isolated nuclei from *Phytophthora megasperma* f. sp. *medicaginis*. Experiment. Mycology **7** (1983): 344—361.

— LOOYEN, D., TURKENSTEEN, L. J., and VAN DER WAL, D.: Occurrence of metalaxyl resistant strains of *Phytophthora infestans* in Dutch potato fields. Netherl. J. Plant Pathol. **87** (1981): 65—68.

DEKKER, J.: Counter measures for avoiding fungicide resistance. In: DEKKER, J., and GEORGOPOULOS, S. G. (Eds.): Fungicide Resistance in Crop Protection. Pudoc, Wageningen 1982, pp. 177—186.

— Non fungicidal compounds, which prevent disease development. Proc. Brit. Crop Protect. Conf. Brighton **1** (1983): 237—248.

— The development of resistance to fungicides. Progress in Pesticide Biochem. Toxicology, **4** (1985): 165—218.

— and GEORGOPOULOS, S. G. (Eds.): Fungicide Resistance in Crop Protection. Pudoc, Wageningen 1982, pp. 273.

— and GIELINK, A. J.: Acquired resistance to pimaricin in *Cladosporium cucumerinum* and *Fusarium oxysporum* f. sp. *narcissi* associated with decreased virulence. Netherl. J. Plant Pathol. **85** (1979a): 67—73.

— — Decreased sensitivity to pyrazophos of cucumber and gherkin powdery mildew. Netherl. J. Plant Pathol. **85** (1979b): 137—142.

DELP, C. J.: Coping with resistance to plant disease control agents. Plant Disease **64** (1980): 652—657.

DE WAARD, M. A.: Negatively correlated cross-resistance and synergism as strategies in coping with fungicide resistance. Proc. Brit. Crop Protect. Conf. **2** (1984): 573—584.

— and VAN NISTELROOY, J. G. M.: An energy-dependent efflux mechanism for fenarimol in a wild-type strain and fenarimol-resistant mutants of *Aspergillus nidulans*. Pesticide Biochem. Physiol. **13** (1980a): 255—266.

— — Mechanism of resistance to pyrazophos in *Pyricularia oryzae*. Netherl. J. Plant Pathol. **86** (1980b): 251—258.

— — Laboratory resistance to fungicides which inhibit ergosterol biosynthesis in *Penicillium italicum*. Netherl. J. Plant Pathol. **88** (1982): 99—112.

FUCHS, A., DE RUIG, S. P., VAN TUIL, J. M., and DE VRIES, F. W.: Resistance to triforine: a non existent problem? Netherl. J. Plant Pathol. **83** (1977) Suppl. **1**: 189—205.

GEORGOPOULOS, S. G.: Development of fungal resistance to fungicides. In: SIEGEL, M. R., and SISLER, H. D. (Eds.): Antifungal Compounds. Vol. 2. Marcel Dekker Inc., New York-Basel 1977, pp. 409—495.

— Cross-resistance. In: DEKKER, J., and GEORGOPOULOS, S. G. (Eds.): Fungicide Resistance in Crop Protection. Pudoc, Wageningen 1982a, pp. 53—59.

— Genetical and biochemical background of fungicide resistance. In: DEKKER, J., and GEORGOPOULOS, S. G. (Eds.): Fungicide Resistance and Crop Protection. Pudoc, Wageningen 1982b, pp. 46—52.

GILPATRICK, J. D.: Case study 2: *Venturia* of pome fruits and *Monilinia* of stone fruits. In: DEKKER, J., and GEORGOPOULOS, S. G. (Eds.): Fungicide Resistance in Crop Protection. Pudoc, Wageningen 1982, pp. 195—206.

GRINDLE, M., and TEMPLE, W.: Sporulation and osmotic sensitivity of dicarboximide resistant mutants of *Neurospora crassa*. Transactions Brit. Mycolog. Soc. **94** (1985): 369—372.

ISHII, H., YANASE, H., and DEKKER, J.: Resistance of *Venturia nashicola* to benzimidazole fungicides. Meded. Fac. Landbouwwet., Rijksuniv. Gent **49/2a** (1984): 163—172.

KABLE, P. F., and JEFFERY, H.: Selection for tolerance in organisms exposed to sprays of biocide mixtures: a theoretical model. Phytopathology **70** (1980): 8—12.

KATO, T., SUZUKI, K., TAKAHASHI, J., and KAMOSHITA, K.: Negatively correlated cross-resistance between benzimidazole fungicides and methyl N-(3,5-dichlorophenyl) carbamate. J. Pesticide Sci. **9** (1984): 489—495.

KOTZÉ, J. M.: Integrated protection in subtropical crops. Proc. 10th Intern. Congr. Plant Protect. November 20—25 (1983): 984—989.

LEROUX, P., and BESSEAT, B.: Pourriture grise: La résistance aux fongicides de *Botrytis cinerea*. Phytoma **359** (1984): 25—31.

LEVY, Y., LEVI, R., and COHEN, Y.: Build up of a pathogen subpopulation resistant to a systemic fungicide under various control strategies: a flexible simulation model. Phytopathology **73** (1983): 1475—1480.

LORENZ, P. H., and EICHHORN, K. W.: *Botrytis cinerea* and its resistance to dicarboximide fungicides. EPPO Bull. **12** (1982): 125—129.

Misato, T., Kakiki, K., and Hori, M.: Mechanism of polyoxin resistance. Netherl. J. Plant Pathol. **83** (1977), suppl. **1**: 253—260.

— and Ko, K.: The development of resistance to agricultural antibiotics. Environmental quality and safety, suppl. **3** (1975): 437—440.

Miura, H., Ito, H., and Takahashi, S.: Occurrence of resistant strains of *Pyricularia oryzae* to kasugamycin as a cause of the diminished fungicidal activity to rice blast. Ann. Phytopathol. Soc. Jap. **41** (1975): 415—417.

Mowery, P. C., Steenkamp, D. J., Ackrell, B. A. C., Singer, T. P., and White, G. A.: Inhibition of Mammalian Succinate Dehydrogenase by Carboxins. Arch. Biochem. and Biophys. **178** (1977): 495—506.

Nishimura, S., Kohmoto, K., and Udagawa, H.: Field emergence of fungicide-tolerant strains in *Alternaria kikuchiama* Tanaki. Rep. Tottori Mycolog. Inst., Jap. **10** (1973): 677—686.

Ogawa, J. M., Manji, B. T., Heaton, C. R., Petrie, J., and Sonoda, R. M.: Methods for detection and monitoring the resistance of plant pathogens to chemicals. In: Georghiou, G. P., and Saito, T. (Eds.): Pest Resistance to Pesticides. Plenum Press, New York-London 1983, pp. 117—162.

Pommer, E. H., and Lorenz, G.: Resistance of *Botrytis cinerea* to dicarboximide fungicides — a literature review. Crop Protect. **1** (1982): 221—230.

Rosenberger, D. A., and Meyer, F. W.: Negatively correlated cross-resistance to diphenylamine in benomyl-resistant *Penicillium expansum*. Phytopathology **75** (1985): 74—79.

Samoucha, Y., and Cohen, Y.: Synergy between metalaxyl-sensitive and metalaxyl-resistant strains of *Pseudoperonospora cubensis*. Phytopathology **74** (1984): 376—378.

Schepers, H. T. A. M.: Changes over a three-year period in the sensitivity to ergosterol biosynthesis inhibitors of *Sphaerotheca fuliginea* in The Netherlands. J. Plant Pathol. **91** (1985), in press.

Schwinn, F. J.: Socio-economic impact of fungicide-resistance. In: Dekker, J., and Georgopoulos, S. G. (Eds.): Fungicide Resistance in Crop Protection. Pudoc, Wageningen 1982, pp. 16—23.

Shephard, M. C., Bent, K. J., Woolner, M., and Cole, M. A.: Sensitivity to ethirimol of powdery mildew from UK barley crops. Proc. Brit. Insecticide Fungicide Conf., Brighton **1** (1475): 59—65.

Skylakakis, G.: The development and use of models describing outbreaks of fungicide resistance. Crop Protect. **1** (1982): 249—262.

Staub, T., and Sozzi, D.: Fungicide resistance: a continuing challenge. Plant Disease **68** (1984): 1026—1031.

Taga, M., Nakagawa, H., Tsuda, M., and Uegama, A.: Ascospore analysis of kasugamycin resistance in the perfect stage of *Pyricularia oryzae*. Phytopathology **68** (1978): 815—817.

Uesugi, Y.: *Pyricularia oryzae* in rice. In: Dekker, J., and Georgopoulos, S. G. (Eds.): Fungicide Resistance in Crop Protection. Pudoc, Wageningen 1982, pp. 207—218.

— and Sisler, H. D.: Metabolism of a phosphoramidate by *Pyricularia oryzae* in relation to tolerance and synergism by a phosphorothiolate and isoprothiolane. Pesticide Biochem. Physiol. **9** (1978): 247—254.

van Tuyl, J. M., Davidse, L. C., and Dekker, J.: Lack of cross-resistance to benomyl and thiabendazole in some strains of *Aspergillus nidulans*. Netherl. J. Plant Pathol. **80** (1974): 165 to 168.

Wade, M.: Antifungal agents with an indirect mode of action. In: Trinci, A. P. J., and Ryley, J. F. (Eds.): In: Mode of Action of Antifungal Agents. Symp. Brit. Mycolog. Soc. Manchester, UK, September 1983. Cambridge Univ. Press, Cambridge 1984, pp. 283—298.

Wild, B. L.: Double resistance by citrus green mould *Penicillium digitatum* to the fungicides guazatine and benomyl. Ann. appl. Biol. **103** (1983): 237—241.

— and Eckert, J. W.: Synergy between a benzimidazole-sensitive and benzimidazole-resistant isolate of *Penicillium digitatum*. Phytopathology **72** (1982): 1329—1332.

Wolfe, M. A.: Trying to understand and control powdery mildew. Plant Pathol. **33** (1984): 451—466.

— Minchin, P. N., and Slater, S. E.: Dynamics of triazole sensitivity in barley mildew, nationally and locally. Brit. Crop Protect. Conf. **2** (1984): 465—470.

Lyr, H. (Ed.): Modern, Selective Fungicides — Properties, Applications, Mechanisms of Action. Longman Group UK Ltd., London, and VEB Gustav Fischer Verlag, Jena, 1987.

Chapter 4

The genetics of fungicide resistance

S. G. Georgopoulos
Athens College of Agricultural Sciences, Athens, Greece

Introduction

All biological phenomena are based on genetic controls and the phenomena of toxicity are certainly no exception. Evolution of resistance to xenobiotics can be considered as a normal part of the overal evolutionary process that has enabled life on earth to be sustained and to achieve levels of amazing complexity. Resistance to toxicants, including those made by man, is the result of either re-adjustment of an organism's existing genetic information or acquisition of additional genes from an external source. This latter way of acquiring resistance is very common in bacteria where the additional genes that are needed are carried on plasmids: cytoplasmic, circular DNA molecules which are transmissible between bacterial cells of the same or of different, even unrelated, species. As a rule, mutation of an existing gene in bacteria modifies the molecular target of the inhibitor, while additional genes usually code for inactivating enzymes (Gale *et al.* 1981).

Cytoplasmic, plasmid-like DNAs are recently being recognised in fungi (Esser *et al.* 1983; Gunge 1983) but no evidence is so far available that such elements may play a role in transmiting resistance to antifungal compounds of agricultural importance. Sensitivity to one type of agricultural fungicide (fentin) has been shown to be controlled by a mitochondrial gene, though not in a plant pathogenic fungus (Lancashire and Griffiths 1971). In all other cases where heritable variation for sensitivity to an agricultural fungicide has been demonstrated, the phenotypic differences were shown to result from differences in chromosomal genes.

To understand the work on the genetics of resistance in the fungi, some knowledge of the genetic features of this group of organisms is essential (Fincham *et al.* 1979). Fungi are eukaryotic organisms with well-defined nuclei, each bounded by an envelope which remains intact during division. In the Fungi imperfecti, a sexual reproductive cycle does not exist. In these organisms, genes can be identified only by the analysis of vegetative (mitotic) segregation: The fungus is normally haploid, but occasional fusion of two nuclei in somatic cells permits the selection of diploid strains by appropriate techniques. Such strains tend to produce sectors showing segregation of markers originally present in the heterozygous condition and resulting mainly from occasional crossing-over between homologous mitotic chromatids or from non-disjunction. This sectoring can be exploited for purposes of genetic mapping.

In members of all other groups, the sexual reproductive cycle consists, as in higher organisms, of a regular alternation of a haploid phase, in which the nuclei contain a single set of chromosomes, and a diploid phase in which a double set of chromosomes is present in each nucleus. The transition from haploidy to diploidy results from fusion (karyogamy) of two haploid gamete nuclei which may be contributed by the same strain in the homothallic but have to be produced by two compatible strains in the heterothallic fungi. The converse transition is accomplished through meiosis which reduces the chromosomes from a double to a single set.

In the Basidiomycetes and the filamentous Ascomycetes the vegetative pathogenic phase is entirely haploid, with meiosis following immediately after karyogamy. In the *Peronosporales,* the most important order of the Oomycetes, the situation is quite different: karyogamy occurs immediately or soon after meiosis so that the organism is predominantly diploid.

In many fungi two or more genetically different nuclei can be carried in the same cytoplasm. The balance of nuclear types in such a heterokaryon may vary from the very stable dikaryons of many heterothallic Basidiomycetes to the multinucleate cells of many Ascomycetes in which the proportions of different nuclei may change in response to selection (DAVIS 1966). It is conceivable that if only some nuclei in a heterokaryon carry genes for resistance to a particular fungicide, their proportion will be increased by exposure to the chemical. To what extent such a change will affect the degree of resistance of the heterokaryon will also depend on the dominance of the genes involved.

If the pathogenic phase is comprised by a haploid homokaryon as in Ascomycetes and most Fungi imperfecti, it is irrelevant whether a resistance gene is dominant or semidominant or recessive. It is also sufficient in such a case to only determine the ratio of the phenotypes in the F_1 from a sensitive $\times$ resistant cross, in order to identify the genes involved. In diploids, such as *Phytophthora,* or dikaryons, such as *Ustilago,* a recessive gene for resistance will not affect the phenotype until it becomes homozygous. A semidominant or dominant gene, however, will be expressed in the heterozygote, so that resistance will be recognised earlier in the laboratory and probably also in the field. On the other hand, if the vegetative phase is diploid or dikaryotic recognition of a Mendelian ratio requires selfing of the F_1 from a sensitive $\times$ resistant cross and examination of the phenotypes of the F_2 generation.

Major-gene control of fungicide resistance

Independently of the test organism, resistance to some of the agricultural fungicides develops in one step as a result of mutation of one gene which has a major effect on the phenotype. One such mutation may achieve the highest level of resistance possible. This does not necessarily mean that only one chromosomal locus controls sensitivity to each type of fungicide. However, if more loci are involved, there is no positive interaction between mutant genes at different loci and a stepwise increase of resistance is not possible. In other words, a mutant allele at one locus is epistatic over wild type alleles at other loci. With this type of genetic control of sensitivity, field populations of sensitive fungi give a discontinuous distribution, i.e. each population consists of at least two distinct subpopulations, one sensitive and one resistant. Because of the lack of overlap between subpopulations, each isolate can be easily classified as unequivocally resistant or sensitive (GEORGOPOULOS 1986).

Benzimidazoles

The genetics of resistance to benzimidazole fungicides has been studied in *Aspergillus nidulans* (HASTIE and GEORGOPOULOS 1971; VAN TUYL 1977), *A. niger* (VAN TUYL 1977), *Ceratocystis ulmi* (BRASIER and GIBBS 1975), *Neurospora crassa* (BORCK and BRAYMER 1974), *Venturia inaequalis* (KATAN *et al.* 1983), *V. pyrina* (SHABI and KATAN 1979), *V. nashicola* (ISHII *et al.* 1984), *Ustilago maydis* (VAN TUYL 1977), the yeast *Schizosaccharomyces pombe* (YAMAMOTO 1980; UMEZONO *et al.* 1983), and the slime mold *Physarum polycephalum* (BURLAND *et al.* 1984). In all of these organisms, highly resistant strains can be obtained by mutation of a single gene. The only case in which a two-step process was claimed is that of *Penicillium italicum* (BERAHA and GARBER 1980). In some species, e.g. *A. nidulans* or *S. pombe,* smaller decreases in sensitivity may result from mutations at other loci. This is shown by the recovery of wild type recombinants from the cross of any two mutants differing in the degree of resistance.

By contrast, in *V. inaequalis* and *V. nashicola* no recombinant phenotypes are found among progenies from crosses between resistant mutants of different phenotypes. This clearly indicates that polymorphism of a single gene causes different resistance levels. In cases of multiple alleles at a single locus, what allele is present may affect also the cross-resistance relationships among benzimidazole derivatives (VAN TUYL 1977) or between benzimidazoles and N-phenylcarbamate compounds (ISHII *et al.* 1984).

In diploid strains of *A. nidulans* heterozygous for benomyl resistance, the sensitive allele was shown to be dominant over the resistant one (HASTIE and GEORGOPOULOS 1971). Similarly, in heterozygous plasmodia of *P. polycephalum* carbendazim caused enlarged nuclei as in the case of sensitive homozygotes, indicating recessivences of the resistant allelomorph (BURLAND *et al.* 1984). In *N. crassa* heterokaryons, however, it was the sensitive gene that was recessive (BORCK and BRAYMER 1974). Benzimidazole resistance is not known to be affected by modifying genes or cytoplasmic components.

Aromatic hydrocarbon group

Genes for resistance to the various members of this group, including the dicarboximides and tolclofos methyl, have been recognized in *A. nidulans* (THRELFALL 1968), *Nectria haematococca* (syn. *Hypomyces solani* — GEORGOPOULOS and PANOPOULOS 1966), *N. crassa* (GRINDLE 1984), *N. sitophila* (WHITTINGHAM 1962), *Penicillium expansum* (BERAHA and GARBER 1966), *P. italicum* (BERAHA and GARBER 1980) and *U. maydis* (ZIOGAS, personal communication). Mutant genes at any one of five chromosomal loci in *N. haematococca* give the same degree of resistance which is not affected by the presence of two or more resistant alleles in the same haploid nucleus. The intraallelic interactions appear to vary, depending on the organism. Thus, in diploids of *A. nidulans,* quintozene resistance behaves as a recessive character, while in *N. crassa* heterokaryons, vinclozolin resistance appeared semidominant. Resistance to sodium orthophenylphenolate was recessive in *P. expansum* and dominant in *P. italicum.* Mutants resistant to fungicides of this group are often found osmotically sensitive. Recent studies (ZIOGAS, personal communication), however, indicate that this depends on the mutated locus.

Carboxamides

Mutations to carboxin resistance in *A. nidulans* map in at least three, freely recombining chromosomal genes (GUNATILLEKE *et al.* 1976). Depending on the carbon source utilized, the three types of mutants may be distinguished by their degree of resistance to the fungicide. In *U. maydis,* low resistance to carboxin is conferred by the *ants* mutation which is more easily recognizable by its effect on the cyanide insensitive respiration (ZIOGAS and GEORGOPOULOS 1984). Intermediate and high resistance, however, are the result of two allelic mutations at the *oxr*-1 locus (GEORGOPOULOS and ZIOGAS 1977). In both *A. nidulans* and *U. maydis,* the resistant gene present may also affect cross resistance to particular oxathiin or thiophene carboxamides (WHITE *et al.* 1978; WHITE and THORN 1980; WHITE and GEORGOPOULOS 1986). With respect to intraallelic interaction, it appears that in both species carboxamide resistance is semidominant, apparently because heterozygous diploid or dikaryotic cells contain a mixture of sensitive and resistant mitochondria (GEORGOPOULOS *et al.* 1975).

Kasugamycin

It is only in *Pyricularia oryzae* among the fungi that kasugamycin resistance has been studied. Genetic experiments have revealed the existence of three unlinked loci for resistance to the antibiotic in this organism. Mutation of only the one of the three genes, however, gives cross resistance to blasticidin S (TAGA *et al.* 1978). Double mutant recombinants from dihybrid crosses did not appear to exhibit higher resistance to kasugamycin than the parental strains.

Kasugamycin is also active against bacteria and several studies on bacterial resistance to this antibiotic have been conducted (GALE *et al.* 1981). In *Escherichia coli*, all three resistance genes, *ksg* A, *ksg* B, and *ksg* C, are chromosomal and it does not seem that plasmid mediated kasugamycin resistance has been reported. In merodiploids heterozygous for the *ksg* A gene, resistance was recessive.

Copper

Although genes for resistance to copper have not been recognized in filamentous fungi and failures of copper fungicides against fungal diseases have not been assigned to resistance (GEORGOPOULOS 1985a), it has been known for many years that copper resistance in the yeast *Saccharomyces cerevisiae* is controlled by the *CUP* 1 chromosomal locus (BRENES-POMALES *et al.* 1955). Further increases of the resistance of mutant cells are the result not of mutation of additional genes, but of gene amplification: the locus copy number is positively correlated with the resistance level of a yeast strain (FOGEL and WELCH 1982; KARIN *et al.* 1984).

Copper resistance is also known in bacteria and, in at least one instance, it has created problems in the control of one bacterial plant pathogen: Copper resistant isolates of *Xanthomonas campestris* pv. *vesicatoria* exist in nature and are not controlled by the amount of Cu^{++} available from fixed copper fungicides. The genetic determinant of this resistance is located in a conjugative plasmid (STALL *et al.* 1981).

Streptomycin

Streptomycin in known as an antibacterial antibiotic, but it inhibits the growth of some fungi, particularly Oomycetes. Several strains of *Phytophthora cactorum* resistant to this antibiotic were obtained by SHAW and ELLIOTT (1968). One of these strains required streptomycin for growth. Resistance and dependence were stable through asexual and sexual reproduction but whether chromosomal or cytoplasmic determinants were involved was not shown.

In bacteria, high level resistance to streptomycin is readily acquired in a single step due to mutation of a chromosomal gene (GALE *et al.* 1981). The three phenotypic responses to the antibiotic (sensitivity, dependence, and high-level resistance) appear to involve multiple alleles of the same gene. In merodiploid heterozygous strains of *E. coli*, sensitivity to streptomycin appears to predominate over resistance. Where, however, resistance genes code for specific enzymes which inactivate antibiotics, they are commonly carried on bacterial plasmids. Thus in the plant pathogenic *Pseudomonas lachrymans*, resistance to dihydrostreptomycin is plasmid-mediated and is due to detoxification of the antibiotic by phosphorylation. The product of the enzymatic inactivation has been identified as dihydrostreptomycin 3-phosphate (YANO *et al.* 1978).

Others

As stated earlier, when stable resistance to fungicides has been observed but not studied genetically, involvement of major genes can be assumed if the sensitivity distributions of field populations show distinct, non-overlapping subpopulations. It appears that this is the case with the acylalanines and the polyoxins (GEORGOPOULOS 1986).

Acylalanine resistance develops apparently in a single step and resistant strains can grow and reproduce on plants treated with much higher doses than doses that are effective against sensitive strains. Distribution is thus distinctly discontinuous and, although details have not yet been published, genetic studies by CRUTE (personal communication) have indeed shown a single locus inheritance of metalaxyl resistance in *Bremia lactucae.*

High level resistance to polyoxins in *Alternaria kikuchiana* appeared suddenly and caused failures in practice. Resistant and sensitive populations do not seem to overlap which indicates involvement of a major gene.

Polygenic control of fungicide resistance

A stepwise development of resistance is observed with some fungicides which do not belong to the previous group. Usually a single gene mutation may have measurable effects on sensitivity (although not necessarily, particularly under field conditions), but many mutant genes are required to achieve the highest level of resistance possible. The various combinations of resistance genes, and often also modifiers, result in a continuous distribution of sensitivities, so that distinct subpopulations cannot be recognized even after long exposures. And if a highly resistant and a wild type strain, differing in many gene pairs, are crossed, a Mendelian ratio of resistant: sensitive progeny cannot easily be recognised. Frequently, the progeny from such a cross will produce a unimodal frequency distribution with no discrete classes with respect to sensitivity.

Dodine

Genes for resistance to dodine have been recognized in *V. inaequalis* (POLACH 1973; YODER and KLOS 1976) and *N. haematococca* (KAPPAS and GEORGOPOULOS 1970). In both organisms, the effects of individual genes are small, so that it is often difficult to separate mutant and wild type strains. In the case of *V. inaequalis,* resistance has caused problems in practice, but it is characteristic that the dodine sensitivity of individuals in a population follows a normal distribution. Treated and untreated populations overlap, with exposure to the fungicide having only a quantitative effect by decreasing mean sensitivity (GILPATRICK 1982). In the case of *N. haematococca* the positive interaction between at least four genes for resistance and the existence of at least two modifiers have been conclusively demonstrated (KAPPAS and GEORGOPOULOS 1970). If more strains had been studied, even higher levels of resistance would have propably been obtained by proper combinations of genes in the same haploid nucleus.

Inhibitors of ergosterol biosynthesis

In *A. nidulans* eight genes for resistance to imazalil were recognised (van Tuyl 1977). With some of these genes pleiotropy and peculiar cross-resistance relationships to unrelated inhibitors were observed. Single gene mutations caused a relatively low level of resistance with a maximum increase of the minimal inhibitory concentration by a factor of 10. This factor could reach 100 by combining proper genes. Modifiers were also recognized. In diploids, four of the resistance genes behaved as semidominant.

In *Erysiphe graminis* f. sp. *hordei*, although much of the variation for triadimenol sensitivity appears to be genetic, individual gene effects were not recognized and the continuous nature of the frequency distributions in progeny populations suggested a complex genetic control (Hollomon *et al.* 1984). However, selection of less sensitive forms in the field does occur, although the least sensitive isolates appear to compete poorly.

Finally, in *N. haematococca*, eight genes for resistance to fenarimol and one modifier have been recognized (Kalamarakis, personal communication).

Cycloheximide

Six chromosomal genes for resistance to cycloheximide in *S. cerevisiae* have been recognized by Wilkie and Lee (1965). While the level of first-step, single gene mutants did not exceed 20 μg of the antibiotic per ml, this level could be increased to 1,000 μg/ml by crossing first-step mutants among themselves or by stepwise selection. Of the six genes, two were recessive, three were semidominant and one was dominant. Polygenic control of cycloheximide resistance has also been observed in *A. nidulans* (van Tuyl 1977) and *N. crassa* (Vomvoyanni 1974). In the latter fungus, cycloheximide resistance is controlled by at least four genes which are also subject to modification by independent genes. In heterokaryons, at least two of the resistant genes are dominant and in double mutant recombinants from dihybrid crosses, the level of resistance can be further increased.

Others

The use of ethirimol against powdery mildew of barley was associated with slightly reduced sensitivity in the surviving populations of the pathogen (Brent 1982). The sensitivity distribution, however, remained unimodal in all cases and if the differences are genetic, a polygenic control must be assumed. An attempt to recognize genes for ethirimol resistance was not successful and it was concluded that many genes of small but additive effects were involved (Hollomon 1981).

In the control of *Cercospora beticola* with fentin fungicides, mean sensitivity may decrease slowly but the distribution remains continuous with considerable overlapping between treated and untreated populations (Giannopolitis 1978). A similar response of field populations of *P. oryzae* to phosphorothiolates has been noticed (Uesugi 1982). It is likely that polygenic control is involved in these two cases also.

Practical implications

To recognize whether genetic variation for sensitivity to a particular type of fungicide is available to target organisms and what is the exact genetic control of such variation is very important in order to make some prediction on the likely useful life of the respective chemicals (Georgopoulos 1986).

From what evidence is so far available, it seems that plant pathogenic fungi possess neither major genes nor polygenic systems for the development of resistance to many of the protectant fungicides, including copper, sulfur, dithiocarbamates, quinones, phthalimides, and chlorothalonil. Mutants of filamentous fungi resistant to these fungicides have never been isolated and there is no data showing selection of less sensitive forms in the field, in spite of long use.

Sudden and complete loss of fungicidal effectiveness has so far taken place only in cases of major-gene resistance. Major genes for resistance do not always lead to complete failures because they may substantially lower fitness (e.g. with kasugamycin). In many cases (e.g. with benzimidazoles or acylalanines), however, fitness is not seriously affected, the response is qualitative and high effectiveness is suddenly followed by complete loss of efficacy.

The quantitative changes which are based on many interacting genes are not likely to lead to sudden and complete failures. Firstly, because with the increase in the number of mutations that are required to achieve agriculturally significant levels of resistance, there is an increased likelihood for a substantial loss of fitness. And secondly, because with the gradually less satisfactory performance there will be ample warning for some changes in the disease control strategies before complete failure.

References

BERAHA, L., and GARBER, E. D.: Genetics of phytopathogenic fungi. XV. A genetic study of resistance to sodium orthophenylphenate and sodium dihydroacetate in *Penicillium expansum*. Amer. J. Bot. **53** (1966): 1041—1047.

— — A genetic study of resistance to thiabendazole and sodium o-phenylphenate in *Penicillium italicum* by the parasexual cycle. Bot. Gaz. **141** (1980): 204—209.

BORCK, K., and BRAYMER, H. D.: The genetic analysis of resistance to benomyl in *Neurospora crassa*. J. Gen. Microbiol. **85** (1974): 51—56.

BRASIER, C. M., and GIBBS, J. N.: MBC tolerance in aggressive and non-aggressive isolates of *Ceratocystis ulmi*. Ann. appl. Biol. **80** (1975): 231—235.

BRENES-POMALES, A., LINDEGREN, G., and LINDEGREN, C. C.: Gene control of copper sensitivity in *Saccharomyces*. Nature **176** (1955): 841—842.

BRENT, K. J.: Case study 4: Powdery mildews of barley and cucumber. In: Dekker, J., and GEORGOPOULOS, S. G. (Eds.): Fungicide Resistance in Crop Protection. Pudoc, Wageningen 1982, pp. 219—230.

BURLAND, T. G., SCHERL, T., GULL, K., and DOVE, W. F.: Genetic analysis of resistance to benzimidazoles in *Physarum:* differential expression of β-tubulin genes. Genetics **108** (1984): 123—141.

DAVIS, R. H.: Mechanisms of inheristance. 2. Heterokaryosis. In: AINSWORTH, G. C., and SUSSMAN, A. S. (Eds.): The Fungi, Vol. 2. Academic Press, New York 1966, pp. 567—588.

ESSER, K., KUCK, U., STAHL, U., and TUDZYNSKI, P.: Cloning vectors of mitochondrial origin for eukaryotes: a new concept in genetic engineering. Curr. Genet. **7** (1983): 239—243.

FINCHAM, J. R. S., DAY, P. R., and RADFORD, A.: Fungal Genetics. 4th edit. Blackwell, Oxford 1979.

FOGEL, S., and WELCH, J. W.: Tandem gene amplification mediates copper resistance in yeast. Proc. Natl. Acad. Sci. (U.S.A.) **79** (1982): 5342—5346.

GALE, E. F., CUNDLIFE, E., REYNOLDS, P. E., RICHMONT, M. H., and WARING, M. J.: The Molecular Basis of Antibiotic Action. 2nd edit. Wiley Intersc., London 1981.

GEORGOPOULOS, S. G.: The genetic basis of classification of fungicides according to resistance risk. EPPO Bull. **15** (1985a): 513—517.

— The development of fungicide resistance. In: WOLFE, M. S., and CATEN, C. E. (Eds.): Populations of Plant Pathogens: Their Dynamics and Genetics. Blackwell Sci. Publ. Oxford 1986.

— and PANOPOULOS, N. J.: The relative mutability of the *cnb* loci in *Hypomyces*. Can. J. Genet. Cytol. **8** (1966): 347—349.

— CHRYSAYI, M., and WHITE, G. A.: Carboxin resistance in the haploid, the heterozygous diploid, and the plant parasitic dicaryotic phase of *Ustilago maydis*. Pesticide Biochem. Physiol. **5** (1975): 543—551.

— and ZIOGAS, B. N.: A new class of carboxin-resistant mutants of *Ustilago maydis*. Netherl. J. Plant Pathol. **83** (1977): 235—242.

GIANNOPOLITIS, C. N.: Occurrence of strains of *Cercospora beticola* resistant to triphenyltin fungicides in Greece. Plant Dis. Reptr. **62** (1978): 205—208.

GILPATRICK, J. D.: Case study 2: *Venturia* of pome fruits and *Monilinia* of stone fruits. In: DEKKER, J., and GEORGOPOULOS, S. G. (Eds.): Fungicide Resistance in Crop Protection. Pudoc, Wageningen 1982, pp. 195—206.

GRINDLE, M.: Isolation and characterization of vinclozolin resistant mutants of *Neurospora crassa*. Trans. Br. mycol. Soc. **82** (1984): 635—643.

GUNATILLEKE, I. A. U. N., ARST, H. N., and SCAZZOCHIO, C.: Three genes determine the carboxin sensitivity of mitochondrial succinate oxidation in *Aspergillus nidulans*. Genet. Res., Camb. **26** (1976): 297—305.

GUNGE, N.: Yeast DNA plasmids. Ann. Rev. Microbiology **37** (1983): 253—275.

HASTIE, A. C., and GEORGOPOULOS, S. G.: Mutational resistance to fungitoxic benzimidazole derivatives in *Aspergillus nidulans*. J. Gen. Microbiology **67** (1971): 371—374.

HOLLOMON, D. W.: Genetic control of ethirimol resistance in a natural population of *Erysiphe graminis* f. sp. *hordei* Phytopathology **71** (1981): 536—540.

— BUTTERS, J., and CLARK, J.: Genetic control of triadimenol resistance in barley powdery mildew. Proc. 1984 Br. Crop Prot. Conf. — Pests and Diseases, Vol. 2: 477—482.

ISHII, H., YANASE, H., and DEKKER, J.: Resistance of *Venturia nashicola* to benzimidazole fungicides. Meded. Fac. Landbouwwet., Rijksuniv., Gent **49** (1984): 163—172.

KAPPAS, A., and GEORGOPOULOS, S. G.: Genetic analysis of dodine resistance in *Nectria haematococca* (Syn. *Hypomyces solani*). Genetics **66** (1970): 617—622.

KARIN, M., NAJARIAN, R., HASLINGER, A., VALENZUELLA, P., and WELCH, J.: Primary structure and transcription of an amplified genetic locus: The *CUP* 1 locus in yeast. Proc. Natl. Acad. Sci. USA **81** (1984): 337—341.

KATAN, T., SHABI, E., and GILPATRICK, J. D.: Genetics of resistance to benomyl in *Venturia inaequalis* from Israel and New York. Phytopathology **73** (1983): 600—603.

LANCASHIRE, W. E., and GRIFFITHS, D. E.: Biocide resistance in yeast: isolation and general properties of trialkyltin resistant mutants. Fed. Europ. Biochem. Soc. Letters **17** (1971): 209—214.

POLACH, F. J.: Genetic control of dodine tolerance in *Venturia inaequalis*. Phytopathology **63** (1973): 1189—1190.

SHABI, E., and KATAN, T.: Genetics, pathogenicity and stability of carbendazim-resistant isolates of *Venturia pyrina*. Phytopathology **69** (1979): 267—269.

SHAW, D. S., and ELLIOTT, C. G.: Streptomycin resistance and morphological variation in *Phytophthora cactorum* J. Gen. Microbiology **51** (1968): 75–84.

STALL, R. E., LOSHKE, D. C., and RICE, R. W.: Conjugational transfer of copper resistance and avirulence to pepper within strains of *Xanthomonas campestris* pv. *vesicatoria*. Phytopathology **74** (1981): 797 (Abstr.).

TAGA, M., NAKAGAWA, H., TSUDA, M., and UEYAMA, A.: Ascospore analysis of kasugamycin resistance in the perfect stage of *Pyricularia oryzae*. Phytopathology **68** (1978): 815—817.

THRELFALL, R. J.: The genetics and biochemistry of mutants of *Aspergillus nidulans* resistant to chlorinated nitrobenzenes. J. Gen. Microbiology **52** (1968): 35—44.

TUYL, J. M. VAN: Genetics of Fungicide Resistance. Meded. Landbouwhogeschool, Wageningen 1977, pp. 77—2.

UESUGI, Y.: *Pyricularia oryzae* in rice. In: DEKKER, J., and GEORGOPOULOS, S. G. (Eds.): Fungicide Resistance in Crop Protection. Pudoc, Wageningen 1982, pp. 207—218.

UMEZONO, K., TAKASHI, T., HAYASHI, S., and YANAGIDA, M.: Two cell division cycle genes *ADA2* and *NDA3* of the fission yeast *Schizosaccharomyces pombe:* control of microtubular organisation and sensitivity to antimitotic benzimidazole compounds. J. Mol. Biol. **168** (1983): 271 to 284.

VOMVOYANNI, V.: Multigenic control of ribosomal properties associated with cycloheximide sensitivity in *Neurospora crassa*. Nature **248** (1974): 508—510.

White, G. A., Thorn, G. D., and Georgopoulos, S. G.: Oxathiin carboxamides highly active against carboxin-resistant succinic dehydrogenase complexes from carboxin selected mutants of *Ustilago maydis* and *Aspergillus nidulans*. Pesticide Biochem. Physiol. **9** (1978): 165—182.

— — Thiophene carboxamide fungicides: Structure activity relationships with succinate dehydrogenase complex from wild-type and carboxin-resistant mutant strains of *Ustilago maydis* Pesticide Biochem. Physiology **14** (1980): 26—40.

— and Georgopoulos, S. G.: Thiophene carboxamide fungicides: Structure-activity relationships with the succinate dehydrogenase complex from wild-type and carboxin-resistant mutant strains of *Aspergillus nidulans* Pesticide Biochem. Physiol. **25** (1986): 188—204.

Whittingham, W. F.: The inheritance of acenaphthene tolerance in *Neurospora sitophila*. Amer. J. Bot. **49** (1962): 866—869.

Wilkie, D., and Lee, B. K.: Genetic analysis of actidione resistance in *Saccharomyces cerevisiae*. Genet. Res., Camb. **6** (1965): 130—138.

Yamamoto, M.: Genetic analysis of resistant mutants to antimitotic benzimidazole compounds in *Schizosaccharomyces pombe*. Molec. gen. Genet. **180** (1980): 231—234.

Yano, H., Fujii, H., Mukoo, H., Shimura, M., Watanabe, T., and Sekizawa, Y.: On the enzymic inactivation of dihydrostreptomycin by *Pseudomonas lachrymans*, the cucumber angular leaf spot bacterium: isolation and structural resolution of the inactivated product. Ann. Phytopath. Soc. Japan **44** (1978): 413—419.

Yoder, K. S., and Klos, E. J.: Tolerance to dodine in *Venturia inaequalis*. Phytopathology **66** (1976): 918—923.

Ziogas, B. N., and Georgopoulos, S. G.: Mitochondrial electron transport in a carboxin-resistant, antimycin A — sensitive mutant of *Ustilago maydis*. Pesticide Biochem. Physiol. **22** (1984): 24—31.

Lyr, H. (Ed.): Modern, Selective Fungicides — Properties, Applications, Mechanisms of Action. Longman Group UK Ltd., London, and VEB Gustav Fischer Verlag, Jena, 1987.

Chapter 5

Aromatic hydrocarbon fungicides

H. Lyr

Institute for Plant Protection Research of the Academy of Agricultural Sciences of the GDR, Kleinmachnow

Introduction

The "Aromatic Hydrocarbon Fungicides" (AHF) include a heterogeneous group of compounds, most of which have been known and used in practise for a long time. Some members of the group have chemical structures (Fig. 5.1) that resemble those of other fungicides (e.g. chlorophenols, chlorothalonil, dichlone), which do not belong to this group because they do not exhibit cross resistance and exert a different mechanism of action. The AHF group was formed after the general cross resistance between these fungicides was recognized (Georgopoulos and Zaracovitis 1967). Therefore, some rather dissimilar compounds with respect to their chemical struture (etridiazole and tolclophosmethyl) are included in this chapter.

Tab. 5.1 summarizes some properties and practical applications of this group of fungicides. In spite of many similarities in the mode of action, the spectrum of anti-

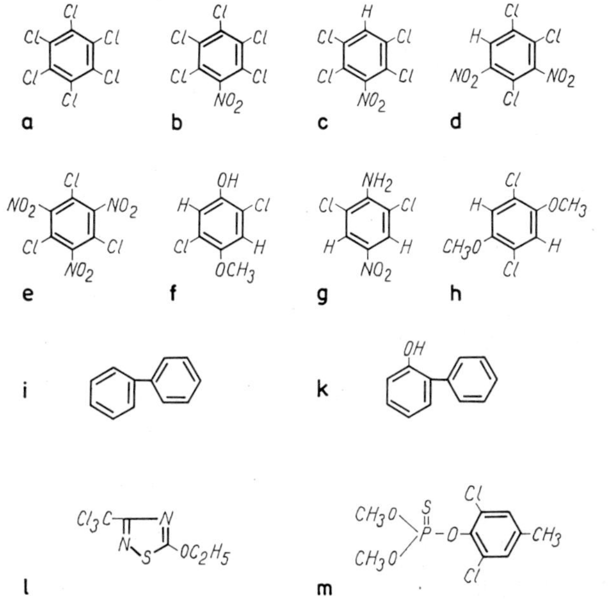

Fig. 5.1 Structure formulae for compounds of the group of aromatic hydrocarbon fungicides. a) Hexachlorobenzene, b) Pentachloronitrobenzene (Quintozene), c) Tetrachloronitrobenzene (Tectacene), d) 1,2,4-trichloro-3,5-dinitrobenzene (Olpisan), e) 1,3,5-trichloro-2,4,6-trinitrobenzene (Phomasan), f) 2,4-Dichloro-3-methoxy-phenol (DCMP), g) Dicloran, h) Chloroneb, i) Diphenyl, k) o-Phenylphenol, l) Etridiazole, m) Tolclophosmethyl.

Table 5.1 Aromatic hydrocarbon fungicides compounds, trade-names, properties and origins

Compound	Trade-names	Usage	Acute oral doses LD_{50} (rats) mg/kg	Vapour pressure mm Hg	Patents	Intro-duction
Hexachlorobenzene		Seed dressing Soil fungicide	10,000	1.09×10^{-5}	—	1945
Quintozene (Pentachloronitrobenzene)	Brassicol Tritisan Folosan	Soil fungicde Seed dressing	12,000	13.3×10^{-3}	IG Farben AG (DRP 682 048)	1930
Tecnazene (Tetrachloronitrobenzene TCNB)	Fosolan Fusarex	Soil fungicide	57	volatile	Bayer AG USP 2 615 801	1946
Trichlorodinitrobenzene	Olpisan	Soil fungicide (*Plasmodiophora brassicae*, a.o. fungi)		moderate volatile		1950
Trichlorotrinitrobenzene	Phomasan	Soil fungicide		moderate volatile		1953
Chloroneb	Demosan	Soil fungicide	11,000	3×10^{-5}	Du Pont de Nemours & Co. (Inc.) USP 3265564	1967
Dicloran	Allisan Bortran	Fruit storage diseases, ornamental *Botrytis* and *Sclerotinia*, Rhizopus diseases	1,500—4,000	1.2×10^{-6}	Boots Co. Ltd. (BP 845 916)	1930
Etridiazole	Terrazol	Soil fungicide Seed dressing	2,000	1×10^{-4}	Olin Chemicals USP 3 260 588 USP 3 260 588 USP 3 260 725	1969
Tolclophosmethyl	Rizolex	Soil, foliar fungicide	5,000		Sumitomo Chemical Co. Ltd.	
2-Phenylphenol (OPP)	Dowicide Nectryl	Fruit storage diseases, Disinfectants	2,480	moderate volatile		1936
Biphenyl		Citrus Citrus storage	3,280	volatile		1944

Table 5.2 Antifungal spectrum of some aromatic hydrocarbon fungicides (ED_{50} values in mg/l for inhibition of radial mycelial growth on malt agar medium). Values for Tolclophosmethyl according to Technical Information "Rizolex" Sumitomo Chem. Corp. Osaka, Japan

Fungus	Chloroneb	Dicloran	Quintozene	Olpisan	Diphenyl	Tolclophosmethyl	Etridiazole
Phytophthora cactorum	2	200	200	35	100	100	0.4
Pythium ultimum	1	200	150	4	100	100	0.3
Mucor mucedo	2	0.5	5	30	30	100	8
Botrytis cinerea	3	1	0.5	1	30	1	20
Sclerotinia sclerotiorum	3	2	5	2	30	1.4	10
Penicillium chrysogenum		0.5	20	—	100	—	90
Penicillium italicum	2	0.4	—	—	1	1.6	—
Aspergillus niger		100	200	—	100	—	100
Fusarium oxysporum	500	200	500	10	100	100	70
Ophiobolus graminearum	70	—	30	—	100	1.4	3
Pseudocercosp. herpotrich.	200	10	60	—	100	—	100
Rhizoctonia solani	15	200	200	20	30	0.1	70
Verticill. albo-atrum	10	8	20	—	—	100	40
Schizophyllum commune	3	100	10	—	100	—	40

fungal activity varies considerably. Utilization in practise is restricted to the control of a few plant pathogens and is further limited by physical properties. Many AHF have been used as soil fungicides because of their high volatility (e.g. chloroneb, PCNB, TCNB, hexachlorbenzene), their low UV-light stability (etridiazole) and their high activity against some soil-born fungi (tolclophosmethyl). Some AHF biphenyl, o-phenylphenol, dichloran are used to control storage diseases. The development of this group of fungicides was stimulated by their relatively simple chemical structure and their low toxicity in mammals, combined with good, but specific, antifungal activity. Therefore, several of the AHF are still in practical use. Cross resistance of this group of fungicides to the modern group of dicarboximide-fungicides is very important from a practical as well as a theoretical point of view (Leroux *et al.* 1977).

The biological conversion of some of these fungicides was summarized by Kaars-Sijpesteijn *et al.* (1977).

Hexachlorobenzene (HCB)

HCB (Fig. 5.1a) is a very effective fungicide of the halogenated benzene class. It was discovered in France (Yersin *et al.* 1946) and exhibits a high degree of vapour-phase activity against some important soil-borne fungi and seed-borne diseases. HCB controls common bunt *(Tilletia foetida* and *T. caries)* and dwarf bunt of wheat *(T. controversa)* (Holton and Purdy 1954; Purdy 1965). It is also effective against the seed borne fungus *Urocystis agropyri* (initiant of flag smut), but is inactive against *Ustilago* spp. Pentachlorobenzene is much less active, about 10 %, than HCB (Stota and Toman 1957)

HCB had been used for more than 20 years as seed dressing and soil fungicide because of its low mammalian toxicity and low price. In Greece it was used since 1958 as a seed dressing agent as an alternative to the organic mercury compounds. However, in 1973 a marked decrease in effectivity against *Tilletia* spp. was observed. Scorda (1977) reported that races of *T. foetida* were cross resistant to HCB, PCNB imazalil and hydantoin. Similar observations were made in Australia (Kuiper 1965).

The use of HCB as a seed dressing has decreased in recent years because it has been replaced by other fungicide combinations with broader spectra of activity against seed-borne pathogens.

Chlorinated nitrobenzenes

The most widely used compound in this group is pentachloronitrobenzene (Fig. 5.1b) (PCNB or quintozene). Quintozene exhibits an antifungal spectrum that is characteristic of this group, combined with low toxicity for mammals (Tab. 5.1 and 5.2). It was extensively used as a soil fungicide to control diseases caused by *Rhizoctonia solani*, *Botrytis* spp., *Sclerotinia* spp., *Sclerotium rolfsii*. However quintozene is completely ineffective against *Pythium*, *Phytophthora* and *Fusarium* all of which cause important soil-borne diseases (Tab. 5.2). The related fungicide TCNB (Tectacene) (Fig. 5.1c) controls *Fusarium coeruleum*, the cause of dry rot in potatoes (Brook and Chesters 1957). PCNB and other chlorinated nitrobenzenes act fungistatically decreasing the mycelium growth of sensitive fungi but without hindering spore germination (Esuruoso *et al.* 1968). Isolates of *Rhizoctonia solani* from diseased cotton plants, differed significantly in their sensitivity to PCNB (Shatla and Sinclair 1963). PCNB stimulated sclerotia production and this phenomenon may result in the selection of resistant strains of *Rhizoctonia* (Shatla and Sinclair 1965). In a comparative study, 19 derivatives of nitrobenzene were evaluated for their fungistatic and phytotoxic properties. Progressive chlorination enhanced fungistatic activity against *Rhizoctonia*, but not against *Pythium*, and decreased phytotoxicity. The 2,3,4,6- and 2,3,4,5-isomers of tetrachloronitrobenzene were more effective against *Rhizoctonia solani* than the 2,3,5,6-isomer (Eckert 1962). Strains resistant to these fungicides can be isolated after culturing sensitive fungi on sublethal concentrations (Kataria and Grover 1974). The strains are cross resistant to all members of this group of compounds (Priest and Wood 1961). The effectiveness of various compounds of this group depends partly on the conditions of testing because very often vapour phase action contributes to overall fungistatic activity. Fungal species may vary in sensitivity to the chloronitrobenzenes. 1,2,4-Trichloro-3,5-dinitrobenzene (Fig. 5.1d) (trade name olpisan) has been used against *Plasmodiophora brassicae* (cause of club root disease) in the soil. A 1,3,5-trichloro-2,4,6-trinitrobenzene (trade name Phomasan) (Fig. 5.1e) had been used against diseases of cucumber and tomato (Thielecke 1963).

Mono-, di- and trichlorinated nitrobenzenes are not very inhibitory to *Rhizoctonia*. These compounds are effective against *Botrytis allii* because their vapour pressure is higher than that of the higher chlorinated compounds (Priest and Wood 1961).

The chlorinated nitrobenzenes affect the growth of higher plants to some degree (Brown 1947; Eckert 1962) and some of these compounds exhibit a systemic activity against *Fusarium* wilt of tomato (Grossmann 1958). Compared to modern benzimidazoles however their innertherapeutic activity is weak.

In general, iodine and bromine 2,5-disubstituted nitrobenzenes are less effective than the corresponding chloro derivatives (Priest and Wood 1961).

The chemical and microbial stability of the chlorinated nitrobenzenes is a major reason for their use as seed and soil fungicides over several decades. Also, the acute mammalian toxicity is quite low, but long term effects have been noted in recent studies. Development of resistance of practical importance in soil fungi seems to require a longer time period compared with leaf pathogens which have a higher sporulation capacity. Therefore the problems of resistance has never become a widespread problem. But the moderate activity of the chloronitrobenzenes against other important soil fungi has limited their practical use (further literature is cited by Corden 1969).

Chloroneb

Chloroneb was developed by Du Pont de Nemours Co. in 1967. It differs from the compounds described above in that it does not contain nitro-groups (Fig. 5.1h). In common with other chlorobenzene fungicides, chloroneb has a low mammalian toxicity (Tab. 5.1) and a significant vapour pressure so that it is used as a soil fungicide, in the culture of beans, cucumber, cotton (FIELDING and RHODES 1967). In contrast to the nitrobenzene compounds, *Phytophthora* spp. are rather sensitive to chloroneb (Tab. 2.3, Tab. 5.2). *Pythium* spp. vary in sensitivity to chloroneb. Because of its low water solubility chloroneb is only very weakly systemic (GRAY and SINCLAIR 1970; SINCLAIR 1975).

Fusarium and *Pseudocercosporella* are not controlled. The spectrum of activity of other chlorinated anisoles is similar in an agar disk test, but the activity of the other compounds is lower than chloroneb which may be due partly to a still lower water solubility (Tab. 5.3). All anisoles have less phytotoxic potential than their demethylation products (phenols) which are strong biocides. Remarkably, the demethylation product of chloroneb, dichlormethoxyphenol (DCMP) (Fig. 5.1f) behaves nearly like chloroneb, but not like chlorinated phenols (WERNER 1980).

Chloroneb has been proposed also as foliar fungicide against *Pythium aphanidermatum* in rye grass (WELLS 1969). Foliar sprays of chloroneb are only weakly active against *Phytophthora infestans* on tomatoes in the greenhouse. This may be partly due to its relatively high vapour pressure.

The main targets of chloroneb are *Rhizoctonia solani*, *Pythium* spp., *Ustilago maydis*, *Typhula* spp. (VARGAS and DEARO 1970), and soil inhibiting *Phytophthora* spp.. Chloroneb is not effective against *Fusarium*, but has a relatively broad spectrum of activity compared with other compounds which are specifically active against Oomycetes (KLUGE 1978). By controlling *Rhizoctonia solani*, by seed-piece or in-furrow applications of chloroneb (LIPE and THOMAS 1979) increased potato yields were produced in Texas.

Like other fungicides of this group, resistant strains can obtained easily in the laboratory and differences in sensitivity seem to exist also in the natural population.

Table 5.3 Activity of some chlorinated anisoles against certain fungi on malt agar medium and their phytotoxicity (measured by the TTC-Test with young tissue of maize and bean hypocotyles). ED_{50}-values in mg/l of mycelial growth inhibition or reduction of TTC, values related to controls) DCMP = Dichloromethoxyphenol

	anisole				phenol		
	Chloro-neb	DCMP	2,4,6-trichloro-	Tetrachloro-	2,4,6-trichloro-	Tetra-chloro-	Penta-chloro-
Phytophthora cactorum	2	3	2	80	2	3	1
Pythium debaryanum	10	40	60	100	2	2	3
Mucor mucedo	2	3	40	25	8	20	2
Botrytis cinerea	5	10	4	6	3	1	0.4
Fusarium oxysporum	100	100	100	100	3	3	5
Colletotrichum lindem.	2	—	30	—	2	0.5	1.5
Cochliobol. carboneum	5	—	30	—	4	2	1.5
Rhizoctonia solani	2	2	20	15	3	0.5	1
Trametes versicolor	2	—	20	—	10	10	2
Phytotoxicity:							
Maize	150	100	1,000	300	30	15	3
Bean	300	200	100	—	3	3	2

Although more effective newer fungicides have displaced the AHF fungicides, in many practical applications these fungicides have been intensively investigated with regard to the mechanism of action because of their simple structure and chemically inert properties.

Dicloran

Dicloran (Botran®) (Fig. 5.1.g) was discovered in 1959 by Boots Co. Ltd. Because of its high activity against *Botrytis, Mucor, Rhizopus,* some *Penicillium* spp., *Monilia fructicola* and its low mammalian toxicity dicloran is used against fruit and vegetable storage diseases, especially in peaches, sweet cherries, grapes, tomatoes, lettuce and cabbage (Ogawa *et al.* 1963; Chastagner and Ogawa 1979; Eckert 1969, 1979).

Tab. 5.2 shows that dicloran has a selective spectrum of antifungal activity similar to that of PCNB. Both fungicides are highly active against *Mucor, Rhizopus* and *Botrytis,* but not active against *Phytophthora, Pythium, Fusarium.* Dicloran may be applied as spray or as fungicide-wax formulation (Chastagner and Ogawa 1979). It has been used as a seed treatment against *Sclerotium cepivorum* on onions (Locke 1965), *Sclerotinia sclerotiorum* on beans (Beckman and Parsons 1965) or on sunflowers. The chemical stability and low volatility of dicloran results in a prolonged disease control on leaves, fruits and in the soil.

Similar to other compounds of this group, dicloran does not inhibit germination of spores but inhibits mycelium growth. Distortion and bursting of the germ tubes have been described.

Resistant isolates were obtained that are cross resistant to other chloronitrobenzene fungicides and to dicarboximide fungicides (Ritchie 1982; Leroux and Gredt 1984), but not to benzimidazol- or triadimefon-fungicides (Bolton 1976). Some properties of dicloran resistant isolates of *Monilinia fructicola* were similar to dicarboximide-resistant isolates (Ritchie 1983). In cross resistant isolates, mycelial growth on agar medium was inhibited more by dicloran than by the dicarboximides. For some practical applications dicloran has been replaced by the more effective dicarboximide fungicides.

Etridiazole (Terrazol®)

Etridiazole produced by OLIN Corp. USA, although of dissimilar chemical structure (Fig. 5.11), it is included into this group of compounds because of its similar mechanism of action and its cross resistance to aromatic hydrocarbon fungicides (Bischoff and Lyr 1980) with which, it is sharing some properties. As Table 5.2 demonstrates, this compound has a pretty broad spectrum of activity and is highly effective against *Phytophthora* and *Pythium* species which can also be seen from Tab. 2.3. Because this compound is pretty sensitive against UV-irradiation (Fig. 5.2) it is mainly used as soil fungicide. Its usage has even increased because it does not so easily produce resistant strains as other compounds of this group. Therefore it is used in various cultures for treatment of seed beds in nurseries, soil treatment for flower bulb crops and other plants which are attacked by *Pythium, Phytophthora* spp., *Botrytis* spp. *Sclerotinia* and similar pathogens (Bakkeren and Olwehand 1970; Wheeler *et al.* 1970).

An unique side effect is its ability to retard soil nitrification (Turner 1979). It has weak insecticidal effects (USP 4,057,639) and can be used as synergist for some inecticides (DP WP A 01N/226475).

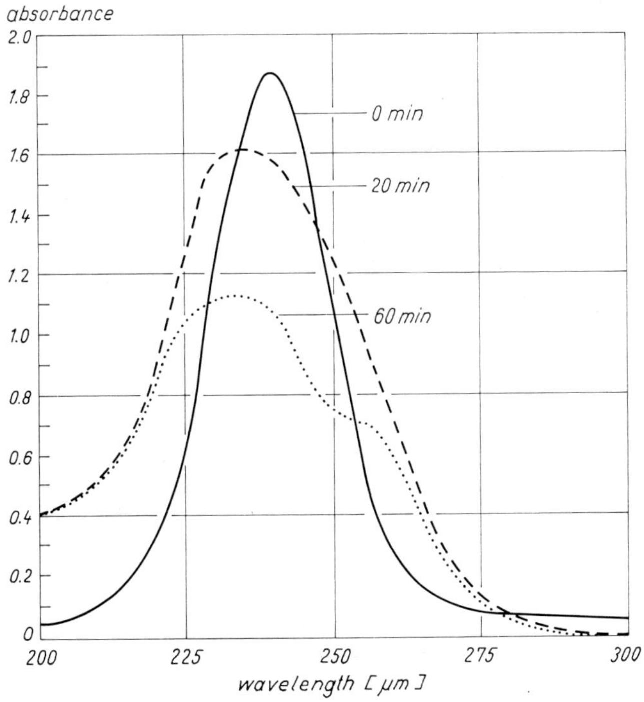

Fig. 5.2 Inactivation of etridiazole by various irradiation times in ethanolic solution (quarz UV-lamp, Hanau). Shifts of the absorption spectrum parallel to biological inactivation. (Biological activity in % inhibition in *Pythium* spec. at 100 mg/l in an agar-mycelial radial growth test.) 0 min: 100 %; 20 min: 30 %; 60 min: 0 %.

Several new derivatives have been produced by various substitutions of the thiadiazole ring. Substances with activities against bean rust, cucumber anthracnose (Katz 1981) or with bactericidal or herbicidal activity have been obtained. The CCl_3-group is essential.

Tolclophos-methyl

This compound, developed in the recent years by Sumitomo Chemical Co., is the newest member of the AHF group (Fig. 5.1m), and the first organo-phosphorus compound (Kato 1983). Its spectrum of activity has been published by the Sumitomo Co. ("A new Fungicide Rizolex"). Some data have been incorporated into Tab. 2.4 and 5.2. This compound is highly active against *Corticium rolfsii*, *Typhula* spp., *Ustilago maydis*, *Botrytis*, *Sclerotinia*, *Penicillium italicum* and especially *Rhizoctonia solani*. *Phytophthora*, *Pythium*, *Fusarium* and *Verticillium albo-atrum* are rather insensitive. Table 5.2 shows the spectrum of activity of this compound. It differs remarkably from other fungicides of this group. Tolclophos-methyl is mainly recommended for control of all diseases where *Rhizoctonia* is involved (seed potatoes, control of stem canker and black scurf). This fungicide is highly effective in controlling this fungus on ornamental crops and vegetables (lettuce). *Sclerotia* of *Rhizoctonia* on potatoes are killed within 30 min after contact with this compound. Such an action cannot be achieved with thiram, mancozeb or iprodione (Barnes et al. 1984).

o-Phenylphenol (OPP)

This compound (Fig. 5.1k) is unique among the phenol group in that it has retained a limited use in plant protection. Its main application is the protection of stored fruits, especially citrus and the disinfection of storage material (Rose *et al.* 1951; Wolf 1956). By vapour action it can protect packed fruits against decay by *Penicillium italicum* and *P. digitatum*, *Diplodia natalenis*, *Botrytis cinerea* and other species.

OPP is more selective than most other free phenols but does produce phytotoxic effects. Sodium o-phenylphenate (SOPP) is less toxic and is used in practice because it is much less phytotoxic to fruits and has a greater water solubility. Tomkins (1963) tried to diminish the phytotoxic effect by esterification of the phenolic group. Acetate and butyrate esters gave efficient control of *Botrytis* on grapes and *Monilinia* on peaches, but esters of o-phenylphenol are easily hydrolyzed into the free phenol (Eckert 1979). The methyl ether was less active against *Diplodia*, and virtually inactive against *Penicillia*. Eckert (1979) demonstrated that *Diplodia* can demethylate about 10% of the anisole, whereas *Penicillium digitatum* does not metabolize this compound. However it is not likely that this small difference can account for the differences in the growth inhibition of these two fungi. Apparently biphenyl anisole which is rapidly accumulated, is itself active against *Diplodia*, but not against *Penicillium*. This resembles strongly the situation with chloroneb, where the first demethylation product (DCMP, dichloromethoxyphenol), has the same spectrum of activity as chloroneb itself and does only partly behave as a phenol (see Tab. 5.3).

Cross resistance of SOPP and diphenyl against *Penicillia* became a practical importance in California in lemon packing houses after continuous use (Eckert *et al.* 1981; Eckert and Wild 1983; Eckert and Ogawa 1985).

Diphenyl (biphenyl)

Diphenyl (Fig. 5.1i) is a vapour-phase fungistat which is used to control postharvest diseases in citrus caused by *Penicillium digitatum* and *P. italicum*, inhibiting mycelial growth and sporulation (Nagy *et al.* 1982). It is impregnated into paper sheets which are added to storage and transportation cartons. Citrus fruits absorb this very stable compound in proportions to the vapour concentration in the surrounding atmosphere which is related to storage temperature and storage time. The official tolerance limit for this compound on citrus fruits is 110 ppm (mg/kg) in the United States, and 70 ppm in Europe and Japan.

Diphenyl, one of the weakest fungicides of this group, because it is highly hydrophobic, is still a part of the protection strategy of citrus shippers.

Conclusions

The short review of this chemically heterogeneous group of fungicides demonstrates, that in spite of the very distinct selectivity of these compounds, many have been successfully used for many years in practice. Reasons are the low mammalian toxicity, low production costs and effectivity against important, mainly soil-born pathogens or postharvest decay fungi. All of these compounds result in the rapid build-up of resistant strains which are cross-resistant to all members of this artificially composed group. This points to a common mechanism of action of this group. The additional cross-resistance to the chemically unrelated group of dicarboximide fungicides (chapter 7) described by LEROUX *et al.* (1977) points to some relationship in their mechanism of action.

I thank Prof. Dr. J. ECKERT (University of California/Riverside) for his kindness to revise this chapter linguistically and for his valuable comments.

References

BAKKEREN, M., and OUWEHAND, I.: Some experiments for control of *Pythium*-caused diseases in flower bulb-crops with 5-ethoxy-3-trichloromethyl-1,2,4-thiadizol. Brit. Crop. Protect. Conf. 70-B 3/2 25 (1970): 241—242.

BARNES, G., HARRIS, R. I., and RUSSEL, P. E.: Tolclophosmethyl, a new fungicide for the control of *Rhizoctonia solani*. Brit. Crop. Protect. Conf. 2 C—S **23** (1984): 460—464.

BECKMANN, K. M., and PARSONS, J. E.: Fungicidal control of *Sclerotinia* wilt in green beans. Plant Disease Rep. **49** (1965): 357—358.

BISCHOFF, G., and LYR, H.: Das Resistenzverhalten von *Mucor mucedo* (L.) Fres. gegenüber Chloroneb und anderen Fungiziden. Arch. Phytopathol. u. Pflanzenschutz, Berlin **16** (1980): 111—118.

BOLTON, A. T.: Fungicide resistance in *Botrytis cinerea*, the result of selective pressure on resistant strains already present in nature. Can. J. Plant Sci. **56** (1976): 861—864.

BROOK, M., and CHESTERS, C. G. C.: The use of tetrachloronitrobenzene isomers on potatoes. Ann. appl. Biol. **45** (1957): 623—634.

BROWN, W.: Experiments on the effect of chlorinated nitrobenzenes on the sprouting potato tubers. Ann. appl. Biol. **34** (1947): 422—429.

CHASTAGNER, G. A., and OGAWA, J. M.: A fungicide-wax treatment to suppress *Botrytis cinerea* and protect fresh-market tomatoes. Phytopathology **69** (1979): 59—63.

CORDEN, M. E.: Aromatic compounds. In: TORGESON, D. C. (Ed.): Fungicides, an Advanced Treatise. Vol. II. Acad. Press Inc. New York-London 1969.

ECKERT, J. W.: Fungistatic and phytotoxic properties of some derivatives of nitrobenzene. Phytopathology **52** (1962): 642—649.

— Chemical treatment for control of postharvest diseases. World Rev. Pest Control Vol. **8** (1969): 116—137.

— Interaction of derivatives of o-phenylphenol with sensitive and tolerant fungi. In: LYR, H., and POLTER, C. (Eds.): Systemic Fungicides. Akademie-Verlag, Berlin 1979.

— Fungicidal and fungistatic agents: Control of pathogenic microorganisms on fresh fruits and vegetables after harvest. In: Food Mycology, G. K. Hall Publ. Co. Boston Chapt. **14**, 1979.

— BRETSCHNEIDER, B. F., and RATNAYAKE, M.: Investigations on new post harvest fungicides for citrus fruits in California. Proc. Int. Soc. Citriculture **2** (1981): 804—810.

— and WILD, B. L.: Problems of fungicide resistance in *Penicillium* rot of citrus fruits. In: GEORGHIOU, G. P., and SAITO, T. (Eds.): Pest Resistance to Pesticides. Plenum Publ. Corp. 1983.

— and OGAWA, J. M.: The chemical control of postharvest diseases: subtropical and tropical fruits. Ann. Rev. Phytopathol. **23** (1985): 421—454.

ESURUOSO, O. F., PRICE, T. V., and WOOD, R. K. S.: Germination of *Botrytis cinerea* conidia in the presence of quintocene, tecnazene and dichloran. Trans. Br. mycol. Soc. **51** (1968): 405 to 410.

FIELDING, M. J., and RHODES, R. C.: Studies with ^{14}C-labeled chloroneb fungicide in plants. Cotton Disease Council Proc. **27** (1967): 56—60.

GEORGOPOULOS, S. G., and ZARACOVITIS, C.: Tolerance of fungi to organic fungicides. Ann Rev. Phytopathol. **5** (1967): 109—130.

GRAY, L. F., and SINCLAIR, J. B.: Uptake and translocation of systemic fungicides by soybean seedlings. Phytopathology **60** (1970): 1486—1488.

GROSSMANN, F.: Untersuchungen über die innertherapeutische Wirkung organischer Fungizide II. Chlornitrobenzole. Z. Pflanzenkrankh. Pflanzenschutz **65** (1958): 594—599.

Holton, C. S., and Purdy, L. H.: Control of soil-borne common bunt of winter wheat in the Pacific Northwest by seed treatment. Plant Disease Rep. **38** (1954): 753—754.

Kaars-Sijpesteijn, A., Dekhuizen, A., and Vonk, J. W.: Biological conversion of fungicides in plants and microorganisms. In: Siegel, M. R., and Sisler, H. D. (Eds.): Antifungal Compounds. Vol. 2. Marcel Dekker Inc., New York 1977.

Kataria, H. R., and Grover, R. K.: Adaption of *Rhizoctonia solani* to systemic and nonsystemic fungitoxicants. Z. Pflanzenkr. Pflanzenschutz **81** (1974): 472—478.

Kato, T.: Mode of action of a new fungicide, tolclophos-methyl. In: Miyamoto, I., and Kearney, P. C. (Eds.): Pesticide Chemistry. Vol. 3. Pergamon-Press, Oxford 1983.

Katz, L. E.: Olin Corp. USP 4263312, 21. April 1981.

Kluge, E.: Vergleichende Untersuchungen über die Wirksamkeit von Systemfungiziden gegen Oomyzeten. Arch. Phytopath. u. Pflanzensch. **14** (1978): 115—122.

Kuiper, J.: Failure of hexachlorobenzene to control common bunt of wheat. Nature **206** (1965): 1219—1220.

Leroux, P., Fritz, R., and Gredt, M.: Etudes en laboratoire de souches de *Botrytis cinerea* Pers. resistantes à la dichlozoline, au dicloran, au quintocene, à la vinchlozoline et au 26019 RP (ou glycophene). Phytopath. Z. **89** (1977): 347—358.

— and Gredt, M.: Resistance of *Botrytis cinerea* Pers. to fungicides. In: Lyr, H., and Polter, C. (Eds.): Systemic Fungicides and Antifungal Compounds. Tagungsber. Akad. Landwirtschaftswiss. DDR **222** (1984): 329—333.

Lipe, W. N., and Thomas, D. G.: Effects of seed-piece and in-furrow fungicide treatments on grade and yield of potatoes in Texas high plains. Texas Agricult. Experim. Station MP-1430 (1979): 1—11.

Locke, S. B.: An improved laboratory assay for testing effectiveness of soil chemicals in preventing white rot of onion. Plant Disease Rep. **49** (1965): 546—549.

Nagy, St., and Wardowski, W. F.: Diphenyl absorption by honey tangerines: the effects of washing and waxing and time and temperature of storage. J. Agr. Food Chem. **29** (1981): 760—763.

— — and Hearn, C. J.: Diphenyl absorption and decay in "Dancy" and "Sunburst" tangerine fruit. J. Am. Hortic. Sci. **107** (1982): 154—157.

Ogawa, J. M., Mathre, J. H., Weber, D. J., and Lyda, St. D.: Effects of 2,6-dichloro-4-nitroaniline on *Rhizopus* species and its comparison with other fungicides on control of *Rhizopus* rot of peaches. Phytopathology **53** (1963): 950—955.

Priest, D., and Wood, R. K. S.: Strains of *Botrytis allii* resistant to chlorinated nitrobenzenes. Ann. appl. Biol. **49** (1961): 445—460.

Purdy, L. H.: Common and dwarf bunts, their chemical control in the Pacific Northwest. Plant Disease Rep. **49** (1965): 42—46.

Renner, G., and Ruckdeschel, G.: Effects of pentachloronitrobenzene and some of its known and possible metabolites on fungi. Appl. Environm. Microbiol. **46** (1983): 765—768.

Ritchie, D. F.: Mycelial growth, peach fruit-rotting capability, and sporulation of strains on *Monilinia fructicola* resistant to dichloran, iprodione, procymidone and vinclozolin. Phytopathology **73** (1983): 44—47.

— Effect of dichloran, iprodione, procymidone, and vinclozolin on the mycelial growth, sporulation, and isolation of resistant strains of *Monilinia fructicola*. Plant Disease Rep. **66** (1982): 484 to 486.

Rose, D. H., Cook, H. T., and Redit, W. H.: Harvesting, handling, and transportation of citrus fruits. U.S. Dept. Agr. Bibliogr. Bull. **13** (1951): 1—178.

Scorda, E. A.: Insensitivity of wheat bunt to hexachlorobenzene and quintocene (Pentachloronitrobenzene) in Greece. Proc. Brit. Crop. Prot. Conf. (1977): 67—71.

Shatla, M. N., and Sinclair, J. B.: Tolerance to pentachloronitrobenzene among cotton isolates of *Rhizoctonia solani*. Phytopathology **53** (1963): 1407—1411.

— — Effect of pentachloronitrobenzene on *Rhizoctonia solani* under field conditions. Plant Disease Rep. **49** (1965): 21—23.

Sinclair, J. B.: Uptake and translocation of systemic fungicides by soybean, creeping bentgras and strawberry. In: Lyr, H., and Polter, C. (Eds.): Systemic Fungicides (Int. Symp. Reinhardsbrunn 1974): Akademie-Verlag, Berlin 1975.

Stota, Z., and Toman, M.: Effect of the constitution of isomers and homologs of benzene on their biological activity. II. Phytotoxicity and fungicidal activity of Cl-substituted derivatives of benzene. Biologia **12** (1957): 683—692.

THIELECKE, H.: Pflanzenschutzmittel. Wiss. Taschenbücher, Akademie-Verlag, Berlin 1963.

TOMKINS, R. G.: Use of paper impregnated with esters of o-phenylphenol to reduce the rotting of stored fruit. Nature **199** (1963): 669—670.

TURNER, F. T.: Soil nitrification retardation by rice pesticides. J. Soil Sci. Soc. Amer. **43** (1979): 955—957.

VAN BRUGGEN, A. H. C., and ARNESON, P. A.: Resistance in *Rhizoctonia solani* to tolklophos-methyl. Neth. J. Plant Path. **90** (1984): 95—106.

VARGAS, J. M., and BEARD, J. B.: Chloroneb, a new fungicide for the control of *Typhula* blight. Plant Disease Rep. **54** (1970): 1075—1080.

WELLS, H. D.: Chloroneb, a foliage fungicide for control of cottony blight of ryegrass. Plant Disease Rep. **53** (1969): 528—529.

WERNER, P.: Zum Wirkungsmechanismus des systemischen Fungizids Chloroneb und zu den möglichen Ursachen erzeugter Resistenz gegenüber *Mucor mucedo* L., Fres., Diss., Martin-Luther-Univ. Halle-Wittenberg 1980.

WHEELER, J. E., HINE, R. B., and BOYLE, A. M.: Comparative activity of dexon and terrazole against *Phytophthora* and *Pythium*. Phytopathology **60** (1970): 561—562.

WOLF, P. A.: Sodium orthophenyl phenate (Dowicide A) combats microbiological losses of fruits and vegetables. Down Earth **12** (1956): 16—19.

YERSIN, H., CHOMETTE, H., BAUMANN, G., and LHOSTE, J.: Hexachlorobenzene, an organic synthetic used to combat wheat smut. Compt. Rend. Acad. Agric. France **31** (1947): 5247 to 5251.

LYR, H. (Ed.): Modern, Selective Fungicides — Properties, Applications, Mechanisms of Action. Longman Group UK Ltd., London, and VEB Gustav Fischer Verlag, Jena, 1987.

Chapter 6

Mechanism of action of aromatic hydrocarbon fungicides

H. Lyr

Institute for Plant Protection Research of the Academy of Agricultural Sciences of the GDR, Kleinmachnow

Introduction

Several investigators have attempted to clarify the mechanism of action of this rather old group of selective fungicides. Although many effects in very different sites of the metabolism have been reported, no clear picture of the primary mechanism of action of these fungicides has emerged (Kaars Sijpestijn 1982; Leroux and Fritz 1984). Some of these effects are summarized in Table 6.1.

However, none of these events are severe enough to explain the primary mechanism of action of this group of fungicides.

Georgopoulos *et al.* (1976) indicated that the genetic effects of Aromatic Hydrocarbon Fungicides (AHF) were the basis of the fungal toxicity of this group as well as that of dicarboximide fungicides, but Beever and Burde (1982) argued that this conclusion is premature. Lyr (1977) classified AHF as compounds with an "unconventional" mechanism of action.

All authors agree that there should be a common mechanism of action in spite of the very different chemical structures (Fig. 5.1), because fungi which developed resistance for one member of this group of fungicides are resistant to other members as well, and also to dicarboximide fungicides, a more recently discovered group of fungicides (cf. chapter 7).

Results on the mechanism of action

With regard to the mechanism of action, the most thoroughly investigated compounds are chloroneb, quintozene (PCNB) and etridiazole, but there is at present no doubt that the other fungicides of this group act by a similar or identical mechanism. The reason for the different antifungal spectra is another problem that is not yet elucidated, but this difference may be due to mechanisms such as detoxification or other processes (chapter 2).

The mechanism of action of chloroneb has been analyzed in detail in *Mucor mucedo* by Werner (1980). Some aspects of these studies have been described by Lyr and Werner (1982), Werner *et al.* (1978).

The ultrastructural effects of chloroneb (Fig. 6.1. and 6.2) are nearly identical with those of PCNB (Fig. 6.3), which has a similar molecular space configuration compared to chloroneb. Therefore, they shall be characterized together. Both compounds induce a lysis of the inner mitochondrial membranes, beginning with a swelling of the cristae. The nuclear envelope vacuolizes, and the cell wall thickness increases dramatically within a few hours in *Mucor mucedo* (Werner *et al.* 1978; Casperson and Lyr 1982; Lyr and Casperson 1982) (Fig. 6.4, 6.7). This stops the hyphal tip (radial) growth and thus hinders the spread of the fungus and a pathogenic attack.

Table 6.1 Some effects of aromatic hydrocarbon fungicides on sensitive fungi which have been described in the literature

	Compounds	Fungus	Author(s)
Inhibition of motility of *Phytophthora* zoospores	biphenyl, chloroneb, dicloran, PCNB, tolclophosmethyl	*Phytophthora capsi*	KATO (1983)
Inhibition of respiration	chloroneb PCNB etridiazole PCNB	*Rhizoctonia Pythium Mucor*	HALOS and HUISMAN (1976b); RADZUHN (1978); WERNER (1980); KATARIA and GROVER (1975)
Thickening of cell walls or abnormal cell wall synthesis	biphenyl chloroneb dicloran PCNB	*Mucor mucedo Actinomucor Mucor*	THRELFALL (1968, 1972), LYR and CASPERSON (1982); FISHER (1977);
Yeast like growth	2,4-dichloro-aniline	*Mucorales*	LYR and CASPERSON (1982)
Effects on cellular or nuclear division	tolclophos-methyl, chloroneb	*Ustilago*	TILLMAN and SISLER (1973); KATO (1983)
Genetic (mutagenic) effects	chloroneb PCNB dicloran and others	*Asp. nidulans*	THRELFALL (1968); GEORGOPOULOS *et al.* (1976) AZEVEDO *et al.* (1977); KAPPAS (1978)
Mitochondrial destruction	etridiazole, PCNB chloroneb DCMP	*Mucor, Botrytis*	CASPERSON and LYR (1982); CASPERSON and LYR (1975)
Protein and nucleic acid synthesis	many compounds	*Ustilago Rhizopus Rhizoctonia Asp. nidulans*	WEBER and OGAWA (1965); TILLMAN and SISLER (1973); THRELFALL (1968); HOCK and SISLER (1969); CRAIG and PEBERDY (1983); FRITZ *et al.* (1977)
Lipid synthesis	dicloran	*Asp. nidulans*	CRAIG and PEPERDY (1983) further literature see LEROUX and FRITZ (1984)

Simultaneously the respiration decreases, but it is not specifically inhibited, because the respiratory quotient of mitochondria remains nearly constant. Dry matter accumulation also decreases.

Chloroneb binds to mitochondrial proteins of sensitive, but not of resistant strains in *Mucor*. The latter do not show changes in the ultrastructure when compared to the control. Comparing the amino acid composition of the proteins of isolated mitochondria of R- and S-strains of *Mucor mucedo* WERNER (1980) found nearly an equal percentage of all amino acids except for tyrosine, which was only 0.2 % in resistant strains com-

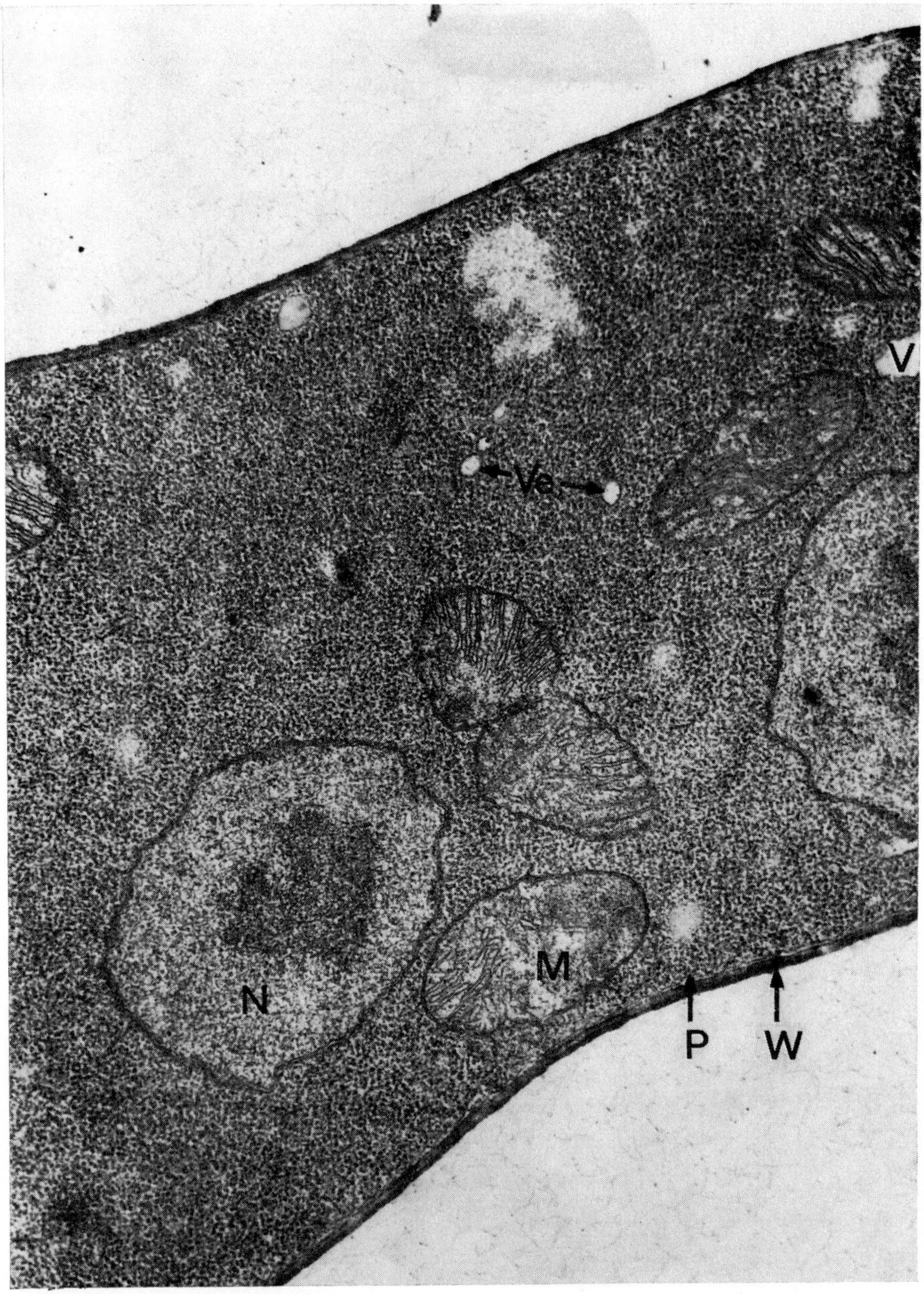

Fig. 6.1 Ultrastructure of *Mucor mucedo,* control. N = nucleus; M = mitochondrium; ER = endoplasmic reticulum; V = vacuole; Ve = vesicle; P = plasmalemma; W = cell wall. 36,000 : 1, phot.: Dr. CASPERSON.

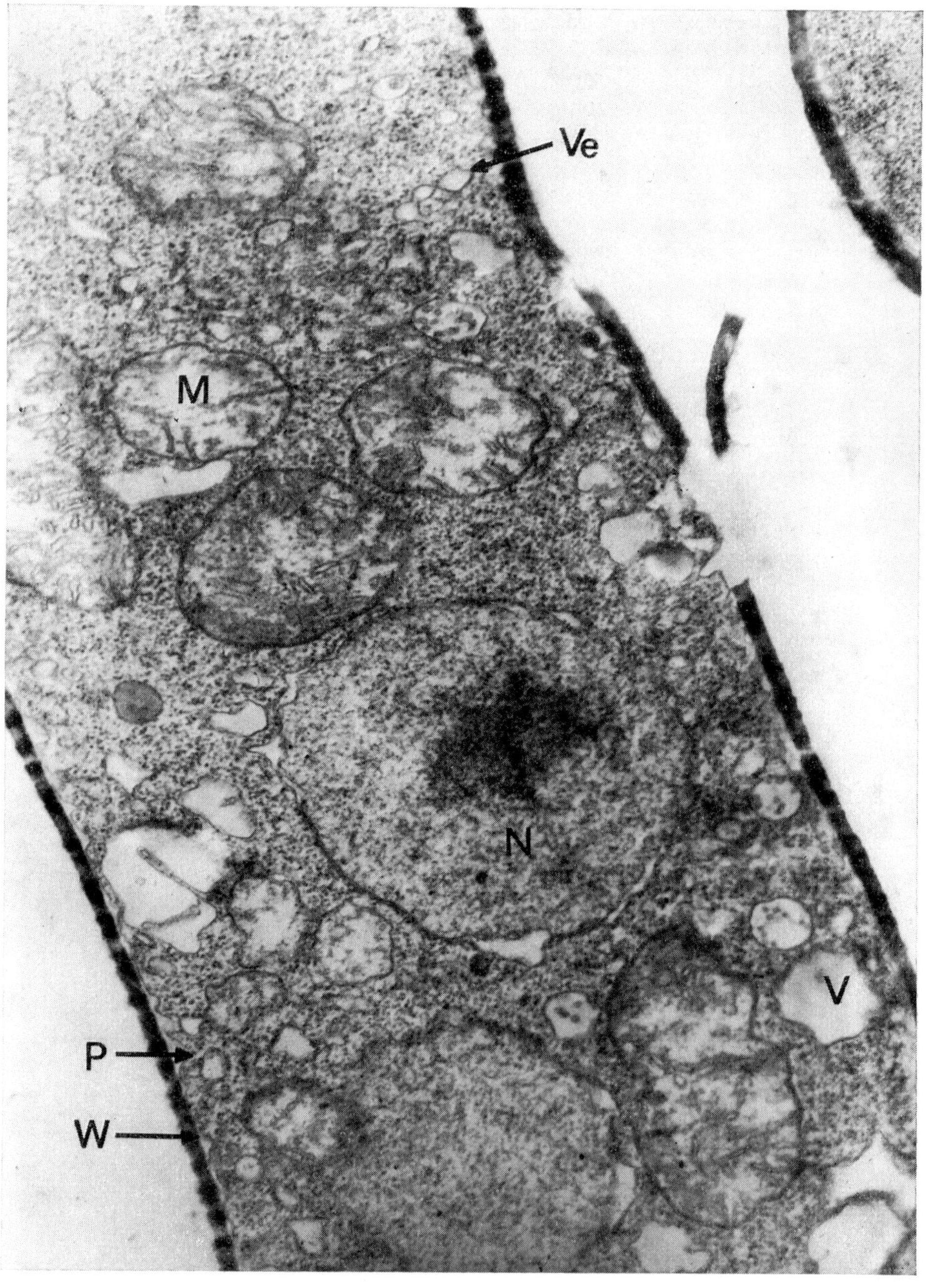

Fig. 6.2 Ultrastructural changes in *Mucor mucedo* 2 hours after application of 20 ppm Chloroneb. Lysis of the cristae in the swollen mitochondria, number of vacuoles and vesicles increased. The perinuclear space is partially enlarged, the plasmalemma is loosened from the thickened cell wall. 36,000 : 1, phot.: Dr. Casperson.

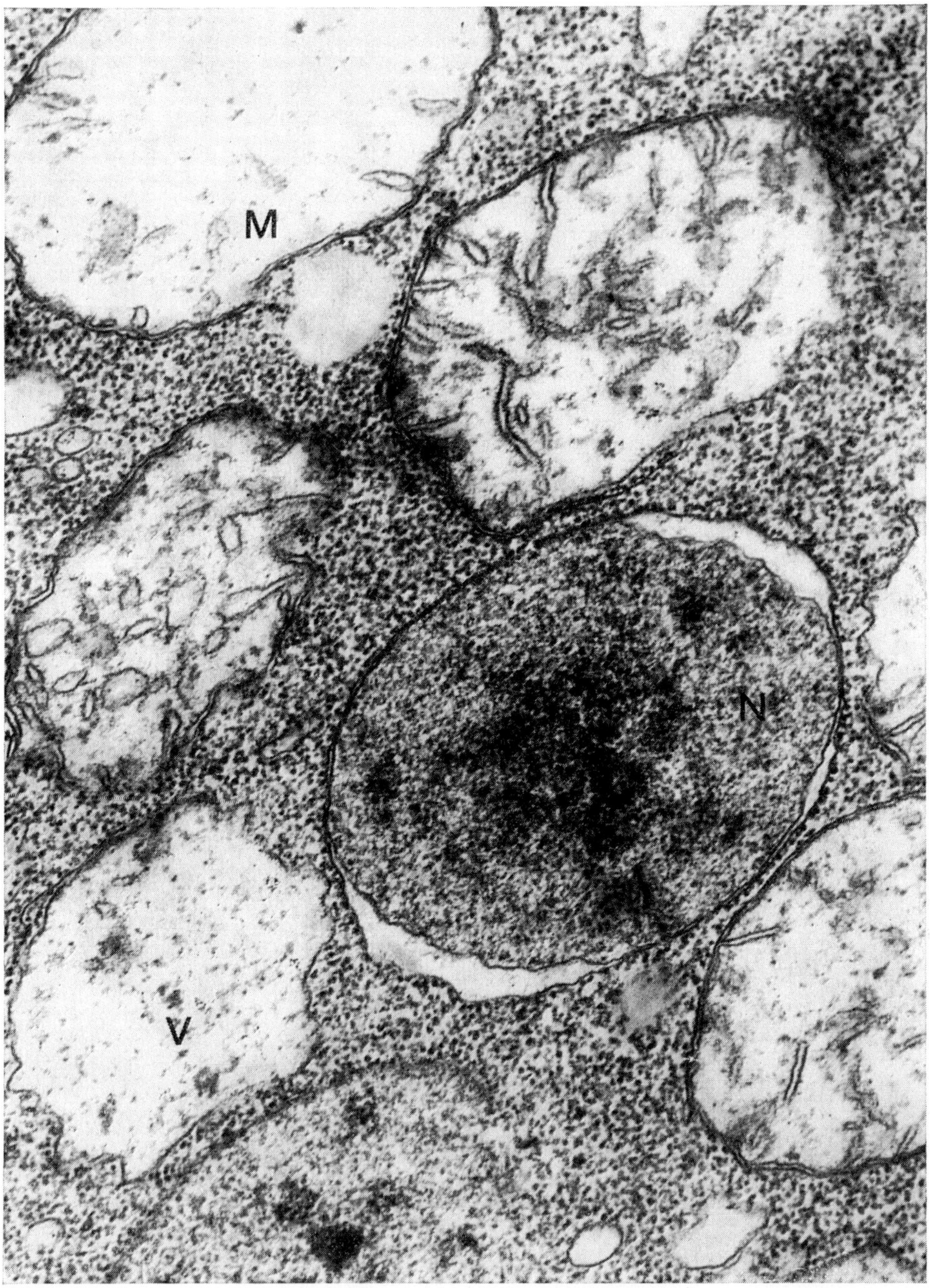

Fig. 6.3 Ultrastructural changes in *Mucor mucedo* 2 hours after application of 20 ppm Pentachloronitrobenzene (PCNB). The perinuclear space is enlarged, in the swollen mitochondria the lysis of the inner membrane is visible. 60,000: 1, phot.: Dr. Casperson.

pared to 2.9% in S-strains. This means, that a tyrosine rich protein (enzyme) has decreased in content or its composition has changed. It may be, that a tyrosine molecule located in the active centre of the enzyme has something to do with the binding capacity for chloroneb. *Neurospora crassa*, a moderately sensitive species, binds significantly more chloroneb in its mitochondria than the insensitive *N. sitophila*. The tyrosine content of mitochondrial proteins of the former species is higher than in the latter (WERNER 1980).

Several biochemical investigations have established that tyrosine and tryptophane play an important role in the binding of flavin coenzymes to the apoenzyme. Tyrosine can form an intermolecular complex with charge-transfer interaction, in contrast to hydroquinones (INOUE *et al.* 1980). Addition of tyrosine to the culture medium did not change the sensitivity of *Mucor* towards chloroneb (in contrast to results of KATARIA and GROVER (1978) in *Rhizoctonia solani*). This indicates that tyrosine synthesis is not blocked but that a specific mutation reducing tyrosine rich sequences seems to have occured in R-strains. According to RUCKPAUL and REIN (1984) tyrosine is also an essential part of the active centre of cytochrome P 450. Therefore, a point mutation at the active centre could abolish the enzyme activity and by this the production of toxic radicals under the influence of AHF. AHF seem to interact with tyrosine sequences in monooxygenases which are capable of hydroxylation of benzene rings. AHF could play the role of substrate analogs which increase the rate of oxidation of NADPH but do not serve as substrates which can be hydroxylated (effector role) (MASSEY *et al.* 1982). Resistance can involve less effective binding or low enzymatic activity.

It should be mentioned here that AESCHBACH *et al.* (1976) described the formation of diphenyl-bridge bonding between hydrophobic tyrosine sequences by a peroxidative

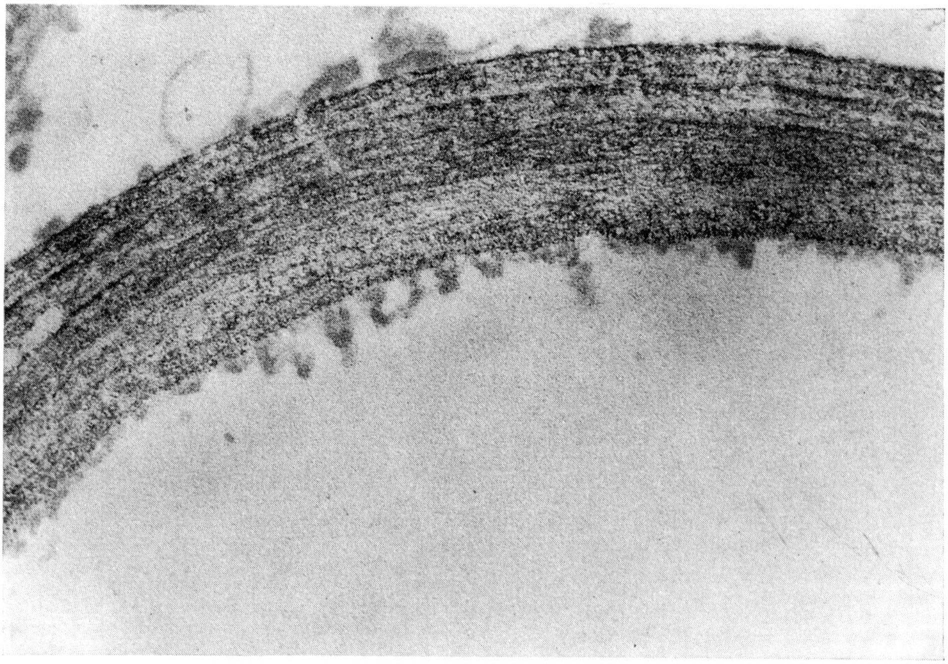

Fig. 6.4 Cell wall of *Mucor mucedo* 2 hours after application of 20 ppm Pentachloronitrobenzene (PCNB). The anomalous thickened cell wall shows a lamellation after PAS-reaction. Dark lamellae indicate mannan and light lamellae glucan and chitin. 60,000 : 1 (According to LYR and CASPERSON 1982).

action. This could be an interesting receptor structure for diphenyl or o-phenylphenol. DCMP, the demethylation product of chloroneb (Fig. 5.1g) behaves nearly identical as chloroneb, but has an additional weak uncoupling effect in isolated mitochondria, as can be expected because of its phenolic structure. But chlorinated phenols to which chloroneb resistant mutants do not show cross resistance (Tab. 6.2)

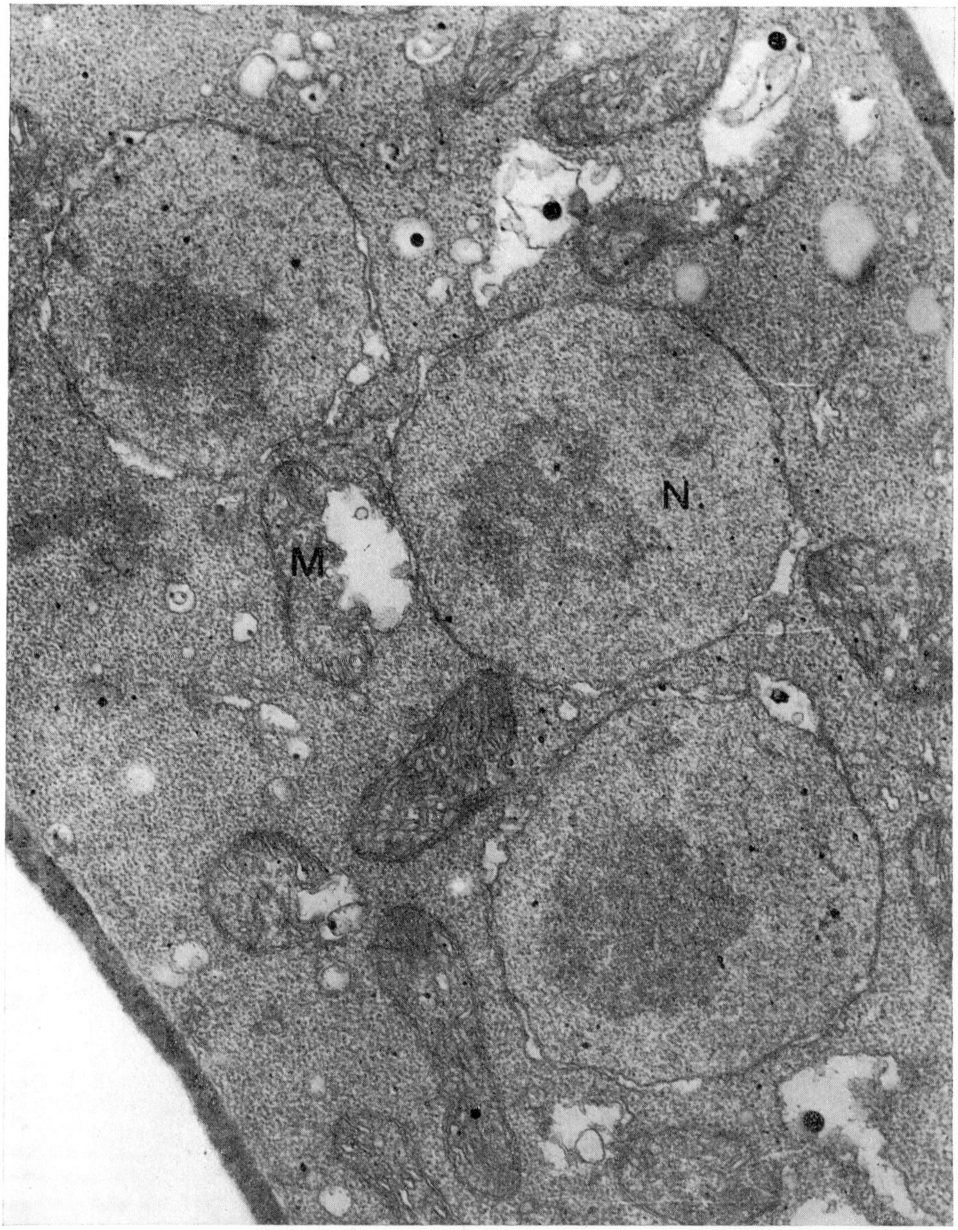

Fig. 6.5 Ultrastructural changes in *Mucor mucedo* 2 hours after application of 10 ppm Ethridiazole. Mitochondria are attacked, the outer membrane is enlarged, the inner membrane shows a partially, local lysis, the nuclear envelope has an enlarged perinuclear space. The number of vesicles is increased, the cell wall is abnormally thickened. 36,000: 1, phot.: Dr. CASPERSON.

have a much stronger uncoupling activity and never produce a cell wall thickening.

Newer investigations by Lyr and Edlich (1984) and Edlich and Lyr (1984) revealed, that chloroneb as well as other members of AHF induce a lipid peroxidation of mitochondrial and endoplasmic membranes which can be nearly nullified by addition of α-tocopherole-acetate, piperonylbutoxide or SKF-525 A. Simultaneously the growth inhibition decreased or is counteracted (Fig. 6.8, Tab. 6.3). The basis of this phenomenon is the interaction of AHF with an enzyme which was preliminarily characterized as cytochrome c-reductase, or with similar NADPH dependent flavin enzymes (chapter 8). Cytochrome c-reductase is totally inhibited by chloroneb and other fungicides of this group, which are the most effective inhibitors of this enzyme presently known. The primary toxic side effect of this inhibition is a stimulated NADPH-oxidation probably *via* a flavin peroxide (Ziegler *et al.* 1980), which results in a lipid peroxidation of the phospholipide of the enzyme. This starts a cascade process in peroxidation of unsaturated fatty acids in the membranes of sensitive fungi, which can be counteracted by external applied radical scavengers, such as α-tocopherole. SKF-525 A or piperonylbutoxide seem to hinder or reverse the binding of chloroneb to the enzyme, and by this avoid an initiation of lipid peroxidation.

According to Kataria and Grover (1978) cysteine, glutathione, hydroquinone and tyrosine decreased or nullify the growth inhibition by chloroneb and PCNB in *Rhizoctonia solani*.

This mechanism of action can sufficiently explain all effects by AHF described until now. A lipid peroxidation by monooxygenases within mitochondria must destroy the structure of the inner membrane system (Fig. 6.2, 6.3) and by this, decrease overall respiration without a specific inhibition of the respiratory chain. It may be that the monooxygenase is normally necessary for the synthesis of ubiquinone (Knoell 1981). This could explain the partial relief of etridiazole inhibition

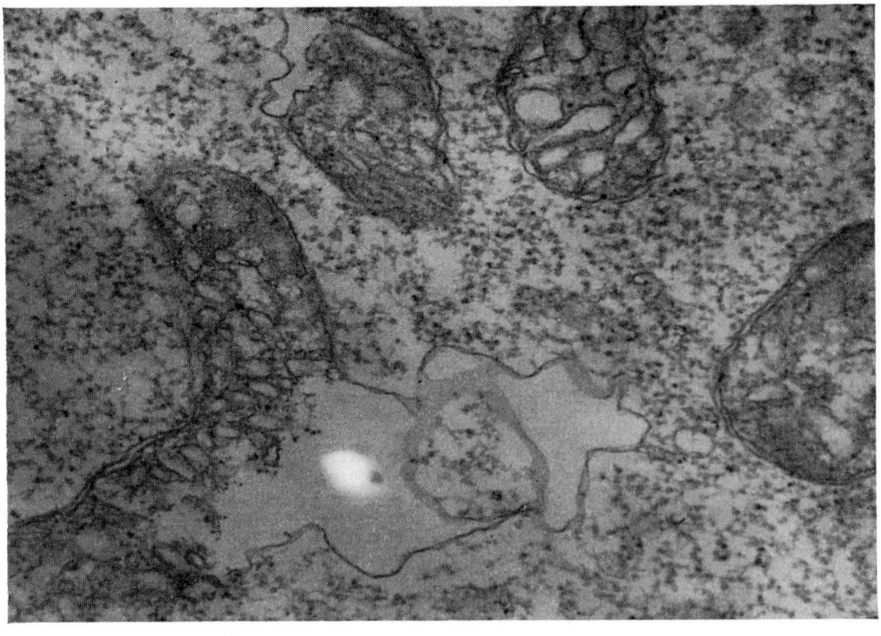

Fig. 6.6 Changes of the ultrastructure of mitochondria in *Mucor mucedo* 2 hours after application of 10 ppm Ethridiazole. The outer membrane is enlarged and the inner membrane shows a beginning destruction. 50,000 : 1 (According to Casperson and Lyr 1975).

of growth and respiration after addition of ubiquinone or menadione, as observed by HALOS and HUISMAN (1976a).

A production of free radicals and lipid peroxidation within the nuclear envelope [the occurrence of a "cytochrome c reductase" was demonstrated in nuclear membranes of mammals by STASIECKI *et al.* (1980)] not only impairs the membrane function and transport of RNA but can affect DNA, leading to strand scissions and chromosome aberrations (BAIRD *et al.* 1980; BRAWN and FRIDOVICH 1981; AMES *et al.* 1982). Desoxynucleosides are very sensitive to singlet oxygen or hydroxyl radicals. This could explain the genetic effects in fungi, observed by GEORGOPOULOS *et al.* (1976).

Cytochrome c reductase is very often located in the endoplasmic reticulum therefore after induction of lipid peroxidation by AHF it is not surprising that protein synthesis at the ribosomal site is impaired.

Membrane bound lipid synthesis can also be disturbed in many ways by lipid peroxidation.

There still remains the phenomenon of pathological cell wall thickening under the influence of AHF as a typical effect (Fig. 6.7), which contributes to the fungistatic effect of AHF.

According to ULANE and CABIB (1974) chitin and glucan synthetase are present behind the hyphal tip in the cytoplasmic membrane as a dormant form under the influence of an inhibitor protein, and can be activated by proteinases decomposing the inhibitor proteins. This was demonstrated also in vitro experiments. AHF also seem to activate chitin synthetase probable allosterically (LYR and SEYD 1979). WATANABE and KONDO (1983) found an activation of proteinases by interaction of lipid peroxidation with a proteinase inhibitor protein. An activations of proteinase by chloroneb was described by WERNER (1980) in *Mucor mucedo*. Therefore it is possible, that under the influence of a moderate lipid peroxidation in the cytoplasmic membrane, chitin- and glucan synthetases are directly or indirectly activated which results in the observed pathological cell wall thickening (Fig. 6.7) and yeast-like growth. Other processes could support this effect (LYR and CASPERSON 1982).

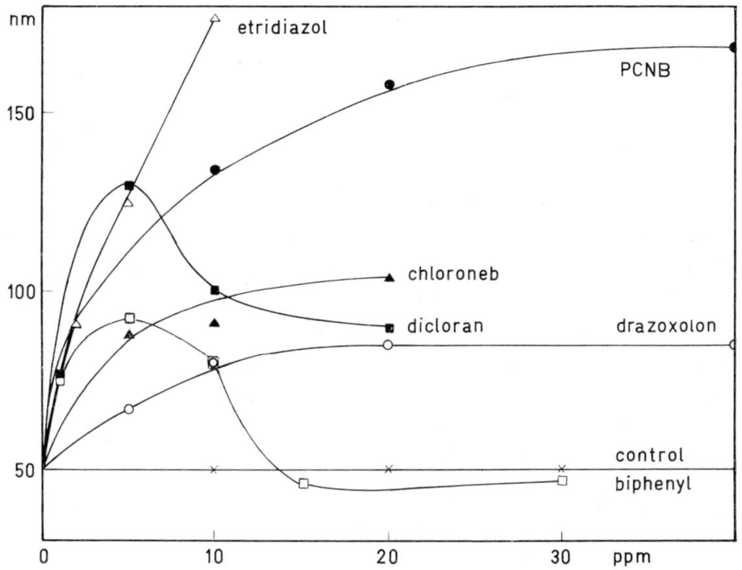

Fig. 6.7 Increase of cell wall thickness in *Mucor mucedo* under the influence of some AHF in dependance of fungicide concentration. (According to CASPERSON and LYR 1982.)

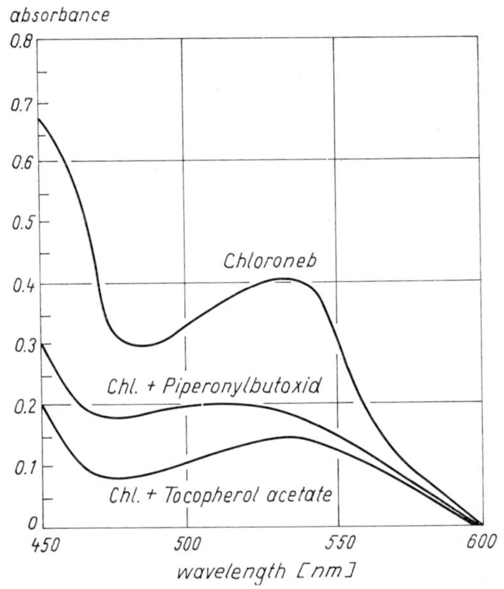

Fig. 6.8 Lipid peroxid content of 16 hours old mycelia of *Mucor mucedo* unter the influence of chloroneb (5×10^{-4} mol) + α-tocopherol acetate or piperonylbutoxide (5×10^{-5} mol). Peroxide content is indicated by the absorbance at 530 nm according to the method of Terao and Matsushita, 1981.

The cytoplasmic membrane itself seems not to be very sensitive to AHF, because it is not lysed and no extreme leakage of cell constituents occurs.

The sensitivity of a membrane to lipid peroxidation is dependent on the presence of radical forming enzymes or radicals, on the content of unsaturated phospholipids, and on the ratio of phospholipids to sterols within the membrane.

In *Mucor mucedo* the rank of sensitivity, appears to occur in the following order: inner mitochondrial membrane > outer membrane > nuclear membrane > EPR > cytoplasmic membrane.

The inner mitochondrial membrane seems to be most sensitive perhaps because of the abundance of radical generating flavin enzymes and/or because of a relatively high content of unsaturated fatty acids combined with a relative low sterol content especially in some fungi (Bottema and Parks 1980).

In the case of etridiazole some differences in the mechanism of action exist, when comparisons are made with other AHF, which make it necessary to discuss this compound separately.

This is demonstrated in the cross resistance values. Etridiazole selected R-strains of *Mucor mucedo* have a high factor of cross resistance with most other AHF, but strains selected on chloroneb exhibit only a cross resistance factor of 3,3 against etridiazole (Tab. 6.2). The ultrastructural changes are somewhat different from those caused by chloroneb or PCNB in *Mucor* (Fig. 6.4, 6.5), especially in regard to the destruction of the inner mitochondrial membrane (Casperson and Lyr 1975). Here a localized lysis of the membrane is typical and this enlarges to a total destruction of the mitochondrial inner membranes (Radzuhn and Casperson 1979). The overall effects on respiration, cell wall thickening, nuclear envelopment and other parameters are rather similar to those of other AHF (Radzuhn and Lyr 1984). The same is true for lipid peroxidation (Lyr and Edlich 1984) (Tab. 6.3).

With etridiazole, in contrast to chloroneb and PCNB, the effect on growth and destruction of mitochondrial membranes can be counteracted by procaine or high

Table 6.2 Cross resistance of *Mucor mucedo* comparing a wild (S-)strain with a resistant (R-strain) selected on chloroneb amended medium. Inhibition of radial growth on malt agar (in percent of controls) by various concentrations of some fungicides (According to Bischoff and Lyr 1980)

		Fungicide concentrations (mg/l)				
		1	3	10	30	100
Chloroneb	S	—	31	70	72	67
	R	—	0	0	0	0
PCNB	S	46	91	96	98	100
	R	0	18	32	47	49
Biphenyl	S	—	2	8	29	70
	R	—	0	0	3	8
Etridiazole	S	—	5	28	73	100
	R	—	4	9	24	71
Dicloran	S	69	85	100	—	—
	R	0	0	0	—	—
Pentachlorophenyl-methylether	S	0	43	83	91	100
	R	0	17	34	52	66
Pentachlorophenol	S	23	42	64	97	100
	R	22	45	65	96	100

Table 6.3 Growth inhibition and lipid peroxidation in *Mucor mucedo* (Lyr and Edlich, 1984) caused by AHF after 16 hours incubation time. Growth inhibition (fresh-weight) is in % related to controls without fungicide. Lipid peroxidation was measured by the method of Terao and Matsushita 1981.

		growth inhibit. %	Lipid peroxide content in %	Lipid peroxide per/g fresh weight µmol
Control	1% DMSO	0	100	6.44
Etridiazole	5 ppm	65	160	10.32
Chloroneb	2 ppm	72	210	13.66
DCMP	2 ppm	64	210	13.66
PCNB	5 ppm	68	146	9.54

doses of calcium ions (Radzuhn 1978). This means, that an activation of a phospholipase is involved in the total process of interaction.

A possible explanation of these deviations can be that a different, but similar enzyme is involved in this process. The molecular mechanism of action seems to differ in some details. The CCl_3-group of etridiazole is essential for its biological activity, replacement by CH_3- or CH_2Cl-results in inactive compounds as our experiments demonstrated. Therefore, one can speculate, that a monoxygenase, such as cytochrome P 450 oxygenase or a cytochrome c reductase can activate the CCl_3-group in a manner described for the insecticide DDT by Baker *et al.* (1984). The produced intermediary $R\text{-}C^{\bullet}Cl_2$ radicals beside initiating a lipid peroxidation can bind to phospholipids and hydrophobic proteins, yielding covalent bondings. This perhaps could explain the lower degree of resistance in etridiazole because by this it would have a multiside effect by its reactivity.

Resistance and cross resistance in AHF has been described by several authors for various fungi (Georgopoulos 1977, 1982; Leroux *et al.* 1984). An open question is, whether the AHF inhibited monooxygenases (target enzymes) are essential for normal

Table 6.4 ED_{50} values (in mg/l), and factors of resistance (RF) in *Mucor mucedo* (S- and R-strains) (from strain P3 of our collection) for various fungicides. The R_{Chl}-strain was selected on chloroneb medium (300 ppm), the R_{ET}-strain on etridiazole medium (according to BISCHOFF and LYR 1980)

Fungicides	S-strain	R_{Chl}-strain	R_{ET}-strain	RF
Chloroneb	3	>300	>300	>100
PCNB	1	100	120	>100
Pentachlorophenol	5	5	5	0
Biphenyl	50	>300	>300	> 6
Hexachlorobenzene	>300	>300	—	—
Drazoxolon	2	2.5	—	0
Etridiazole	20	65	70	3.3
Fenarimol	180	260	—	1.4

growth and are still operating in the R-strains, though perhaps with small structural changes. Resistant strains in most cases have similar growth rates as S-strains, and sporulation can be quite normal.

VAN TUYL (1977) observed, that a chloroneb resistant strain of *Penicillium expansum* was unable to grow and to sporulate normally and could not produce the typical green colour of mature spores. Addition of chloroneb or PCNB restored growth and sporulation to normal. WERNER (1980) confirmed these results and found, that dichlorohydroquinone, p-dichlorobenzene, 2,4,5-trichlorophenol, 2,4,5-trichloroanisole, hexachlorobenzene, dicloran or diphenyl did not reverse this defect, whereas chloroneb, DCMP and PCNB at 100 and 1,000 ppm induced normal sporulation and a green colouring of the spores. This means that the R-mutant needs chloroneb for induction of the target enzymes or for some peroxidative reactions necessary for normal development. On the other hand, these experiments could reflect differences in the receptor structure for the various AHF. The basis of this phenomenon is not yet elucidated.

Conclusions

Summarizing it is now possible to state, that the primary toxic effect of AHF is an induced lipid peroxidation, which is performed especially in the inner mitochondrial membrane, in the nuclear membrane, in the endoplasmatic reticulum and probably also to a small degree in the cytoplasma membrane of sensitive fungi.

The reason is the interaction of AHF with flavin enzymes, e.g. cytochrome c reductase, or similar monooxygenases which are located within the membrane systems. Their distribution seems to be correlated at least in part, with the intensivity of lipid peroxidation. AHF block the electron transport from flavin to the substrate and induce, probably by generation of free radicals a pathological peroxidation of the surrounding phospholipids which can start a cascade process of membrane peroxidation.

Whereas chloroneb and most other AHF are only indirectly involved by binding ability in the form of pseudosubstrate complexes, etridiazole may be activated to a free radical form which can bind to phospholipids or hydrophobic proteins. Additionally, a phospholipase is involved which can be inhibited by procain or high concentrations of calcium ions.

α-Tocopherol acetate, cysteine, and other compounds may act as scavengers, depressing the lipid peroxidation and the growth inhibition. SKF 525 and piperonylbutoxide prevent the binding of AHF to the receptor structure, which could be a tyrosine residue in the active centre of the flavin enzymes or the isoalloxazin ring system itself and by this suppress lipid peroxidation.

Resistance may be caused by a decreased activity of the toxifying enzyme(s), by changed receptor structures or by increased protecting measurements.

All other effects of AHF described in the literature can be sufficiently explained by the lipid peroxidation mechanism.

References

Aeschbach, R., Amado, R., and Neukom, H.: Formation of dityrosine cross-links in proteins by oxidation of tyrosine residues. Biochem. Biophys. Acta **439** (1976): 292—299.

Ames, B. N., Hollstein, M. C., and Cathcart, R.: Lipid peroxidation damage to DNA. In: Yagi, K. (Ed.): Lipid Peroxides in Biology and Medicine. Academ. Press. Inc., New York 1982, pp. 339—351.

Azevedo, J. L., Santana, E. P., and Bonatelli, R.: Resistance and mitotic instability to chloroneb and 1,4-oxathiin in *Aspergillus nidulans*. Mutation Res. **48** (1977): 163—172.

Baird, M. B., Birnbaum, L. S., and Speir, G. T.: NADPH-driven lipid peroxidation in rat liver nuclei and nuclear membranes. Arch. Biochem. Biophys. **200** (1980): 108—115.

Baker, M. T., and van Dyke, R. A.: Metabolism-dependent binding of the chlorinated insecticide DDT and its metabolite, DDD, to microsomal protein and lipids. Biochem. Pharmacology **33** (1984): 255—260.

Beever, R. E., and Byrde, J. W.: Resistance to the dicarboxamid fungicides. In: Dekker, J., and Georgopoulos, S. G. (Eds.): Fungicide Resistance in Crop Protection. Centre for Agricult. Publ. and Document., Wageningen 1982, pp: 101—117.

Bischoff, G., and Lyr, H.: Resistenzverhalten von *Mucor mucedo* (L.) Fres. gegenüber Chloroneb und anderen Fungiziden. Arch. Phytopath. Pflanzenschutz **16** (1980): 111—118.

Bottema, C. K., and Parks, L. W.: Sterol analysis of inner and outer mitochondrial membranes in yeast. Lipids **15** (1980): 987—992.

Brawn, K., and Fridovich, I.: DNA strand scission by enzymatically generated oxygen redicals. Arch. Biochem. Biophys. **206** (1981): 414—419.

Casperson, G., and Lyr, H.: Wirkung von Terrazol auf die Ultrastruktur von *Mucor mucedo*. Z. Allg. Mikrobiol. **15** (1975): 481—488.

— and Lyr, H.: Wirkung von Pentachlornitrobenzol (PCNB) auf die Ultrastruktur von *Mucor mucedo* und *Phytophthora cactorum*. Z. Allg. Mikrobiol. **22** (1982): 219—226.

Craig, G. D., and Peperdy, J. F.: The mode of action of s-benzyl-0,0-di-isopropyl phosphorothionate and of dicloran on *Aspergillus nidulans*. Pesticide Sci. **14** (1983): 17—24.

Edlich, W., and Lyr, H.: Occurrence and properties of a Cytochrome c-reductase in *Mucor* and its interaction with some fungicides. In: Lyr, H., and Polter, C. (Eds.): Systemic Fungicides and Antifungal Compounds. Tagungsbericht Akad. Landwirtschaftswiss. DDR 222 (1984): 203—206.

Fisher, D. J.: Induction of yeast-like growth in *Mucorales* by systemic fungicides and other compounds. Transact. Brit. Mykol. Soc. **68** (1977): 397—402.

Fritz, R., Leroux, P., and Gredt, M.: Mécanisme de l'action fongitoxique de la promidone (260 19 RP ou glycophène), de la vinclozoline et du dichloran sur *Botrytis cinerea* Pers. Phytopatholog. Z. **90** (1977): 152—163.

Georgopoulos, S. G.: Development of fungal resistance to fungicides. In: Siegel, M. R., and Sisler, H. D. (Eds.): Antifungal Compounds. Vol. 2, Marcel Dekker Inc., New York 1977, pp. 439—495.

— Cross-resistance. In: Dekker, J., and Georgopoulos, S. G. (Eds.): Fungicide Resistance in Crop Protection. Centre for Agricult. Publ. and Document., Wageningen 1982, pp. 53—59.

— Kappas, A., and Hastie, A. C.: Induced sectoring in diploid *Aspergillus nidulans* as a criterion of fungitoxicity by interference with hereditary processes. Phytopathology **66** (1976): 217—220.

Halos, P. M., and Huisman, O. C.: Mechanism of tolerance of *Pythium* species to ethazol. Phytopathology **66** (1976a): 152—157.

Halos, P. M.: Inhibition of respiration in *Pythium* species by ethazol. Phytopathology **66** (1976b): 158—164.

Hock, W. K., and Sisler, H. D.: Specificity and mechanism of antifungal action of chloroneb. Phytopathology **59** (1969): 627—632.

Inoue, M.; Shibata, M., and Ishida, T.: X-ray cristal structure of 7,8-dimethyl-isoalloxazine-10-acetic acid: tyramine (1 : 1) tetrahydrate complex. Biochem. Biophys. Res. Commun. **93** (1980): 415—419.

Kaars Sipesteijn, A.: Mechanism of action of fungicides. In: Dekker, I., and Georgopoulos, S. G. (Eds.): Fungicide Resistance in Crop Protection. Centre for Agricult. Publ. and Document. Wageningen 1982, pp. 32—45.

Kappas, A.: On the mechanisms of induced somatic recombination by certain fungicides in *Aspergillus nidulans*. Mutation Res. **51** (1978): 189—197.

Kataria, H. R., and Grover, R. K.: Effect of chloroneb (1,4-dichloro-2,5-dimethoxybenzene) and pentachloronitrobenzene on metabolic activities of *Rhizoctonia solani* Kühn. Ind. J. Exp. Biology **13** (1975): 281—285.

— — Reversal of toxicity of some systemic and non-systemic fungitoxicants by chemicals. Z. Pflanzenkrankh. Pflanzenschutz **85** (1978): 76—83.

Kato, T.: Mode of antifungal action of a new fungicide, tolclofosmethyl. Miyamoto, J., and Kearney, P. C. (Eds.): Pesticide Chemistry. Vol. 3, Pergamon Press, Oxford 1983, pp. 153 to 157.

Knoell, H.-E.: On the nature of the monooxygenase system involved in ubiquinone-8 synthesis. FEMS Microbiol. Letters **10** (1981): 63—65.

Leroux, P., and Fritz, R.: Antifungal activity of dicarboximides and aromatic hydrocarbons and resistance to these fungicides. In: Trinci, A. P. J., and Ryley, J. F. (Eds.): Mode of Action of Antifungal Agents. Brit. Mycol. Soc. (1984): 207—237.

Lyr, H.: Effect of fungicides on energy production and intermediary metabolism. In: Siegel, M. R., and Sisler, H. D. (Eds.): Antifungal Compounds. Vol. 2, Marcel Dekker Inc., New York 1977, pp. 301—332.

— and Casperson, G.: Anomalous cell wall synthesis in *Mucor mucedo* (L.) Fres. induced by some fungicides and other compounds related to the problem of dimorphism. Z. Allg. Mikrobiol. **22** (1982): 245—254.

— and Edlich, W.: On the molecular mechanism of action of the fungicides chloroneb and etridiazole. In: Lyr, H., and Polter, C. (Eds.): Systemic Fungicides and Antifungal Compounds. Tagungsbericht Akad. Landwirtschaftswiss. DDR 222 (1984): 59—64.

— and Seyd, W.: Die Wirkung einiger Fungizide und anderer Verbindungen auf die Chitin-Synthese *in vitro*. In: Lyr, H., and Polter, C. (Eds.): Systemic Fungicides. Abh. Akad. Wiss. DDR N 2 (1979): 151—157.

— and Werner, P.: On the mode of action of the fungicide Chloroneb. Pesticide Biochem. Physiol. **18** (1982): 69—78.

Massey, V., Claiborne, Al., Detmer, K., and Schopfer, L. M.: Comparative aspects of flavo protein monooxygenases. In: Oxygenases and oxygen metabolism. Acad. Press Inc., New York 1982, pp. 185—195.

Radzuhn, B.: Zum Wirkungsmechanismus des systemischen Fungizides Terrazol. Diss. Humboldt-Univ. Berlin (1978): 1—81.

— and Casperson, G.: Zum Wirkungsmechanismus von Terrazol. In: Systemic Fungicides. V. Intern. Symp., Abh. Akad. Wiss. DDR N 2. Lyr, H., and Polter, C. (Eds.): Akademie-Verlag, Berlin 1979, pp. 195—206.

— and Lyr, H.: On the mode of action of the fungicide etridiazole. Pesticide Biochem. Physiol. **22** (1984): 14—23.

Ruckpaul, K., and Rein, H.: Cytochrome P 450. Akademie-Verlag, Berlin 1984.

Stasiecki, P., Oesch, F., Bruder, G., Jarasch, E.-D., and Franke, W. W.: Distribution of enzymes involved in metabolism of polycyclic aromatic hydrocarbons among rat liver endomembranes and plasma membranes. Eur. J. Cell Biol. **21** (1980): 79—92.

Terao, J., and Matsushita, S.: Thiobarbituric acid reaction of methyl arachidonate monohydroperoxide isomers. LIPIDS **16** (1981): 98—101.

Threlfall, R. J.: The genetics and biochemistry of mutants of *Aspergillus nidulans* resistant to chlorinated nitrobenzenes. J. Gen. Microbiol. **52** (1968): 35—44.

— Effect of pentachloronitrobenzene (PCNB) and other chemicals of sensitive and PCNB resistant strains of *Aspergillus nidulans*. J. Gen. Microbiol. **71** (1972): 173—180.
Tillman, R. W., and Sisler, H. D.: Effect of chloroneb on the growth and metabolism of *Ustilago maydis*. Phytopathology **63** (1973): 219—225.
Ulane, R. E., and Cabib, E.: The activating system of chitin synthetase from *Saccharomyces cerevisiae*. J. Biol. Chem. **249** (1974): 3418—3422.
Van Tuyl, J. M.: Genetics of fungal resistance to systemic fungicides. Mededelingen Landbouwhogeschool (Wageningen) **77** (1977): 1—136.
Watanabe, T., and Kondo, N.: The change in leaf protease and protease inhibitors activities after supplying various chemicals. Biol. Plant **25** (1983): 100—109.
Weber, J. R., and Ogawa, J. M.: The mode of action of 2,6-dichloro-4-nitroaniline in *Rhizopus arrhizus*. Phytopathology **55** (1965): 159—165.
Werner, P.: Zum Wirkungsmechanismus des systemischen Fungicides Chloroneb und zu den möglichen Ursachen erzeugter Resistenz gegenüber *Mucor mucedo* L. Fres. Diss. Martin-Luther-Univ. Halle–Wittenberg (1980): 1—142.
— Lyr, H., and Casperson, G.: Die Wirkung von Chloroneb, seinen Abbauprodukten sowie von chlorierten Phenolen auf das Wachstum und die Ultrastructur verschiedener Pilzarten. Arch. Phytopathol. Pflanzenschutz, Berlin **14** (1978): 301—312.
Ziegler, D. M., Poulsen, L. L., and Duffel, M. W.: Kinetic studies on mechanism and substrate specificity of the microsomal flavin-containing monooxygenases. In: Microsomes, Drug Oxidations, and Chemical Carcinogenesis. Acad. Press Inc., New York 1980, pp. 637 to 645.

Lyr, H. (Ed.): Modern, Selective Fungicides — Properties. Applications, Mechanisms of Action. Longman Group UK Ltd., London, and VEB Gustav Fischer Verlag, Jena, 1987.

Chapter 7

Dicarboximide fungicides

E.-H. Pommer and Gisela Lorenz
BASF Aktiengesellschaft, Limburgerhof, FRG

Introduction

The antimicrobial activity of 3-phenyl-oxazolidine-2,4-diones and related compounds was reported by Fujinami, Ozaki and Yamamoto in 1971. In their studies on structure-activity relationships, these authors were able to show the effect on the fungicidal activity of, firstly, substituents in the benzene ring and, secondly, substituents in the 5-position of the oxazolidine ring. A powerful fungicidal effect was only produced by a dichloro-substitution in the 3,5-position of the benzene ring; a dimethyl-substitution in the 5-position of the oxazolidine ring gave the best activity. Their studies were carried out using *Sclerotinia sclerotiorum* as test fungus. The first active compound which arose out of this class of compounds, now designated dicarboximides, was dichlozoline (Fig. 7.1f). On account of toxicological problems (Nakaya *et al.* 1969), dichlozoline was not pursued further although it possessed excellent fungicidal activity. A particular feature of dichlozoline is its strong activity towards *Botrytis and Sclerotinia* species (Menager *et al.* 1971). Dimethachlor (Fig. 7.1g) resulted from investigations into substituted 1-phenylpyrrolidine-2,5-diones (Fujinami *et al.* 1972). This active compound found, however, no wide application and is currently of only minor importance.

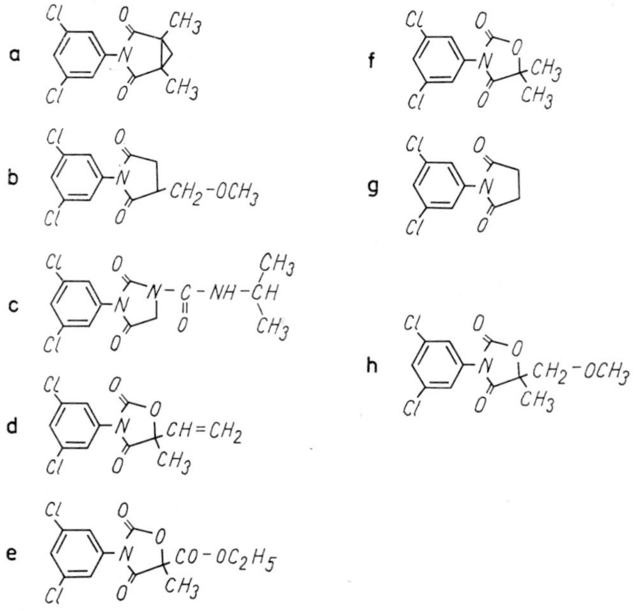

Fig. 7.1 Structure formulae of dicarboximide fungicides. a) Procymidone, b) Metomeclan, c) Iprodione, d) Vinclozolin, e) Chlozolinate, f) Diclozoline, g) Dimethachlor, h) Myclozoline.

Table 7.1 Development of dicarboximides as fungicides for commercial use

common name code number	product name on introduction	year of introduction by
dichlozoline CS 8890	Sclex	1967 Sumitomo/Hokko withdrawn
dimethachlor S-47 127	Ohric	1969 Sumitomo withdrawn
iprodione 26 019 RP	Rovral	1974 Rhone-Poulenc
vinclozolin BAS 352 F	Ronilan	1975 BASF
procymidone S-7131	Sumilex Sumisclex	1976 Sumitomo
chlozolinate M 81 64	Serinal	1980 Montedison/Farmoplant
metomeclan Co-6054	Drawifol	1984 Wacker

The search for further cyclic imides of high fungicidal activity and substituted on the nitrogen atom by a 3,5-dichlorophenyl group was successful. Within three years, three new dicarboximide fungicide were introduced on the market: iprodione (LACROIX *et al.* 1974), vinclozolin (POMMER and MANGOLD 1975) and procymidone (HISADA *et al.* 1976). These active compounds are used world-wide. Although the interest in new active compounds out of this group has been greatly reduced on account of the resistance to dicarboximides in *Botrytis cinerea* since the end of the 70's, two further active compounds have been developed into marketed products: chlozolinate (DI TORO *et al.* 1980) and metomeclan (VULIE *et al.* 1984) (Fig. 7.1).

Amongst the development products that have emerged in this group of compounds, mention should be made of myclozoline, experimental code BAS 436 F, in which the active compound is 3-(3,5-dichlorophenyl)-5-(methoxymethyl)-5-methyl-2,4-oxazolidinedione (POMMER and ZEEH 1982).

Dicarboximides which have established themselves on the market are listed in Table 7.1. Table 7.2 lists those dicarboximides currently in use with details of the active compounds they contain: only limited information is currently available for chlozolinate and metomeclan.

In ornamental plant cultivation, dicarboximides are also employed in bulb flower and other flower crops for disinfection against *Botrytis* (bulb fire) in tulips: *Sclerotinia bulborum* (black slime) in hyacinths; *Botrytis, Curvularia* (dry rot) and *Stromatinia* (dry rot) in gladioli. The fungicides are applied by dipping.

There are no indications to be found in the literature about serious problems with respect to plant tolerance when dicarboximides are used in horticultural or agricultural crops, however the manufacturers' recommendations should be followed.

In order to broaden the spectrum of activity of dicarboximides, mixtures containing other fungicides have been developed into commercial products:

iprodione + maneb + sulphur
vinclozolin + maneb
vinclozolin + metiram
vinclozolin + thiram
vinclozolin + thiophanate methyl
vinclozolin + sulphur

Table 7.2 Dicarboximides in commerical use as fungicides

Structure	Chemical name	Patent No.	LD_{50} rat p.o. mg/kg
Procymidone 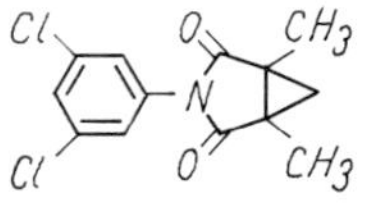	N-(3,5-dichloro-phenyl)-1,2-dimethylcyclo-propane-1,2-dicarboximide	US 3 903 090 (Sumitomo)	6800
Metomeclan 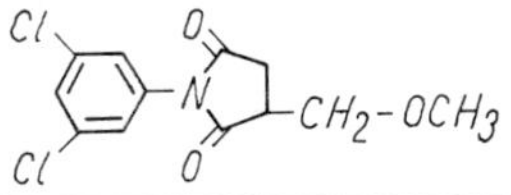	1-(3,5-dichloro-phenyl)-3-methoxy-methyl-2,5-pyrrolidindione	EP 46 274 (Wacker)	>10000
Iprodione 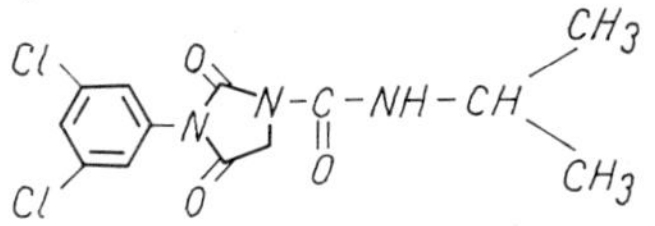	3-(3,5-dichloro-phenyl)-N-isopropyl-2,4-dioxoimidazolidine-1-carboxamide	FR 2 120 222 (Rhone-Poulenc)	3500
Vinclozolin 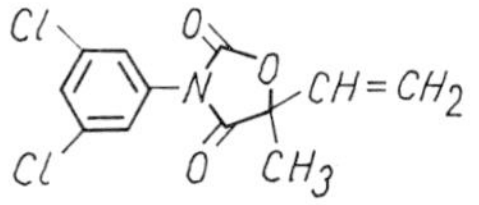	3-(3,5-dichloro-phenyl)-5-ethenyl-5-methyl-2-4-oxazolidine-dione	DE 2 207 576 (BASF)	>10000
Chlozolinate 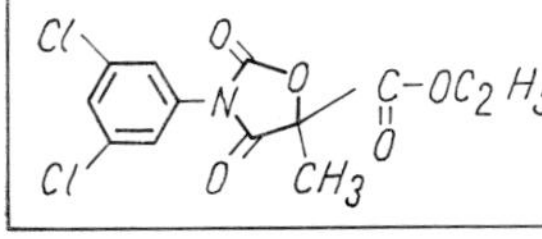	(I)-3-(3,5-dichlorophenyl)-5-methyl-2,4-dioxo-oxazolidine-5-carboxylate	GB 874 406 (Montedison)	>4500

Fungicidal spectrum of activity and the use of dicarboximides in horticultural or agricultural crops

A general feature of dicarboximides is an essentially protective activity against representatives of the following genera of fungi: *Botrytis, Sclerotinia, Monilinia, Alternaria, Sclerotium* and *Phoma. Helminthosporium, Rhizoctonia* and *Corticium* should also be mentioned. Since the cyclic imide component in the dicarboximides can be an oxazolidine-dione (vinclozolin or chlozolinate), a succinimide (procymidone or metomeclan) or a hydantoin (iprodione) there are understandable differences in their degree of activity against the pathogens listed or shifts in the spectrum of activity. Dicarboximides are mainly employed for the control of *Botrytis cinerea* and *Scleotinia sclerotiorum* or *Sclerotinia minor* in agricultural crops. Controllable plant diseases and the most important crops are listed in Table 7.3.

Table 7.3 The plant diseases of important crops which can be controlled with dicarboximide (fungicides)

Pathogen	Disease	Main crops
Botrytis cinerea *Botrytis* spp.	grey mould	vines, strawberries, lettuce, capsicum, aubergines, beans, onions, tomatoes, ornamentals
Sclerotinia sclerotiorum	*Sclerotinia* rot *Sclerotinia* wilt *Sclerotinia* stem rot	lettuce, endives, chicory, cucurbits, celery, peanuts beans, rape, turf, lawn
Sclerotinia minor	*Sclerotinia* rot	lettuce
Sclerotinia homoeocarpa	dollar spot	turf, lawn
Monilinia laxa	blossom wilt	peaches, apricots, cherries
Monilinia fructigena *Monilinia fructicola*	brown rot	apples, pears, peaches, apricots
Alternaria brassicae	dark leafspot	cabbage, rape
Alternaria radicina	black rot	celery
Sclerotium bataticola	stem rot	sunflower
Phoma betae	black rot	beet
Phoma lingam	black leg	cabbage, rape
Didymella lycopersici	stem and fruit rot	tomatoes
Didymella bryoniae	black rot	cucumber
Rhizoctonia solani	wirestem, bottom rot, head rot	radish, black radish, kohlrabi, potatoes
Laetisaria fuciformis (*Corticium fuciforme*)	black rot, red thread	turf, lawn

Effect on the development stages of fungi

Chapter 8 is concerned with the biochemical aspects of the mechanism of activity of dicarboximides. This section, therefore, will deal with the extent of our knowledge of the effects of the active compounds on the various development stages of fungi:

Depending on the concentration of active compound present, the conidial germination in *Botrytis cinerea* is inhibited less strongly than the mycelial development (Buchenauer 1976; Hisada and Kawase 1977; Hisada *et al.* 1978; Pappas and Fisher 1979). Should conidial germination occur, the germ tubes remain short and stumpy, swell up and burst; in the case of an already developed mycelium, the cell wall structure is altered in the region of the hyphal tips that are still growing. Disturbances occur in the cell wall synthesis which is associated with an outflow of cytoplasm (Buchenauer 1976; Hisada and Kawase 1977; Eichhorn and Lorenz 1978; Albert 1979). Using ^{14}C-labelled procymidone, Hisada and Kawase (1977) established that the compound was rapidly bound to the hyphal cell wall. The binding process was reversible and more than 95 % of the bound procymidone was removed from the cell walls through washing. Hyphal growth, which was completely suppressed under the influence of the compound, recovered after washing. Hagan and Larsen (1979) reported a slight effect on the germination of conidia after using iprodione for the control of *Bipolaris (Drechslera) sorokiniana* on *Poa pratensis*, but the growth of germ tubes and the formation of appressoria was substantially suppressed. Conidia of *Drechslera sorokiniana* germinated in in vitro tests in the presence of iprodione, but they proceeded to swell and burst (DANNENBERGER and VARGAS 1982). Mycelial growth of *Corticium rolfsii* was strongly inhibited by iprodione but not by procymidone or vinclozolin, and all three fungicides suppressed the formation of sclerotia. GEORGOPOULOS et al. (1979) observed in *Aspergillus nidulans* that procymidone, iprodione and vinclozolin increased the frequency of mitotic recombination in diploid colonies. This effect was concentration dependent.

Although the dicarboximides are classed amongst the contact fungicides which, as a rule, are applied prophylactically, various papers are devoted to the translocation of these active compounds. The significance of these results for practical use is, however, still not clear as these were essentially model experiments. COOKE *et al.* (1979) improved the activity of procymidione against *Botrytis cinerea* on strawberries by treating both the blossoms and soil rather than a blossom treatment alone. In tests with ^{14}C-labelled procymidone, MIKAMI *et al.* (1984) were not able to detect any uptake of the active compound via cucumber leaves (in contrast to HISADA *et al.* 1977); in bean seedlings active compound uptake from a nutrient solution occurs via the roots. HISADA *et al.* (1977) also worked with a ^{14}C-labelled active compound. They observed a leaf uptake in cucumbers and a migration of the active compound to the stem, with subsequent translocation both upwards and slightly downwards. Iprodione was taken up by potato plants growing in different soils, but the amount of fungicide detected in the aerial parts was dependent on the soil type (CAYLEY and HIDE 1980). After inoculation with *Phoma exigua* var. *foveata*, development of lesions was prevented. Application of iprodione to the sprouts of seed potato tubers prior to planting decreased the incidence and severity of infection by *Rhizoctonia solani* and *Polyscytalum pustulans* on stem bases.

Controversely, POMMER and MANGOLD (1975) established that the uptake of vinclozolin via the roots of beans and green peppers was slight and that transfer of the active compound in the stem tissue only takes place over short distances.

Effect on yeast flora and the course of fermentation

The treatment of vines against grape-*Botrytis* is an important field for the use of dicarboximides. The effect which dicarboximides have on the yeast flora and the course of fermentation in grape juice consequently forms the subject of numerous papers. Considerable attention has also been given to the behaviour of residues in wine.

In in vitro tests, the addition of 0.025—0.075 % vinclozolin and procymidone to a nutrient agar did not inhibit the growth of *Kloeckera apiculata, Saccharomyces oviformis* and *S. ellipsoideus* (STOJANOVIC and VUKMIROVIC 1979). Similar results were obtained by VOJTEKOVA *et al.* (1983) with *Saccharomyces oviformis* and *S. cerevisiae* after adding vinclozolin to grape must. No changes in the composition of the natural yeast flora of the grape or the must resulted in field trials with vinclozolin (BENDA 1978: BARBERO and GAIA 1979). Non-sulphited, non-defecated grape juice treated with iprodione or vinclozolin was recovered and allowed to ferment spontaneously. Yeast development during the fermentation was identical to that in juice from grapes which had not been treated with these fungicides (SAPIS-DOMERCQ *et al.* 1977). In contrast, in France in 1977, a dry year, SAPIS-DOMERCQ *et al.* (1978) observed a slight effect on the course of fermentation after the use of vinclozolin and procymidone in viticulture. The authors attributed this to the residual effect of the fungicides applied to the grapes. According to BENDA (1983), the auxiliary substances used for the formulation of the active compounds iprodione and procymidone can inhibit the fermentation process: the auxiliary substances in the vinclozolin formulation did not affect the fermentation.

The degradation of the dicarboximides chlozolinate, iprodione, procymidone and vinclozolin has been studied in wine at pH 3.0 and 4.0 at 30 °C (CABRAS *et al.* 1984). The authors reached the conclusion that chlozolinate is degraded very rapidly. Procymidone and vinclozolin, which have a greater stability than chlozolinate, showed degradation times that were shorter than the regular maturation time of wines. Iprodione exhibited a high stability even after 92 days. Metabolites found were 3-(3,5-dichlorophenyl)-5-methyloxazolidine-2,4-dione (from chlozolinate) and 3′,5′-dichloro-2-hydroxy-2-methylbut-3-enanilide (from vinclozolin). The second of these probably results by hydrolytic opening of the heterocyclic part of the molecule. 3,5-dichloroaniline was not detected as a degradation product of any of the compounds. FLORI and ZORINI (1984) were able to demonstrate that residues of vinclozolin, iprodione and procymidone were partially removed by cold clarification of the must or by treatment of the wine with active charcoal. Grapes treated with vinclozolin, iprodione and procymidone were also examined for the presence of residues. The residues detected were considerably below legal tolerances (FIMA and WOMASTEK 1983).

Behaviour in the soil

Several lines of enquiry have been followed with respect to the behaviour of dicarboximides in soils. Only limited information is available on the degradation and migration of the active compounds in soil. WALKER *et al.* (1984) reported an enhanced degradation of iprodione in soils previously treated with this fungicide. ^{14}C-labelled iprodione was incubated in two soils which had not been treated previously. After 45 days' incubation, 18—20 % of the iprodione applied initially was recovered from the previously treated soils, whereas recovery from one of the soils not treated previously was 40 % of the initial dose, from the other it was 80 %. In a second experiment with four paired soil samples, iprodione degraded at least twice as rapidly in the previously treated sample, compared with the previously untreated one.

The migration and degradation of iprodione and vinclozolin in Woodstown loamy sand (a peanut soil) and in Lod. loam (not a peanut soil) has been investigated by ELMER and STIPES (1985) using a bioassay. They found that both fungicides were washed 25—35 cm deep into a soil, the fungicide mobility being greatly affected by the soil type. Both fungicides disappeared slightly faster after incubation at 28 °C than at 21 °C. The time required to reduce the fungicides to trace levels was a function of the initial concentration.

Evidence suggests that dicarboximides only have a slight effect on the microbial flora and microbiological activity of the soil.

10 and 100 mg iprodione per kg soil had little effect on soil respiration. Similarly, 1 and 10 mg/kg soil did not affect ammonification or nitrification; a slight retardation was observed with the addition of 100 ppm (HELWEG 1983).

Vinclozolin added at a concentration of 300 ppm to a humous sandy soil produced no adverse effects on the soil microflora or respiration; there was a slight negative effect on various enzyme activities, such as dehydrogenase, amylase and protease, and also on nitrification (POMMER and MANGOLD 1975).

Only very limited information is available about the effect of dicarboximides on plant mycorrhizas. According to RHODES and LARSEN (1981), iprodione has a reducing activity. Iprodione reduced mycorrhizal development on creeping bentgrass when applied in the spring to golf course greens or 4—8 weeks after bentgrass was seeded and inoculated with *Glomus fasciculatus* in a glasshouse.

In connection with the use of dicarboximides in lawns and turf or in soils, important observations were made by Roark and Dale (1979), according to which iprodione did not cause a reduction in longevity of earthworms.

Resistance to dicarboximides

Several reviews on resistance to dicarboximides have been published in the past few years which have in part laid emphasis on different aspects (Beever and Byrde 1982; Pommer and Lorenz 1982; Leroux and Fritz 1983).

The following section deals essentially with the resistance situation in vitro and in vivo, the biological properties of resistant strains, their population dynamics and the consequences arising therefrom (resistance management) for the practical use of this group of fungicides.

Induction of resistance in vitro

The relevant literature indicates that it is comparatively easy to produce resistant strains of fungi in appropriate laboratory experiments, even without pretreatment with mutagens (Table 7.4). Naturally occurring and very considerable fluctuations

Table 7.4 List of dicarboximide-resistant species of fungis, as determined by *in vitro* induction of resistance

Fungus	Source
Alternaria alternata	McPhee 1980
Alternaria kikuchiana	Kato *et al.* 1979
Aspergillus nidulans	Leroux *et al.* 1977
	Beever 1983
Botrytis cinerea	Leroux *et al.* 1977
	Lorenz and Eichhorn 1978
	Schüepp and Küng 1978
	Albert 1979
	Gullino and Garibaldi 1979
	Hisada *et al.* 1979
	Maraite *et al.* 1980
	Davis and Dennis 1981 (a)
Botrytis squamosa	Presley *et al.* 1982
Botrytis tulipae	Chastagner and Vassey 1979
Cochliobolus miyabeanus	Kato *et al.* 1979
Monilinia fructicola	Sztejnberg and Jones 1978
	Ritchie 1983
Monilinia laxa	Katan and Shabi 1982
Neurospora crassa	Grindle 1984
Penicillium expansum	Leroux *et al.* 1978
	Rosenberger *et al.* 1979
	Beever 1983
Rhizoctonia sclani	Kato *et al.* 1979
Rhizopus nigricans	Leroux *et al.* 1979
Sclerotinia minor	Brennemann and Stipes 1983
	Porter and Phipps 1985
Sclerotinia sclerotiorum	Kato *et al.* 1979
Sclerotium cepivorum	Littley and Racke 1984
Ustilago maydis	Leroux *et al.* 1978

in mutation rates are found, these depending on the fungus and pretreatment. In general the values range from 1×10^{-6} to 1×10^{-8} and for *Botrytis cinerea*, the fungus most frequently and most thoroughly investigated on account of its practical importance, lie between 1×10^{-6} an 1×10^{-7} (see references in BEEVER and BYRDE 1982; POMMER and LORENZ 1982; LEROUX and FRITZ 1983).

Using mycelium as the starting material, resistant strains of fungi can be obtained either by adaptation to slowly increasing concentrations of fungicide or by the further propagation of resistant sectors that can develop spontaneously at fairly high concentrations of fungicide (LORENZ and EICHHORN 1978; SCHÜEPP and KÜNG 1978).

The occurrence of cross-resistance has been investigated in particular detail for *Botrytis cinerea*. As a rule, cross-resistance covers all dicarboximides introduced (see Tab. 7.2), irrespective of the compound used to induce resistance (see references in LEROUX and FRITZ 1983; KATAN 1982). Cross-resistance can, however, also appear to fungicides from the aromatic hydrocarbon group (dichlorane or quintozene) and other aromatic ring compounds (see references in LEROUX and FRITZ 1983).

It is interesting to note that resistant strains of fungi selected on nutrient media containing ergosterol biosynthesis inhibitors are occasionally also resistant to dicarboximides, although similar mechanisms of action can be excluded (FUCHS *et al.* 1984; LEROUX and FRITZ 1983; LORENZ unpublished results).

No cross-resistance exists between MBC fungicides and dicarboximides. The fact that in the majority of cases field isolates of *Botrytis cinerea* that are resistant to dicarboximides are also MBC-resistant (double resistance) is due to the circumstance that in virtually all cases MBC fungicides were used for *Botrytis* control before dicarboximides. MBC-resistant strains of *Botrytis* are thus widely distributed; in addition, MBC-resistance is extremely stable.

The development of resistance under practical conditions

The intensive observation of tests plots (vine/*Botrytis*) treated with dicarboximides in the period 1973—1978, and also the deliberate attempts to induce resistance in the field or in a glasshouse have shown that the selection of resistant *(Botrytis)* strains under practical conditions does not proceed with the rapidity that would perhaps have been expected from the laboratory results described above. For example, in an experimental plot of vines that had been treated with dicarboximides up to 5 times a year since 1973 (as specified in the official recommendations for use), it was still not possible to isolate any resistant *Botrytis* strains ever in 1978 (SPENGLER *et al.* 1979). Similarly, no resistant strains could be isolated after two years in a field trial (vine/*Botrytis*) in which vinclozolin and iprodione had been used in gradually increasing concentrations (5—1,000 ppm) up to eleven times a year (LORENZ and EICHHORN 1978). Nor could HISADA *et al.* (1979) achieve the selection of resistant *Botrytis* isolates on glasshouse roses in spite of applying procymidone nineteen times in the course of three years.

The first dicarboximide-resistant *Botrytis* field isolates were found in a vineyard in the Mosel growing area at the end of 1978 (HOLZ 1979). In 1979 isolates of this type were found in all other wine-growing areas of West Germany (SPENGLER *et al.* 1979; LORENZ and EICHHORN 1980), and in 1980 they were also found in vines in Switzerland (SCHÜEPP *et al.* 1982) and in France (LEROUX *et al.* 1982). Since, for the lack of usable alternatives, the dicarboximides were also employed for *Botrytis* control in the following years in the vine-growing areas and countries mentioned, the proportion of resistant strains increased further. Since 1981 the proportion of resistant strains has settled to a level of about 60—80 % in West Germany (LÖCHER *et al.* 1985). Despite this relatively high percentage, total losses of activity, such as occurred very rapidly following the initial appearance of MBC resistance, has not yet been observed in West Germany.

Further reports relating to the appearance of dicarboximide-resistant *Botrytis* strains are mainly concerned with strawberries (Davis and Dennis 1979; Hunter *et al.* 1979; Maraite *et al.* 1981) and glasshouse crops of a wide variety of types (Katan 1981, 1982; Panagiotaku and Malathrakis 1981; Pappas 1982; Beaver and Brien 1983; Hartill *et al.* 1983).

In the case of other species of fungi, for example *Fusarium nivale* and *Sclerotinia homoeocarpa* in lawns (Chastagner and Vassey 1983; Detweiter *et al.* 1983) and also *Monilia fructicola* on stone fruit (Penrose *et al.* 1985), resistance to dicarboximides has hitherto only been found sporadically.

Despite the presence of resistant strains in field crops, dicarboximides are still recommended and also employed, with certain limitations, as before (Leroux and Fritz, 1983; Löcher *et al.* 1985). With glasshouse crops, however, the occurrence of resistance is more critically assessed (Panagiotaku and Malathrakis 1981; Katan 1982). This applies particularly to the Mediterranean area where vegetables, such as cucumbers and tomatoes, are cultivated under plastic film during the winter months. The conditions which prevail under the plastic film are extremely favourable for the mass propagation of *Botrytis* and it is necessary to employ botryticides in a frequency appropriate to the situation. This can have the result that populations consisting solely of resistant individuals develop and also maintain themselves, as a result of the high infestation and selection pressure, and also with no entry of conidia from outside. In such cases severe losses of activity can result (Katan 1981).

The biological characteristics of dicarboximide-resistant strains of fungi

The vast majority of investigations to characterise dicarboximide-resistant fungal isolates have been carried out with *Botrytis cinerea* (see references in Beever and Byrde 1982; Pommer and Lorenz 1982). Hence the results discussed here also relate to this fungus, especially as investigations on other fungi, for example *Sclerotinia homoeocarpa* (Detweiter *et al.* 1983), *Monilia fructicola* (Sztejnberg and Jones 1979; Ritchie 1983), *Monilia laxa* (Katan and Shabi 1982), *Sclerotium cepivorum* (Littley and Rahe 1984), *Botrytis squamosa* (Presley *et al.* 1982) and *Neurospora crassa* (Grindle 1984; Grindle and Temple 1985), have not led to fundamentally different conclusions.

In general, dicarboximide-resistant strains of fungi are less vigorous than sensitive ones. This manifests itself particularly when adapted isolates are compared with the appropriate sensitive strains (Lorenz and Pommer 1985). However it is possible for overlapping to occur due to the range of variation that exists in all characteristics, which is especially pronounced in *Botrytis cinerea* and which is to be found in both sensitive and resistant strains. This natural variation, in addition to the method chosen, probably also accounts for the contradictory results obtained in some cases by various authors (see references in Pommer and Lorenz 1982).

Mycelium growth rate and sporulation are, as a rule, lower in resistant than in sensitive strains; it has been repeatedly pointed out, however, that resistant isolates can also be found which are comparable to sensitive strains in these respects (see references in Pommer and Lorenz 1982). Our investigations with sensitive strains and the appropriate adapted strains have indicated that mycelium growth rates are insignificantly different between the two groups; on the other hand, conidial formation in resistant strains was, with only one exception, considerably less than that of the sensitive strains. In the one case mentioned the original isolate itself had a very weak rate of sporulation (Lorenz and Pommer 1985).

It is very evident that the method chosen to test the pathogenicity of resistant strains has a pronounced effect on the results obtained. Thus, for example, where mycelium is used as inoculum and/or separated plant parts are used as the host material, differences in pathogenicity between resistant and sensitive strains are, if at all present, only relatively slight (Lorenz and Eichhorn 1978; Pappas *et al.* 1979; Hartill *et al.* 1983). On the other hand, when conidial suspensions and entire plants are used, dicarboximide-resistant isolated of *Botrytis* are, as a rule, less pathogenic than sensitive strains (Hisada *et al.* 1979; Gullino and Garibaldi 1979, 1981; Lorenz and Pommer 1982, 1985).

An increased osmotic sensitivity of resistant strains is frequently quoted as an explanation for their lower pathogenicity (Beever 1983; Beever and Brien 1983; Leroux and Fritz 1983). The mycelial growth of such isolates on nutrient media which contain an increased concentration of NaCl (0.68 M — Beever 1983) or various sugars (10 mg/ml — Leroux and Fritz 1983) is strongly inhibited. Leroux and Gredt (1982a) were able to establish a direct correlation between osmotic sensitivity and the degree of resistance in strains they used. Similarly, Beever and Brien (1983) found that all their resistant isolates were more sensitive to increased osmotic values than sensitive strains; this was, however, substantially independent from the degree of resistance. In investigations with *Neurospora crassa*, Grindle and Temple (1985) were able to show that dicarboximide resistance and osmotic sensitivity are not necessarily correlated. This agrees with our own investigations in which only one of five isolates had a higher osmotic sensitivity after adaptation than the appropriate sensitive strain (Lorenz and Pommer 1985). In addition to this the testing of 50 sensitive and 70 resistant field isolates (Tab. 7.5) yielded no decisive differences in the range of variation of osmotic sensitivity between the two groups. The proportion of isolates having an average or higher osmotic sensitivity was the same in both groups.

The degree of resistance to dicarboximides exhibited by *Botrytis* strains is generally relatively low with ED_{50} values of <5—<10 ppm a.i. This applies to both field isolates (Maraite *et al.* 1981; Leroux and Gredt 1982b; Lorenz and Pommer 1985) as well as to isolates from glasshouse crops (Panagiotaku and Malathrakis 1981; Katan 1982; Pappas 1982). Under laboratory conditions, on the other hand, it is possible to obtain strains which exhibit very much higher degrees of resistance (ED_{50} values >100—>500 ppm a.i.; (Leroux *et al.* 1977). In laboratory experiments of this kind the isolates used have generally been in culture for a prolonged period of

Table 7.5 Osmotic sensitivity of sensitive and dicarboximide-resistant field-isolates of *Botrytis cinerea* grown on media amended with 0.68 M an 1.0 M (figures in brackets) respectively

	Total number of Isolates tested	% Isolates with ... osmotic sensitivity			
		no	low	medium	high
sensitive Isolates	50	86 % (20 %)	10 % (67 %)	2 % (3 %)	2 % (10 %)
resistant Isolates	70	77 % (8 %)	19 % (77 %)	3 % (11 %)	1 % (4 %)

no osmotic sensitivity >60 % control
low osmotic sensitivity 30—60 % control
medium osmotic sensitivity 11—29 % control
high osmotic sensitivity <11 % control

time, which would suggest that there is a correlation between progressive homokaryotisation and degree of resistance. In our own investigations we too could only obtain daughter isolates having a high degree of resistance ($ED_{50} > 500$ ppm a.i.) by adaptation from a very homogeneous, stable laboratory strain, whereas with relatively freshly isolated field strains the low degree of resistance characteristic of such isolates was retained (LORENZ and POMMER 1985).

It is very evident that the question whether mainly heterokaryotic or homokaryotic strains have been used is of importance in interpreting the various results obtained on the stability of resistance to dicarboximides. With respect to the characteristic "resistance to dicarboximides", homokaryotic isolates only produce resistant conidia whereas heterokaryotic strains produce both sensitive and resistant conidia. In the case of the latter, the proportion of sensitive or resistant conidia, respectively, increases very rapidly under certain circumstances depending on whether selection pressure caused by the appropriate fungicide is absent or present (SUMMERS *et al.* 1984). The continued application of selection pressure in relatively closed systems, which exist, for example, under laboratory conditions and also to a certain extent under glasshouse conditions, can finally result in the overwhelming homokaryotisation of individual isolates, and also entire populations (KATAN 1982; SUMMERS *et al.* 1984). In the field, on the other hand, as a result of constant mixing of populations, predominantly heterokaryotic populations can be expected. If the results obtained by various authors are considered from this aspect, then a clear correlation between the origin of the isolates and the stability of resistance does in fact manifest itself. In laboratory isolates and strains from glasshouse crops, resistance to dicarboximides has generally proved stable (LEROUX *et al.* 1977; LORENZ and EICHHORN 1978; KATAN 1982). With field isolates that have not been cultivated on nutrient media containing fungicides for any length of time, the proportion of resistant conidia decreased very rapidly; occasionally such isolates regained a complete sensitivity (DENNIS and DAVIS 1979; LORENZ and EICHHORN 1980, 1982; MARAITE *et al.* 1981; LORENZ and POMMER 1982). The latter does not, however, represent the rule. Testing isolates from vine plots in which no further applications of dicarboximides had been carried out for four years, has shown that a certain, even if only a small proportion of resistant strains is retained in the population (LORENZ and POMMER 1985). A similar result was obtained from infection tests (Tab. 7.6) with resistant strains over several passages in a glasshouse (LORENZ and POMMER 1985). This, and also the studies by SUMMERS *et al.* (1984) on heterokaryotic and homokaryotic *Botrytis* isolates support the hypothesis formulated by DEKKER (1976) in which the heterokaryotic state which exists in multinucleate fungi "may serve as a mechanism in maintaining low levels

Table 7.6 Competitive ability of a resistant isolate of *Botrytis cinerea* (Nr. 920) in mixture with conidia from the corresponding sensitive isolates

Mixture sensitive : resistant	% Resistant isolates after reisolation			
	I	II	III	IV
100 % sensitive	0	0	0	0
80 % sensitive : 20 % resistant	44 %	0	0	0
50 % sensitive : 50 % resistant	50 %	0	10 %	0
20 % sensitive : 80 % resistant	33 %	30 %	0	10 %
100 % resistant	80 %	37 %	10 %	14 %

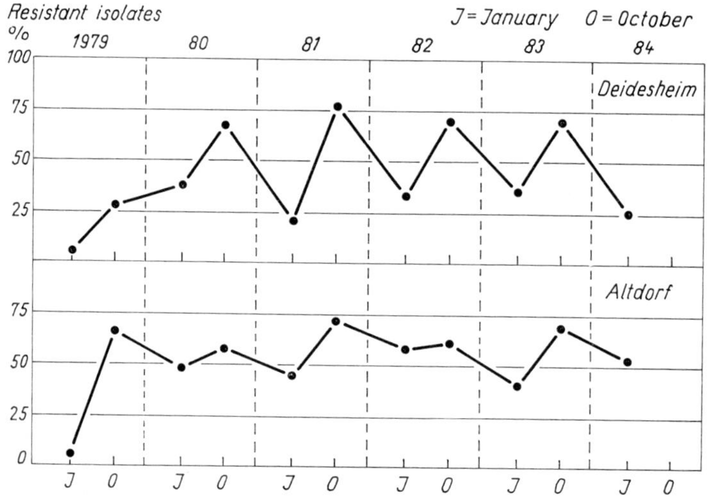

Fig. 7.2 Population dynamics of dicarboximide-resistant strains in two vineyards of the Rhenish-Palatinate area (Deidesheim and Altdorf) during the years 1979—1984 (dicarboximides were used continously at least three times per season).

of fungicide tolerance in populations in the presence and absence of selection pressure from the fungicide".

The competitive ability of resistant *Botrytis* isolates is markedly inferior to that of sensitive isolates. It has been repeatedly possible to demonstrate this in a very wide variety of test situations, both under glasshouse conditions (Hisada *et al.* 1979; Lacroix and Gouot 1981; Lorenz and Pommer 1985) and in the field (Davis and Dennis 1981; Gullino and Garibaldi 1981). Field observations also indicate a low competitive ability. Leroux and Gredt (1982a) observed a considerable reduction in resistant strains in vine plots when the selection pressure was absent during the vegetation period. The decline of resistant strains was particularly pronounced at high infection pressures. In our own tests (Löcher and Lorenz 1985) the proportion of resistant strains was, as a rule, highest after the use of dicarboximides in the autumn (at harvest), but decreased decisively during the winter and early spring (Fig. 7.2). The explanation for this is surely to be found in the interaction of various factors, namely the weaker production of conidia, the reduced pathogenicity and the less stable resistance, at least under field conditions.

References

Albert, G.: Wirkungsmechanismus und Wirksamkeit von Vinclozolin bei *Botrytis cinerea* Pers. Diss. Rheinische Friedrich-Wilhelms-Univ., Bonn 1979.

— Sphäroplastenbildung bei *Botrytis cinerea*, hervorgerufen durch Vinclozolin. Z. Pflanzenkrankh. u. Pflanzenschutz **88** (1981): 337—342.

Barbero, L., and Gaia, P.: Enological consequence of the use of Ronilan in grape cultivation. Vini Ital. **21** (119) (1979): 95—100.

Beever, R. E.: Osmotic sensitivity of fungal variants resistant to dicarboximide fungicides. Transactions of the Brit. Mycolog. Soc. **80** (1983): 327—331.

— and Brien, H. M. R.: A survey of resistance to the dicarboximide fungicides in *Botrytis cinerea*. New Zealand J. Agric. Res. **26** (1983): 391—400.

— and BYRDE, R. J. W.: Resistance to the dicarboximide fungicides. In: DEKKER, J., and GEORGOPOULOS, S. G. (Eds.): Fungicide Resistance in Crop Protection., Pudoc, Wageningen 1982, pp. 101—117.

BENDA, J.: Mikrobiologische Untersuchungen über den Einfluß des Fungizides Ronilan auf die Hefeflora der Trauben und des Weines. Weinwissenschaft **33** (1978): 153—158.

— Botrytizide, ihre Wirkstoffe und Formulierungsmittel in der mikrobiologischen Prüfung. Weinwissenschaft **38** (1983): 41—50.

BRENNEMAN, T. P., and STIPES, R. J.: Sensitivity of *Sclerotinia minor* from peanut to dicloran, iprodione and vinclozolin. Phytopathology **73** (1983): 964.

BUCHENAUER, H.: Preleminary studies on the mode of action of vinclozolin. Meded. Fak. Landbouwwet. Rijksuniv. Gent **41** (1976): 1509—1519.

CABRAS, P., MELONI, M., PIRISI, F. P., and PIRISI, M.: Degradation of dicarboximide fungicides in wine. Pesticide Sci. **15** (1984): 247—252.

CAYLEY, G. R., and HIDE, G. A.: Uptake of iprodione and control of diseases on potato stems. Pesticide Sci. **11** (1980): 15—19.

CHASTAGNER, G. A., and VASSEY, W. E.: Tolerance of *Botrytis tulipae* to glycophene and vinclozolin. Phytopathology **69** (1979): 914.

— — Occurrence of iprodione tolerant *Fusarium nivale* under field conditions. Plant Dis. **66** (1983): 112—114.

COOKE, B. K., PAPPAS, A. C., JORDAN, V., and WESTERN, N. M.: Translocation of benomyl prochloraz and procymidone in relation to control of *Botrytis cinerea* in strawberries. Pesticide Sci. **10** (1979): 467—472.

DANNEBERGER, T. K., and VARGAS, J. M.: Systemic activity of iprodione in *Poa annua* and post-infection activity for *Drechslera sorokiniana*. Plant Dis. **66** (1982): 914—915.

DAVIS, R. P., and DENNIS, C.: Use of dicarboximides on strawberries and potential problems of resistance in *Botrytis cinerea*. Proc., 1979 Brit. Crop Protect. Conf., Brighton, Pests and Diseases **1** (1979): 193—201.

— — Properties of dicarboximide-resistant strains of *Botrytis cinerea*. Pesticide Sci. **12** (1981): 521—535.

— — Studies on the survival and infective ability of dicarboximide-resistant strains of *Botrytis cinerea*. Ann. appl. Biol. **98** (1981): 395—402.

DEKKER, J.: Acquired resistance to fungicides. Ann. Rev. of Phytopathol. **14** (1976): 405—428.

DENNIS, C., and DAVIS, R. P.: Tolerance of *Botrytis cinerea* to iprodione and vinclozolin. Plant Pathol. **28** (1979): 131—133.

DETWEITER, A. R., VARGAS, J. M., and DANNEBERG, T. K.: Resistance of *Sclerotinia homeocarpa* to iprodione and benomyl. Plant Dis. **67** (1983): 617—630.

DI TORO *et al.*: Atti Giornate Fitopatologico: 1980.

EICHHORN, K. W., and LORENZ, D. H.: Untersuchungen über die Wirkung von Vinclozolin gegenüber *Botrytis cinerea* in vitro. Z. für Pflanzenkrankh. u. Pflanzenschutz **85** (1978): 449—460.

ELMER, W. H., and STIPES, R. J.: Movement and disappearance of dicloran, iprodione and vinclozolin in peanut and nonpeanut soils. Plant Dis. **69** (1985): 292—294.

FIMA, P., and WOMASTEK, R.: Residues in/on grapes and in wine. Mitt. Klosterneuburg **33** (1983): 253—256.

FLORI, P., and ZORINI, R.: Anti-botrytis fungicide residues and ecological technology. Vignevini **11** (1984): 47—50.

FUCHS, A., DE VRIES, F. W., and DE WAARD, M. A.: Simultaneous resistance in fungi to ergosterol biosynthesis inhibitors and dicarboximides. Netherl. J. Plant Pathol. **90** (1984): 3—11.

FUJINAMI, A., OZAKI, K., NODERA, T., and TANAKA, K.: Studies on biological activity of cyclic imide compounds. Part II. Antimicrobial activity of 1-phenylpyrolidine-2,5-diones and related compounds. Agric. Biol. Chem. **36** (1972): 318.

— — and YAMAMOTO, S.: Studies on biological activity of cyclic amide compounds. Part I. Antimicrobial activity of 3-phenyloxazolidine-2,4-diones and related compounds. Agric. Biol. Chem. **35** (1971): 1707—1719.

GEORGOPOULUS, S. G., SARRIS, M., and ZIOGAS, B. N.: Mitotic instability in *Aspergillus nidulans* caused by the fungicides iprodione, procymidone and vinclozolin. Pesticide Sci. **10** (1979): 389—392.

GRINDLE, M.: Isolation and characterization of vinclozolin resistant mutants of *Neurospora crassa*. Transactions of the Brit. mycolog. Soc. **82** (1984): 635—643.

— and TEMPLE, W.: Sporulation and osmotic sensitivity of dicarboximide-resistant mutants of *Neurospora crassa*. Transactions of the Brit. mycolog. Soc. **84** (1985): 369—372.

GULLINO, M. L., and GARIBALDI, A.: Osservazioni sperimentali sulla resistenza di isolamenti Italiani di *Botrytis cinerea* a vinclozolin. La difesa delle piante **6** (1979): 341—350.

— — Biological properties of dicarboximide-resistant strains of *Botrytis cinerea* Pers. Phytopathol. Mediterranea **20** (1981): 117—122.

HAGAN, A., and LARSEN, P. O.: Effect of fungicides on conidium germination, germ tube elongation, and appressorium formation by *Bipolaris sorokiniana* on Kentucky bluegrass. Plant Dis. Rep. **63** (1979): 474—478.

HARTILL, W. F. T., TOMPKINS, G. R., and KLEINSMAN, P. J.: Development in New Zealand of resistance to dicarboximide fungicides in *Botrytis cinerea*, to acylalanines in *Phytophthora infestans* and to guazatine in *Penicillium italicum*. New Zealand J. Agric. Res. **26** (1983): 261—269.

HELWEG, A.: Influence of the fungicide iprodione on respiration, ammonification and nitrification in soil. Pedobiologia **25** (1983): 87—92.

HISADA, Y., KATO, T., and KAWASE, Y.: Systemic movement in cucumber plants and control of cucumber gray mould by a new fungicide, S-7131. Netherl. J. Plant Pathol. **83** (1977): 71—78.

— — — Mechanism of antifungal action of procymidone in *Botrytis cinerea*. Ann. Phytopathol. Soc. Japan **44** (1978): 509—518.

— and KAWASE, Y.: Morphological studies of antifungal action of N-(3′5′-dichlorophenyl)-1,2-dimethylcyclopropane-1,2′-dicarboximide on *Botrytis cinerea*. Ann. Phytopathol. Soc. Japan **43** (1977): 151—158.

— — Reversible binding of procymidone to a sensitive fungus *Botrytis cinerea*. J. Pesticide Sci. **5** (1980): 559—564.

— MAEDA, K., TOTTORI, N., and KAWASE, Y.: Plant disease control by N-(3,5-dichlorophenyl)-1,2-dimethyl-cyclopropane-1,2-dicarboximide. J. Pesticide Sci. **1** (1976): 145—149.

— TAKAKI, H., KAWASE, J., and OZAKI, T.: Difference in the potential of *Botrytis cinerea* to develop resistance to procymidone *in vitro* and in the field. Ann. Phytopathol. Soc. Japan **45** (1979): 283—290.

HOLZ, B.: Über eine Resistenzerscheinung von *Botrytis cinerea* an Reben gegen die neuen Kontaktbotrytizide im Gebiet der Mittelmosel. Weinberg u. Keller **26** (1979): 18—25.

HUNTER, T., JORDAN, V. W. L., and PAPPAS, A. C.: Control of strawberry fruit rots caused by *Botrytis cinerea* and *Phytophthora cactorum*. Proc., 1979 Brit. Crop. Protect. Conf., Brighton-Pests and Dis. **1** (1979): 177—183.

KATAN, T.: Resistance to dicarboximide fungicides in *Botrytis cinerea* from cucumbers, tomatoes, strawberries and roses. Netherl. J. Plant Pathol. **87** (1981): 244.

— Resistance to 3,5-dichlorophenyl-N-cyclic imide (dicarboximide) fungicides in the grey mould pathogen *Botrytis cinerea* on protected crops. Plant Pathol. **31** (1982): 133—141.

— and SHABI, E.: Characterization of a dicarboximide-resistant laboratory idolate of *Monilinia laxa*. Phytoparasitica **10** (1982): 241—245.

KATO, T., HISADA, I., and KAWASE, I.: Nature of procymidone-tolerant *Botrytis cinerea* strains obtained in vitro. In: Proc. of Seminar on Pest resistance in Pesticides, Palm Spring, USA 1979.

LACROIX, L., BIC, C., BURGAUD, L., GUILLOT, M., LEBLANC, R., RIOTTOT, R., and SAULI, M.: Etude des propriétés antifongiques d'une nouvelle famille de dérivés de l'hydantoine et en particulier du 26 019 R.P.: Phytiatrie-Phytopharmacie **23** (1974): 165—174.

— and GOUOT, J. M.: Investigations on the resistance of *Botrytis cinerea* to iprodion in greenhouses. Meded. Fak. Landbouwwet. Rijksuniv., Gent **46** (1981): 979—989.

LEROUX, P., and FRITZ, R.: Antifungal activity of dicarboximides and aromatic hydrocarbons and resistance to these fungicides. In: TRINCI, A. P. J., and RYLEY, J. F. (Eds.): Mode of Action of Antifungal Agents. Cambridge Univ. Press, Cambridge 1983, pp. 207—237.

— — and GREDT, M.: Etudes en laboratoire de souches de *Botrytis cinerea* Pers. résistantes à la Dichlozoline, au Dicloran, au Quintozène, à Vinclozoline et au 26 019 RP (Glycophéne). Phytopathol. Z. **89** (1977): 347—358.

— — — Cross-resistance between 3,5-dichlorophenyl cyclic imide fungicides and various aromatic compounds. In: LYR, H., and POLTER, C. (Eds.): Systemische Fungicide und Antifungale Verbindungen. VI. Internationales Symposium. Akademie-Verlag, Berlin 1982, S. 79—88.

— and GREDT, M.: Effets d'alcools primaires, de polyols, de sels mineraux et de sucres sur des souches de *Botrytis cinerea* sensibles ou resistantes à l'iprodione et à la procymidone. Comptes rendus de l'Academie des Sci., Paris, série III, **294** (1982a): 53—56.

— — Phénomènes de resistance de *Botrytis cinerea* aux fongicides. La Défense des Végétaux **213** (1982b): 3—17.

— — and FRITZ, R.: Etudes en laboratoire de souches de quelques champignons phytopathogènes, résistantes à la vinclozoline et à divers fongicides aromatique. Meded. Fak. Landbouwwet. Rijksuniv., Gent **43** (1978): 881—889.

— — — Resistance to 3,5-dichlorophenyl-N-cyclic imide fungicides. Netherl. J. Plant Pathol. **87** (1981): 242.

— LAFON, R., and GREDT, M.: La resistance de *Botrytis cinerea* aux benzimidazoles et aux imides cycliques: situation dans les vignobles alsaciens, bordelais et champenois. OEPP/EPPO Bull. **12** (1982): 137—143.

LITTLEY, E. R., and RAHE, J. E.: Specific tolerance of *Sclerotium cepivorum* to dicarboximide fungicides. Plant Dis. **68** (1984): 371—374.

LÖCHER, F. J., BRANDES, W., LORENZ, G., HUBER, W., SCHILLER, R., and SCHREIBER, B.: Development of a strategy to maintain the efficacy of the dicarboximides in the presence of resistant strains of *Botrytis* in grapes. Gesunde Pflanzen **37** (1985): 3—8.

— and LORENZ, G.: Resistance management strategies for dicarboximides in grapes — results of six years trialwork (1985): in press.

LORENZ, D. H., and EICHHORN, K. W.: Untersuchungen zur möglichen Resistenzbildung von *Botrytis cinerea* an Reben gegen die Wirkstoffe Vinclozolin und Iprodione. Wein-Wissensch. **33** (1978): 2—10.

— — Vorkommen und Verbreitung der Resistenz von *Botrytis cinerea* gegen Dicarboximid-Fungizide im Anbaugebiet der Rheinpfalz. Wein-Wissensch. **35** (1980): 199—210.

— — *Botrytis* and its resistance against dicarboximide fungicides. OEPP/EPPO Bull. **12** (1982): **125**—129.

LORENZ, G., and POMMER, E.-H.: Studies on pectolytic and cellulolytic enzymes of dicarboximide-sensitive and -resistant strains of *Botrytis cinerea*. OEPP/EPPO Bull. **12** (1982): 145—149.

— — Morphological and physiological characteristics of dicarboxmide-sensitive and -resistant isolates of *Botrytis cinerea*. OEPP/EPPO Bull. **15** (1985): 353—360.

MARAITE, H., GILLES, G., MEUNIER, S., WEYNS, J., and BAL, E.: Resistance of *Botrytis cinerea* Pers. ex Pers. to dicarboximide fungicides in strawberry fields. Parasitica **36** (1981): 90—101.

— MEUNIER, S., POURTOIS, A., and MEYER, J. A.: Emergence *in vitro* and fitness of strains of *Botrytis cinerea* resistant to fungicides. Meded. Fak. Landbouwwet. Rijksuniv. Gent **45** (1980): 159—167.

MCPHEE, W. J.: Some characteristics of *Alternaria alternata* strains resistant to iprodione. Plant Dis. **64** (1980): 847—849.

MENAGER, TISSIER, COMELLI, and ALLUIS: La dichlozoline. Phytiatrie-Phytopharmacie **20** (1971): 169—172.

MIKAMI, N., YOSHIMURA, J., YAMADA, H., and MIYAMOTO, J.: Translocation and metabolism of procymidone in cucumber and bean plants. Nippon Noyaku Gakkaishi **9** (1984): 131—136.

NAKAYA, S., KITAYAMA, H., UEDA, M., KITAYAMA, M., SAITO, M., and KITSUTAKA, T.: Subacute toxicity of a new pesticide, 3-(3,5-dichlorophenyl)-5,5-dimethyl-2,4-oxazolidinedione (DDOD). Hokkaidoritsu Eisei KENKYUSHOHO **19** (1969): 132—138.

PANAGIOTAKU, M. G., and MALATHRAKIS, N. E.: Resistance of *Botrytis cinerea* Pers. to dicarboximide fungicides. Netherland J. Plant Pathol. **87** (1981): 242.

PAPPAS, A. C.: Unzureichende Bekämpfung des Grauschimmels auf Alpenveilchen mit Dicarboximid-Fungiziden in Griechenland. Z. Pflanzenkrankh. u. Pflanzenschutz **89** (1982): 52—58.

— COOKE, B. K., and JORDAN, V. W. L.: Insensitivity of *Botrytis cinerea* to iprodione, procymidone and vinclozolin and their uptake by the fungus. Plant Pathol. **28** (1979): 71—76.

— and FISHER, D. J.: A comparison of the mechanismen of action of vinclozolin, procymidone and prochloraz against *Botrytis cinerea*. Pesticide Sci. **10** (1979): 239—246.

PENROSE, L. J., HOFFMANN, W., and NICHOLLS, M. R.: Field occurrence of vinclozolin resistance in *Monilinia fructicola*. Plant Pathol. **34** (1985): 228—234.

POMMER, E.-H., and LORENZ, G.: Resistance of *Botrytis cinerea* Pers. to dicarboximide fungicides — a literature review. Crop Protect. **1** (1982): 221—230.

— and MANGOLD, D.: Vinclozolin (BAS 352 F), ein neuer Wirkstoff zur Bekämpfung von *Botrytis cinerea*. Meded. Fak. Landbouwwet. Rijksuniv. Gent **40** (1975): 713—722.

POMMER, E. H., and ZEEH, B.: Myclozolin, ein neuer Wirkstoff aus der Klasse der Dicarboximide. Meded. Fak. Landbouwwet. Rijksuniv. Gent (1982): 935—942.

PORTER, D. M., and PHIPPS, P. M.: Effects of three fungicides on mycelial growth, *sclerotium* production and development of fungicide-tolerant isolates of *Sclerotinia minor*. Plant Dis. **69** (1985): 143—146.

PRESLEY, A. H., and MAUDE, R. B.: Tolerance in *Botrytis squamosa* to iprodione. Ann. appl. Biol. **100** (1982): 117—127.

RHODES, L. H., and LARSEN, P. O.: Effects of fungicides on mycorrhizal development of creeping bentgrass. Plant Dis. **65** (1981): 145—147.

RITCHIE, D. F.: Mycelial growth, peach fruit-rotting capability and sporulation of strains of *Monilinia fructicola* resistant to dichloran, iprodione, procymidone and vinclozolin. Phytopathology **73** (1983): 44—47.

ROARK, J. H., and DALE, J. L.: The effect of turf fungicides on earthworms. Proc. Arkansas Academy of Sci. **33** (1979): 71—74.

ROSENBERGER, D. A., MEYER, F. W., and CECILIA, C. V.: Fungicide strategies for control of benomyl-tolerant *Penicillium expansum* in apple storage. Plant Dis. Rep. **63** (1979): 1033 to 1037.

SAPIS-DOMERCQ, S., BERTRAND, A., MUR, FR., and SARRE, C.: Effects of fungicides residues from grapevines on fermentating microflora, 1976 Experiments. Connaissance Vigne Vin **11** (1977): 227—242.

— — JOYEUX, A., LUCMARET, V., and SARRE, C.: Study of the effect of vine treatment products on grape and vine microflora. 1977 Experiments. Connaissance Vigne Vin **12** (1978): 245—275.

SCHÜEPP, H., and KÜNG, M.: Gegenüber Dicarboximid-Fungiziden tolerante Stämme von *Botrytis cinerea* Pers. Ber. Schweizer. Bot. Ges. **88** (1978): 63—71.

— — and SIEGFRIED, W.: Développement des souches de *Botrytis cinerea* résistentes aux dicarboximides dans les vignes de la Suisse alémanique OEPP/EPPO Bull. **12** (1982): 157—161.

SPENGLER, G., SCHERER, M., and POMMER, E. H.: Untersuchungen über das Resistenzverhalten von *Botrytis cinerea* gegenüber Vinclozolin. Mitt. Biol. Bundesanstalt für Land- und Forstwirtschaft, Berlin-Dahlem, **191** (1979): 236.

STOJANOVIC, M., and VUKMIROVIC, M.: Effect of some fungicides on fermentation yeasts. Mikrobiologija **16** (1979): 39—49.

SUMMERS, R. W., HEANEY, S. P., and GRINDLE, M.: Studies on a dicarboximide resistant heterokaryon of *Botrytis cinerea*. Proc. Brit. Crop Protect. Conf. 1984, Pests and Diseases Vol. 2 (1984): 453—458.

SZTEJNBERG, A., and JONERS, A. L.: Resistance of the brown rot fungus *Monilinia fructicola* to iprodione, vinclozolin and procymidone. Phytoparasitica **7** (1979): 46.

VOJTEKOVA, G., VODOVA, J., and JURKACKOVA, M.: Effect of several fungicides on the fermentation of grape must. Vinohrad (Bratislava) **21** (1983): 207—210.

VULIE, M., EBERLE, O., and HÄBERLE, N.: Metomeclan, ein neuer Dicarboximid-Wirkstoff mit breitem fungizidem Wirkungsspektrum. Meded. Fak. Landbouwwet. Rijksuniv. Gent **49** (1984): 293—301.

WALKER, A., ENTWISTLE, A. R., and DEARNALEY, N. J.: Evidence for enhanced degradation of iprodione in soils treated previously with this fungicide. Monograph-British Crop Protect. Council **27** (Soils Crop Protect. Chemistry) (1984): 117—123.

LYR, H. (Ed.): Modern, Selective Fungicides — Properties, Applications, Mechanisms of Action. Longman Group UK Ltd., London, and VEB Gustav Fischer Verlag, Jena, 1987.

Chapter 8

Mechanism of action of dicarboximide fungicides

W. EDLICH and H. LYR

Institute for Plant Protection Research of the Academy of Agricultural Sciences of the GDR, Kleinmachnow

Introduction

Dicarboximide fungicides (DCOF) represent a group of highly active and selective fungicides as described in chapter 7.

In contrast to other groups of fungicides, there is a lack of knowledge concerning the mechanism of action of the dicarboximide group. Various effects have been described (Tab. 8.1) but they do not indicate a primary action which could be regarded as the basis for the very different effects observed in various parts of the fungal cell metabolism. The detection of a general cross resistance between the dicarboximide and the Aromatic Hydrocarbon Fungicides (AHF) was surprising (LEROUX *et al.* 1978).

Table 8.1 Some effects of dicarboximides on fungal cell structure and metabolism

Affected Process	Symptoms	References
cell division	mitotic instability	GEORGOPOULOS *et al.* 1979
DNA synthesis	somatic segregation	LEROUX *et al.* 1978
	low inhibition	FRITZ *et al.* 1977
	strong inhibition	HISADA and KAWASE 1978
	without effect	BUCHENAUER 1976
RNA and protein synthesis	low inhibition	PAPPAS and FISHER 1979
cell wall synthesis	without effect	BUCHENAUER 1976
	increased precursor insertion	HISADA and KAWASE 1977
	low inhibition	BUCHENAUER 1976
	strong inhibition	ALBERT 1981
metabolism of sterols	low effects	BUCHENAUER 1976
		PAPPAS and FISHER 1979
metabolism of lipids	increased level of free fatty acids	PAPPAS and FISHER 1979
		BUCHENAUER 1976
		FRITZ *et al.* 1977
K^+-efflux	without effect	HISADA and KAWASE 1977
respiration	without effect	BUCHENAUER 1976
		HISADA and KAWASE 1977
		FRITZ *et al.* 1977
		PAPPAS and FISHER 1977
cell structure	swelling of mitochondria and their lysis; vesiculation of ER	MÜLLER 1981

Morphological effects

DCOF inhibit spore germination and growth of mycelia, but inhibit the latter much more efficiently (BUCHENAUER 1976; FRITZ *et al.* 1977; PAPPAS and FISHER 1979). In liquid nutrient media germ tubes of *Botrytis cinerea* swell rapidly after treatment with DCOF and may burst when certain concentrations of DCOF are used (HISADA and KAWASE 1977). At sublethal doses, the growth of the hyphal tips is disturbed and anomalous branching occurs (chapter 7) similar to that which occurs with some AHF (chapter 6).

But morphological changes are rather unspecific in relation to the mechanism of action of a fungicide, because very different agents (inhibitors of protein synthesis, antitubulins and others) can induce analogous abnormalities. Such effects have been described for example, with AHF (SHARPLES 1961; GEORGOPOULOS *et al.* 1967; MACRIS and GEORGOPOULOS 1973; THRELFALL 1972; TILLMANN and SISLER 1973), with sorbose (MOORE 1981), or with cytochalasin (BETTINA *et al.* 1972; ALLEN *et al.* 1980).

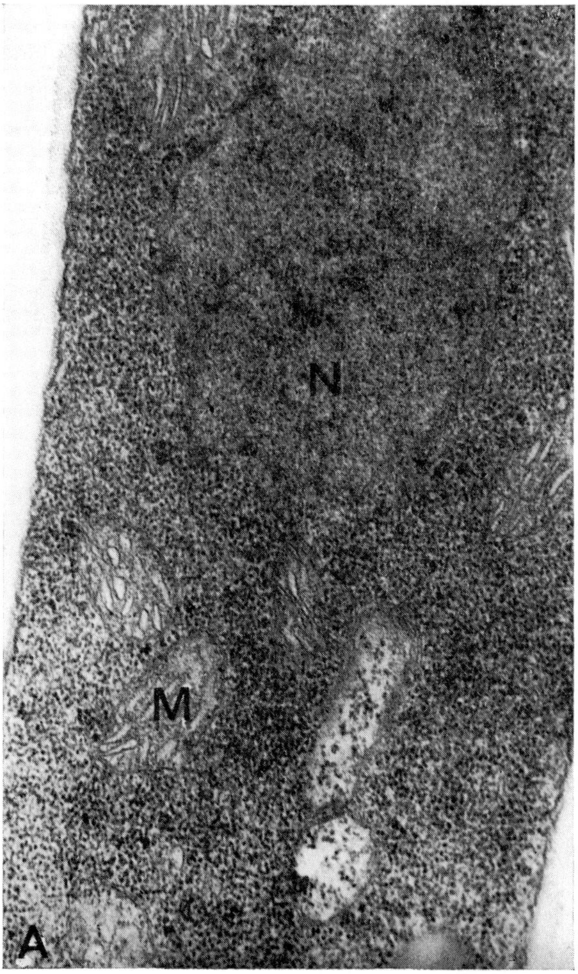

Fig. 8.1 *Botrytis cinerea*, untreated control in submers culture 8 h after spore germination. ×35,000 (phot. Dr. H. M. MÜLLER).

M — mitochondria, ER — endoplasmatic reticulum, V — vesicles, N — nucleus, Z — cell wall.

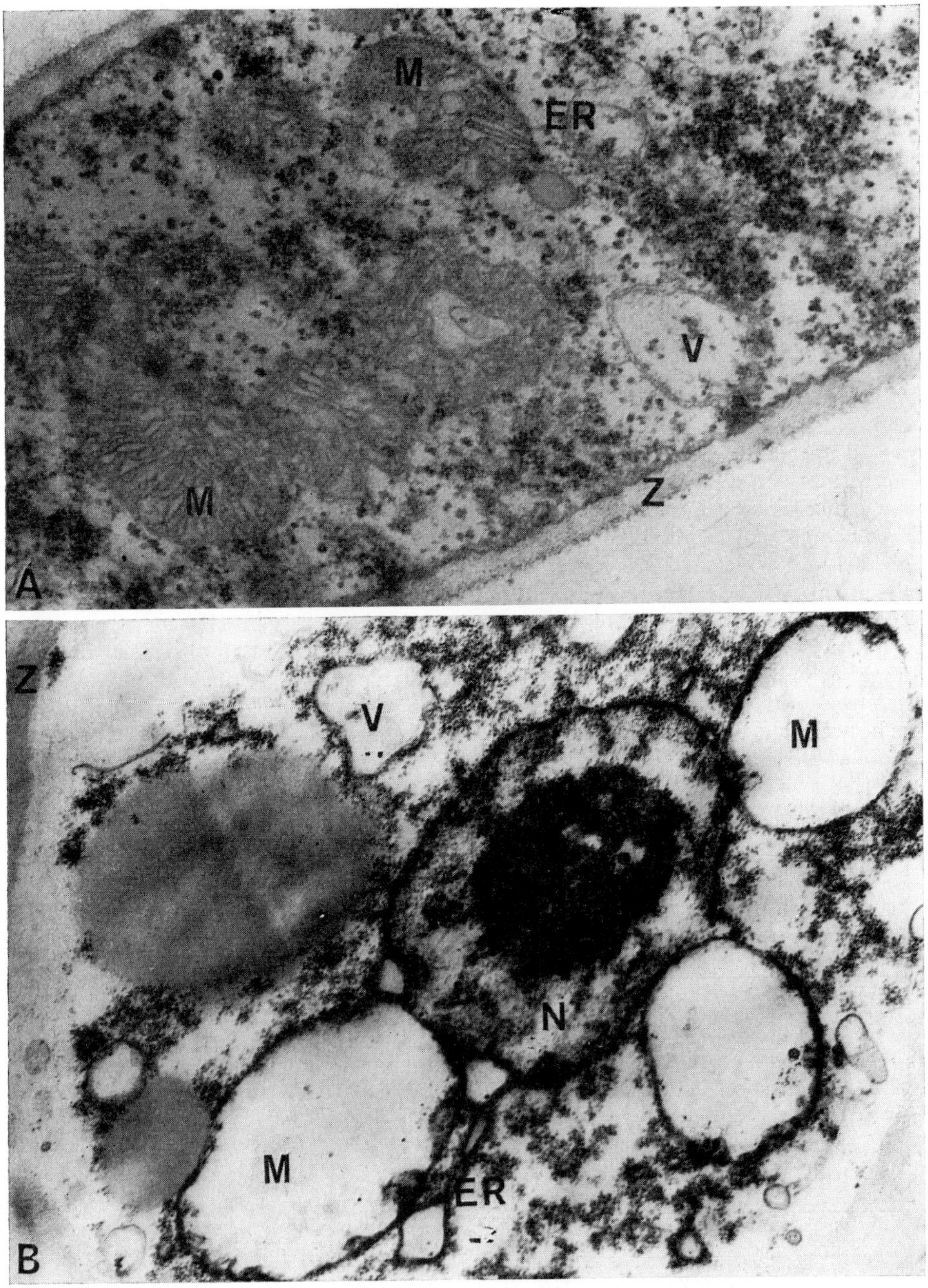

Fig. 8.2 *Botrytis cinerea*, treated by 1 mg · l^{-1} vinclozolin. A — 2 h after treatment, B — 4 h after treatment. Figures show swelling of mitochondria, loss of their matrix and strong reduction of christae, vesiculation of cytoplasma ×30,000 (phot. Dr. H. M. MÜLLER).

Hisada and Kawase (1977) working with fungal protoplasts did not find an influence of DCOF on cell wall synthesis, whereas Albert (1979, 1981) in similar investigations could inhibit cell wall synthesis with DCOF. But this may have been favoured by the rather high osmotic pressure (up to 2.3 M) in the culture medium. The strong pathological cell wall thickening induced by chloroneb and other AHF (chapter 6) is not produced by DCOF.

Physiological and biochemical effects

As with the AHF the main metabolic pathways are not, or only to a small degree disturbed by DCOF. Some authors found minor changes of the synthesis of proteins, nucleic acids and lipids, whereas in other experiments no severe influence on catabolic or anabolic pathways was observed (Buchenauer 1979; Fritz *et al.* 1977; Hisada and Kawase 1978; Pappas and Fisher 1979).

DNA-synthesis seems to be relatively strongly impaired by iprodione (Pappas and Fisher 1979). But all authors agree that the observed effects are only the consequence of an unknown primary action. Georgopoulos *et al.* (1979) and Fritz *et al.* (1977) described irregularities in the course of the cell division under the influence of DCOF.

The results demonstrate that the evaluated effects are not directly related to the primary mechanism of action. Apparently the conventionally used analytical procedures are not sufficient to explore the effects which are caused by dicarboximide fungicides and which lead to the final death of the fungal cell. Stimulated by our results with Aromatic Hydrocarbon Fungicides, we investigated the effect of DCOF on the lipid peroxidation ability in sensitive fungi, mainly in *Mucor mucedo* and *Botrytis cinerea*. Electron microscopic pictures (Müller 1981) demonstrated damage in fungal cells in the mitochondrial inner membrane and in membranes of the endoplasmic reticulum (EPR) caused by DCOF (Fig. 8.1 and 8.2). There exist clear differences between the effects produced by DCOF and those produced by AHF (chapter 6). The cell wall is not thickened in cells treated with DCOF, and lysis of mitochondrial

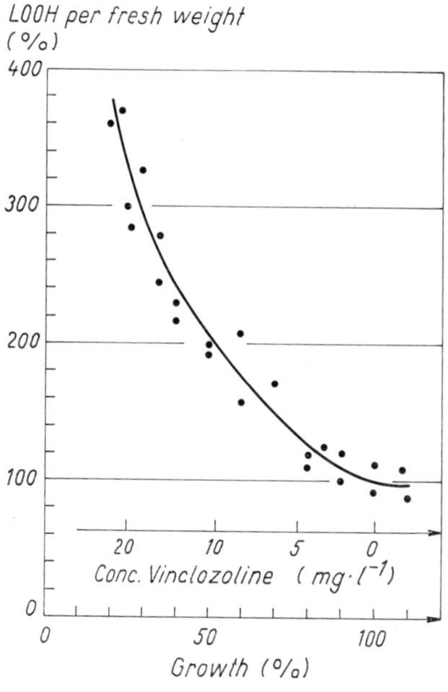

Fig. 8.3 Correlation between growth of mycel and level of lipid peroxides in cells of *Mucor mucedo* in dependence of concentration of vinclozolin. Growth is expressed as fresh weight of mycel in relationship to a control without fungicide. Level of lipid peroxides (LOOH) were measured by using the thiobarbituric acid method.

membranes is a bit different in those treated with AHF. It seems possible however, that DCOF initiate effects similar to AHF by inducing pathological oxidative processes in sensitive fungal cells. Indeed we found a correlation between inhibition of hyphal growth in *Mucor mucedo* and *Botrytis* and the level of intracellular lipid peroxidation caused by DCOF application (Fig. 8.3). Addition of α-tocopherol acetate not only antagonized the growth inhibition but also decreased the level of peroxides to the level of the untreated controls. Among other compounds which we investigated, only piperonylbutoxide and SKF 525 A significantly inhibited the lipid peroxidation which is induced by DCOF. These compounds were active at concentrations of 5×10^{-6} mol $\times$ l^{-1} (1—2 ppm). Other fungicides such as tridemorph, carbendazim, or chlorothalonil had little or no effect on this reaction, whereas the effect of chloroneb, dichloran and PCNB was similar to that of DCOF.

Mechanism of action

There is a question, of course, concerning the mechanism by which DCOF initiates lipid peroxidation. Experiments with *Mucor mucedo* and *Botrytis cinerea* revealed, that a membrane bound NADPH dependent flavin enzyme of the type of a "cytochrome-c-reductase" (EC 1.6.2.) (as well as some other, but not all flavin enzymes) are specifically inhibited by vinclozolin and other DCOF *in vitro*. The inhibition begins at concentrations of 5×10^{-6} mol $\times$ l^{-1} (1—2 ppm). This reaction is not sensitive to CO, therefore a participation of a cytochrome P-450 dependent enzyme can be excluded. The involvement of a cytochrome P-450 dependent enzyme has been proposed by Leroux *et al.* (1983) and is discussed in a recent paper by Gullino and Sisler (1985). They concluded this indirectly from an antagonism of some cytochrome P-450 inhibitors like piperonylbutoxide, SKF 525 A, metyrapone a.o.

Indeed, according to our results the effect of DCOF on lipid peroxidation can be counteracted *in vivo* and *in vitro* by α-tocopherol acetate, piperonylbutoxide and partly by SKF 525 A. Piperonylbutoxide is in practical use as an insecticide synergist and the assumption is that it inhibits cytochrome P-450 enzymes which detoxify insecticides. Apparently it interacts with cytochrome c reductase which can, but must not be a part of the P-450 enzyme complex, i.e. it can also exist as a distinct enzyme. SKF 525 A has also been described as an inhibitor of cytochrome P-450 enzymes. In contrast, α-tocopherol acetate acts as general radical scavenger in membranes and by this mechanism, suppresses lipid peroxidation.

Therefore the antagonizing effects of these two types of a forementioned substances, which can be demonstrated *in vitro* as well as *in vivo* in mycelial growth tests (Tab. 8.2. and Fig. 8.3), are realized by different mechanisms.

DCOF even at relative high concentrations do not very strongly inhibit to NADPH independent endogenous monoxygenases such as xanthine oxidase, whereas some AHF, such as dicloran, PCNB and chloroneb are more inhibitory towards this enzyme (Tab. 8.3). This demonstrates that the two groups of fungicides, AHF and DCOF, have many common features in their mechanism of action but do differ in some properties.

There is at present no doubt, that the pathological oxidative processes observed *in vitro* are identical with those occurring *in vivo*. DCOF seems to interact with the flavin enzyme cytochrome-c-reductase in such a manner that the normal electron flow from NADPH to cytochrome c is blocked. As a consequence NADPH as well as the essential phospholipids surrounding the active centre of the enzyme are oxidized probably by a peroxide intermediary product of the flavin enzyme (Ziegler *et al.* 1980) and/or by free radicals.

Table 8.2 Effects of some antidotes acting against vinclozolin induced growth inhibition of *Botrytis cinerea* (S-strain) in nutrition solution in relation to the level of lipid peroxides

Compounds	Conc. $mg \times l^{-1}$	Dry weight decrease %	Level of lipid* peroxides %
Without	—	0	100
Vinclozolin	0.1	67	180
	0.5	97	295
	10	100	—
α-Tocopherol acetate	100	0	96
	500	0	98
Piperonyl butoxide	0.1	4	103
	1.0	8	107
SKF 525A	1	1	93
	10	14	99
Vinclozolin + α-Tocopherol acetate	0.5 + 100	77	210
	0.5 + 500	50	175
	1.0 + 100	82	235
	1.0 + 500	67	180
Vinclozolin + Piperonyl butoxide	0.5 + 0.1	82	175
	0.5 + 1.0	76	155
Vinclozolin + SKF 525A	0.5 + 1.0	88	220
	0.5 + 10	87	215

* Lipid peroxide content was measured by using the thiobarbituric acid method.

Table 8.3 Inhibitory effects of some fungicides on xanthine oxidase and D-amino acid oxidase

Fungicide	Xanthine oxidase I_{50} $mol \times l^{-1}$	D-Amino acid oxidase I_{50} $mol \times l^{-1}$
Iprodione	2×10^{-4}	3×10^{-4}
Procymidone	10^{-3}	4×10^{-4}
Vinclozolin	10^{-3}	3×10^{-4}
Chloroneb	10^{-4}	7×10^{-4}
Dicloran	4×10^{-4}	9×10^{-4}
Etridiazole	10^{-3}	8×10^{-4}
PCNB	3×10^{-4}	4×10^{-4}
Tolclophosmethyl	10^{-3}	7×10^{-4}

* I_{50} — Concentration for 50% inhibition of enzyme activity

Such a pathological side reaction has been described for monoxygenases under the influence of pseudosubstrates which cannot be hydroxylated (Massey *et al.* 1982). The lipid peroxidation can spread in a cascade process if the natural cellular defence systems are overrun by an overproduction of free radicals. Their nature in connection with the action of DCOF is not yet established, but they could be single oxygen, hydroxyl (·OH) or superoxide (O_2·) radicals which by their very good solubility in

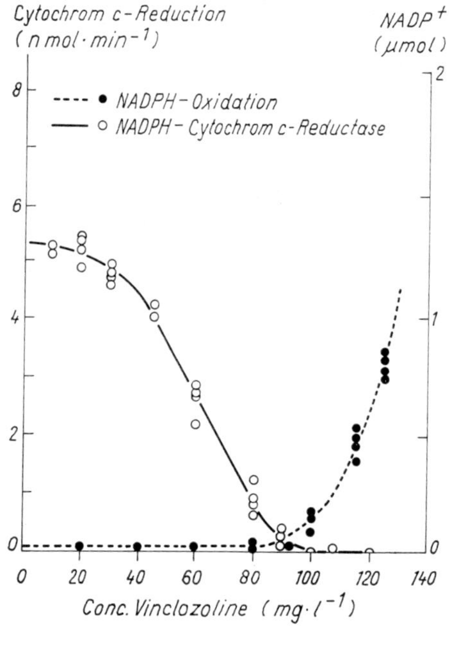

Fig. 8.4 Influence of vinclozolin on both NADPH dependent reduction of cytochrome c and NADPH oxidation *in vitro*. Solubilized 100,000 g pellets of homogenates of *Botrytis cinerea* serve as enzyme sources. Results represent from 4 independent experiments. (correction: Conc. Vinclozolin ...)

Table 8.4 Actions induced by active kinds of oxygen

	References
Peroxidation of polyunsaturated fatty acids	Bryan *et al.* 1982
Disintegration of membranes	Halliwell 1984
Inactivation of enzymes	Halliwell and Gutteridge 1984
Lignin degradation	Bes *et al.* 1983
Single strain breaks and disintegration of DNA	Mello Filho *et al.* 1984
Destruction of protease inhibitors	Halliwell and Gutteridge 1984
Oxidation of sulfhydryl groups	Hall *et al.* 1984
Oxidation of cytosine	Hazra and Steenken 1983
Destruction of chromosomes	Hassan and Moody 1984
Inhibition of respiration	Halliwell and Gutteridge 1984

lipids oxidatively attack membranes and other cell structures. That has been described for several systems in the literature (Tab. 8.4). Protective systems within the cells are enzymes, such as superoxide dismutase (SOD), catalase and peroxidases, or scavengers of various kinds (tocopherols and mannitol). When there is an overproduction of aggressive radicals, these systems are inactivated and a collapse of cell structures or a damage of DNA results, as has been demonstrated in numerous biological systems (Flamingini *et al.* 1982; Mitchell and Morrison 1983; Mello Filho *et al.* 1984; Iwata *et al.* 1984; Hall *et al.* 1984; Klebanoff *et al.* 1984; Norkus *et al.* 1983; Reiner and Kazura 1982).

Due to their high reactivity active oxygen radicals can attack various structures. Unsaturated fatty acids are peroxidized, SH-groups in enzymes are oxidized (Bhuyan *et al.* 1982; Hall *et al.* 1984) and biopolymers such as nucleic acids (Gutteridge and Halliwell 1982; Mitchell and Morrison 1983; Mello Filho *et al.* 1984; Iwata *et al.* 1984), proteins (Brunori and Rotilo 1984; Halliwell and Gutteridge 1984) and cell wall components (Bes *et al.* 1983; Gold *et al.* 1983) are attacked. The consequences are numerous, such as an inactivation of enzymes (Karagezyan 1982), activation of proteolysis (Watanabe and Kondo 1983) strand scission of nucleic acids (Hassan and Moody 1984; Hazra and Steenken 1983), destruction of cell wall material (Green and Gould 1983) and destruction of membranes (Esterbauer 1982).

Several of these effects could be the consequence of the action of dicarboximide fungicides. The inner membranes of mitochondria (Fig. 8.2) are especially sensitive which is probably due to their high content of unsaturated fatty acids. The plasmalemma, in contrast is much more stable, therefoɪe, it does not lyse or allow cell constituents to leave the cell at an early stage of cell damage.

The good correlation of lipid peroxidation and growth inhibition under the influence of DCOF supports this theory. Support is also found in the simultaneous depression of lipid peroxidation and alleviation of growth inhibition by antagonists of the radical production or propagation such as piperonylbutoxide, SKF 525 A or α-tocopherol.

It should be mentioned that cytochrome P-450 produces free radicals also under certain circumstances (Ruckpaul and Rein 1984). It seems, that piperonylbutoxide inhibits specifically the cytochrome-c-reductase and as a consequence of course, the cytochrome P-450 enzymes (mfo).

Selectivity

The reason for the selectivity of dicarboximide fungicides is not yet clear, because the mechanism of action described here could be a common biocidal principle. *Botrytis cinerea*, *Sclerotinia* and *Monilia* are especially sensitive. They could be distinguished by a high activity of the toxifying enzyme, by an especially sensitive membrane composition, by weak protection mechanisms, or by a combination of these properties. This remains to be investigated.

Resistance

Another problem is that of resistance to DCOF. The practical observations regarding *Botrytis cinerea* are contradictory. But more or less resistant strains can be isolated, either in the laboratory or from the field and from glasshouse. Sporulation and growth rates are often weaker compared to wild strains (Pommer and Lorenz 1982). But there exist also R-strains with normal or high pathogenicity (Lorenz and Pommer 1982), which demonstrates the great genetic variability of *Botrytis cinerea* and indicates, that resistance is not firmly linked genetically to other properties as fitness and pathogenicity (chapter 7).

The infection biochemistry of *Botrytis* is not very clear. Active oxygen species seem to play, among other factors, an important role in the virulence (Rist and Lorbeer 1984; Edlich unpubl.). In some investigations an increased osmolability in R-strains has been observed (Beever and Brien 1983; Edlich and Lyr 1986) (Fig. 8.5) which indicates changes in the membrane or cell wall structure. However, investigation of more than 100 resistant isolates showed no clear correlation between these parameters (Lorenz pers. communication) (chapter 7). The surprising observation of a general

Table 8.5 Cross resistance of *Botrytis cinerea*

Fungicide	Strain	Concentration [mg × l⁻¹]		
		0.1	1.0	10
Iprodione	S	25	75	100
	R	5	7	12
Vinclozolin	S	40	97	100
	R	5	5	14
Carbendazim	S	77	100	100
	R	66	100	100
Chloroneb	S	—	10	70
	R	—	2	10
Dicloran	S	2	30	100
	R	0	5	24
PCNB	S	—	43	82
	R	—	8	20
Tridemorph	S	22	57	92
	R	19	68	90

Cross resistance, measured by the inhibition of radial growth on malt agar petri dishes. Results are expressed in relationship to untreated control.

S — dicarboximide sensitive strain

R — dicarboximide resistant strain

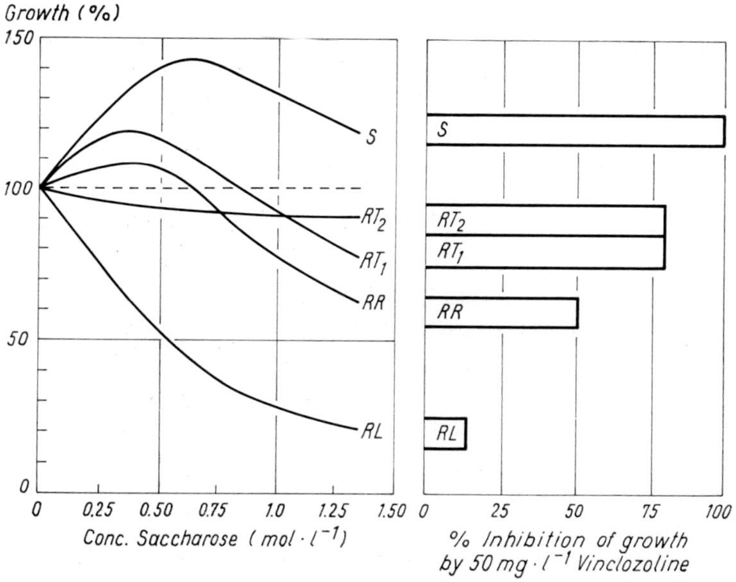

Fig. 8.5 Correlation between osmotic sensitivity and tolerance against vinclozoline in mycelia of various isolates of *Botrytis cinerea*. Growth is expressed as diameters of colonies on malt agar dishes related to controls without saccharose or fungicide. S — dicarboximide sensitive isolate, RT — dicarboximide tolerant isolates from bunches, RR — dicarboximide tolerant isolate from vine, RL — dicarboximide resistant labour strain. Tolerant field isolates of *Botrytis cinerea* were obtained from Dr. Lorenz. (correction = right part, read: by 50 mg · l^{-1} Vinclozolin)

cross resistance between dicarboximide and AHF (LEROUX et al. 1978), that there exist equal resistance mechanisms. Their nature is at present not identified. It can be concluded from the results of BEEVER and BYRDE (1982) that there is not an absolute identity, because some strains isolated from glycophen media are resistant to DCOF, but not to dicloran and PCNB. A reverse effect was found with strains isolated from dicloran media which are resistant to dicloran and PCNB, but sensitive to DCOF.

This demonstrates that in spite of many similarities in their general mechanism of action (inducing an increased, pathological lipid peroxidation) DCOF and AHF differ in some respects regarding their detailed mechanism of action. Perhaps there are also some differences in fungal mechanisms of resistance to them. This is shown for a special case in Table 8.3. This point remains to be elucidated. But in spite of that, all data, which can only partly be mentioned here, agree with the general mechanism of action which is described for DCOF in this chapter.

Acknowledgements: We are indebted to Dr. POMMER, Dr. MANGOLD and Dr. LORENZ for stimulating discussions and Prof. SISLER for his kind help in linguistic respect.

References

ALBERT, G.: Wirkungsmechanismus und Wirksamkeit von Vinclozolin bei *Botrytis cinerea*. Inaug. Diss. Landw. Fak. Rhein. Friedr. Wilh. Univ., Bonn 1979.

— Sphäroblastenbildung bei *Botrytis cinerea*, hervorgerufen durch Vinclozolin. Z. Pflanzenkr. Pflanzenschutz **88** (1981): 337—342.

ALLEN, G., R. AIUTO, R., and SUSSMAN, A. S.: Effect of cytochalasin on *Neurospora crassa*. Protoplasma **102** (1980): 63—75.

BEEVER, R. E., and BRIEN, H. M. R.: A survey of resistance to the dicarboximide fungicides in *Botrytis cinerea*. N. Z. J. Agric. Res. **26** (1983): 391—400.

— and BYRDE, R. J. W.: Resistance to the dicarboximide fungicides. In: DEKKER, J., and GEORGOPOULOS, S. G. (Eds.): Fungicide Resistance in Crop Protection. Pudoc, Wageningen 1982.

BES, B., RANJEVA, R., and BOUDET, A. M.: Evidence for the involvement of activated oxygen in fungal degradation of lignocellulose. Biochemie (Paris), **65** (1983): 283—290.

BETINA, V., MICEKOVA, D., and NEMEC, P.: Antimicrobial properties of cytochalasin and their alternation of fungal morphology. J. Gen. Microbiol. **71** (1972): 343—349.

BHUYAN, K. C., BHUYAN, D. K., KUCK, J. F. R., KUCK, K. D., and KERN, H. L.: Increased lipid peroxidation and altered membrane functions in binory mouse cataract. Curr. Eye Res. **2** (1982/1983): 597—606.

BRUNORI, M., and ROTILO, G.: Biochemistry of oxygen radical species. In: COLOWICK, S. P., and KAPLAN, N. O. (Eds.): Methods of Enzymology. Vol. 105. Academic Press, New York 1984.

BUCHENAUER, H.: Preliminary studies on the mode of action of vinclozolin. Meded. Fac. Landbouwwet. Rijksuniv. Gent **41** (1976): 1509—1519.

EDLICH, W., and LYR, H.: Occurrence and properties of a cytochrome c-reductase in *Mucor* and its interaction with some fungicides. Tagungsber. Akad. Landwirtschaftswiss. DDR **222** (1984): 37—40.

— — Biochemical and physiological properties of dicarboximide resistant strains of *Botrytis cinerea*. Z. Allg. Mikrobiol.: in press.

ESTERBAUER, H.: Aldehydic Products of lipid peroxidation. In: MCBRIEN, D. C. H., and SLATER, T. F. (Eds.): Free Radicals, Lipid Peroxidation and Cancer. Academic Press, London 1982, pp. 101—108.

FLAMINGHINI, F., GUANIERI, C., TONI, R., and CALDARERA, C. M.: Effect of oxygen radicals on heart mitochondrial function in tocopherol deficient rabbits. Int. J. Vitam. Nutr. Res. **52** (1982): 402—406.

FRITZ, R., LEROUX, P., and GREDT, M.: Mecánisme de l'action fongitoxique de la promidione (26019 RP ou glycophene), de la vinclozolin et du dicloran sur *Botrytis cinerea*. Phytopathol. Z. **90** (1977): 152—163.

Georgopoulos, S. G., Zafiratos, C., and Georgiadis, E.: Membrane functions and tolerance to aromatic hydrocarbon fungitoxicants in conidia of *Fusarium solani*. Physiol. Plant **20** (1967): 373—381.

— Sarris, M., and Ziogas, B.: Mitotic instability in *Aspergillus nidulans* caused by the fungicides iprodione, procymidone and vinclozolin. Pesticide Sci. **10** (1979): 389—392.

Gold, M. H., Kutsuki, H., and Morgan, M. A.: Oxidative degradation of lignin by photochemical and chemical radical generating systems. Photochem. Photobiol. **38** (1983): 647—652.

Green, R. V., and Gould, J. M.: Substrate-induced hydrogen peroxide production in mycelia from the lignin-degrading fungus *Phaenerochaete chrysosporium*. Biochem. Biophys. Res. Commun. **117** (1983): 275—281.

Gullino, M. L., and Sisler, H. D.: Antagonism of iprodione toxicity to *Botrytis cinerea* by mixed function oxidase inhibitors. (1985): in press.

Gutteridge, J. M. C., and Halliwell, B.: The role of superoxide and hydroxyl radicals in the degradation of DNA and deoxyribose induced by a copper-phenanthroline complex. Biochem. Pharmacol. **31** (1982): 2801—2806.

Hall, N. D., Maslen, C. L., and Blake, D. R.: The oxidation of serum sulfhydryl groups by hydrogen peroxide secreted by stimulated phagocytic cells in rheumatoid arthritis. Rheumatol. Int. **4** (1984): 35—38.

Halliwell, B., and Gutteridge, J. M. C.: Role of iron in oxygen radical reactions. In: Colowick, S. P., and Kaplan, N. O. (Eds.): Methods in Enzymology. Vol. 105. Academic Press, New York 1984.

Hassan, H. M., and Moody, C. S.: Determination of the mutagenity of oxygen free radicals using microbial systems. In: Colowick, S. P., and Kaplan, N. O. (Eds.): Methods in Enzymology. Academic Press, New York 1984.

Hazra, D. K., and Steenken, S.: Pattern of hydroxyl radical addition to cytosine and 1-, 3-, 5- and 6 substituted cytosines. J. Am. Chem. Soc. **105** (1983): 4380—4386.

Hisada, Y., and Kawase, Y.: Morphological studies on antifungal action of N-(3′,5′-dichlorophenyl)-1,2-dimethyl-cyclopropane-1,2-dicarboximide on *Botrytis cinerea*. Ann. Phytopathol. Soc. Jap. **43** (1977): 151—158.

Iwata, K., Shibuya, H., Ohkawa, V., and Inui, N.: Chromosomal aberrations in V79 cells induced by superoxide radical generated by the hypoxanthine-xanthine oxidase system. Toxicol. Letters **22** (1984): 75—82.

Karagezyan, K. G.: Phospholipid-phospholipid correlations and the dynamics of free radical oxidation of lipids in biological membranes in alloxan diabetes. Voprosij medicinoi khimii **28** (1982): 56—60.

Klebanoff, S. J., Waltersdorph, A. M., and Rosen, H.: Antimicrobial activity of myeloperoxidase. In: Colowick, S. P., and Kaplan, N. O. (Eds.): Vol. 105 Academic Press, New York 1984.

Leroux, P., Fritz, R., and Gredt, M.: Cross resistance between 3,5-dichlorophenyl cyclic imide fungicides (Dichlozolin, Iprodion, Procymidone, Vinclozolin) and various aromatic compounds. In: Lyr, H., and Polter, C. (Eds.): Systemische Fungizide und Antifungale Verbindungen. Abh. Akad. d. Wiss. DDR, Akademie-Verlag, Berlin 1983, pp. 79—88.

— Gredt, M., and Fritz, R.: Etudes en laboratoire de souches de quelques champions phytopathogenes resistantes a la dichlozoline, a la dicyclidine, a l'iprodione, a la vinclozolin et a divers fondicides aromatiques. Med. Fac. Landbouw. Rijksuniv. Gent **43** (1978): 881—889.

— — — Resistance to 3,5-dichlorophenyl-N-cyclic imide fungicides. Neth. J. Plant Pathol. **87** (1981): 244—249.

Lorenz, G., and Pommer, E.-H.: Pectolytic and cellolytic enzymes of dicarboximide sensitive and resistant strains of *Botrytis cinerea*. EPPO Bull. **12** (1982): 145—149.

Macris, B., and Georgopoulos, S. G.: Reduced hexosamine content of fungal cell wall due to the fungicide pentachloronitrobenzene. Z. Allg. Mikrobiol. **13** (1973): 415—423.

Massey, V., Palmer, G., and Ballou, D.: On the reaction of reduced flavins and flavoproteins with molecular oxygen. In: Slater, E. C. (Ed.): Flavins and Flavoproteins. BBA Library, Vol. 8, 1966.

Mello Filho, A. C., Hoffmann, M. E., and Meneghini, R.: Cell killing and DNA damage by hydrogen peroxide are mediated by intracellular iron. Biochem. J. **218** (1984): 273—276.

Mitchell, R. E., and Morrison, D. P.: A comparison between rates of cell death and DNA damage during irradiation of *Saccharomyces cerevisiae* nitrogen and nitrous oxide. Radiat. Res. **96** (1983): 374—379.

Moore, D.: Effects of hexose analogues on fungi, mechanism of inhibition and of resistance. New Phytol. **87** (1981): 487—515.

Müller, H. M.: Zytopathologische Veränderungen bei *Botrytis cinerea* unter dem Einfluß verschiedener Fungizide. Tag.-Ber. 10. Tagung „Elektronenmikroskopie", Ges. Topochem. und Elektronenmikroskopie DDR, Leipzig 1981.

Norkus, E. P., Kuenzig, W., and Conney, A. H.: The mutagenic activity of ascorbic acid in vitro and in vivo. Mutat. Res. **117** (1983): 183—191.

Pappas, A. C., and Fisher, D. H.: A comparison of the mechanism of action of vinclozolin, procymidone, iprodione and prochloraz against *Botrytis cinerea*. Pesticide Sci. **10** (1979): 239—246.

Pommer, E.-H., and Lorenz, G.: Resistance of *Botrytis cinerea* Pers. to dicarboximide fungicides — literature review. Crop Protect. **1** (1982): 221—230.

Reiner, N. E., and Kazura, J. W.: Oxidant-mediated damage of *Leishmania donovani* promastigotes. Infect. Immun. **36** (1982): 1023—1027.

Rist, D. L., and Lorbeer, J. W.: Moderate dosages of ozone enhance infection of onion *(Allium cepa)* leaves by *Botrytis cinerea* but not by *Botrytis squamosa*. Phytopathology **74** (1984): 1217—1220.

Ruckpaul, K., and Rein, H.: Cytochrome P-450. Akademie-Verlag, Berlin 1984.

Sharples, R. O.: The fungitoxic effects of dicloran on *Botrytis cinerea*. Proc. Brit. Insect. Fung. Conf. London 1961: 327—336.

Threlfall, R. J.: Effect of pentachloronitrobenzene (PCNB) and other chemicals on sensitive and PCNB-resistant strains of *Aspergillus nidulans*. J. Gen. Microbiol. **71** (1972): 173—180.

Tillman, R. W., and Sisler, H. D.: Effect of chloroneb on the growth and metabolism of *Ustilago maydis*. Phytopathology **63** (1973): 219—225.

Watanabe, T., and Kondo, N.: The change in leaf protease and protease inhibitor activities after supplying various chemicals. Biol. Plant **25** (1983): 100—109.

Ziegler, D. M., Poulsen, L. L., and Duppel, M. W.: Kinetic studies on mechanism of the microsomal flavin-containing monooxygenase. In: Microsomes, Drug Oxidations and Chemical Carcinogenesis (1980): 637—645.

Lyr, H. (Ed.): Modern, Selective Fungicides — Properties, Applications, Mechanisms of Action Longman Group UK Ltd., London, and VEB Gustav Fischer Verlag, Jena, 1987.

Chapter 9

Carboxin fungicides and related compounds

M. KULKA*) and B. VON SCHMELING**)

*) Uniroyal Ltd. Research Laboratories Guelph/Ontario, Canada

**) Uniroyal Chemical Company, World Headquaters, Middleburry CT 06749, USA

Introduction

The protectant and eradicant fungicides, which have been available for more than a century, are effective only in controlling fungal pathogens on the surface of the plant. To control internal pathogens of plants it was necessary to find a systemic fungicide which could be absorbed by the host plant and be transported via the xylem or phloem systems to the site of the pathogen where eradication could take place. Although a search for such fungicides continued for many years, it was not until the nineteen sixties that a great deal of progress was made.

In 1966 von SCHMELING and KULKA reported the discovery of systemic fungicides carboxin (5,6-dihydro-2-methyl-N-phenyl-1,4-oxathiin-3-carboxamide) (Vitavax) (carbathiin in Canada) and oxycarbox in (5,6-dihydro-2-methyl-N-phenyl-1,4-oxathiin-3-carboxamide, 4,4-dioxide) (Plantvax) (Fig. 9.1). They announced that these chemicals were effective in controlling plant pathogenic fungi such as wheat leaf rust, *Puccinia rubigo-vera tritici* (Eriks) Carleton; bean rust, *Uromyces phaseoli typica* Arth.; loose smut of barley, *Ustilago nuda* (Jens) Rostr. and damping off, *Rhizoctonia solani* Kühn.

In the initial experiments, bean seeds *(Phaseolus vulgaris)* were treated with carboxin and oxycarboxin at the rate of 0.25% chemical per weight of seed. The seeds were planted in the greenhouse and the resulting plants were inoculated at 1 and 2 week intervals with uredospores of bean rust *Uromyces phaseoli*. Both chemicals controlled the development of rust symptoms on the primary leaves when inoculated 7 days after planting. In the experiment of the 2 week interval between planting and inoculation, oxycarboxin gave complete control whereas carboxin failed to control the rust. Barley seed *(Hordeum vulgare)* infected with loose smut, *Ustilago nuda*, when treated with carboxin (0.125% chemical per weight of seed) and planted in the field, yielded two months later smut free barley plants. Oxycarboxin in a similar experiment was only moderately effective.

Systemic nature of carboxin and oxycarboxin

The systemic nature of carboxin and oxycarboxin was demonstrated by a number of methods. The control of rust disease on the leaves of bean plants following soil treatment as well as control of the systemic loose smut fungus by seed treatment (v. SCHMELING and KULKA 1966; v. SCHMELING *et al.* 1966, 1968, 1969, 1970; MATHRE 1968) is evidence that the fungitoxic compound entered the plant and was transported to the site of the pathogen in sufficient quantity to cause eradication.

More direct evidence for the systemicity of carboxin and oxycarboxin was obtained (SNEL and EDGINGTON 1968) by labelling these chemicals with ^{14}C and then applying them and tracing their movements in the plant by autoradiography and liquid

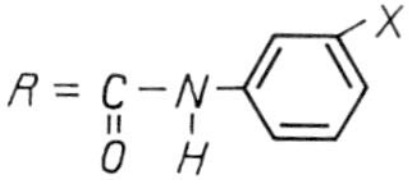

X = H (except in h, i and j.)

			patents	introduction
a	Carboxin (Oxathiin) (Vitavax)	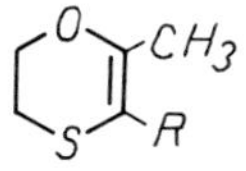	Uniroyal Inc. US 3 249 499 3 393 202 3 454 391	1966
b	Oxycarboxin (F 461) (Plantvax)	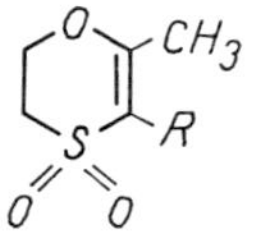	Uniroyal Inc. US 3 399 214 3 402 241 3 454 391	
c	Carboxin Sulfoxide	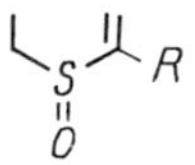	(inactive)	
d	Pyracarbolid (HOE 2989) (Sicarol)	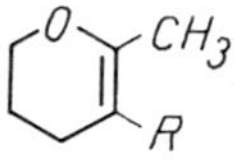	Hoechst AG DE 1 668 790	1970
e	Salicylanilid (Shirlan)	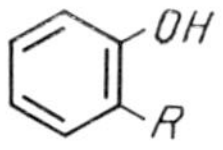	Shirley Institute	1930
k	Fenfuram (WL 22 361) (Panoram)	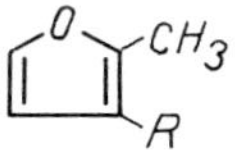	Shell Res. Ltd. GB 1 215 066 Uniroyal Inc. Canada P. 893, 676	1974 1972
l	Methfuroxam (UBI H 719) Furavax	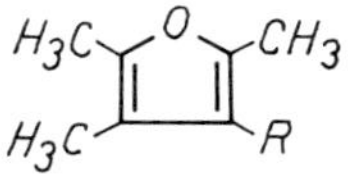	Uniroyal Inc. Canada P. 893 676 DE 2 006 471 US 3,959,481 4,054,585	1972
m	Furcarbanil (BAS 3191 F)	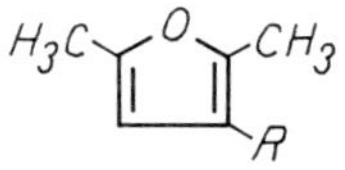	BASF AG DE 1 768 686	1970
n	2, 4-Dimethyl-thiazole-5-carboxanilide G 696, F847	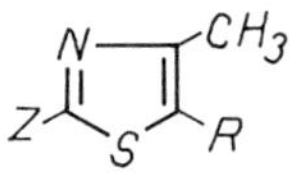 Z = CH_3 Z = NH_2	Uniroyal Inc. US 3,547,917 3,505,055 3,709,992 3,725,427	1970
o	2-Methylthiophene-2-carboxanilide	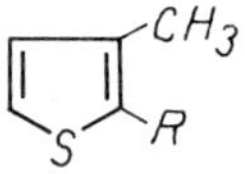		1975
p	1,3,5-Trimethyl-pyrazole-4-carboxanilide	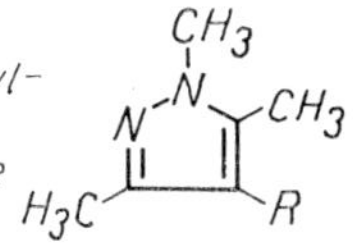		1976

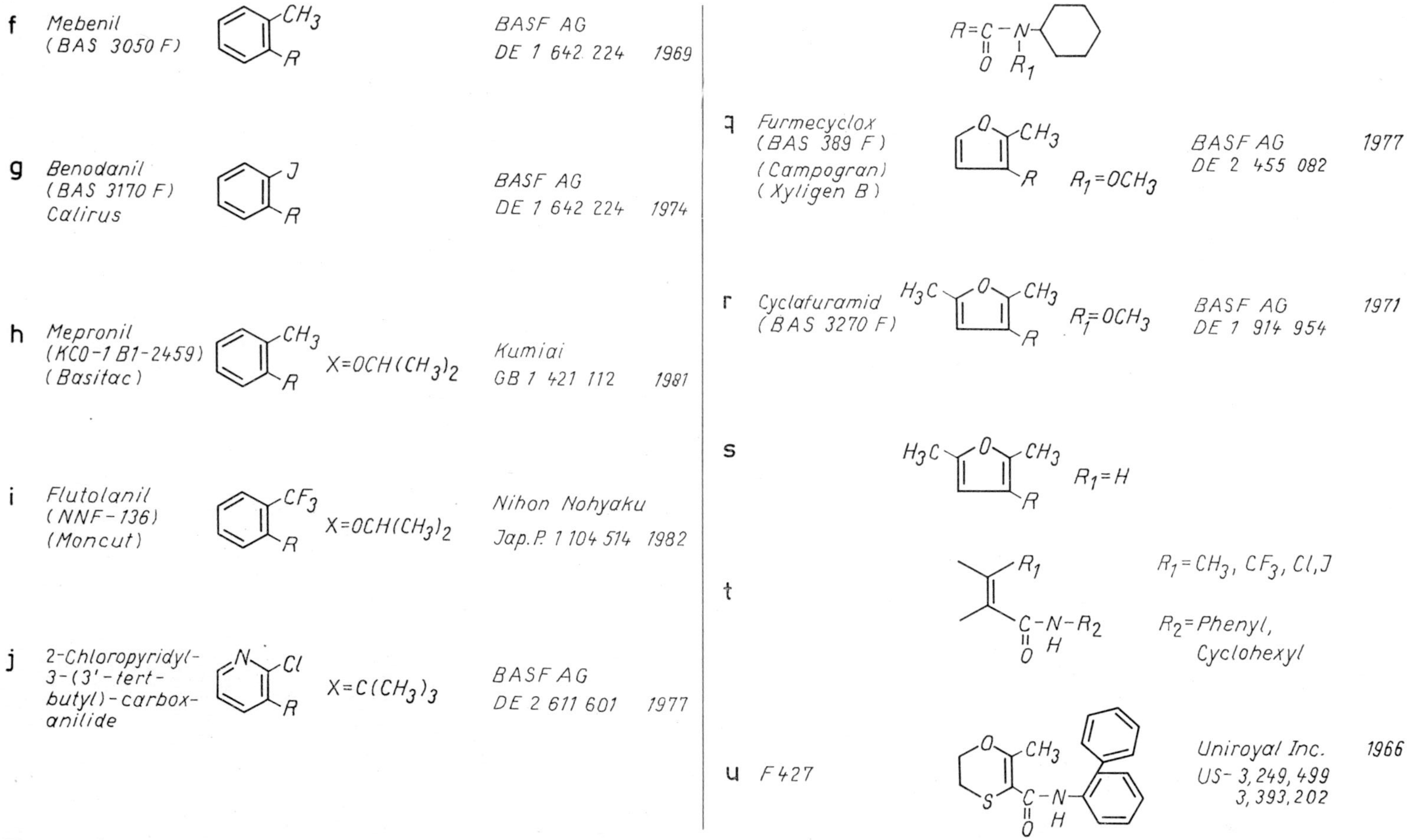

Fig. 9.1 Structure relationships of carboxins and related carboxamides. (corrections: in n: F 849, instead of F 847; in o: 3-methylthiophene-, instead of 2-methylthiophene-)

scintillation. Snel and Edginton (1968) found that these chemicals were translocated in bean plants upward from roots and shoots and accumulated in the margins of the leaves. Similarly, it was found (Kirk *et al.* 1969) that the ^{14}C labelled carboxin translocated from treated seeds or soil to stems and leaves in cotton plants *(Gossypium hirsutum)*.

Applications in agriculture

Great interest was shown in carboxin and oxycarboxin following the disclosure of their systemic properties. They were evaluated all over the world in different climates and over seven hundred papers were published confirming the results of the original disclosure and extending applications to other fungal pathogens. This extensive research also showed that carboxin and oxycarboxin are specific systemic fungicides controlling mainly pathogens of the Basidiomycetes — a class of fungi which include such important pathogens as smuts, bunts and rusts of cereal grains and the soil fungi *Rhizoctonia solani*. The economic benefits of carboxin and oxycarboxin became evident. The treatment of seed or plants with these chemicals results in higher yields of crops not only because of disease control but also because of growth stimulation. Varieties of grain seed that yield well but could not be used in agriculture because of their high susceptibility to diseases, could now be used after treatment with these systemic fungicides. Erwin (1970) reviewed the development of the systemic fungicides up to 1970. An exhaustive review of the many publications which followed 1970 is beyond the scope of this chapter.

Carboxin was first registered for use on grains in France in 1969 and in Canada in 1970. Other countries followed and today carboxin and oxycarboxin are used in many countries of the world to control such diseases as smuts, bunts and rusts of cereal grains and damping-off diseases on cotton. These systemic fungicides may be applied to seed, soil or plants but seed application is the most practical. Carboxin is applied as a seed dressing at the rate of 0.4 g to 1 g per kg of cereal seed and 1 to 2 g per kg of cotton seed and these applications are sufficient to control the diseases for the growing season. For the control of rust diseases of grain, in addition to seed treatment, one or more foliar applications of oxycarboxin usually as 0.1 % spray is required. Some pathogens such as onion smut require considerably heavier dosages of carboxin for complete control (Edgington and Kelly 1966). The diseases of rust on ornamentals are controlled by a few 0.1 % spray applications of oxycarboxin to the growing plants.

Carboxin is used either alone or in combinations with other fungicides such as thiram and copper 8-hydroxyquinolinate. The purpose of this is to increase the efficacy of carboxin and to broaden the spectrum of fungal control. Also synergistic effects of one fungicide on another can increase efficacy. A synergistic action between carboxin and copper 8-hydroxy quinolinate has been reported (Richard and Vallier 1969).

Another method used for increasing the efficacy of the systemic fungicides is through formulation techniques. Thus oil dispersible formulations are better than wettable powder formulations mainly because oil formulations penetrate the host and pathogen better. In comparing the efficacy of different formulations in the control of oat leaf rust *(Puccinia coronata avenae)* with oxycarboxin, it was found that the emulsifiable concentrates and oil dispersible concentrates had lower ED values than wettable powder formulations (Chin *et al.* 1975).

Tables 9.1 and 9.2 list the pathogens which are controlled by carboxin and oxycarboxin.

Table 9.1 Fungal pathogens controlled by carboxin

Fungal pathogen	Host	References
Rhizoctonia solani (damping off)	cotton, bean	AL-BELDAWI and PINCĆARD (1970); SHARMA and SOHI (1982)
	conifer seeds	BELCHER and CARLSON (1968)
Ustilago nuda (loose smut)	barley	MAUDE and SHURING (1969); REINBERGS *et al.* (1968)
Ustilago nigra (black loose smut)	barley	BATALOVA *et al.* (1978)
Ustilago tritici (loose smut)	wheat	MAUDE and SHURING (1969); TYLER (1969)
Ustilago avenae (loose smut)	oats	TOLLENAAR *et al.* (1969)
Ustilago kolleri (covered smut)	oats	PATHAK *et al.* (1971); RICHARD and VALLIER (1969); WALLACE (1969)
Ustilago hordei (covered smut)	barley	PATHAK *et al.* (1971); WALLACE (1969)
Tilletia foetida (bunt)	wheat	PATHAK *et al.* (1971); WALLACE (1969)
Tilletia caries (bunt)	wheat	PATHAK *et al.* (1971); WALLACE (1969)
Sphacelotheca panici-miliacei (smut)	millet	KOISHIBAEV (1974)
Sphacelotheca sorghi (covered kernel smut)	sorghum	DUSCHANOV (1983); POPOV and SILAEV (1978)
Sphacelotheca reilianum (head smut)	sorghum, corn	SIMPSON and FENWICK (1971); POPOV and SILAEV (1978)
Urocystis agropyri (flag smut)	wheat	METCALFE and BROWN (1969)
Urocystis cepulae (onion smut)	onion	EDGINGTON and KELLY (1966)
Exobasidium vexans (blister blight)	tea	VENKATA RAM (1969)
Exobasidium vaccinii (red leaf disease)	blueberry	LOCKHART (1969)
Sclerotium rolfsii (white mold)	peanuts	DIOMANDE and BEUTE (1977)
Tolyposporium penicillariae (kernel smut)	pearl millet	WELLS (1967)
Helminthosporium gramineum (leaf stripe)	barley	KINGSLAND (1969); NAVUSHTANOV (1978)
Helminthosporium sativum (root rot)	barley	SMIRNOVA *et al.* (1977)
Cochliobolus sativus (seedling blight)	barley	MILLS and WALLACE (1970); WALLACE (1969)
Drechslera sorokiniana	barley	HAMPTON (1978)
Corticium sasakii (sheath blight)	rice	LAKSHMANAN *et al.* (1980)
Ustilago maydis (smut)	corn	MATHRE (1968)

Structure-activity relationships of carboxin and related carboxamides

Diligent search by the inventors and by others, in the area of structure-activity relations followed soon after the discovery of carboxin. The aim was to find out what structural modifications could be made to carboxin without losing its unique biological properties. Controlled oxidation of carboxin (Fig. 9.1a) yielded the sulfoxide (Fig. 9.1c) (also formed in plants and animals) and this is many times less active than the parent compound. Further oxidation of carboxin (Fig. 9.1a) or (Fig. 9.1c) yielded

Table 9.2 Fungal pathogens controlled by oxycarboxin

Fungal pathogen	Host	References
Uromyces phaseoli typica (rust)	bean	Rolim *et al.* (1981)
Puccinia recondita (leaf rust)	wheat	Hagborg (1971); Rowell (1967)
Puccinia graminis (stem rust)	wheat	Hagborg (1971); Rowell (1967)
Puccinia coronata avenae (leaf rust)	oats	Chin *et al.* (1975)
Puccinia hordei Otth. (brown rust)	barley	Udeogalanya and Clifford (1982)
Puccinia striiformis (stripe, yellow rust)	wheat K. bluegrass	Powelson and Shaner (1966) Hardison (1967, 1971)
Puccinia arenariae (rust)	dianthus	Umgelter (1969)
Puccinia antirrhini (rust)	snapdragon	Umgelter (1969)
Puccinia malvacearum (rust)	mallow	Umgelter (1969)
Puccinia horiana (white rust)	chrysanthemum	Umgelter (1969); Zadoks *et al.* (1969)
Puccinia obscura (rust)	daisy	Umgelter (1969)
Puccinia carthami (rust)	safflower	Zimmer (1967)
Puccinia helianthi (rust)	sunflower	Bhowmik and Singh (1979)
Urocystis agropyri (flag smut)	wheat, K. bluegrass	Hardison (1967, 1971); Metcalfe and Brown (1969)
Ustilago striiformis (stripe smut)	cr. bent grass c. orchard grass K. bluegrass	Hardison (1967, 1971)
Hemileia vastatrix (rust)	coffee	Figueiredo *et al.* (1981)
Phragmidium mucronatum (rust)	rose	Umgelter (1969); Shattock and Rahbar Bhatti (1983)
Coleosporium campanulae (rust)	bluebell	Umgelter (1969)
Uromyces transveralis (rust)	gladiolus	Lopes *et al.* (1983)
Uromyces dianthi (rust)	carnation	Spencer (1979)
Puccinia pelargonii-zonalis (rust)	geranium	Harwood and Raabe (1979)
Melampsora lini (flax rust)	flax	Froiland and Littlefield (1972)

oxycarboxin (Fig. 9.1b) which was found to be less active on smut diseases but more active on the rust diseases. Of the many carboxins substituted in the phenyl ring which were prepared, only a few had significant fungicidal activity. The 3′-methyl- and the 3′-methoxyderivatives of carboxin were at least as active as carboxin.

The 2′-phenyl-derivative of carboxin (F427) (Fig. 9·1u) is unique in that it exhibits a broader spectrum of activity than carboxin, controlling some species of Deuteromycetes and of Phycomycetes as well as those of the Basidiomycetes (Edgington and Barron 1967).

The existence of the 1,4-oxathiin ring in carboxin does not appear to be important since its replacement by other rings does not cause a loss of the systemic fungicidal activity. Thus two thiazolecarboxamides, 2-amino-4-methyl-, (Fig. 9.1n), and 2,4-dimethyl- (Fig. 9.1n) N-phenyl-5-thiazole-carboxamide are systemic fungicides with biological properties similar to that of carboxin (Harrison *et al.* 1970, 1971, 1973). The thiophenecarboxamide, (Fig. 9.1o) also possesses fungicidal activity (White and Thorn 1980). The replacement of the 1,4-oxathin ring in carboxin with a furan ring creates little change in biological activity. Thus 2,5-dimethyl-N-phenyl- and 2,4,5-trimethyl-N-phenyl- (Fig. 9.1l, m) 3-furan-carboxamide (Davis *et al.* 1972, 1973, 1976, 1977) are systemic fungicides with activity similar to that of carboxin and oxycarboxin, but (Fig. 9.1l) is somewhat more effective than carboxin. The replacement of sulfur in carboxin with a CH_2 group resulted in 5,6-dihydro-2-methyl-N-phenyl-4H-pyran 3-carboxamide (Fig. 9.1d) and this is also active against smuts, rusts and *Rhizoctonia* spp. although (Fig. 9.1d) is somewhat more phytotoxic than carboxin (JANK and GROSSMANN 1971).

The simple compound mebenil (2-methyl-N-phenylbenzamide) (Fig. 9.1f) which contains the 2-butenamide moiety (Fig. 9.1t) common to all the systemic fungicides (Fig. 9.1a) to (Fig. 9.1u) above controls rust in cereals (POMMER and KRADEL 1969). Thus it appears that the main structural feature responsible for systemic fungicidal activity is N-phenyl-2-butenamide (Fig. 9.1t). The contributions of the methyl group, the double bond, the carboxanilide function and stereoisomerism to systemic fungicidal activity are discussed in detail by TEN HAKEN and DUNN (1971).

Among the benzoic acid anilides substituted in position 2, the 2-hydroxy-benzoic acid anilide (salicylic acid anilide) (Fig. 9.1e) introduced in 1930 is worth mentioning This was used at first for material and textile protection (FARGHER *et al.* 1930). It is still used in limited amounts in plant protection for the control of *Cladosporium fulvum* in tomatoes. The 2-methyl derivative Mebenil (Fig. 9.1f), was not used commercially in spite of good activity against cereal rusts and *Rhizoctonia solani*. By replacement of the methyl-group by halogen, especially iodine, new compounds with a specific activity against Basidiomycetes were found (POMMER *et al.* 1974). It should be mentioned, that 2-jodobenzoic acid anilide (Fig. 9.1g) (benodanil) has excellent photostability and can be used as fungicide for the control of *Puccinia* species in cereals (POMMER *et al.* 1973; FROST *et al.* 1973). Benodanil exhibits good activity against diseases caused by Basidiomycetes in turf and lawn, such as *Corticium fuciforme, Puccinia* spp. and *Marasmius oreades.* Among the fungicidally active benzanildes substituted in the aniline moiety, two are distinguished by their usefulness for controlling sheath blight in rice: mepronil (Fig. 9.1h) (DOI 1981) and flutolanil (Fig. 9.1i) (ARAKI and YABUTANI 1981). Both compounds are also active as are the other benzoic acid anilides against *Rhizoctonia solani* in potatoes (black scurf) and *Typhula* snow blight in wheat.

In regard to the substituted furanilides and furamides apparently none came to market as leaf fungicides, in spite of good results in the greenhouses. The main reasons seem to be the insufficient UV-stability and the correlated low persistence (BUCHENAUER 1975). Fenfuram (Fig. 9.1k) (TEN HAKEN and DUNN 1971; JORDOW 1978), methfuroxam (Fig. 9.1l) (ALCOCK 1978) and Furmecyclox (Fig. 9.1q) are used mainly in seed dressing formulations for *Tilletia-* and *Ustilago* spp. in cereals. For enlarging the fungicidal spectrum of activity, i.e. for *Drechslera graminea* and *Fusarium nivale,* various seed dressing combinations have been developed:

Fenfuram + Guaratine
Fenfuram + Guazatine + Imazalil
Fenfuram + Quintocene + Thiabendazol
Methfuroxam + Thiabendazol
Furmecyclox + Imazalil

Methfuroxam (Fig. 9.1l) and Furmecyclox (Fig. 9.1q) are quite suitable for the control of *Rhizoctonia solani* in cotton, potatoes and flower bulbs. Furmecyclox is formulated as a soluble wood preservative. This compound gives excellent control of the economically important wood destroying *basidiomycetes* (POMMER and REUTHER 1978). Pyracarbolid (Fig. 9.1d) described by STINGL *et al.* (1970) was experimentally used against *Ustilago* and *Tilletia* spp. and *Rhizoctonia solani*. It also exhibits excellent activity against *Hemileia vastatrix* in coffee.

Animal toxicology

Carboxin has a low toxicity to animals. Albino rats suffered no detectable symptoms when fed 600 ppm of carboxin in their diets for two years. The acute oral LD_{50} values for white rats, quail and partridge are 3,200, 5,600 and 2,000 mg per kg body weight, respectively. The acute dermal LD_{50} for rabbits is $> 8,000$ mg per kg. The formulation known as Vitavax 250 which consists of 25.3 % carboxin and 74.7 % diluent, anti-freeze and dispersants has the following toxicity values: acute oral LD_{50} (rats) $>5,000$ mg/kg; acute dermal LD_{50} (rabbits) $> 23,600$ mg/kg; inhalation LC_{50} (rat) > 20 mg/L. The acute oral LD_{50} for hens is 24,400 mg/kg with cumulation coefficient > 5. However when administered repeatedly to hens at 1 % LD_{50} it caused some symptoms of poisoning (Padalkin 1978).

The formulation Plantvax 75W which consists of 75 % oxycarboxin and 25 % carrier and surfactants has an oral LD_{50} to rats of 2,570 mg/kg body weight, dermal LD_{50} to rabbits of $> 8,000$ mg/kg and inhalation LC_{50} to rats of > 6.5 mg/L.

Carboxin is not toxic to bees. When carboxin was blown into the holes of the nests of alfalfa leafcutting bees once a week, the chalkbrood disease of *Megachile rodundata* was reduced and the fungicide had no effect on the nesting or mortality of the bees (Parker 1984).

Fate of carboxin and oxycarboxin in soil, plants and animals

Carboxin is rapidly oxidized to the sulfoxide, in soil, plants and animals (Chin *et al.* 1969, 1970). It loses its fungicidal effectiveness in soil and plants three weeks after application mainly because of conversion to the sulfoxide, which is much less active than the parent compound. Eventually the sulfoxide, disappears and is believed to be bound to lignin by complex formation. No residue was present in wheat *(Triticum aestivum)*, barley or cotton seed harvested from ^{14}C-carboxin-treated seed. The residues of carboxin and its sulfoxide, were also determined in wheat, barley, oats *(Avena)*, peanuts *(Arachis hypogaea)*, sorghum *(Holcus gramineae)*, flax *(Phormium tenax)*, and cotton seed (each harvested from carboxin-treated seeds), by extraction, base hydrolysis and chromatographic determination of the cleavage product aniline (Sisken and Newell 1971). No residues were detected. This easily met the tolerance level requirement of 0.2 ppm set by the U.S. Environmental Protection Agency.

More recent metabolic studies (Larson and Lamoureux 1984) of carboxin have shown that in peanuts not only oxidation takes place but also some amide cleavage occurs and in barley plants the hydroxylation product of carboxin, namely, 5,6-dihydro-2-methyl-N-(4-hydroxyphenyl)-1,4-oxathiin-3-carboxamide forms and was detected and identified (Briggs *et al.* 1974). This phenol is believed to be bound to lignin.

Dogs fed carboxin did not accumulate it in the body. Instead they excreted it in the faeces and urine together with the sulfoxide (Fig. 9.1c) (Chin *et al.* 1969, 1970).

Carboxin and oxycarboxin are also degraded by bacteria. The bacterium *Pseudomonas aeruginosa* isolated from red sandy loam soil, when perfused with a solution of carboxin, first oxidized carboxin to oxycarboxin. Then the oxycarboxin was converted to 2-aminophenol and 2-(2-hydroxyethylsulfonyl) acetic acid (Balasubramanya *et al.* 1980). The latter was apparently formed via the intermediate 2-(vinylsulfony) acetanilide with hydroxylation of the phenyl group and addition of water to the vinyl group occurring at some stage. Somewhat similar degradation of carboxin occurs in cultures of *Rhizopus japonicus* (Wallnoefer *et al.* 1972).

Chemical degradation of oxycarboxin in base follows the same pattern and 2-(vinylsulfonyl) acetanilide has been isolated. It forms from the intermediate 3-oxo-N-phenyl-2-(vinylsulfonyl) butanamide through de-acetylation. 2-(Vinylsulfonyl) acetanilide is a reactive compound and readily undergoes intra-moleculqr cyclization to form N-phenyl-3-thiomorpholinone, 1,1-dioxide but in the presence of nucleophilic reagents, addition to the vinyl group occurs to form 2-(2-substituted-ethylsulfonyl) acetanilide.

Development of resistance

Resistance development of fungal organisms to specific systemic fungicides would be expected to take place rapidly. However, so far, no widespread development of resistance of fungal pathogens to carboxin and oxycarboxin has been observed in the field. Strains of *Ustilago maydis* resistant to carboxin were first obtained in 1970 (Georgopoulos and Sisler 1970) by UV irradiation of *Ustilago* sporidia placed on carboxin-containing media. In 1981 (Grouet *et al.*) in France a *Puccinia horiana* strain resistant to oxycarboxin was isolated in the greenhouse from chrysanthemums treated repeatedly for one year with oxycarboxin sprays. In the absence of further treatment with oxycarboxin, the resistance decreased but was still noticeable three months later after the last treatment. At about the same time, similar cases of resistance development of the same fungal pathogen were observed in the Netherlands (Dirkse *et al.* 1982) and in Taiwan (Pai and Sun 1981). The LC_{50} for the resistant strains of *P. horiana* was > 100 ppm while the LC_{50} for the sensitive strain was only 1.4 ppm.

In Israel (Ben-Yephet *et al.* 1974) the tolerance of UV-induced mutants of *Ustilago hordei* to carboxin was demonstrated in smut inoculated and fungicide treated barley plants. These mutants were also resistant to oxycarboxin.

Bochow *et al.* (1971) found that when cultivated in repeated passages on agar in the presence of the systemic fungicides, the pathogens *Sclerotinia sclerotiorum* and *Fusarium solani pisi* developed resistance to carboxin and oxycarboxin but *Rhizoctonia solani* did not.

Growth stimulation and other side effects

Carboxin and oxycarboxin when applied to seeds or plants not only control diseases but they also stimulate growth of the plants (v. Schmeling and Kulka 1969). Thus when pinto beans *(Phaseolus vulgaris)* with or without disease were sprayed with a 125 ppm suspension of carboxin and then allowed to grow in the greenhouse, growth stimulation was evidenced by such effects as increased height of the plants, increased number of leaves, increased length of internodes, increased weight and dark green colour of the leaves as compared to controls. Seed treatment of barley seed with carboxin and oxycarboxin resulted in production of greener plants and in an increase of the number of seed heads as compared to controls.

Concern was expressed that these systemic fungicides might interfere with nitrogen fixation bacteria in the soil in view of the fact that carboxin and oxycarboxin are bactericides effective against *Staphylococcus aureus* Rosenbach on Petri plates. However, while high concentrations of carboxin and oxycarboxin in the soil significantly inhibited nitrogen fixation on white clover, there was no effect on growth or nodulation at lower concentrations, such concentrations still being much greater than would be normally encountered in the field (Fisher and Hayes 1981). There could be no accumulation of these fungicides in the soil through repeated seasonal applications because of rapid degradation.

Carboxin and oxycarboxin also protect plants against air pollutants (Hager 1973). Thus pinto beans growing in soil treated with 12 ppm carboxin showed no ozone damage as compared to controls. Tobacco, cotton, soybean and tomato plants were also protected by carboxin and oxycarboxin against injury by ozone in the air.

Conclusion

Carboxin and oxycarboxin are systemic fungicides effective in the control of such fungal pathogens as smuts, bunts and rusts of cereal grains, ornamentals and vegetables and the soil fungi *Rhizoctonia solani* of cotton. They have low animal toxicity and are quickly degraded in the soil, plants and animals and leave no residue in crops. These systemic fungicides provide an added benefit in that they also stimulate the growth of plants resulting in crop yield increases. While carboxin and oxycarboxin are effective against some bacteria they do not interfere with the nitrogen fixation bacteria in the soil when applied at the recommended rates.

Acknowledgement: We are indebted to Dr. E.-H. Pommer for contributing the last two paragraphs to the section on structure-activity relationships of carboxin and related carboxamides.

References

Al-Beldawi, A. S., and Pinckard, J. A.: Control of *Rhizoctonia solani* on cotton seedings by means of a derivative of 1,4-oxathiin. Plant Dis. Rep. **54** (1970): 524—528.

Alcock, K. T.: Field evaluation of 2,4,5-trimethyl-N-phenyl-3-furancarboxamid (UBI-H 719) against cereal smut diseases in Australia. Plant Dis. Rep. **62** (1978): 854—858.

Araki, F., and Yabutani, K.: α,α,α-Trifluoro-3′-isopropoxy-o-toluanilide (NNF-136), a new fungicide for the control of diseases caused by Basidiomycetes. Proc. Brit. Crop Protect. Conf. — Pests and Diseases (1981): 3—10.

Balasubramanya, R. H., Patil, R. B., Bhat, M. V., and Negendrappa, G.: Degradation of carboxin and oxycarboxin by *Pseudomonas aeruginosa* isolated from soil. J. Environ. Sci. Health, Part B B **15** (1980): 485—505.

Batalova, T. S., Tyulina, L. R., and Mal'tseva, A. I.: Disinfection of barley seeds for the elimination of black loose smut. Khim. Sel'sk Khoz. **16** (1978): 33—34.

Belcher, J., and Carlson, L. W.: Seed treatment fungicides for control of conifer damping off. Can. Plant Dis. Surv. **48** (1968): 47—52.

Ben-Yephet, Y., Henis, Y., and Dinoor, A.: Genetic studies on tolerance of carboxin and benomyl at the asexual phase of *Ustilago hordei*. Phytopathol. **64** (1974): 51—56.

Bhowmik, T. P., and Singh, A.: Evaluation of certain fungitoxicants for the control of sunflower rust. Indian Phytopathol. **32** (1979): 443—444.

Bochow, H., Luc, L. H., and Sung, Ph. O.: Development of tolerance by phytopathogenic fungi to systemic fungicides. Acta Phytopathol. **6** (1971): 399—414.

Briggs, D. E., Waring, R. H., and Hackett, A. M.: Metabolism of carboxin in growing barley. Pesticide Sci. **5** (1974): 599—607.

Buchenauer, H.: Differences in light stability of some carboxylic acid anilide fungicides in relation to their applicability for seed and foliar treatment. Pesticide Sci. **6** (1975): 525—535.

Chin, M. Y., Edgington, L. V., Bruin, G. C. A., and Reinburgs, E.: Influence of formulations on efficacy of three systemic fungicides for control of oat leaf rust. Can. J. Plant Sci. **55** (1975): 911—917.

Chin, W. T., Stone, G. M., Smith, A. E., and von Schmeling, B.: Fate of carboxin in soil, plants and animals. Proc. 5th Brit. Insecticide Fungicide Conf. (1969) 322—327: J. Agric. Food Chem. **18** (1970): 709—712; 731—732.

Davis, R. A., von Schmeling, B., Kulka, M., and Felauer, E.: Furan-3-carboxamides. U.S. Patent 3,959,481 (1976); U.S. Patent 4,054,585 (1977); Can. Patent 893,676 (1972); Can. Patent 932, 334 (1973).

Diomande, M., and Beute, M. K.: Comparison of soil plate fungicide screening and field efficacy in control of *Sclerotium rolfsii* on peanuts. Plant Dis. Rep. **6** (1977): 408—412.

Dirkse, F. B., Dil, M., Linders, R., and Rietstra, I.: Resistance in white rust (*Puccinia horiana* P. Hennings) of chrysanthemum to oxycarboxin and benodanil in the Netherlands. Meded. Fac. Landbouwwet., Rijksuniv. Gent, **47** (1982): 793—800.

Doi, S.: Basitac (Mepronil). Japan Pesticide Inf. **38** (1981): 17—20.

Duschanov, I. D.: Characteristics of spore germination by kernel smut of sorghum. Khim. Sel'sk. Khoz. (1983): 32—34.

Edgington, L. V., and Barron, G. L.: Relation of structure to fungitoxic spectrums of oxathiin fungicides. Can. Phytopathol. Soc. Proc. **33** (1967): 18—19.

— and Kelly, C. B.: Chemotherapy of onion smut with oxathiin systemic fungicides. Phytopathology **56** (1966): 876.

Erwin, D. C.: Progress in the development of systemic fungitoxic chemicals for control of plant diseases. FAO Plant Protect. Bull. **18** (1970): 73—82.

Fargher, R. G., Galloway, L. D., and Probert, M. E.: The inhibitory action of certain substances on the growth of mould fungi. Mem. Shirley Inst. **9** (1930): 37.

Figueiredo, P., Mariotto, P. R., Bonini, R., De Oliveira, F. N. L., and Oliveira, D. A.: Effect of pyrocarbolid and oxycarboxin applied alone, in mixture, and alternated with copper fungicide in the control of coffee rust *(Hemileia vastatrix)*. Biologico (Brazil) **47** (1981): 239 to 244.

Fisher, D. J., and Hayes, A. L.: Effects of some fungicides used against cereal pathogens on the growth of *Rhizobium trifolii* and its capacity to fix nitrogen in white clover. Ann. appl. Biol. **98** (1981): 101—108.

Froiland, G. E., and Littlefield, L. J.: Systemic protectant and eradicant chemical control of flax rust. Plant Dis. Rep. **56** (1972): 737—739.

Frost, A. J. P., Jung, K. V., and Bedford, J. L.: The timing of application of benodanil (BAS 3170 F) for the control of cereal rust diseases. Proc. 7th Brit. Insecticide and Fungicide Conf. **2** (1973): 105—110.

Georgopoulos, S. G., and Sisler, H. D.: Gene mutation eliminating antimycin A-tolerant electron transport in *Ustilago maydis*. J. Bacteriol. **103** (1970): 745—750.

Grouet, D., Montfort, F., and Leroux, P.: Presence of a strain of *Puccinia horiana* resistant to oxycarboxin in France. Phytiatr-Phytopharm. **30** (1981): 3—12.

Hagborg, W. A. F.: Oxycarboxin emulsifiable concentrate in the control of leaf and stem rusts in wheat. Can. J. Plant Sci. **51** (1971): 239—241.

Hager, F. M.: 5,6-Dihydro-2-methyl-1,4-oxathiin-3-carboxamide as plant protectant against air pollution. Ger. Offen. 2,238,053 (1973).

Hampton, J. G.: Seed treatments for the control of *Drechslera sorokiniana* in barley. N.Z.J. Exp. Agric. **6** (1978): 85—89.

Hardison, J. R.: Chemotherapeutic control of stripe smut *(Ustilago striiformis)* in grasses by two derivatives of 1,4-oxathiin. Phytopathol. **57** (1967): 242—245; **61** (1971): 731—735.

Harrison, W. A., Kulka, M., and von Schmeling, B.: Thiazolecarboxamides U.S. Patent 3,547,917 (1970); Can. Patent 873,888 (1971); U.S. Patent 3,505,055 (1970); Can. Patent 837,517 (1970); U.S. Patent 3,709,992 (1973); Can. Patent 985,901 (1970); U.S. Patent 3,725,427 (1973).

Harwood, C. A., and Raabe, R. D.: The disease cycle and control of geranium rust. Phytopathology **69** (1979): 923—927.

Jank, B., and Grossmann, F.: 2-Methyl-5,6-dihydro-4H-pyran-3-carboxylic acid anilide, a new systemic fungicide against smut diseases. Pesticide Sci. **2** (1971): 43—44.

Jordow, E.: Panoram-chemical for fungus control on cereals. Vaextskyddsrapporter (Vaextskyddskonferensen) **4** (1978): 118—123.

Kingsland, G. C.: Barley leaf stripe control and *in vitro* inhibition of *Helminthosporium sorokinianum* obtained with carboxin in South Carolina. Phytopathology **59** (1969): 115.

KIRK, B. T., SINCLAIR, J. B., and LAMBREMONT, E. N.: Translocation of C_{14}-labelled chloroneb and DMOC in cotton seedlings. Phytopathology **59** (1969): 1473—1476.

KOISHIBAEV, M.: Effectiveness of new fungicides in controlling millet smut. Khim. Sel'sk. Khoz. **12** (1974): 765—766.

LAKSHMANAN, P., NAIR, M. C., and MENNON, M. R.: Comparative efficacy of certain fungicides on the control of sheath blight of rice. Pesticides (India) **14** (1980): 31—32.

LARSON, J. D., and LAMOUREUX, G. L.: Comparison of the metabolism of carboxin in peanut plants and peanut cell suspension cultures. J. Agric. Food Chem. **32** (1984): 177—182.

LOCKHART, C. L.: Control of red leaf in lowbush blueberry. Pesticide Res. Rep. (1969): 235—236.

LOPES, L. C., BARBOSA, J. G., and FILHO, J.: Effects of fungicides on the control of rust *Uromyces transversalis* (Thueman, Winter) on gladiolus. Rev. Ceres, **30** (1983): 366—374.

MATHRE, D. E.: Uptake and binding of oxathiin systemic fungicides by resistant and sensitive fungi. Phytopathology **58** (1968): 1464—1469.

MAUDE, R. B., and SHURING, C. G.: Seed treatments with carboxin for the control of loose smut of wheat and barley. Ann. appl. Biol. **64** (1969): 259—263.

METCALFE, P. B., and BROWN, J. F.: Evaluation of nine fungicides in controlling flag smut of wheat. Plant. Dis. Rep. **53** (1969): 631—633.

MILLS, J. T., and WALLACE, H. A. H.: Effect of Fungicides on *Cochliobolus sativus* and other fungi on barley seed in soil. Can. J. Plant Sci. **49** (1969): 543—548; **50** (1970): 129—136.

NAVUSHTANOV, S.: Results of tests with some fungicides for control of barley stripe disease *(Helminthosporium gramineum)*. Rastenievud. Nauki, **15** (1978): 134—140 (Bulgaria).

PADALKIN, I. Ya.: Toxocity of Vitavax to hens. Veterinariya (Moscow) **6** (1978): 85—86.

PARKER, F. D.: Effect of fungicide treatments on incidence of chalkbrood disease in nests of the alfalfa leaf cutting bee *(Hymenoptera: Megachilidae)*. J. Econom. Entomol. **77** (1984): 113 to 117.

PATHAK, K. D., SHARMA, R. C., and JOSHI, L. M.: Seed treatment for control of some important cereal smuts. Plant Dis. Rep. **55** (1971): 544—545.

PEI, C. L., and SUN, S. K.: Study on fungicide-tolerant strain of pathogenic fungi in Taiwan. Oxycarboxin-resistance of *Puccinia horiana* P. Hennings, the white rust of chrysanthemum. Chih Wu Pao Hu Hsueh Hui Hui K'an **23** (1981): 221—227.

POMMER, E.-H., GIRGENSOHN, B., KÖNIG, K.-H., OSIEKA, H., and ZEEH, B.: Development of new systemic fungicides with carboxanilide structure. Kemia-Kemi **1** (1974): 617—618.

— JUNG, K., HAMPEL, M., and LOECHER, F.: BAS 3170 F (2-Jodbenzoesäureanilid), ein neues Fungizid zur Bekämpfung von Rostpilzen in Getreide. Mitt. Biol. Bundesanstalt f. Land- und Forstwirtschaft **151** (1973): 204.

— and KRADEL, J.: Mebenil (BAS 3050 F) new compound with specific action against some Basidiomycetes. Proc. 5th Brit. Insecticide Fungicide Conf. **2** (1969): 563—568.

— and REUTHER, W.: Furmetamid, a new active ingredient for the control of wood-destroying Basidiomycetes. Proc. 4th Intern. Biodeterioration Symp. (1978): 67—70.

POPOV, V. I., and SILAEV, A. I.: Effectiveness of dressing sorghum seed against two types of smut. Nauchn. Tr. Leningrad S-kh. Inst. **351** (1978): 85—87.

POWELSON, R. L., and SHANER, G. E.: An effective chemical seed treatment for systemic control of seedling infection of wheat by stripe rust *(Puccinia striiformis)*. Plant. Dis. Rep. **50** (1966): 806—807.

REINBERGS, E., EDGINGTON, L. V., METCALFE, D. R., and BENDELOW, V. M.: Field control of loose smut in barley with systemic fungicides Vitavax and Plantvax. Can. J. Plant Sci. **48** (1968): 31—35.

RICHARD, G., and VALLIER, J. P.: Treatment of cereal seed by combination of carboxin with copper quinolinate. Proc. 5th Brit. Insecticide Fungicide Conf. (1969): 45—53.

ROLIM, P. R. R., NETO, F. B., ROSTON, A. J., and OLIVEIRA, D. A.: Chemical controls of bean (*Phaseolus vulgarus* L.) diseases Biologico (Brazil) **47** (1981): 201—205.

ROWELL, J. B.: Control of leaf and stem rust on wheat by an 1,4-oxathiin derivative. Plant Dis. Rep. **51** (1967): 336—339; **52** (1968): 856—858.

SCHMELING, B. VON, and KULKA, M.: Systemic fungicidal activity of 1,4-oxathiin derivatives. Science, **152** No. 3722 (1966): 659—660.

— — Regulation of plant growth. Can. Patent 828,771 (1969); U.S. Patent 3,454,391 (1969).

— — THIARA, D. S., and HARRISON, W. A.: U.S. Patents 3,249,499 (1966); 3,393,202 (1968); 3,399,214 (1968); and 3,402,241 (1968). Canadian Patents 787,893 (1968); 791,151 (1968); 842,243 (1970) and 825,665 (1969).

SHARMA, S. R., and SOHI, H. S.: Effect of different fungicides against *Rhizoctonia* root rot of French bean *(Phaseolus vulgaris)* Indian J. of Mycology and Plant Pathol. (1982): 216—220.
SHATTOCK, R. C., and RAHBAR BHATTI, M. H.: Fungicides for control of *Phragmidium mucronatum* on *Rosa laxa hort.* Plant Pathol. **32** (1983): 67—72.
SIMPSON, W. R., and FENWICK, H. S.: Suppression of corn head smut with carboxin seed treatments. Plant Dis. Rep. **55** (1971): 501—503.
SISKEN, H. R., and NEWELL, J. E.: Determination of residues of Vitavax and its sulfoxide in seeds. J. Agric. Food Chem. **19** (1971): 738—741.
SMIRNOVA, K. F., ANDREEVA, E. I., PROCHENKO, T. S., and PETROVA, L. P.: Injury-bearing root root and some results of a study of fungicides against its basic agents with *in vitro* experiments. Kihm. Primen. Pestits. Prep. (1977): 81—86.
SNEL, M., and EDGINGTON, L. V.: Fungitoxicity, uptake and transportation of two oxathiin systemic fungicides in bean. Phytopathology **58** (1968): 1068; **60** (1970): 1708—1716.
SPENCER, D. M.: Carnation rust caused by *Uromyces dianthii*. In: Annual Report of the Glasshouse Crops Research Institute. Res. Inst. P. (1979): **136**. Plant Pathol. **28** (1979): 10—16.
STINGL, H., HÄRTEL, K., and HEUBACH, G.: HOE 2989, ein neues systemisches Fungizid, wirksam gegen verschiedene Basidiomyceten (Rost- und Brandpilze sowie *Rhizoctonia*). VII. Int. Congr. Plant Protect., Paris (1970): 205.
TEN HAKEN, P., and DUNN, C. L.: Structure-activity relationships in a group of carboxanilides systemically active against broad bean rust. *Uromyces fabae* and wheat rust *(Puccinia recondita)*. Proc. 6th Brit. Insecticide-Fungicide Conf. (1971): 455—462.
TOLLENAAR, H., BERATTO, E. M., and NAREA, G. C.: Preliminary note on the control of smut of oats in Chile. Agr. Tec. (Santiago de Chile) **29** (1969): 32—33.
TYLER, L. J.: Treatment of winter wheat seed with systemic fungicides for control of loose smut. Plant Dis. Rep. **53** (1969): 733—736.
UDEOGALANYA, A. C. C., and CLIFFORD, B. C.: Control of barley brown rust, *Puccinia hordei* Orth. by benodanil and oxycarboxin in the field and the effects on yield. Crop Protect. **1** (1982): 299—308.
UMGELTER, H.: Experience with Plantvax for control of rust diseases of ornamentals and vegetable crops. Gesunde Pflanzen **3** (1969): 53—60.
VENKATA RAM, C. S.: Systemic control of *Exobasidium vexans* on tea with 1,4-oxathiin derivatives. Phytopathology **59** (1969): 125—128.
WALLACE, H. A. H.: Cooperative seed treatment trials — 1969 Can. Plant Dis. Surv. **49** (1969): 49—53.
WALLNOEFER, P. R., KOENIGER, M., SAFE, S., and HUTZINGER, O.: Metabolism of the systemic fungicide carboxin by *Rhizopus japonicus*. Int. J. Environ. Anal. Chem. **2** (1972): 37—43.
WELLS, HOMER D.: Effectiveness of two 1,4-oxathiin derivatives for control of *Tolyposporium* smut of pearl millet. Plant Dis. Rep. **51** (1967): 468—469.
WHITE, G. A., and THORN, G. D.: Thiophenecarboxamide fungicides: structure-activity relationships with the succinate dehydrogenase complex from wild-type and carboxin-resistant mutant strains of *Ustilago maydis*. Pesticide Biochem. Physiol. **14** (1980): 26—40.
ZADOKS, J. C., KODDLE, A., and HOOGKAMER, W.: Effect of derivatives of 1,4-oxathiin on *Puccinia horiana* in *chrysanthemum morifolium*. Netherl. J. Plant Pathol. **75** (1969): 193—196.
ZIMMER, D. E.: Efficacy of 1,4-oxathiin fungicides for control of seedling safflower rust. Plant Dis. Rep. **51** (1967): 586—588.

LYR, H. (Ed.): Modern, Selective Fungicides — Properties, Applications, Mechanisms of Action. Longman Group UK Ltd., London, and VEB Gustav Fischer Verlag, Jena, 1987.

Chapter 10

Mechanism of action of carboxin fungicides and related compounds

T. Schewe*) and H. Lyr**)

*) Institute of Biochemistry of the Humboldt-University Berlin, **) Institute for Plant Protection Research Kleinmachnow of the Academy of Agricultural Sciences of the GDR

Introduction

The interesting systemic features of carboxin and oxycarboxin and their selectivity for Basidiomycetes (Edgington and Barron 1967) were unique for fungicides at the time of their discovery. These compounds stimulated progress towards new systemic fungicides.

The primary site of attack of these fungicides is in the mitochondria. This observation was confirmed by several groups (Mathre 1968, 1970, 1971; Ragsdale and Sisler 1970; Lyr *et al.* 1971). Other effects such as inhibition of nucleic acid synthesis and of protein synthesis proved to be secondary events. Such a target was somewhat unexpected in light of the generally accepted conclusion that the mitochondrial system is an ancient and conservative one. But studies with carboxins have revealed differences in the sensitivity of the mitochondria of higher plants and of fungi as well as differences even among the fungi.

Biochemistry of the mechanism of action

Mathre (1970) was the first who recognized that the oxidation of succinate is the critical site of the fungicidal action of the carboxins. Studies on the mechanism of action in the succinate oxidase system have revealed that the carboxins act on complex II (succinate-ubiquinone oxidoreductase system) of the mitochondrial electron transfer chain (White 1971; Lyr *et al.* 1972; Ulrich and Mathre 1972; Schewe *et al.* 1973; Mowery *et al.* 1976) (Fig. 10.1). This conclusion was also confirmed by genetical studies; carboxin-resistant mutants of *Aspergillus nidulans*, *Ustilago maydis* or *Ustilago hordii* exhibited diminished sensitivity of succinate-ubiquinone oxidoreductase or succinate oxidase activities towards carboxin and related compounds as well as altered enzymatic properties and stability of complex II activity (Georgopoulos *et al.* 1972; Georgopoulos *et al.* 1975; Ben-Yephet *et al.* 1975; Gunatilleke *et al.* 1975; Georgopoulos and Ziogas 1977; White *et al.* 1978). It should be emphasized that the action of carboxins is not restricted to fungal mitochondria. A carboxin-sensitivity was also shown for the complex II activities of the mitochondria of rat liver (Mathre 1971; Schewe *et al.* 1973), bovine heart (Schewe *et al.* 1973; Mowery *et al.* 1976) higher plants (Mathre 1971; Schewe *et al.* 1974; Day *et al.* 1978) as well as for respiring membranes of *Micrococcus denitrificans* (Tucker and Lillich 1974).

There are, however, differences in the concentration of carboxins needed for half-inhibition of the enzymatic activity. The relative tolerance of the succinate oxidase activity of higher plants may be one of the prerequisites for the systemic efficacy against fungal pathogens with low toxicity to the host organism.

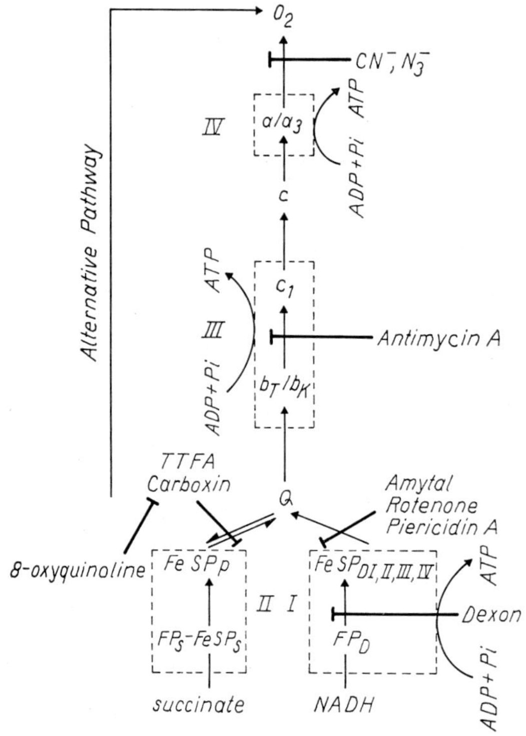

Fig. 10.1 Scheme of the respiratory pathway and sites of attack of some inhibitors, at site I (amytal, rotenone, piericidin A, dexon), site II (carboxin, TTFA), site III (antimycin A), site IV (cyanide, azide), including the possible relation to the alternative pathway of respiration (O_2-consumption).

More precise knowledge on the mechanism of action of carboxins arose from the contemporary progress in the field of the chemistry and enzymology of succinate dehydrogenase and its interaction with the other parts of the mitochondrial electron transfer system, especially from that of bovine heart. The use of mammalian mitochondrial systems as models for the target of the fungicidal action of carboxins appears to be possible or even advantageous for the following reasons;

i) These mitochondria show a sensitivity comparable to those of carboxin-sensitive fungi

ii) Many carboxin-sensitive fungi possess alternative pathways of succinate oxidation, which are absent in mammalian mitochondria, so that a serious complication in the studies with fungal mitochondria is alleviated

iii) Bovine heart mitochondria are the only source until now from which reconstitutively active succinate dehydrogenase could be extracted from untreated and carboxin-treated mitochondria. SDH of the carboxin-sensitive fungus *Trametes versicolor* did not differ significantly (Lyr *et al.* 1972). These observations rule out succinate dehydrogenase *per se* as site of action of carboxins. The conclusion is further supported by observations with bovine heart submitochondrial particles.

Interaction with receptor structures

Only 50% of the succinate-phenazine methosulfate oxidoreductase activity (PMS reductase) is inhibited, i.e. that part which includes ubiquinone (coenzyme Q), whereas the other part of the PMS reductase activity being insensitive, is due to a direct interaction with the succinate dehydrogenase (MOWERY *et al.* 1976).

The reduction of the Fe—S centres S_1 and S_3 of the succinate dehydrogenase by succinate is not inhibited by carboxins (ACKRELL *et al.* 1977), whereas real succinate dehydrogenase inhibitors such as malonate should do so.

By contrast, EPR measurements revealed that carboxins inhibit the reoxidation of the two Fe—S centres S_1 and S_3 after having been reduced by succinate (ACKRELL *et al.* 1977). Assuming a linear electron flow through the chain histidylflavin$\rightarrow S_1 \rightarrow S_3 \rightarrow Q$ this behaviour implies that carboxins affect the electron transfer from the reduced form of S_3 to Q. An identical site of action is likely also for the succinate oxidase inhibitor thenoyltrifluoroacetone (TTFA). The effects of carboxin and TTFA resemble each other in many respects, although differences have been reported (ULRICH and MATHRE 1972; SCHEWE *et al.* 1973; MOWERY *et al.* 1977). From binding experiments it had been suggested that both inhibitors compete for the same specific binding site, but these experiments were complicated by the obvious simultaneous presence of unspecific binding sites not giving rise to inhibition of the enzymic activity (COLES *et al.* 1978; RAMSEY *et al.* 1981). These authors demonstrated also the reversibility of the carboxin action at least for complex II preparation from bovine heart. The site of action of carboxins is analogous to that of rotenone and piericidin A with respect to the fact that the latter inhibitors also interrupt the electron transfer between an Fe—S-cluster and ubiquinane.

It has been established that the succinate-ubiquinone oxidoreductase system is composed of four polypeptides having molecular masses of approximately 70 Da, 30 kDa, 15 kDa and 7.5 kDa. The 70 kDa subunit contains the histidylflavin prosthetic group and an iron-sulfur moiety including the 2 Fe—2 S cluster S_1 which gives rise to a typical EPR signal on reduction of succinate dehydrogenase by succinate. The 30 kDa subunit is another Fe—S component containing centre S_3. It was formerly believed to be a HIPIP-type cluster (High-Potential Iron Protein) because it exhibits an EPR signal in the oxidized state as is the case with HIPIP's from bacteria (for review see BEINERT and ALBRACHT 1982). Recently it was shown that the oxidized species of centre S_3 is a 3 Fe cluster (ACKRELL *et al.* 1984) which is apparantly converted to an EPR-silent 4 Fe—4 S cluster on reduction by succinate. Moreover, these authors discuss the possibility that the centre S_2 which appears in EPR spectroscopy only upon addition of sodium dithionite, may represent a superreduced state of this 4 Fe—4 S-cluster. The two large polypeptides represent the soluble succinate dehydrogenase which transfers electrons from succinate to artificial acceptors (phenazine methosulfate, ferricyanide, Wurster's blue) but not to ubiquinone, its physiological electron acceptor. Recombination of soluble succinate dehydrogenase with the two small polypeptides confers succinate-ubiquinone oxidoreductase activity (ACKRELL *et al.* 1980; VINOGRADOV *et al.* 1980). The 15 kDa polypeptide has the properties of a ubiquinone-binding protein (YU and YU 1981). It is assumed that the ubiquinone-binding proteins enable the ubiquinone to react with the proximate electron carriers of the respiratory chain and protect the ubisemiquinone radicals, which appear to be obligatory intermediates of electron transfer, from dismutation yielding oxidized and reduced ubiquinone.

Carboxins do not inhibit soluble succinate dehydrogenase from various sources (ULRICH and MATHRE 1972; LYR *et al.* 1972; SCHEWE *et al.* 1973; MOWERY *et al.* 1976; GEORGOPOULOS *et al.* 1975). Apparently the interaction of Fe—S clusters with ubiquinone is a critical step in the electron transfer. Unfortunately this electron transfer is not yet understood from the chemical point of view.

Ubiquinone occurs in excess over the other electron carriers of the respiratory chain and is thought to exert a "pool" function in the electron transfer chain, but only a small proportion of it is rapidly reduced by succinate or NADH. This behaviour may be rationalized in the light of the recently discovered Q-binding proteins (Yu and Yu 1981). Distinct Q-binding proteins have been proposed to function in the complex I, II and III of the electron transfer system (Fig. 10.1). Their interaction with Q may predispose the latter to specific electron transfer reactions. The obvious absence of Q-protein(s) in soluble succinate dehydrogenase preparations appears to be the reason for the lack of both Q reductase activity and carboxin sensitivity. This hypothesis was experimentally supported by the work of Vinogradov and coworkers, who isolated a protein fraction consisting of three small polypeptides (molar mass 13 kDa) from submitochondrial particles depleted of succinate dehydrogenase. Addition of this protein fraction to reconstitutively active succinate dehydrogenase resulted in a highly active, carboxin-sensitive succinate-Q oxidoreductase activity as measured with 2,6-dichlorophenolindophenol as acceptor (Vinogradov *et al.* 1980).

The reconstitution also greatly increased the stability of the enzyme. The involvement of protein(s) conferring Q reductase activity and carboxin-sensitivity in fungi is supported by the studies of Georgopoulos *et al.* (1972, 1975), in which they showed that carboxin-resistance in *Ustilago maydis* is due to a single gene mutation which gives rise to a fairly labile succinate dehydrogenase activity which is resistant to low concentrations of carboxin. Such a lability is also typical for soluble succinate dehydrogenase preparations, where it is due to a destruction of the Fe—S centre S_3 by oxygen. Apparently the Q proteins protect this Fe—S cluster from aerobic inactivation which may be favoured by the complicated mechanism of redox change (Ackrell *et al.* 1984). Apparently only the complex between succinate dehydrogenase and Q-protein(s) is able to interact with carboxins, which is also evident from studies of Coles *et al.* (1978) and Ramsay *et al.* (1981). These authors showed that extraction of soluble dehydrogenase from complex II by perchlorate abolishes the specific binding of a radiolabelled carboxin derivative. In a photoaffinity labelling experiment using a labelled azido derivative of carboxin a sizable covalent binding to the small polypeptides of complex II was observed.

All of the aforementioned data indicate that the complex of the Fe—S cluster S_3 and specific small polypeptide(s), called QP_S according to the proposal of Yu and Yu, constitute the carboxin receptor in which the Fe—S cluster must be reduced. The

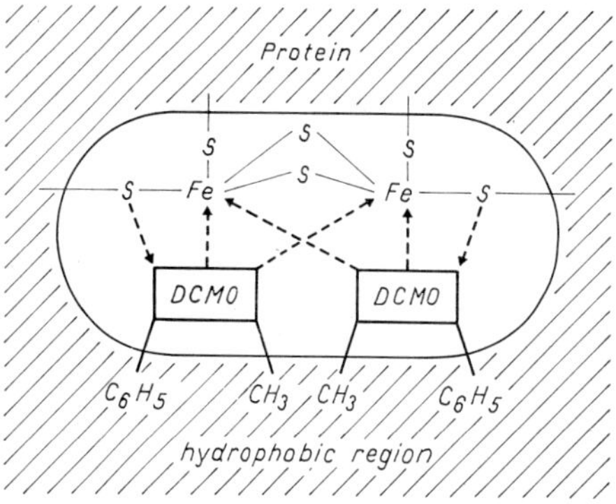

Fig. 10.2 Scheme of the interaction of carboxin (DCMO) with the active centre of the succinodehydrogenase complex (site II) of the respiratory chain.

reduced form represents presumably a 4 Fe—4 S cluster (ACKRELL *et al.* 1984). Unfortunately its interaction with carboxin cannot be studied by EPR spectroscopy, since it is diamagnetic in this redox state.

It should be emphasized that the lack of drastic effects of carboxin on the EPR signal of oxidized S_3 (ACKRELL *et al.* 1977) does not exclude its interaction with reduced cluster S_3. It has been reported, however, that carboxin triggers the oxidative destruction of the cluster S_3 by dioxygen (ref. 13 in MOWERY *et al.* 1976).

This effect may be rationalized by an interference with the stabilization of cluster S_3 afforded by QP_S so that cluster S_3 becomes susceptible to degradation by oxygen as it is in the case with soluble succinate dehydrogenase as well as with the particulate preparations from the carboxin-resistant mutants reported by GEORGOPOULOS *et al.* (1972, 1975).

Since carboxins act on the complex between reduced S_3 and QP_S, they should possess a binding affinity for at least one of these components. Ten years ago we proposed a model for the interaction of carboxin with an Fe—S cluster (LYR *et al.* 1975; SCHEWE *et al.* 1979) which meets the following requirements:

i) the present knowledge on the spatial structure of 2 Fe—2 S or 4 Fe—4 S ferredoxins
ii) X-ray structural data of the carboxin molecule
iii) concordance with the structure-activity relationship of the inhibitory effects of a broad spectrum of carboxin congeners.

This model (Fig. 10.2) is still valid in the light of the recent progress in the carboxin research.

The elucidation of the mechanism of action of carboxins were substantially supported by synthesis of various chemical analogues, which demonstrated, that only a certain central configuration of the molecule, namely N-phenylbutenamid, or according to TEN HAKEN and DUNN (1971) "cis-crotonanilide" is required for activity.

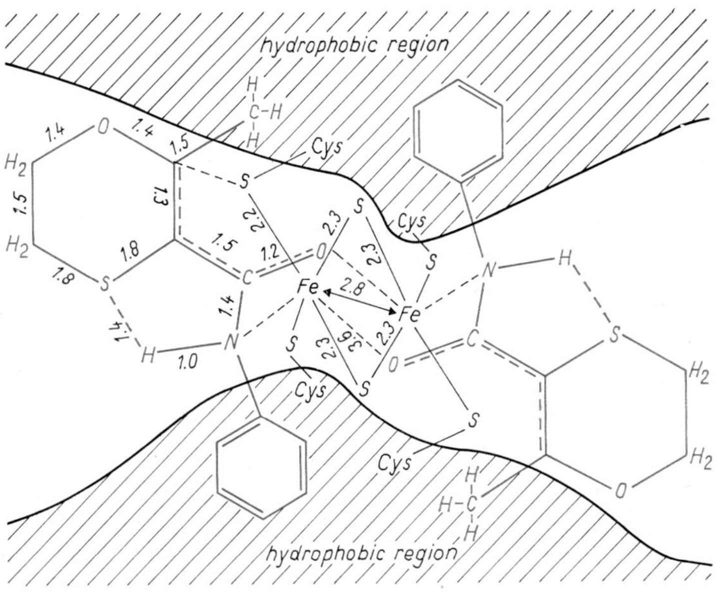

Fig. 10.3 Receptor model for carboxin at the binding site II in the mitochondrial respiratory chain of sensitive fungi. Red: carboxin structure, black: receptor structure (2 carboxin molecules are tentatively used).

The 1,4-oxathiin ring could be replaced by different planar or nonplanar ring systems such as benzene, dihydropyran, pyridine, furan, pyrazole, thiophene and triazole. This means, that this part of the molecule does not show special structural requirements for the binding to the specific receptor. The common basic structure can be described as indicated in Figure 9.1t.

Since the methyl group can be replaced by halogens (POMMER *et al.* 1974) such as iodine in benodanil (LÖCHER *et al.* 1974) or chlorine in 2-chloropyridyl-3-(3′-tert-butyl)-carboxanilide (ZEEH *et al.* 1977), it follows that it is not involved in the specific interaction with the receptor region. Comparing all structures only the very general expression "carboxamide-fungicides" is adequate.

It is surprising that a replacement of the benzene ring of the aniline moiety by cyclohexane does not appreciably alter the spectrum of activity. Extensive structure-activity comparisons have been made by HARDISON (1971), MATHRE (1971), TEN HAKEN and DUNN (1971), WHITE and THORN (1975), MÜLLER *et al.* (1977), HUPPATZ *et al.* (1984), SNEL *et al.* (1970). A considerable change in the spectrum of activity appeared only with compound F 427 from the Uniroyal Comp. in which the anilide ring is substituted in the o-position by a phenyl group (Fig. 9.1u). According to EDGINGTON and BARRON (1967) this substitution extended the fungitoxic activity to some, but not all Deuteromycetes (*Aspergillus, Botrytis, Drechslera* and others) and to some Phycomycetes such as *Thamnidium elegans, Cunninghamella echinulata,* whereas Basidiomycetes as *Rhizoctonia solani* and *Polyporus giganteus* remained sensitive as with other oxathiins.

In summary it follows that for the interaction with the receptors the following parts of the molecule are involved:

i) the hydrophobic ring at the amide nitrogen
ii) the cis-methyl (or halogen) group in 2-position of the oxathiin ring
iii) the vinylogous carbonyl grouping including both an electrophilic centre at C—2 and a nucleophilic one at the oxygen atom
iv) a weakly nucleophilic amide nitrogen.

Accordingly a multi-centre binding is accomplished (SCHEWE *et al.* 1979). The presence of sulfur may favour stabilization of the agent-receptor complex via a hydrogen bond with the amide group. Such hydrogen bonds are unusual but are favoured by both the hydrophobic environment in the receptor region and the formation of a five-membered ring.

If the carboxin molecule is allowed to react with an Fe—S cluster, a complex may be formed as shown in Figure 10.3. For simplicity a 2 Fe—2 S cluster is depicted instead of the 4 Fe—4 S cluster, since the former cluster corresponds to one area of the cube-like structure of the Fe—S bonds. The presence of two carboxin molecules in the receptor complex is tentative; another possibility is the occupation of the second binding site by a ubiquinone molecule which shows some structural similarities with carboxin. According to this model the electrophilic carbon atom of the α, β-unsaturated carbonyl grouping should interact with a cysteine sulfur of the cluster, whereas the oxygen and nitrogen atoms are coordinatively liganded with two different Fe atoms of the cluster. In line with this assumption is the fact that α, β-unsaturated compounds will react readily with mercaptans under physiological conditions (cf. addition of methyl mercaptane to acrolein at 37° without catalyst). The nucleophilic activity of the sulfur should be even higher in the Fe—S cluster owing to the charge influence of the iron. The coordination of iron via a carbonyl oxygen is strongly supported by the fact that thenoyltrifluoroacetone which acts at the same specific binding site (COLES *et al.* 1978) is known to form such complexes with inorganic ferric iron. The two essential nonpolar groups of the carboxin molecule may serve as anchors for the fixation in hydrophobic regions of the receptor. These hydrophobic regions may be formed by QP_S or other small polypeptides of complex II. The photoaffinity labelling of the small but not of the large polypeptides by a carboxin analogue possessing a azido group at the phenyl residue (WHITE *et al.* 1983) strongly supports such an assumption. Phospholipids may be excluded with regard to an involvement in the specific receptor binding, since the carboxin-sensitivity of soluble succinate dehydrogenase is conferred solely by the addition of the small polypeptide fraction (VINOGRADOV *et al.* 1980). On the other hand, a large proportion of unspecific binding of carboxin may involve binding to the phospholipids (COLES *et al.* 1978; WHITE *et al.* 1983). One may speculate that the formation of the specific complex with the receptor is preceded by an accumulation of the fungicide in the phospholipid bilayer.

The specificity of carboxins for some groups of fungi may be based on the specific steric assembly of the receptor site. The changed selectivity of F 427 (o-phenyl derivative) indicates that the hydrophobic orientation caused by hydrophobic amino acid sequences in the active centre might be different in Basidiomycetes and Ascomycetes.

Carboxin action and alternative pathway of respiration

The observation that many carboxin-sensitive fungi possess an alternative cyanide- and antimycin A — resistant respiratory pathway raises the question whether the two processes are related. Carboxins inhibit the oxidation of succinate via both cytochrome chain and alternative respiratory pathway (SHERALD and SISLER 1972; LYR and SCHEWE 1975). An indispensable role of the alternative respiratory pathway for the action of carboxins must be however excluded since mammalian mitochondria are highly carboxin-sensitive, but do not possess cyanide-insensitive respiration. The nature of the alternative oxidase has been the subject of extensive studies within the last two decades. A spectrally detectable component has not been observed. Therefore it is reasonable to assume that alternative respiration may be due to the autoxidation of an obligatory member of the respiratory chain. In an earlier report we proposed that the Fe—S centre S_3 may function as "alternative oxidase" (LYR and SCHEWE 1975). Later studies of RICH and BONNER (1978) have indicated a more likely role of ubisemiquinone radicals. As mentioned above their sizable occurrence presupposes the presence of ubiquinone-binding proteins, which prevent spontaneous dismutation. It appears likely that a special type of Q proteins renders the ubisemiquinone radicals susceptible to the direct reaction with dioxygen forming oxidized ubiquinone and superoxide anion radical that is in turn converted by mitochondrial superoxide dismutase to hydrogen peroxide. Inhibitors of the main respiratory pathway such as antimycin A or cyanide trigger this process by enhancing the level of ubisemiquinone radicals, since their oxidation via the cytochrome chain is interrupted.

In this manner the special Q proteins may act as "ubisemiquinone oxidases". It is not yet established whether a particularly modified QP_S itself is the alternative oxidase or whether it is another Q protein. The selective inhibition of the alternative pathway by certain hydroxamates argues in favour of the latter possibility. The observation that succinate is in many cases a much better substrate for the alternative pathway than NADH or NAD^+-dependent substrates may indicate a direct functional interaction of QP_S and "ubisemiquinone oxidase" without involvement of the Q pool. One has to take into consideration the occurrence of more than one alternative oxidase (AINSWORTH *et al.* 1980). Such a heterogeneity would explain the conflicting results in the recent literature (DAY *et al.* 1978; ZIOGAS and GEORGOPOULOS 1979; ZIOGAS and GEORGOPOULOS 1984) concerning the action of carboxin and thenoyltrifluoroacetone on the alternative pathway.

Possible mechanisms of side effects

It may be of interest, that carboxin can be oxidized to the inactive sulfoxide (Fig. 9.1c) by mitochondria of sensitive and insensitive fungi as well as by riboflavins and flavin enzymes. Light strongly increases the conversion. Catalase has a protective effect. Only a small amount of the sulfoxide is further oxidized to the biologically active oxycarboxin (Fig. 9.1b) (Lyr *et al.* 1974).

This implies that carboxin can act as oxygen radical scavenger. One may speculate, that this could be the background of the growth-stimulating effect described by von Schmeling and Kulka (1969) and the reason for a reduction of ozone injury (Hofstra *et al.* 1978). For further literature on the carboxins, see the article by Kuhn (1984).

Acknowledgement: We thank Prof. Dr. H. D. Sisler for linguistic corrections.

References

Ackrell, B. A. C., Ball, M. B., and Kearney, E. B.: Peptides from complex II active in reconstitution of succinate-ubiquinone reductase. J. Biol. Chem. **255** (1980): 2761—2769.

— Kearney, E. B., Coles, C. J., Singer, T. P., Beinert, H., Wan, Y. P., and Folkers, K.: Kinetics of the reoxydation of succinate dehydrogenase. Arch. Biochem. Biophys. **182** (1977): 107—117.

— — Mims, W. B., Peisach, J., and Beinert, H.: Iron-sulfur cluster 3 of beef heart succinate-ubiquinone oxidoreductase is a 3-iron cluster. J. Biol. Chem. **259** (1984): 4015—1018.

Ainsworth, P. J., Ball, A. J. S., and Tustanoff, E. R.: Cyanide-resistant respiration in yeast II. Characterization of a cyanide-insensitive NAD (P)H oxidoreductase. Arch. Biochem. Biophys. **202** (1980): 187—200.

Beinert, H., and Albracht, S. P. J.: New insights, ideas and unanswered questions concerning iron-sulfur clusters in mitochondria. Biochim. Biophys. Acta **683** (1982): 245—277.

Ben-Yephet, Y., Dinoor, A., and Henis, Y.: The physiological basis of carboxin sensitivity and tolerance in *Ustilago hordei.* Phytopathology **65** (1975): 936—942.

Coles, Ch. J., Singer, T. P., White, G. A., and Thorn, G. D.: Studies on the binding of carboxin analogs to succinate dehydrogenase. J. Biol. Chem. **253** (1978): 5573—5578.

Day, D. A., Arron, G. P., and Laties, G. G.: The effect of carboxins on higher plant mitochondria. FEBS Lett. **85** (1978): 99—102.

Edgington, L. V., and Barron, G. L.: Fungitoxic spectrum of oxathiin compounds. Phytopathology **57** (1967): 1256.

Georgopoulos, S. G., Alexandri, E., and Chrysayi, M.: Genetic evidence for the action of oxathiin and thiazole derivatives on the succinic dehydrogenase system of *Ustilago maydis* mitochondria. J. Bacteriol. **110** (1972): 809—817.

— Chrysayi, M., and White, G.: Carboxin resistance in the haploid, the heterozygous diploid, and the plant parasitic dicaryotic phase of *Ustilago maydis.* Pesticide Biochem. Physiol. **5** (1975): 543—551.

— and Ziogas, B. N.: A new class of carboxin-resistant mutants of *Ustilago maydis.* Netherl. J. Plant Pathol. **83** (1977): 235—242.

Gunatilleke, I. A. U. N., Arst, H. N., and Scazzocchio, C.: Three genes determine the carboxin sensitivity of mitochondrial succinate oxidation in *Aspergillus nidulans.* Genet. Res., Camb. **26** (1975): 297—305.

Haken, P. ten, and Dunn, C. L.: Structure-activity relationsships in a group of carboxanilides systemically active against broad bean rust *(Uromyces fabae)* and wheat rust *(Puccinia recondita).* Proc. 6th Brit. Insecticide Fungicide Conf. (1971): 453—463.

Hardison, J. R.: Relation ships of molecular structure of 1,4-oxathiin fungicides to chemotherapeutic activity against rust and smut fungi in grasses. Phytopathology **61** (1971): 731—735.

Hofstra, G., Littlejohns, A., and Wukasch, R. T.: The efficacy of the antioxidant ethylenediurea (EDU) compared to carboxin and benomyl in reducing yield losses from ozone in navy bean. Plant Dis. Rep. **62** (1978): 350—352.

Huppatz, J. L., Phillips, J. N., and Witrzens, B.: Structure-activity relationships in a series of fungicidal pyrazole carboxanilides. Agricult. Biol. Chem. **48** (1984): 45—50.

Kuhn, P. J.: Mode of action of carboxamides. In: Trinci, A. P. J., and Ryley, J. F. (Eds.): Mode of Action of Antifungal Agents. Brit. Mykol. Soc. (1984): 155—183.

Löcher, F., Hampel, M., and Pommer, E.-H.: Ergebnisse bei der Rostbekämpfung mit Benodil (BAS 317 00 F). Meded. Fac. Landbouwwet. Rijksuniv. Gent **39** (1974): 1079—1089.

Lyr, H., Luthardt, W., and Ritter, G.: Wirkungsweise von Oxathiin-Derivaten auf die Physiologie sensitiver und insensitiver Hefe-Arten. Z. Allg. Mikrobiol. **11** (1971): 373—385.

— Ritter, G., and Casperson, G.: Zum Wirkungsmechanismus des systemischen Fungizids Carboxin. Z. Allg. Mikrobiol. **12** (1972): 271—280.

— — and Banasiak, L.: Detoxication of carboxin. Z. Allg. Mikrobiol. **14** (1974): 313—320.

— and Schewe, T.: On the mechanism of the cyanide-insensitive alternative pathway of respiration in fungi and higher plants and the nature of the alternative terminal oxidase. Acta Biol. med. germ. **34** (1975): 1631—1641.

— — Müller, W., and Zanke, D.: Zum Problem der Selektivität sowie der Struktur-Rezeptorbeziehungen von Carboxin und seinen Analogen. In: Lyr, H., and Polter, C. (Eds.): Systemic Fungicides (Int. Symp. Reinhardsbrunn 1974). Akademie-Verlag, Berlin 1975, pp. 153—166

Mathre, D. E.: Uptake and binding of oxathiin systemic fungicides by resistant and sensitive fungi. Phytopathology **58** (1968): 1464—1469.

— Mode of action of oxathiin systemic fungicides. I. Effect of carboxin and oxycarboxin on the general metabolism of several basidiomycetes. Phytopathology **60** (1970): 671—676.

— Mode of action of oxathiin systemic fungicides III. Effect on mitochondrial activities. Pesticide Biochem. Physiol. **2** (1971): 216—224.

Mowery, P. C., Ackrell, B. A. C., Singer, T. P., White, G. A., and Thorn, G. D.: Carboxins, powerful selective inhibitors of succinate oxidation in animal tissues. Biochem. Biophys. Res. Comm. **71** (1976): 354—361.

— Steenkamp, D. J., Ackrell, B. A. C., Singer, T. P., and White, G. A.: Inhibition of mammalian succinate dehydrogenase by carboxins. Arch. Biochem. Biophys. **178** (1977): 495—506.

Müller, W., Schewe, T., Lyr, H., and Zanke, D.: Wirkungsmechanismus der Atmungshemmung durch die Systemfungizide der Carboxingruppe. Wirkung von Oxathiinderivaten und -analoga auf nicht phosphorylierende submitochondriale Partikeln aus Rinderherz sowie *Trametes versicolor* und *Trichoderma viride*. Z. Allg. Mikrobiol. **17** (1977): 359—372.

Pommer, E.-H., Girbensohn, B., König, K.-H., Osieka, H., and Zeeh, B.: Development of new systemic fungicides with carboxanilide structure. Kemia-Kemi **1** (1974): 617—618.

Ragsdale, N. N., and Sisler, H. D.: Metabolic effects related to fungitoxicity of carboxin. Phytopathology **60** (1970): 1422—1427.

Ramsay, R. R., Ackrell, B. A. C., Coles, C. J., Singer, T. P., White, G. A., and Thorn, G. D.: Reaction site of carboxanilides and of thenoyltrifluoroacetone in complex II. Proc. Nation. Acad. Sci. USA **78** (1981): 825—828.

Rich, P. R., and Bonner, W. D.: The nature and location of cyanide and antimycin resistant respiration in higher plants. In: Degn, H., *et al.* (Ed.): Functions of Alternative Terminal Oxidases Vol. 49, Proc. 11th FEBS Meeting Copenhagen. Pergamon Press, Oxford 1978, pp. 149—158.

Schewe, T., Hiebsch, Ch., Garcia Parra, M., and Rapoport, S.: Effect of some respiratory inhibitors on the respiratory enzymes of the mitochondria from the cauliflower *(Brassica oleracea)*. Acta Biol. Med. Germ. **32** (1974): 419—426.

— Müller, W., Lyr, H., and Zanke, D.: Ein molekulares Rezeptormodell für Carboxin. In: Lyr, H., and Polter, C. (Eds.): Systemic Fungicides (Symposium 1977). Abh. Akad. Wiss. DDR 2 N, Berlin 1979, pp. 241—251.

— Rapoport, S., Böhme, G., and Kunz, W.: Zum Angriffspunkt des Systemfungicids Carboxin in der Atmungskette. Acta Biol. Med. Germ. **31** (1973): 73—86.

Schmeling, B. v., and Kulka, M.: Regulation of plant growth. Can. Patent 828, 771 (1969); U.S. Patent **3,454,391** (1969).

Sherald, J. L., and Sisler, H. D.: Selective inhibition of antimycin A — insensitive respiration in *Ustilago maydis* and *Ceratocystis ulmi*. Plant and Cell Physiol. **13** (1972): 1039—1052.

Snel, M., Schmeling, B. v., and Edgington, L. V.: Fungitoxicity and structure-activity relationships of some oxathiin and thiazole derivatives. Phytopathology **60** (1970): 1164—1169.

Tucker, A. N., and Lillich, T. T.: Effect of the systemic fungicide carboxin on electron transport function in membranes of *Mikrococcus denitrificans*. Antimicrob. Agents and Chemotherapy **6** (1974): 522.

Ulrich, J. T., and Mathre, D. E.: Mode of action of oxathiin systemic fungicides. V. Effect on electron transport of *Ustilago maydis* and *Saccharomyces cerevisiae*. J. Bacteriol. **110** (1972): 628—632.

Vinogradov, A., Gavrikov, V. G., and Gavrikova, E. V.: Studies on the succinate dehydrogenating system II. Reconstitution of succinate-ubiquinone reductase from the soluble components. Biochim. Biophys. Acta **592** (1980): 13—27.

White, G. A.: A potent effect of 1,4-oxathiin systemic fungicides on succinate oxidation by a particulate preparation from *Ustilago maydis*. Biochem. Biophys. Res. Comm. **44** (1971): 1212—1219.

— and Thorn, G. D.: Structure-Activity relationships of carboxamide fungicides and the succinic dehydrogenase complex of *Crytococcus laurentii* and *Ustilago maydis*. Pesticide Biochem. and Physiol. **5** (1975): 380—395.

— — and Georgopoulos, S. G.: Oxathiin carboxamides highly active against carboxin-resistant succinic dehydrogenase complexes from carboxin — selected mutants of *Ustilago maydis* and *Aspergillus nidulans*. Pesticide Biochem. Physiol. **9** (1978): 165—182.

— Thorn, G. D., Ackrell, B. A. C., Kearny, E. B., Ramsey, R. R., and Singer, T. P.: Site of action of carboxamides in mitochondrial complex II. In: Miyamoto, J., *et al.* (Ed.): IUPAC Pesticide Chemistry. Pergamon Press, Oxford 1983, pp. 141—146.

Yu, Ch., and Yu, L.: Ubiquinone-binding proteins. Biochim. Biophys. Acta **639** (1981): 99—128.

Zeeh, B., Linhart, F., and Pommer, E.-H.: Nicotinic anilides. German Offenlegungsschrift 26 11 601 (1977): 1—17.

Ziogas, B. N., and Georgopoulos, S. G.: The effect of carboxin and of thenoyltrifluoroacetone on cyanide-sensitive and cyanide-resistant respiration of *Ustilago maydis* mitochondria. Pesticide Biochem. Physiol. **11** (1979): 208—217.

— — Mitochondrial electron transport in a carboxin-resistant, antimycin A-sensitive mutant of *Ustilago maydis*. Pesticide Biochem. Physiol. **22** (1984): 24—31.

Lyr, H. (Ed.): Modern, Selective Fungicides — Properties, Applications, Mechanisms of Action. Longman Group UK Ltd., London, and VEB Gustav Fischer Verlag, Jena, 1987.

Chapter 11

Morpholine fungicides

E.-H. POMMER
BASF Aktiengesellschaft, Limburgerhof, FRG

Introduction

In 1965 KÖNIG, POMMER and SANNE reported the fungicidal activity of N-substituted tetrahydro-1,4-oxazines or morpholine derivatives. They were able to demonstrate that morpholine derivatives with major N-positioned cycloaliphatic or long-chain alkyl groups exhibited good to very good fungicidal activity, particulary against the powdery mildew fungi. As a result of the systematic approach to this new class of compounds, the active compounds dodemorph (POMMER and KRADEL 1967a) and tridemorph (KRADEL *et al.* 1968) were produced and which were introduced into the market in 1965 and 1969 respectively. Whereas dodemorph found favour in the control of powdery mildew fungi in ornamental plant cultivation (KRADEL and POMMER 1967a, b), tridemorph gained importance in the first instance as a fungicide for the control of barley mildew, which had become a serious problem in barley growing in Western Europe at the end of the 1950's. Over the past few years this compound has also been increasingly used in other crops such as bananas and rubber. The active compound aldimorph, which belongs to the same class of substances as dodemorph

Table 11.1 Development of morpholine derivatives as fungicides for commercial use

Common name Code number	Tradename at the time of introduction Formulation	Year of introduction (by)	Main area of application
Dodemorph BAS 238 ... F	BASF-Mehltau-mittel EC 400 g/l	1965 (BASF)	Powdery mildew on ornamentals
Tridemorph BAS 200 ... F 220 ... F	Calixin ED 750 g/l	1969 (BASF)	Powdery mildew on cereals (barley); sigatoka disease (bananas)
Aldimorph	Falimorph EC 662 g/l	1980 (Fahlberg-List)	Powdery mildew on cereals (barley)
Fenpropimorph BAS 421 ... F Ro 14—3169	Corbel/Mistral EC 750 g/l	1980 (BASF/Maag)	Powdery mildew and rust on cereals (wheat and barley)
Trimorphamide VUAgT-866 BAS 463 ... F	Fademorf EC 200 g/l	1981 (Vysk. Ustav Agrochem. Technol.)	Powdery mildew on grapes and apples

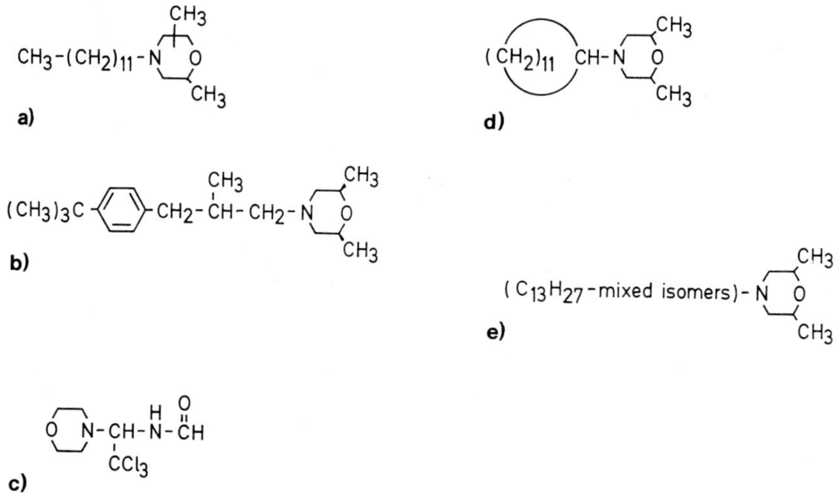

Fig. 11.1 Structure formulae of morpholine fungicides. a) Aldimorph, b) Fenpropimorph, c) Trimorphamide, d) Dodemorph, e) Tridemorph.

Table 11.2 Morpholine derivatives in commercial use as fungicides

Structure	Chemical name	Patent No.	LD_{50} rat p.o. mg/kg
Dodemorph 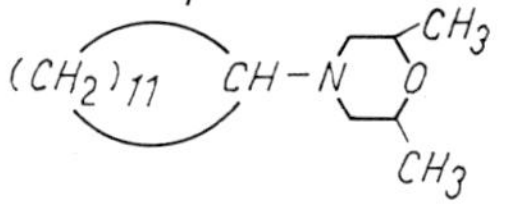	4-cyclododecyl-2,6-dimethyl-morpholine	DE 1 198 125	1800
Tridemorph 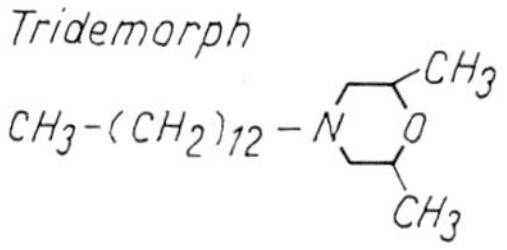	4-tridecyl-2,6-dimethyl-morpholine	DE 1 164 152	860
"Aldimorph" 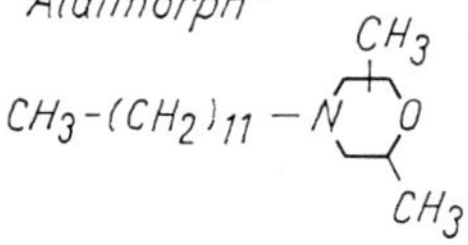	N-n-dodecyl-2,5/2,6-dimethyl-morpholine	DD 140 412	3500
Fenpropimorph 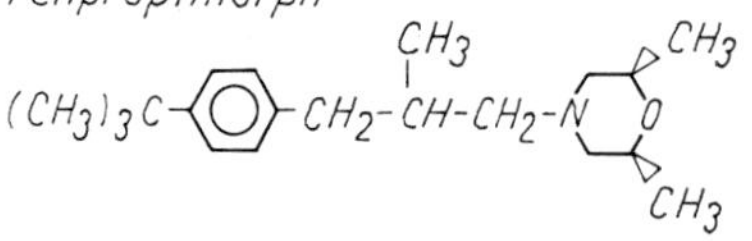	cis-4-3-(4-tert-butyl-phenyl)-2-methylpropyl-2,6-dimethyl-morpholine	DE 26 56 747 DE 27 52 096	3515
Trimorphamide 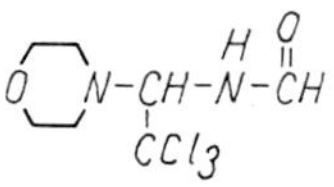	N-2,2,2-tri-chloro-1-(4-morpholinyl)-ethyl-formamide	CS 193 659	2820

and tridemorph, was introduced by JUMAR and LEHMANN in 1982 as a cereal mildew fungicide.

A substantial improvement in the field of morpholine derivatives with a fungicidal activity was achieved through the study of substances on the basis of arylalkylamines. The result of these studies was the active compound fenpropimorph; POMMER and HIMMELE (1979) and BOHNEN and PFIFFNER (1979) reported its fungicidal properties and possibilities of its use in the control of cereal mildew and rust fungi.

The active compound trimorphamide, which is mainly active against powdery mildew fungi (DEMECKO and JARAS 1975), also belongs to the N-substituted tetrahydro-1,4-oxazines. The morpholine derivatives which have been introduced as commercial products are listed in Table 11.1.

The structural formulae (Fig. 11.1) and the chemical names of these five active compounds together with their patent numbers and the LD_{50} values for the rat given in Table 11.2.

Uptake and transport in the plant

The first investigations into the systemic properties of morpholine derivatives were carried out with dodemorph (KRADEL and POMMER 1967c). Soil treatment experiments made using emulsions containing 250 ppm active ingredient showed that barley plants, artificially inoculated with *Erysiphe graminis*, could be kept free of infection over a period of 16 days. Barley plants readily take up tridemorph via their roots and transport this acropetally with the transpiration stream into the leaves. It was possible to measure 7.2 ppm of this substance in leaf extracts of plants 4 h after their roots had been submerged in a nutrient solution containing 50 ppm tridemorph; after 48 h an equilibrium had been established.

In comparison with the root uptake, absorption of tridemorph by the leaf surface occurs more slowly. After the active substance has penetrated the leaf, however, further transport within the tissue occurs relatively rapidly (POMMER *et al.* 1969). Similar results were obtained by OTTO and POMMER (1973) in experiments with ^{14}C-labelled tridemorph. When 5 ml of a suspension which contained 0.15 % active ingredient were pipetted into the leaf sheaths of barley seedlings at the 1-leaf growth stage, acropetal transport of the active ingredient could be detected in an autoradiogram as soon as 8 h after application.

Due to the predominantly upward transport of the compound, a certain degree of impoverishment of the compound in the lower leaf parts was observed 54 h after application. When the labelled active compound was applied to a barley plant in the 1-node growth stage into a leaf sheath above this node, then within 2 days transport of the active compound into all leaves above the node could be detected, plant parts which developed later did not contain any radioactivity; a basipetal transport does not occur. In experiments with cucumber plants the same authors found that labelled tridemorph applied to the leaves was distributed in the entire growing plant. Here, the tendency of the active compound to accumulate in the marginal zones of the leaves could prove to be disadvantageous if as a result the concentration of the active compound in the leaf centres were too greatly reduced. In bananas an equilibrium in concentration between treated and non-treated leaves occurs very slowly due to the low rate at which the active compound is absorbed through the leaf surfaces. In contrast to barley, in wheat tridemorph is taken up better by the morphological upper leaf surface than by the lower (POMMER and KRADEL 1971). Since the wheat upper leaf surfaces are considerably more densely covered with stomata than the lower leaf surfaces, the stomata may possibly be regarded as important gates of entry for tridemorph. The fact that the leaf surfaces are more densely covered with fine hairs may play a role in the lower uptake of the active compound, since the small air bubbles trapped between the hairs may greatly impede the wetting of the epidermal cells by tridemorph. DEYMANN (1981) observed that wheat plants take up lower quantities of tridemorph than barley plants. The degradation of the active compound is accelerated during the first week following treatment. With fenpropimorph no statistically reliable differences have been observed between barley and wheat plants. After treatment with equal quantities of active compound, the amount of fenpropimorph found in barley and wheat plants is greater than that of tridemorph. Furthermore, the author found that the distribution of the active compound varied depending upon the substance used and the growth stage of the plant. A treatment with tridemorph led in younger and older plants to an accumulation of the active substance in the leaf tips such that the lower parts of the leaves became impoverished (cf. OTTO and POMMER 1973). In fenpropimorph treatments it was found that plants at the heading growth stage showed a prolonged retention of the active compound, although here, too, the active compound accumulated in the leaf tips. Uptake of tridemorph and fenpropimorph is lower in older cereal plants than in younger ones, and their degradation occurs more slowly. A basipetal transport of fenpropimorph has not been detected (POMMER and HIMMELE 1979; BOHNEN and PFIFFNER 1979; DEYMANN 1981).

Fenpropimorph is characterized by a pronounced vapour phase which can lead to a secondary distribution of the active compound, as has been demonstrated in glasshouse experiments with *Erysiphe graminis* on infected wheat and barley plants (BOHNEN and PFIFFNER 1979; POMMER and HIMMELE 1979). When infected, untreated wheat and barley plants were placed together with plants treated with fenpropimorph, then the infection could be stopped at a distance of up to one metres between the plants. According to DEYMANN (1981), the activity of fenpropimorph via the vapour phase does not play an important role in practice.

JUMAR and LEHMANN (1982), working with *Hordeum distichon* and *Sinapis alba*, established that relatively high quantities of aldimorph were taken up by the plants both after leaf as well as root application. According to these authors, the non-metabolized compound shows only a low rate of translocation. However, the fungicidally active metabolites are more mobile and are responsible for the systemic activity.

Despite repeated inoculation with *Erysiphe cichoracearum*, cucumber plants could be kept completely free of infection over a period of 29 days following an initial soil treatment with 250 mg trimorphamide (HUDECOVA *et al.* 1978). After 41 days the fungicide still provided 40 % protection.

Influence of morpholine derivatives on pathogenesis

This section deals with the influence of morpholine derivatives on the pathogenesis of plant pathogenic fungi. Their mode of action will be described in chapter 12.

SCHLÜTER and WELTZIEN (1971) investigated the effect of dodemorph and tridemorph on the pathogenesis of barley mildew by means of cytological and cytochemical methods. In these studies they observed an only incomplete haustoria formation, i.e. the growth of the finger-shaped outgrowths was impeded. At the same time the authors found an irreversible inhibition in sporulation. As an inhibitor of haustoria development, fenpropimorph interferes relatively late in the development of an infection; it also exhibits a directly damaging and germination inhibiting effect on conidia of *Erysiphe graminis* on barley and wheat. In this respect fenpropimorph is more active than tridemorph (DEYMANN and WELTZIEN 1980).

Bohnen and Pfiffner (1979) could demonstrate with barley mildew that 30—36 hours after a curative application of fenpropimorph, the young hyphae growing out from the periphery of existing infection centres grew in a hump-backed fashion away from the epidermis. The tips of these hyphae ballooned to 2—5 times their original size, finally collapsing and drying up. The production of conidia was interrupted. The mycelium which had grown prior to the treatment, as well as the conidia which had already been formed lose their turgor and die. Saur and Löcher (1984) established that fenpropimorph had a good stopping effect against *Erysiphe graminis* on wheat at low as well as high temperatures.

Zobrist *et al.* (1982) investigated the morphological and microstructural effect of fenpropimorph on the germination of uredospores, as well as the formation of appressoria and uredospores of oat crown rust. Following leaf applications of 750 ppm fenpropimorph, the germination of uredospores and the infection process proceeded as on untreated plants. However, the pathogen died after penetrating the leaf tissue before uredospores could be formed. After treatment with 300 ppm active compound *Puccinia coronata* developed up to the formation of uredospores. However, the spores were for the most part collapsed or had only developed rudimentary cell walls.

Fluorescence microscopy investigations on the effect of fenpropimorph on *Ustilago maydis* and *Penicillium italicum* (Kerkenaar and Barug 1984) have indicated that this substance caused morphological changes in the sporidia of *Ustilago maydis* and in the germinating conidia of *Penicillium italicum*. The sporidia of *Ustilago maydis* appeared swollen, distorted, multicellular and sometimes branched; conidia of *Penicillium italicum* swelled in size and the extension of germ tubes was strongly inhibited. The mycelium of *Penicillium italicum* also exhibited a much enlarged hyphal diameter and relatively short distances between the septae.

The behaviour of morpholine derivatives in soil and plant

In 1973 Otto and Drescher investigated the behaviour of tridemorph in the soil using tridemorph which had been radioactively labelled on the C-atoms in positions 2 and 6 of the morpholine ring. They were able to demonstrate that tridemorph is strongly adsorbed onto the soil particles (loamy sand and peaty soil) such that leaching onto deeper soil layers and a subsequent contamination of the surface water is prevented. In addition to this they were able to establish that tridemorph is not persistent in the soil and is biologically degraded. The initial step in this process is the formation of tridemorph-N-oxide; 2,6-dimethylmorpholine and CO_2 have been detected as further degradation products.

The persistence of aldimorph in the soil was established by using ^{14}C-labelled active compound (Sieber *et al.* 1981). In laboratory experiments, the half-life values of aldimorph, after the addition of 20.6 ppm active compound to two experimental soils, amounted to 18 and 26 days. Investigations carried out under field conditions also produced favourable results. Under these conditions it was possible to establish a half-life value of 10 days following irrigation. Aldimorph showed a great affinity to the soil. After irrigation with 200 ml precipitation, the active compound was retained in the upper 5 cm layer of soil.

Fenpropimorph is not persistent in the soil; it is degraded relatively quickly. The half-life was investigated in field experiments and found to be around 21 days. Leaching of fenpropimorph into the soil does not occur (Anonym 1982).

Investigations into the effect of fenpropimorph on the soil microflora have revealed that the addition of 10 and 100 ppm active compound to a humus sandy soil and a loamy soil had no influence on the total number of viable cells, their respiration, various enzyme activities nor nitrification (Pommer and Himmele 1979).

The metabolism of aldimorph in treated barley plants starts relatively quickly. During this process polar, water soluble metabolites are mainly formed which contain the intact morpholine ring (Jumar and Lehmann 1982). Experimental results from these authors indicate that a probable primary hydroxylation of the terminal methyl group of the alkyl side-chain or of one of the methyl groups attached to the ring takes place followed by a conjugation with glucose. In addition to this it is very probable that N-carboxymethyl-2,6-dimethylmorpholine is formed by beta-oxidation.

At present there is little information available with respect to the metabolism of tridemorph and fenpropimorph in plants. Experiments carried out with radioactively labelled fenpropimorph in wheat have revealed that the active compound is extensively degraded in the plant and metabolized largely to four identified degradation products; the half-life value was found to equal approximately 5 days. In the grain the main part of the radioactivity is built into starch, which indicates that the degradation of the active compound is complete (Anonym 1984).

The use of morpholine derivatives in agriculture

Whilst describing the use of morpholine derivatives in agriculture it must be taken into consideration that tridemorph, for example, is a fungicide that has been known for many years and, consequently, there is a large amount of information available on the use of this compound. In contrast, aldimorph and trimorphamide are relatively young products such that to date there is only a limited amount of literature available. Fenpropimorph is predominantly used in cereal crops such that experiments with this compound have been concentrated on its use in these crops. Finally, dodemorph has established itself as a fungicide for the control of powdery mildew fungi in ornamental plants; consequently, the literature available describing the experiences gained with the use of this compound is not extensive.

Dodemorph

In commercial products (as EC formulations), the active compound dodemorph is not used in its free base form but as acetate. The two cis- and trans-stereoisomers of the 4-cyclodecyl-2,6-dimethylmorpholine are present in an approximately equal ratio (Kradel and Pommer 1967a). Using barley and wheat powdery mildew it could be demonstrated that the two stereoisomers applied separately or as a cis-, trans-stereoisomeric mixture were equally effective (Pommer 1984). Kradel *et al.* (1969) reported experiments conducted over a period of several years in which dodemorph (cyclomorph had been originally proposed as the common name) was compared with tridemorph for the control of barley mildew. Similar results were obtained by using application rates of 1.0—1.5 kg dodemorph/ha and 0.4—0.6 kg tridemorph/ha. Since dodemorph exhibits a good degree of activity against mildew, but also because it is sufficiently well tolerated by a number of crops, even after repeated application, this fungicide was developed for the use in ornamental plant cultivation (Kradel and Pommer 1967a; Kradel *et al.* 1967; Kradel *et al.* 1970). A favoured area of application is roses (Frost and Pattissen 1971). In the control of *Podosphaera leucotricha* on apples severe fruit russetting developed in the sensitive varieties Cox's Orange and Golden Delicious; the activity against mildew was equal to that of dinocap. Good results against *Erysiphe cichoracearum* were achieved on cucumbers and pumpkins.

Occasionally a yellowing of the leaf margins was observed when the compound was applied under glass (Kradel *et al.* 1967).

In order to extend the range of activity and also to intensify its efficacy against mildew, the following mixtures with dodemorph have been developed into commercial products:

dodemorph + dodine

dodemorph + nitrothal-isopropyl

Aldimorph

The active compound aldimorph is formulated as an emulsifiable concentrate and is recommended for the control of *Erysiphe graminis* in spring and winter barley. Sieber *et al.* (1981) recommended 0.8 kg a.i./ha as the application rate to be used. On account of its low phytotoxicity, this compound can also be applied to other sensitive crops, such as wheat. According to Jumar and Lehmann (1982), the active compound aldimorph is not a uniform substance but consists chiefly of a mixture of 4-n-dodecyl-2,6-dimethylmorpholine (cis-, trans-mixture) and a smaller share of 4-n-dodecyl-2,5-dimethylmorpholine (also a cis-, trans-mixture), which operate synergistically.

Tridemorph

The active compound tridemorph is not a uniform substance; it is a reaction mixture of C_{11}—C_{14}-4-alkyl-2,6-dimethylmorpholine (cis-, trans-mixture) homologues containing 60—70 % 4-tridecyl-isomers and 26 % 4-dodecyl-isomers (Anonym 1984). Tridemorph was initially developed in particular for the control of barley and wheat mildew (Pommer and Kradel 1967a; Kradel *et al.* 1970; Kradel and Pommer 1971). The rates of app ication lie between 375 and 560 g a.i./ha cereal. Tridemorph is distinguished by its plant tolerance in barley. Wheat exhibits varying reaction tridemorph; Siddiqui and Haahr (1971) established genetically based differences in the sensitivity of *Triticum aestivum* mutants. Hampel and Lang (1972) pointed out that with strong sunshine leaf burning occasionally occurs, which, however, does not cause any damage. In order to be able to investigate the tolerance of winter wheat varieties to tridemorph, Bergmann *et al.* (1983) developed a test using 2,3,5-triphenyl tetrazolium chloride (TTC) to determine the phytotoxic effect. It was thus possible to detect a clear dependence on the concentration. In agreement with investigations on entire plants in the glasshouse, an increasing phytotoxic effect in the following order was observed: barley < wheat < cucumber. Tridemorph exhibits a wide range of activity against pathogens belonging to the classes Ascomycetes and Basidiomycetes as well as out of the group of the Fungi imperfecti.

In addition to cereal mildew, positive results are available for the control of the following fungi (Tab. 11.3).

When tridemorph is used for the control of *Mycosphaerella* spp. causing leaf diseases in bananas, spray oil is added to the fungicide mixture in order to increase the penetration into the leaf tissue and to reduce the rate of vaporization of the active compound (Hamm 1983).

Tran van Canh (1984) reported a new method for the direct control of *Rigidoporus lignosus*, the causal organism of white root disease in *Hevea* (rubber), with tridemorph. By applying 10 g a.i. tridemorph (formulated as EC) per tree every 6 months, *Hevea* is well protected against an infection by *Rigidoporus*. In trials over three years the drenching has proved to be very effective. Mappes and Hiepko (1984) reported good

Table 11.3 In addition to cereal mildew, positive results in the control of the following fungi are available

Pathogen	Disease	Crop	Source
Erysiphe cichoracearum } *Sphaerotheca fuliginea*	Powdery mildew	Cucurbits	
Erysiphe cichoracearum	Powdery mildew	Tobacco	
Erysiphe polygoni	Powdery mildew	Legumes	Munshi (1983)
Erysiphe betae	Powdery mildew	Sugar beet	Meeus (1977)
Mycosphaerella musicola	Yellow sigatoka	Banana	Hampel and Pommer (1971)
Mycosphaerella fijiensis var. *difformis*	Black sigatoka	Banana	Mappes (1979)
Mycosphaerella fijiensis	Black leaf streak	Banana	Mappes (1979)
Cladosporium musae	Leaf speckle	Banana	Mappes (1979)
Corticium salmonicolor	Pink disease	*Hevea* (rubber)	Yeoh and Tan (1974)
Ceratocystis fimbriata	Moldy rot	*Hevea*	
Oidium hevea	Powdery mildew	*Hevea*	Lim (1976)
Fomes lignosus (*Rigidoporus lignosus*)	White root disease	*Hevea*	Tran van Canh (1982)
Fomes noxius (*Phellinus noxius*)	Stem rot	*Hevea*	
Ganoderma philippii	Rod root disease	*Hevea*	
Geotrichum candidum	Sour rot	*Citrus*	
Exobasidium vexans	Blister blight	Tea	Venkata Ram (1974)
Oidium mangiferae	Powdery mildew	Mango	

results achieved against *Rigidoporus lignosus*, *Phellinus noxius* and *Ganoderma* sp. with a tridemorph-bitumen formulation which had been applied to exposed, infected roots.

A treatment of tea leaves with a fungicide emulsion containing 0.11 % tridemorph inhibited sporulation of *Exobasidium vexans*; at a concentration of 0.375 % it eradicated the fungal infection in situ. In the field, satisfactory disease control that was achieved with tridemorph at 225 and 420 g/ha under mild and moderate rain fall conditions broke down under severe conditions (Venkata Ram 1974).

The pink disease of *Hevea*, caused by *Corticium salmonicolor*, could be effectively controlled under field conditions with a natural rubber latex formulation which contained 2 % tridemorph (Yeoh and Tan 1974).

Lim (1976) described a low volume procedure for the control of *Oidium* secondary leaf fall in *Hevea* with a formulation of tridemorph plus a refined white oil. Stem canker and black scurf of potato, caused by *Rhizoctonia solani*, was controlled by misting with 20 g tridemorph/l (Chand *et al.* 1981).

The following combinations of tridemorph with other fungicides, to extend the range of activity, are on the market:

tridemorph + maneb
tridemorph + carbendazim + maneb
tridemorph + propiconazol
tridemorph + triademifon

Side effects of tridemorph

In laboratory experiments tridemorph emulsion was toxic to the pink mite *(Acaphylla theae)*, purple mite *(Calacarus carinatus)* and scarlet mite *(Brevipalpus australis)* on tea leaves (CHANDRASEKRAN 1980).

Field applications of tridemorph had no effect on the aphid fungal pathogen *Entomophthora* on wheat (ZIMMERMANN and BASEDOW 1980).

ZIMMERMANN (1975) carried out in vitro experiments to study the effect of systemic fungicides on the germination of spores and mycelial growth of beneficial entomopathogenic Fungi imperfecti *Beauveria bassiana, B. tenella, Metarrhizium anisopliae, Paecilomyces farinosus* and *P. fumosoroseus*. At a concentration of 0.22 % tridemorph he observed an impairment of mycelial growth.

The antibacterial activity of tridemorph was detected against a wide spectrum of bacterial species (OROS and SULE 1983) *Corynebacterium* species were found to be the most sensitive (MIC values = 10—100 ppm), whereas *Agrobacterium* and *Erwinia* species were generally the most resistant (MIC values = > 1,000 ppm). Electrolyte studies have indicated that irreversible membrane damage is involved in the sensitivity of bacteria to tridemorph.

Tridemorph does not negatively effect the symbiotic N-fixation by *Rhizobium trifolii* in root nodules of white clover (FISHER 1976). In a further study, FISHER (1981), established that in the presence of higher tridemorph concentrations in the soil, the formation of root nodules in white clover was reduced. However, after application of the recommended rates of tridemorph under field conditions no influence on the N-fixation is to be expected.

The treatment of winter wheat with a mixture of compounds, which contained, amongst others, 750 g tridemorph/ha, had no influence on the decomposition of the straw. In laboratory experiments to investigate the destruction of cellulose, SCHROEDER (1979) showed that tridemorph influenced the rate of degradation.

Fenpropimorph

It was possible at a very early stage in the development of fenpropimorph to establish that the configuration characteristics of the 2,6-dimethylmorpholine component are important for the fungicidal activity. Of the two stereoisomeric forms, the cis-dimethylmorpholine compound exhibited considerably better activity against *Puccinia recondita* and *Erysiphe graminis* on wheat than the corresponding trans-form (POMMER and HIMMELE 1979; HIMMELE and POMMER 1980). The marked activity of fenpropimorph against mildew and rust fungi in cereals, in addition to its good compatibility in wheat have been confirmed in field trials (BOHNEN and PFIFFNER 1979; HAMPEL *et al.* 1979; SAUR *et al.* 1979). Under laboratory and glasshouse conditions (BOHNEN *et al.* 1979) determined the range of activity indicated in Table 11.4.

Spectrum of activity

Strong fungicidal side effects against species of *Uromyces, Hemileia, Rhizoctonia, Pyricularia* and *Rhynchosporium* have been detected. To control these and other pathogens the concentrations given in Table 11.4 and 11.5 are necessary under laboratory conditions.

In seed dressing tests with cabbage seed naturally infected with *Alternaria brassicola, A. brassicae* and *Phoma lingam*, fenpropimorph showed at an application rate of 2.5 g a.i./kg seed an activity comparable to that of iprodione (MAUDE *et al.* 1984).

Table 11.4 Main activities under laboratory, glasshouse or climate chamber conditions (Bohnen *et al.* 1979)

Pathogen	Host plant	ED_{75} values (MIC)
Foliar treatment		mg (a. i.)/l spray
Erysiphe graminis	barley	4
Erysiphe graminis	wheat	70
Puccinia triticina	wheat	300
Puccinia dispersa	rye	200
Puccinia graminis	wheat	250
Puccinia coronata	oat	75
Puccinia sorghi	maize	400
Puccinia striiformis	wheat	300
Uromyces spp.	bean	275
Seed treatment		mg (a.i.)/kg seed
Tilletia tritici	wheat	250
Ustilago avenae	oat	400
Pyrenophora graminea	barley	350
Rhizoctonia solani	cotton	400

Table 11.5 Fungicidal side effects under laboratory, glasshouse or climate chamber conditions (Bohnen *et al.* 1979)

Pathogen	Host plant	ED_{75} values (MIC) in mg a.i./l spray mixtures, kg seed or agar culture-medium
Foliar treatment		*in vivo*
Cercospora arachidicola	groundnut	400
Sphaerotheca pannosa	roses	200
Uncinula necator	vine	100
Septoria nodorum	wheat	~1,000
Agar culture medium		*in vitro*
Poria vaporaria		0.1
Lenzites trabea		0.2
Ceratocystis ulmi		0.2
Ustilago nuda		1
Septoria avenae		2
Ustilago maydis		3

Both seed treatments were highly effective and produced healthy transplants and later healthy crops.

Fenpropimorph, formulated as an emulsifiable concentrate, is currently primarily used for the control of cereal diseases caused by *Erysiphe graminis*, *Puccinia* species and *Rhynchosporium secalis;* the rate of application is 750 g a.i./ha (Hampel *et al.* 1979; Atkin *et al.* 1981; Goebel 1983). Trials on a range of winter wheat and winter and spring barley cultivars, under conditions of differing disease intensity, have confirmed that fenpropimorph gives very good efficacy against powdery mildew with good control persisting, at some sites, for up to 45 days after a single application in the spring (Atkin *et al.* 1981). In connection with the control of rust fungi, Rathmell and Skidmore (1982) emphasize the eradicative and curative activity of fenpropimorph; they also mention the redistribution of the active compound via the vapour phase. A persistence of 4—5 weeks is stated.

The stop-effect of fenpropimorph is emphasized by various authors (Bohnen and Pfiffner 1979; Pommer and Himmele 1979; Hopp *et al.* 1984). Geobel (1983) could demonstrate that even if 5—10 % of the leaf surface was covered with mildew, it was still possible to use fenpropimorph without disadvantage to its efficacy. At low temperatures and low rates of evaporation the wheat mildew in the crop stand has reportedly been stopped and killed even in cases where the infected leaf surface was considerably larger (Frahm 1983). Comprehensive investigations to clarify environmental influences on the efficacy of fenpropimorph against *Erysiphe graminis* on wheat have been carried out by Saur and Löcher (1984). They investigated the effect of varying day temperatures and precipitation after fungicide application (750 g a.i./ha in 400 l of water) with beginning to medium (3.8 % of the leaf area infected) mildew attack. According to these investigations, fenpropimorph exhibits at low as well as at higher temperatures a good stop effect and persistence. The tests for resistance to wash-off indicated that a rapid penetration of fenpropimorph into the plant occurred after application. If the crop was irrigated immediately after application of the compound, then no losses of active compound occurred when the plants were "dry" when they were treated; when they were dew-or rain-wet at the time of treatment then the losses were only negligible. Further experiments under field condition produced almost identical results.

To widen the range of activity of fenpropimorph, the following combinations have been developed into commercial products:

fenpropimorph + carbendazim
fenpropimorph + chlorothalonil

Trimorphamide

The active compound trimorphamide is used either as a wettable powder or as an emulsifiable concentrate. Its main area of application is in the control of powdery mildew fungi, however, the substance also exhibits a pronounced activity against *Venturia inaequalis* on apples (Seidl and Lansky 1981).

Good results have been obtained under field tests against the following organisms:

— *Erysiphe graminis* on barley and wheat (Klimach 1981)
— *Erysiphe cichoracearum* on cucumbers (Demecko 1975)
— *Sphaerotheca mors-uvae* on gooseberries
— *Sphaerotheca pannosa* on roses (Swiech 1981)
— *Podosphaera leucotricha* on apples (Demecko and Pandlova 1977)
— *Uncinula necator* on grapes (Gaher and Hudecova 1978).

Malenin and Todorova (1983) treated grape vines at 2-weekly intervals with 0.06% trimorphamide emulsions and observed a very good curative as well as prophylactic activity.

Resistance

Triazole derivatives have been widely use for a number of years to control cereal mildew. As a consequence of the increased selection pressure, mildew populations, e.g. in barley, have appeared which show a reduced sensitivity to this type of active compound (Fletcher and Wolfe 1981; Hollomon 1982; Buchenauer 1984).

Morpholines, like triazoles and pyrimidines, belong to the ergosterol biosynthesis inhibitors; they interfere, however, in different ways in the ergosterol biosynthesis (see Chapter 12 and 14). HOLLOMON (1982) investigated the effects of tridemorph on barley powdery mildew. He compared these to those of cycloheximide (protein synthesis inhibitor), triadimenol (sterol biosynthesis inhibitor) and ethirimol (inhibitor of adenosine metabolism). Tridemorph, like triadimenol, stopped the mildew development after the first haustorium had been formed. No strong cross-sensitivity was observed between tridemorph and the other fungicides, although triadimenol and ethirimol showed some evidence of negative cross-insensitivity.

Successive sowings of glasshouse-grown barley plants were treated with 50 g to 2,000 mg/l tridemorph and inoculated with an initially fungicide-sensitive isolate of *Erysiphe graminis* f. sp. *hordei*. The time required for symptoms to appear, compared with that for untreated plants, decreased with successive sowings. This was interpreted as evidence for the increase in the frequency of fungicide-tolerant propagules in the pathogen population. Effective mildew control was, however, obtained by the use of tridemorph in trial plots and field crops (WALMSLEY-WOODWARD *et al.* 1979a). The same authors (1979b) found considerable variation among the isolates of *Erysiphe graminis* f. sp. *hordei,* both in pathogenicity and in the level of tolerance to tridemorph. Tolerance was detected as the mean number of pustules on all treated material expressed as a percentage of the number on the untreated plants.

In model experiments with *Botrytis cinerea*. BISCHOFF (1983) isolated two strains that had different levels of resistance, different growth rates and different cross resistances. The author supposed that a series of genetic variations within the fungal cell was responsible for this. Cross resistance to dodine and carboxin persisted, but not that toward drazoxolon. This would appear to indicate that tridemorph resistance depends on variations in the respiratory chain at site I.

Laboratory isolates of *Penicillium italicum* with varying levels of resistance to fenarimol showed cross resistance to other fungicides which inhibit ergosterol biosynthesis (bitertanol, etaconazole, fenapanil and imazalil) but not to fenpropimorph (DE WAARD *et al.* 1982; DE WAARD and VAN NISTELROY 1982). In contrast, all isolates with a relatively high degree of resistance to ergosterol biosynthesis inhibitors exhibited increased sensitivity to fenpropimorph (negatively correlated cross resistance).

In glasshouse experiments over a two year period (POMMER and HIMMELE 1979) or of more than three years (BUTTERS *et al.* 1984) it has not been possible to induce an adaptive resistance to fenpropimorph. There are no indications to date that the sensivity of powdery mildew populations alters following the application of fenpropimorph.

LORENZ and POMMER (1984) reported a fenpropimorph monitoring programme designed to establish the fungicide sensitivity of wild populations, the variation in the range of these values and whether the sensitivity of powdery mildew populations to triazole fungicides is influenced by the use of fenpropimorph. In examples drawn from three different regions, the fenpropimorph values were found to be independent of treatment and lie in the range of values found in ten wild populations.

References

Anonym: Corbel Getreidefungizid. BASF AG u. Dr. Maag AG Sept. 1982.

— Calixin-Fungizide for the control of Sigatoka and other diseases in bananas. BASF AG 1984.

ATKIN, J. C., PARSONS, R. G., RICLEY, C. E., WATERHOUSE, S., and SIEGLE, H.: Control of cereal diseases with fenpropimorph and fenpropimorph mixtures in the U. K. Proc. 1981 Brit. Crop Protect. Conf. — Pests and Diseases, 307—316.

BERGMANN, H., MÜLLER, R., and NEUHAUS, W.: Bestimmung der phytotoxischen Wirkung von Calixin auf Kulturpflanzen. In: LYR, H., and POLTER, C. (Eds.): System. Fungizide u. Antifungale Verbindungen. Abh. Akad. Wiss. DDR Abt. Math., Naturwiss., Techn., N 1. Akademie-Verlag, Berlin 1983, S. 355—359.

BISCHOFF, G.: Tridemorph resistance in *Botrytis cinerea*. In: LYR, H., and POLTER, C. (Eds.): System. Fungizide u. Antifungale Verbindungen. Abh. Akad. Wiss. DDR Abt. Math., Naturwiss., Techn., N 1. Akademie-Verlag, Berlin 1983, S. 361—364.

BOHNEN, K., and PFIFFNER, A.: Fenpropemorph, ein neues, systemisches Fungizid zur Bekämpfung von echten Mehltau- und Rostkrankheiten im Getreidebau. Meded. Fak. Landbouwwet. Gent. **44/2** (1979): 487—497.

— SIEGLE, H., and LÖCHER, F.: Further experiences with a new morpholine fungicide for the control of powdery mildew and rust diseases on cereals. Proc. 1979 Brit. Crop Protect. Conf. — Pests and Diseases: 541—548.

BUCHENAUER, H.: Stand der Fungizidresistanz bei Getreidekrankheiten am Beispiel der Halmbruchkrankheit und des echten Mehltaus. Gesunde Pflanzen **36** (1984): 161—170.

BUTTERS, J., CLARK, J., and HOLLOMON, D. W.: Resistance to inhibitors of sterol biosynthesis in barley powdery mildew. Lecture held at XXXVI Intern. Symp. on Crop Protect., Gent, 8. 5. 1984.

CHAND, T., COPELAND, R. B., and LOGAN, C.: Fungicidal control of stem canker and black scurf of potato. Tests Agrochem. Cultivat. **2** (1981): 22—23.

CHANDRASEKARAN, R.: Acaricidal action of calixin against tea mites. Pestic. **14/5** (1980): 16—17.

DEMECKO, J., and JARAS, A.: Biological properties of N-(1-formamido-2,2,2-trichlorethyl) morpholine. Dokl. Soobshch.-Mezhdunar. Kongr. Zashhita Rast., Moskva **3** (1975): 178—183.

— and PANDLOVA, M.: Effect of trimorfamid on *Podosphaera leucotricha salm.* Agrochemia **17** (1977): 314—316.

DE WAARD, M. A., GROENEWEG, H., and VAN NISTELROOY, J. G. M.: Laboratory resistance to fungicides which inhibit ergosterol biosynthesis in *Penicillium italicum*. Netherl. u. Plant Pathol. **88** (1982): 99—112.

— and VAN NISTELROOY, J. G. M.: Toxicity of fenpropimorph to fenarimol-resistant isolates of *Penicillium italicum*. Netherl. J. Plant Pathol. **88** (1982): 231—236.

DEYMANN, A.: Untersuchungen zu Aufnahme und Transport der systemischen Fungizide Tridemorph und Fenpropimorph und ihrer Wirkungsweise gegen den echten Mehltau an Getreide *Erysiphe graminis* D.C. Marchal. 1981. Rheinische Friedrich-Wilhelm-Univ., Landw. Fak., Bonn-Poppelsdorf, Diss.

— and WELTZIEN, H. C.: Wirkung von Tridemorph und Fenpropimorph auf den echten Mehltau an Getreide *Erysiphe graminis* D.C. Marchal. Meded. Fak. Landbouwwet. Gent. **45/2** (1980): 129—136.

FISHER, D.: Effects of some fungicides on *Rhizobium trifolii* and its symbiotic relationship with white clover. Pesticide Sci. **7 (1)** (1976): 10—18.

— Effects of some fungicides used against cereal pathogens on the growth of *Rhizobium trifolii* and its capacity to fix nitrogen in white clover. Ann. appl. Biol. **98 (1)** (1981): 101—107.

FLETCHER, J. T., and WOLFE, M. S.: Insensitivity of *Erysiphe graminis* f. sp. hordei to triadimefon, triadimenol and other fungicides. Proc. Brit. Crop Protect. Conf. Pest and Diseases (1981): 633—640.

FRAHM, J.: Lehren aus einem schwierigen Mehltaujahr. Top Agrar. **10** (1983): 52—54.

FROST, A., and PATTISSEN, N.: Some results obtained in the United Kingdom using dodemorph for the control of powdery mildews in roses and other ornamentals. Proc. 6th Brit. Insecticide and Fungicide Conf. (1971): 349—354.

GAHER, S., and HUDECOVA, D.: Practical use of the fungicidal properties of trimorfamid. Agrochemia. **18** (1978): 185—189.

GOEBEL, G.: Corbel-systemisches Fungizid für Getreide mit neuer Wirkungsart. Gesunde Pflanzen **35** (1983): 243—247.

HAMM, R.: On the behaviour of Calixin in bananas. BASF Symposium, Central Amer. (1983): 15—24.

HAMPEL, M., and LANG, H.: Erfahrungen bei der Bekämpfung des Getreidemehltaus *(Erysiphe graminis)* in den Jahren 1970 und 1971. BASF-Mitt. Landbau April 1972.

— LÖCHER, F., and SAUR, R.: Two year field trial results with fenpropimorph in cereals. Meded. Fak. Landbouwwet. Gent **44** (1979): 511—518.

Hampel, M., and Pommer, E.-H.: Results of trials with Tridemorph on the control of *Mycosphaerella musicola* on Bananas. Proc. 6th Brit. Insecticide and Fungicide Conf. (1971): 842—847.

Himmele, W., and Pommer, E.-H.: 3-phenylpropylamines, a new class of systemic fungicides. Angew. Chemie, Internat. Ed. in English. **19** (1980): 184—189.

Hollomon, D. W.: The effects of tridemorph on barley powdery mildew: its mode of action and cross sensitivity relationships. Phytopathol. Z. **105** (1982): 279—287.

Hopp, H., Siegle, H., Saur, R., and Risch, H.: Wirkungsabhängigkeit der Getreidemehltaufungizide von unterschiedlichen Temperaturen und verschieden starkem Ausgangsbefall unter besonderer Berücksichtigung von Fenpropimorph. Mitt. aus der Biologischen Bundesanstalt für Land- und Forstwirtschaft Berlin-Dahlem Heft 223—244. Dt. Pflanzenschutz-Tag. Gießen (1984): 214.

Hudecova, D., Kovac, J., and Gaher, S.: Biological activity of trimorfamide against *Erysiphe cichoracearum* on cucumbers. Agrochemia. **18** (1978): 361—363.

Jumar, A., and Lehmann, H.: The special properties of fungicidal N-n-alkyldimethyl morpholines synthesized by an unorthodox route. Proc. 5th Intern. Congr. of Pesticide Chem., Kyoto 1982.

Kerkenaar, A., and Barug, D.: Fluorescence microscope studies of *Ustilago maydis* and *Penicillium italicum* after treatment with imarzalil or fenpropimorph. Pesticide Sci. **15 (2)** (1984): 199—205.

Klimach, A.: Effectiveness of Fademorf EK-20 against *Erysiphe graminis* DC. in winter wheat and spring barley. Agrochemia. **21** (1981): 51—53.

König, K.-H., Pommer, E.-H., and Sanne, W.: N-substituierte Tetrahydro-1,4-oxazine, eine neue Klasse fungizider Verbindungen. Angew. Chemie. **77** (1965): 327—333.

Kradel, J., Effland, H., and Pommer, E.-H.: Mehrjährige Ergebnisse bei der Bekämpfung des Getreidemehltaus *(Erysiphe graminis)* an Sommergerste. Meded. Fak. Landbouwwet. Gent. **33** (1968): 997—1004.

— — — Response of barley varieties to the control of powdery mildew with cyclomorph and tridemorph. Proc. 5th Brit. Insecticide and Fungicide Conf. (1969): 16—19.

— and Pommer, E.-H.: Vierjährige Versuche mit Cyclododecyl-2,6-dimethyl-morpholinacetat, einem neuen Wirkstoff gegen echte Mehltaupilze. BASF Mitt. Landbau Mai 1967, (1967 a).

— — Erfahrungen mit einem neuen Wirkstoff gegen Echte Mehltaupilze. Gartenwelt. **67** (1967 b): 185—186.

— — Some remarks and results on the control of powdery mildew in cereals. Proc. 4th Brit. Insecticide and Fungicide Conf. (1967 c): 170—175.

— — and Will, H.: Langjährige Erfahrungen bei der Mehltaubekämpfung in Zierpflanzen mit Dodemorph. BASF Mitt. Landbau. Juni 1970.

Lim, T. M.: Low-volume spray of an oil-based systemic fungicide for controlling *Oidium* secondary leaf fall. Proc. Rubber Res. Inst. Malays. Plant Conf. (1976): 231—242.

Lorenz, G., and Pommer, E.-H.: Investigations into the sensitivity of wheat powdery mildew populations towards fenpropimorph. Brit. Crop Protect. Conf. — Pests and Diseases. (1984): 489—494.

Malinin, I., and Todorova, M.: New fungicides for the control of powdery mildew. Lozar, Vinar. **32** (1983): 28—31.

Mappes, D.: Results of new trials with tridemorph against *Mycosphaerella* spp. on bananas. 9th Intern. Congr. Plant Protect. and 71th Ann. Meeting Amer. Phytopathological Soc., Washington (1979): Ref. No. 606.

— and Hiepko, G.: New possibilities for controlling root diseases of plantation crops. Meded. Fak. Landbouwwet. Gent **49 (2 a)** (1984): 283—292.

Maude, R. B., Humpherson-Jones, F. M., and Shuring, C. G.: Treatments to control *Phoma* and *Alternaria* infections of brassica seeds. Plant Pathol. **33** (1984): 525—535.

Meeus, P.: Fungicidal protection against powdery mildew in sugar beets. Driemaand. Publ., Belg. Institute Verbetering Beit **45 (3)** (1977): 131—141.

Munshi, G. D.: Efficacy of fungicides for the control of powdery mildew of pea. J. Res. (Punjab Agric. Univ.) **20 (4)** (1983): 485—488.

Oros, G., and Sule, S.: Antibacterial activity of calixin on phytopathogenic bacteria. In: Lyr, H., and Polter, C. (Eds.): System. Fungizide u. Antifungale Verbindungen. Abh. Akad. Wiss. DDR Abt. Math., Naturwiss., Techn., N 1. Akademie-Verlag, Berlin 1983, S. 417—420.

Otto, S., and Drescher, N.: The behaviour of tridemorph in soil. Proc. 7th Brit. Insecticide and Fungicide Conf. (1973): 57—64.

— and POMMER, E.-H.: Über das systemische Verhalten von Tridemorph (Untersuchungen zur Translokation von ^{14}C-markiertem Tridemorph an Gerste, Gurken und Bananen). Meded. Fak. Landbouwwet. Gent **38** (1973): 1493—1506.

POMMER, E.-H.: Chemical structure-fungicidal activity relationships in substituted morpholines. Pesticide Sci. **15** (1984): 285—295.

— and HIMMELE, W.: Die Bekämpfung wichtiger Getreidekrankheiten mit Fenpropemorph, einem neuen Morpholinderivat. Meded. Fak. Landbouwwet. Gent **44**/2 (1979): 499—510.

— and KRADEL, J.: Substituierte Dimethylmorpholin-Derivate als neue Fungizide zur Bekämpfung echter Mehltaupilze. Meded. Fak. Landbouwwet. Gent **32** (1967 a): 735—744.

— — Beiträge zur Wirkungsweise von Tridemorph gegen *Erysiphe graminis*. Meded. Fak. Landbouwwet. Gent **36**/1 (1971): 120—125.

— OTTO, S., and KRADEL, J.: Some results concerning the systemic action of tridemorph. Proc. 5th Brit. Insecticide and Fungicide Conf. (1969): 347—353.

RATHMELL, W. G., and SKIDMORE, A. M.: Recent advances in the chemical control of cereal rust diseases. Outlook on Agriculture **11** (1982): 37—43.

SAUR, R., HOPP, H., SIEGEL, H., and RISCH, W.: Fenpropimorph, ein neues Fungizid zur Bekämpfung von Getreidekrankheiten — dreijährige Versuchsergebnisse aus dem Freiland. — Mitt. Biol. Bundesanstalt für Land- und Forstwirtsch., Berlin-Dahlem. **191** (1979): 183.

— and LÖCHER, F.: Untersuchungen von Umwelteinflüssen auf die Anwendung von Fenpropimorph gegen *Erysiphe graminis* an Weizen. Mitt. Biol. Bundesanstalt für Land- und Forstwirtsch. Berlin-Dahlem Heft 223—244. Dt. Pflanzenschutz-Tag. Gießen (1984): 213.

SCHLÜTER, K., and WELTZIEN, H. C.: Ein Beitrag zur Wirkungsweise systemischer Fungizide auf *Erysiphe graminis*. Mededelingen Fak. Landbouwwet. Gent **36** (1971): 1159—1164.

SCHROEDER, D.: Effect of agrochemical compounds on the decomposition of straw and allulose in soil. Z. Pflanzenernähr. u. Bodenkunde **142** **(4)** (1979): 616—625.

SEIDL, V., and LANSKY, M.: Subsidiary effects of the antioidial fungicide Fademorph EC-20. Ved. Prace Ovocnarske. **8** (1981): 117—128.

SIDDIQUI, K. A., and HAAHR, V.: Different reactions of wheat mutants to a systemic fungicide. Naturwissenschaften **58** (1971): 415—416.

SIEBER, K., LINK, V., and KÖNNIG, M.: Zum Rückstandsverhalten des Mehltaufungizids Falimorph. Nachr.-Bl. Pflanzenschutz DDR (1981): 135—137.

SWIECH, T.: Effectiveness of Fademorf EK-20 against *Sphaerotheca pannosa* Lev. var rosae Wor. and *Diplocarpon rosae* Wolf in roses. Agrochemia. **21** (1981): 53—55.

TRAN VAN CANH: Lutte contre le *Fomes:* Nouvelle methode d'etude. Caoutchoucs et Plastiques Nr. 617/618 (1982).

— A new method of direct control of *Rigidoporus lignosus* causal agent of white root disease of *Hevea*. Intern. Rubber Conf., 17.—19.th Sept. 1984, Colombo.

VENKATA RAM, C. S.: Calixin, a systemic fungicide effective against blister blight. Pesticides **8** (5) (1974): 21—25.

WALMSLEY-WOODWARD, D., LAWS, F. A., and WHITTINGTON, W. J.: Studies on the tolerance of *Erysiphe graminis* f. sp. hordei to systemic fungicides. Ann. appl. Biol. **92** **(2)** (1979 a): 199 to 202.

— — — The Characteristics of isolates of *Erysiphe graminis* f. sp. hordei varying in response to tridemorph and ethirimol. Ann. appl. Biol. **92** **(2)** (1979 b): 211—219.

YEOH, C. S., and TAN, A. M.: Natural rubber latex formulation for controlling pink disease. Proc. Rubber Res. Institute Malays. Plant Conf. (1974): 171—177.

ZIMMERMANN, G.: Effect of systemic fungicides on different entomopathogenic fungi imperfecti in vitro. Nachr.-Bl. dt. Pflanzenschutzdienstes **27** **(8)** (1975): 113—117.

— and BASEDOW, T.: Field tests on the effect of fungicides on the mortality of cereal aphids caused by *entomophthoraceae* (Zygomycetes). Z. Pfl.-Krankh. u. Pflanzenschutz **87** **(2)** (1980): 65—72.

ZOBRIST, P., COLOMBO, V. E., and BOHNEN, K.: Action of fenpropimorph on exterior structures of *Puccinia coronata* on oat as revealed by scanning electron microscopy. Phytopathol. Z. **105** (1982): 11—19.

Chapter 12

Mechanism of action of morpholine fungicides

A. KERKENAAR
Netherlands Organization of Applied Scientific Research, TNO Institute of Applied Chemistry, Zeist, The Netherlands

Introduction

Nowadays it has been generally accepted that morpholine fungicides like tridemorph dodemorph, fenpropimorph and Ro 14—4767/002 act primarily on sterol biosynthesis in fungi but also in higher plants.

Many different modes of action have been proposed before. The earliest work on the mode of action of tridemorph indicated a primary effect on the respiratory electron transport chain (BERGMANN 1979; BERGMANN *et al.* 1977; MÜLLER and SCHEWE 1975, 1976; RAPOPORT 1974). However, this has been questioned by several authors (FISHER 1974; GARG and MEHROTRA 1975; KERKENAAR and KAARS SIJPESTEIJN 1979; KATO *et al.* 1980). Primary effects have been proposed furthermore on cell membrane function in fungi (KAARS SIJPESTEIJN 1972) and higher plants (BUCHENAUER 1975), on protein synthesis (FISHER 1974) and lipid biosynthesis (POLTER and CASPERSON 1979).

The first indications that tridemorph might be an inhibitor of sterol biosynthesis came from the observation that the compound had many features in common with the sterol demethylation inhibitors. Similarities were found in the antimicrobial spectrum, the morphology, the alleviating effects of unsaturated lipophilic compounds and the alterations in the neutral lipid pattern (KERKENAAR *et al.* 1979), together with the occurrence of cross-resistance between tridemorph and sterol C-14 demethylation inhibitors (BARUG and KERKENAAR 1979, 1984). Inhibition of sterol biosynthesis was already reported by LEROUX and GREDT (1978) but this effect was attributed to inhibition of protein synthesis.

This chapter considers the various morpholines in relation to growth, morphology and biochemical target sites in fungi and higher plants and their effects on organelles, cell-free systems and enzymes.

Effects on growth, morphology and biochemical activities in whole cells and organisms

A. Fungi

Growth

Morpholine fungicides act against many fungi, but are particularly effective against powdery mildews where they interfere with haustorial development (HOLLOMON 1982; SCHLÜTER and WELTZEIN 1971). In non-obligate fungi *in vitro* morpholines have little or no effect on fungal growth until after one or more cell doublings have occurred. Spore germination is not inhibited by concentrations which are ultimately lethal or strongly inhibitory to hyphal growth (Kato *et al.* 1980; Kerkenaar *et al.* 1979, 1984; Kerkenaar 1983; Kerkenaar and Barug 1984). In *Ustilago maydis* only a moderate reduction of dry-weight increase is evident after 6—8 h of exposure to tridemorph or fenpropimorph, whereas striking evidence for inhibition of sporidial multiplication is obtained after 2—4 h (Kerkenaar *et al.* 1979; Kerkenaar 1983). Dilution of the terminal sterol, e.g. ergosterol, fucosterol or brassicasterol depending on the species, is probably required before the effect on growth becomes evident. The inhibition of dry-weight increase is accompanied by the first visible irregular deposition of cell wall material (Kerkenaar and Barug 1984).

Morphology and cytology

Treatment of *Ustilago maydis* sporidia in liquid cultures with morpholine fungicides gives rise to production of large sporidia which are multicellular and frequently branched (Kerkenaar *et al.* 1979; Kerkenaar 1983). Each cell contains only one nucleus (Kerkenaar unpublished results).

Treatment of *Botrytis cinerea*, *Botrytis allii* and *Penicillium italicum* (Kato *et al.* 1980; Kerkenaar *et al.* 1979; Kerkenaar 1983) lead to short excessively branched germ tubes. Mycelium of *Penicillium italicum*, treated with fenpropimorph, shows much enlarged hyphal diameters and relatively short distances between septa (Kerkenaar and Barug 1984).

Fenpropimorph causes an irregular deposition of β-1,3 and β-1,4 polysaccharides, probably chitin, in *Ustilago maydis* and *Penicillium italicum*. However, this phenomenon shows up much later than the accumulation of sterols different from the terminal sterol, ergosterol.

Respiration

Tridemorph does not inhibit respiration of whole cells of *Ustilago maydis*, *Saccharomyces cerevisiae*, *Torulopsis candida*, *Botrytis allii*, *Cladosporium cucumerinum* and *Botrytis cinerea* (Kato *et al.* 1980; Kerkenaar *et al.* 1979). Fenpropimorph does not inhibit respiration of *Ustilago maydis* (Kerkenaar *et al.* unpublished results). In previous publications it has been claimed that tridemorph inhibits fungal respiration (Bergmann *et al.* 1975; Bergmann 1979). However, Kerkenaar and Kaars Sijpesteijn (1979) demonstrated that these effects were probably obtained with the formulated product Calixin, at least in tests with whole cells. The synergistic effect of antimycin A or rotenone and tridemorph, observed by Bergmann *et al.* (1979) and by Kerkenaar *et al.* (unpublished results), is at least partly due to its tenside character. In ETP of beef hearts NADH-oxidase and succinate cytochrome c reductase are strongly inhibited by pure tridemorph in confirmation of the results of Müller and Schewe (1975) (unpubl. results).

Protein, RNA and DNA syntheses

Tridemorph has little or no effect on protein or RNA syntheses in either *Botrytis cinerea* or *Ustilago maydis* at concentrations which strongly inhibit ergosterol biosynthesis (Kato *et al.* 1980; Kerkenaar *et al.* 1979). Moreover, tridemorph does not inhibit protein synthesis in a cell-free system prepared from powdery mildew conidia (Hollomon 1982). Inhibition of DNA synthesis by tridemorph has been reported within 2 h of treatment (Kerkenaar and Kaars Sijpesteijn 1979).

Lipid synthesis

Most of the literature concerns the effects of morpholines on fungal sterol biosynthesis. Because of the complexicity of the material each compound will be dealt with separately.

Tridemorph

Treatment of *Botrytis cinerea* with tridemorph hardly inhibits incorporation of [^{14}C]acetate into the various lipid fractions (Kato *et al.* 1980). Analysis of the 4-desmethylsterol fraction reveals large differences between control and treated mycelium. The amounts of ergosterol and episterol were reduced and there was an accumulation of fecosterol, ergosta-8,22,24(28)-trien-3β-ol and ergosta-8,22-dien-3β-ol (Fig. 12.1). Inhibition of $\Delta^8 \rightarrow \Delta^7$ isomerization causes the accumulation of these sterols retaining the C8 (9) double bond.

Treatment of *Ustilago maydis* with tridemorph inhibits the incorporation of [^{14}C] acetate into desmethyl sterols and increases the incorporation into C4-desmethyl

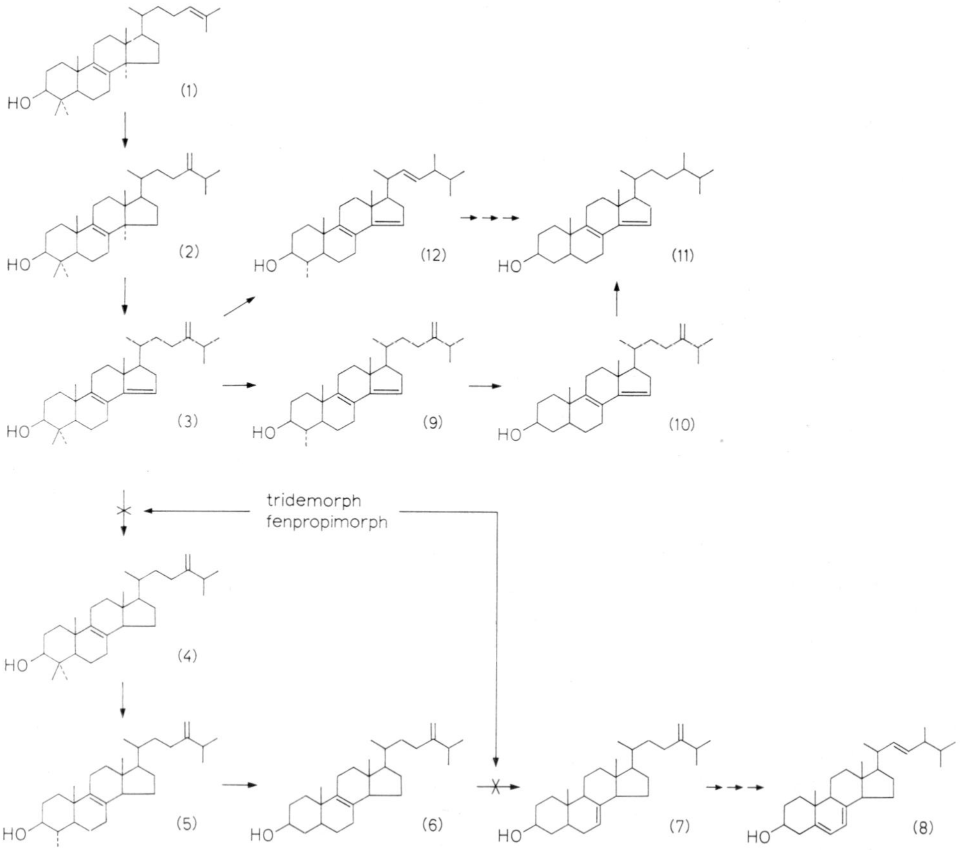

Fig. 12.1 Biosynthetic pathway of ergosterol indicating the presumed sites of inhibition by tridemorph and fenpropimorph. (1) Lanosterol, (2) 24-methylenedihydrolanosterol, (3) 4,4-dimethylergosta-8,14,24(28)-trien-3β-ol, (4) 4,4-dimethylergosta-8,24(28)-dien-3β-ol, (5) 4α-methylergosta-8,24(28)-dien-3β-ol, (6) fecosterol, (7) episterol, (8) ergosterol, (9) 4α-methylergosta-8,14,24(28)-trien-3β-ol, (10) ergosta-8,14,24(28)-trien-3β-ol, (11) ignosterol. (12) 4α-methylergosta-8,14,22-trien-3β-ol.

sterols and free fatty acids (Kerkenaar *et al.* 1979). Accumulation of sterols retaining the Δ^8 as well as the Δ^{14} double bonds leads to the concept of inhibition of sterol Δ^{14} reductase (Kerkenaar *et al.* 1981). Because the Δ^8 sterols accumulating in *Botrytis cinerea* have been identified only by mass spectroscopy, and misinterpretation due to the similarities in mass spectra of fecosterol and ignosterol is rather easy, it has been suggested that their identification was wrong. Moreover, evidence has been presented that inhibition of Δ^{14} reductase is more likely to be harmful to fungal growth than inhibition of the $\Delta^8 \rightarrow \Delta^7$ isomerase (Kerkenaar *et al.* 1981). The picture is further complicated by the observations that in *Saccharomyces cerevisiae* treatment with tridemorph leads to the accumulation of probably fecosterol (Kato 1983). The identity of fecosterol was confirmed by proton-NMR spectroscopy. The proton absorption of the olefinic hydrogen at C-15 is absent (Kato 1983). This result indicates that the $\Delta^8 \rightarrow \Delta^7$ isomerase is inhibited. However, in a recent study of inhibition of sterol biosynthesis in tridemorph-treated *Saccharomyces cerevisiae* it is shown that about 40% of the total sterol present retains the Δ^8 as well a the Δ^{14} double bonds, while about 30% represents fecosterol. These results indicate inhibition of the sterol Δ^{14} reductase as well as of the $\Delta^8 \rightarrow \Delta^7$ isomerase in *Saccharomyces cerevisiae* (Baloch *et al.* 1984a, 1984b). On the other hand, in *Ustilago maydis* tridemorph only causes the accumulation of Δ^8 sterols; sterols retaining the Δ^8 as well as the Δ^{14} double bonds are absent (Baloch *et al.* 1984a, 1984b). Leroux and Gredt (1983) give similar results for *Ustilago maydis*. For *Saccharomyces cerevisiae, Botrytis cinerea* and *Penicillium expansum*, after tridemorph treatment, they also report the accumulation of only Δ^8 sterols. In budding fungi, like *Ustilago maydis* and *Saccharomyces cerevisiae*, fecosterol and ergosta-8-en-3β-ol accumulate, whereas in filamentous fungi, like *Botrytis cinerea* and *Penicillium expansum*, fecosterol and ergosta-8,22,24(28)-trien-3β-ol accumulate.

Tridemorph treatment of *Saprolegnia ferax* leads to inhibition of the terminal Δ^5 sterols: fucosterol, 24-methylenecholesterol, desmosterol and cholesterol. The following sterols accumulate: zymosterol, fecosterol and stigmasta-8,24(28)-dien-3β-ol (Berg *et al.* 1983; Berg 1983). Similar results are obtained using another four species from the order of the *Saprolegniales: Achlya americana, Dictyuchus monosporus, Apodachlyella completa* and *Lagenidium callinectes* (Berg 1983). *Achlya radiosa*, treated with tridemorph, accumulates mainly stigmasta-8,24(28)-dien-3β-ol (Kerkenaar *et al.* unpublished results).

Finally, tridemorph treatment of *Pyricularia oryzae* causes inhibition of sterol biosynthesis and leads to accumulation of ergosta-5,8-dien-3β-ol (Berg *et al.* 1984; Berg 1984).

These results indicate inhibition of the $\Delta^8 \rightarrow \Delta^7$ isomerase as the most frequently cited site of inhibition by tridemorph.

Fenpropimorph

Fenpropimorph treatment leads to the accumulation of ignosterol in *Ustilago maydis* and of ergosta-8,14,24(28)-trien-3β-ol in *Penicillium italicum* (Kerkenaar 1983; Kerkenaar *et al.* 1984). This suggests that fenpropimorph inhibits the $\Delta^{8,14}$ sterol reductase. However, when *Ustilago maydis* is treated with fenpropimorph at low oxygen pressure, Δ^8 sterols accumulate (Kerkenaar, 1983). This suggests that under these conditions the $\Delta^8 \rightarrow \Delta^7$ isomerase can also be a target of fenpropimorph.

In *Ustilago maydis* like in *Saccharomyces cerevisiae* fenpropimorph has been reported to inhibit the $\Delta^8 \rightarrow \Delta^7$ isomerase, but not the sterol Δ^{14} reductase, due to the accumulation of only fecosterol and ergosta-8-en-3β-ol. The same applies to treatment by fenpropimorph of *Botrytis cinerea* and *Penicillium expansum*, where accumulation of fecosterol and probably ergosta-8,22,24(28)-trien-3β-ol takes place (Leroux and

GREDT 1983). However, BALOCH *et al.* (1984) have reported that fenpropimorph inhibits Δ^{14} reductase along with the $\Delta^8 \rightarrow \Delta^7$ isomerase in *Saccharomyces cerevisiae* and *Ustilago maydis* (BALOCH *et al.* 1984a, 1984b). Accumulation of 4,4-dimethylcholesta-8,14,24-trien-3β-ol, 4,4-dimethylcholesta-8,14-dien-3β-ol and ignosterol in *Saccharomyces cerevisiae* leads to the concept of inhibition of sterol Δ^{14} reduction. More than 60% in the fenpropimorph-treated cells retains the Δ^8 as well as the Δ^{14} double bonds; only 2% retains the Δ^8 double bond in the form of fecosterol. In *Ustilago maydis* about 30% of the sterols retains the Δ^8 as well as the Δ^{14} double bonds in the form of ignosterol, whereas about 52% of the desmethyl sterols is ergost-8-en-3β-ol and ergosta-8,22-dien-3β-ol (BALOCH *et al.* 1984a, 1984b).

In *Pyricularia oryzae* fenpropimorph treatment causes the accumulation of ignosterol indicating the inhibition of the sterol Δ^{14} reductase (BERG 1983; BERG *et al.* 1983).

These results indicate inhibition of the sterol Δ^{14} reductase as the most frequently cited site of inhibition by fenpropimorph.

Dodemorph

Only one publication deals with the effect of dodemorph on sterol biosynthesis. This morpholine compound inhibits sterol biosynthesis, causing the accumulation of fecosterol and ergosta-8-en-3β-ol in *Ustilago maydis* and *Saccharomyces cerevisiae* and fecosterol and ergosta-8,22,24(28)-trien-3β-ol in *Botrytis cinerea* and *Penicillium expansum* (LEROUX and GREDT 1983). These results indicate the inhibition of $\Delta^8 \rightarrow \Delta^7$ isomerase.

Ro 14—4767/002

Ro 14—4767/002 induces an increase of the total sterol content of cells of *Candida albicans,* due to accumulation of sterols not present in control cells. The main sterol that accumulates is ignosterol indicating an inhibition of Δ^{14} reductase. However, also inhibition of the $\Delta^8 \rightarrow \Delta^7$ isomerase is assumed (POLAK-WYSS *et al.* 1985).

Other fungicides containing a morpholine ring

Besides the above mentioned fungicides, the following fungicides contain a morpholine ring in their structure: trimorphamide and aldimorph. Data on the mode of action of these compounds are lacking. However, trimorphamide is probably a C-14-demethylation inhibitor (LEROUX personal communication) and aldimorph, being closely related to tridemorph, is considered to act in the same way as tridemorph. It will be interesting to study the effect of trimorphamide in a mutant with an insensitive C-14-demethylation site.

B. Plants and tissue cultures

Growth and miscellaneous effects excluding lipid synthesis

Growth of bramble cells *(Rubus fruticosus)* is only slightly reduced after tridemorph (10 μg/ml) treatment (SCHMITT *et al.* 1981). At lower concentrations growth of cells is not significantly reduced (SCHMITT *et al.* 1981, 1982). Fenpropimorph is completely ineffective (BENVENISTE *et al.* 1984). Tridemorph and dodemorph induce a drastic leakage of betacyanin from discs of beet roots and an efflux of electrolytes from discs of bean leaves and sections of barley leaves (BUCHENAUER 1975). A pretreatment of barley seedlings by root drench with tridemorph and dodemorph results in an increased leakage of electrolytes from sections of primary leaves (BUCHENAUER 1975).

A mitodepressive effect of tridemorph on the meristematic cells of *Allium cepa* has been observed. This effect is dosage dependent. Tridemorph has been revealed as a strong C-mitotic agent (CORTÉS *et al.* 1982).

Tridemorph as Calixin has been reported to cause phytotoxicity in barley, wheat and cucumber, depending on the concentration (BERGMANN *et al.* 1982). Calixin causes chlorosis in wheat (*Triticum aevistivum*), varieties Indus 66, Nayab and C 591 (SIDDIQUI and HAAHR 1971). Tridemorph and fenpropimorph treatment of maize seedlings lead to growth inhibition even at the lowest concentration (1 μg/ml). Higher concentrations resulted in higher inhibition. Spikes and ears are formed after 2—3 months, but they are smaller than the same parts from control plants (BENVENISTE *et al.* 1984).

Growth as measured by dry-weight production in carrot, tobacco and soybean cultures is inhibited by tridemorph (HOSOKAWA *et al.* 1984).

Lipid synthesis

Tridemorph

In bramble cell suspensions tridemorph (10 μg/ml) induces an increase of the tota quantity of sterols. A strong accumulation of 9β,19-cyclopropyl sterols and the disappearance of Δ^5 sterols, normally present in untreated cells, is observed. The major sterols accumulating in treated cells are cycloeucalenol and 24-methylenepollinastanol (SCHMITT *et al.* 1981). Tridemorph inhibits probably the cycloeucalenol-obtusifoliol isomerase in bramble cells (Fig. 12.2).

Treatment with tridemorph modified drastically the sterol composition of maize seedlings (BLADOCHA and BENVENISTE 1983). Various 9β,19-cyclopropyl sterols accumulate after treatment with 20 μg/ml of tridemorph. The major sterols of the treated plants are 24-methylpollinastanol, 24-dehydrocycloeucalenol and cycloeucalenol. When the fungicide is applied at 5 μg/ml, the initial Δ^5 sterols of the leaves are replaced by both cyclopropyl sterols and Δ^8 sterols (Fig. 12.2) indicating that besides the cycloeucalenol-obtusifoliol isomerase also the $\Delta^8 \rightarrow \Delta^7$ isomerase is inhibited (BLADOCHA and BENVENISTE 1983; BENVENISTE *et al.* 1984).

HOSOKAWA *et al.* (1984) reported the accumulation of sterols with a cyclopropane ring upon treatment of carrot, tobacco and soybean cultures; about 33 % of total sterol in carrot, 58 % in tobacco and 49 % in soybean. An increase in the concentration of cyclopropyl sterols results from the accumulation of cycloeucalenol, 24-methylenecycloartanol, 24-methylenepollinastanol and 24-methylpollinastanol in tridemorph-treated cultures. Tridemorph appears to be an effective inhibitor of the cycloeucalenol-obtusifoliol isomerase in carrot, tobacco and soybean cells.

The other major tridemorph-induced modification related to the sterol composition is that tridemorph causes a reduction of sterols with a 10-carbon side chain and an increase of sterols with a 9-carbon side chain. In addition, the level of 24-methylene sterols increases in the fungicide-treated cultures, especially in carrot and tobacco cultures. Tridemorph appears to be an effective inhibitor of the second alkylation at carbon 24 (HOSOKAWA *et al.* 1984).

A minor effect of tridemorph may be to prevent the removal of 14α-methyl groups in soybean cultures. The other secondary effect of the fungicide is on the accumulation of sterols with $\Delta^{5,24(28)}$ double bonds, indicating the inhibition of the $\Delta^{24(28)}$ double bond reduction (HOSOKAWA *et al.* 1984).

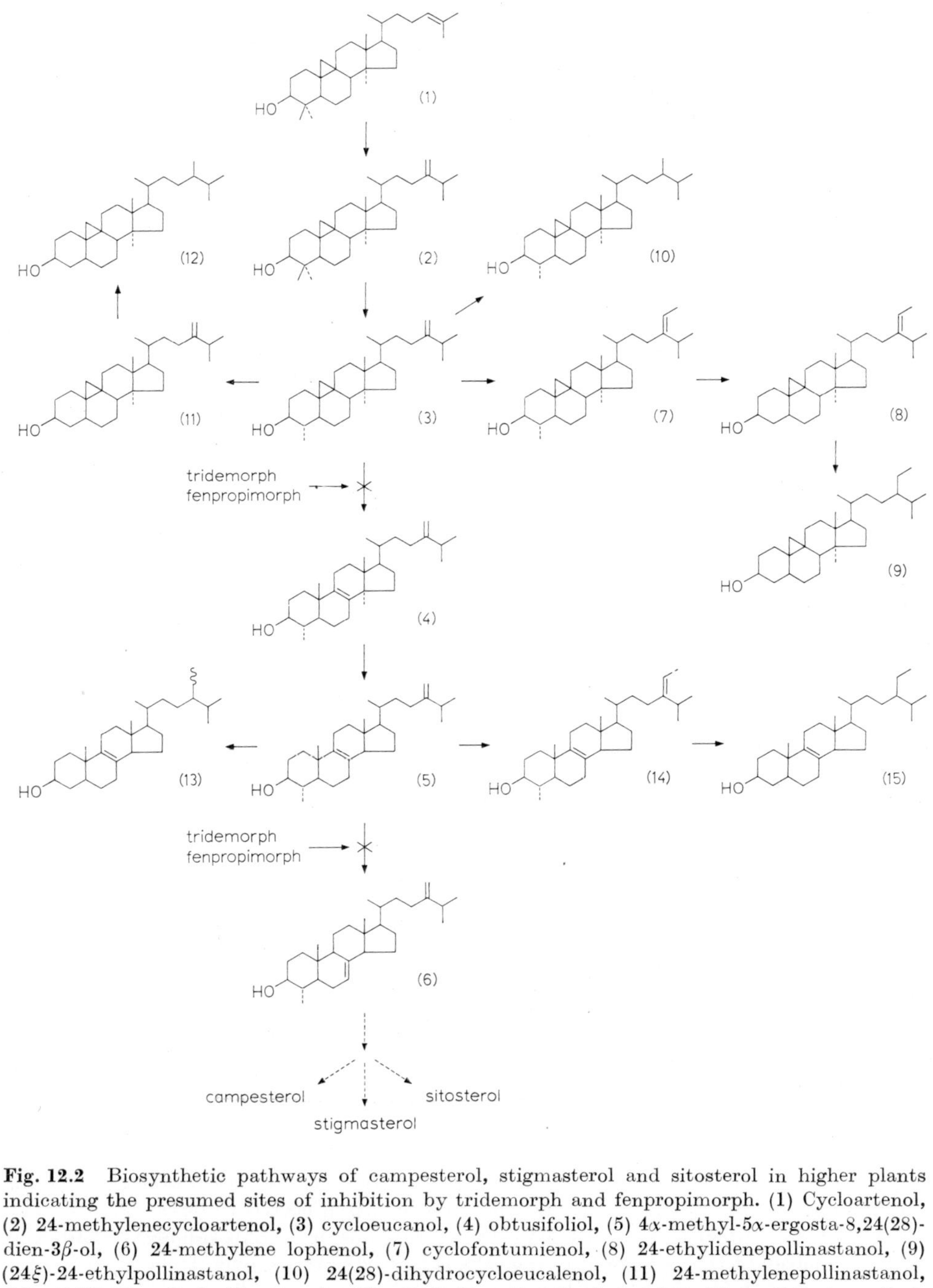

Fig. 12.2 Biosynthetic pathways of campesterol, stigmasterol and sitosterol in higher plants indicating the presumed sites of inhibition by tridemorph and fenpropimorph. (1) Cycloartenol, (2) 24-methylenecycloartenol, (3) cycloeucanol, (4) obtusifoliol, (5) 4α-methyl-5α-ergosta-8,24(28)-dien-3β-ol, (6) 24-methylene lophenol, (7) cyclofontumienol, (8) 24-ethylidenepollinastanol, (9) (24ξ)-24-ethylpollinastanol, (10) 24(28)-dihydrocycloeucalenol, (11) 24-methylenepollinastanol, (12) (24**R**)-24-methylpollinastanol, (13) (24ξ)-24-methyl-5α-cholest-8-en-3β-ol, (14) 4α-methyl-5α-stigmasta-8,Z-24(28)-dien-3β-ol, (15) (24**R**)-24-ethyl-5α-cholest-8-en-3β-ol.

Fenpropimorph

Fenpropimorph treatment of maize seedlings causes the accumulation of 9β,19-cyclopropyl sterols in roots and leaves (Benveniste *et al.* 1984). When plants are treated with low concentrations (1 μg/ml) of fenpropimorph, Δ^8 sterols are detected in addition to cyclopropyl sterols. Thus fenpropimorph, like tridemorph, acts on the cycloeucalenol-obtusifoliol isomerase and the $\Delta^8 \rightarrow \Delta^7$ isomerase (Benveniste *et al.* 1984).

Fenpropimorph treatment of maize seedlings results in the presence of 9β,19-cyclopropyl sterols containing over 90 % of 24-methyl and only 3 % of 24-ethyl sterols in both roots and leaves. In both organs of control plants over 75 % of the 24-ethyl sterols is present in the Δ^5 sterols (Benveniste *et al.* 1984).

Effects on organelles and enzymes of fungi and plants

The effects of tridemorph on three membrane fractions, prepared from roots of maize seedlings after treatment, are reported by Benveniste *et al.* (1984). One fraction consists of crude mitochondria, the other two are considered to be enriched in endoplasmic reticulum and plasmalemma, respectively. More than 98 % of the total sterols in membranes from control plants are Δ_5 sterols, whereas in membranes from treated plants less than 7 % is present in this form. On the other hand, cyclopropyl sterols which are not present in control membranes, accounted for more than 90 % of the total sterols in membranes from treated plants. In addition, the sterol profile was almost identical in the three membrane fractions from treated plants and very similar to the sterol profile of the whole tissue (Benveniste *et al.* 1984).

When microsomes from maize seedlings are incubated in the presence of cycloeucalenol with various amounts of tridemorph, the fungicide strongly inhibits the cycloeucalenol-obtusifoliol isomerase ($I_{50} = K_I = 0.5$—1 μM; K_m for cycloeucalenol = 100 μM) (Rahier *et al.* 1983; Schmitt *et al.* 1982). The cycloeucalenol-obtusifoliol isomerase has a higher affinity for tridemorph than for its substrate cycloeucalenol. Fenpropimorph also strongly inhibits the cycloeucalenol-obtusifoliol isomerase in a cell-free extract from maize seedlings (Benveniste *et al.* 1984; Rahier *et al.* 1983).

Tridemorph has been reported to inhibit both the NADH-oxidase and the succinate cytochrome C oxydoreductase system of non-phosphorylating electron-transfer particles from bovine heart mitochondria (Müller and Schewe 1976). Reinvestigations with tridemorph, double distilled, and investigations with fenpropimorph using this system reveal that especially the NADH-oxidase is fairly sensitive, although sterol synthesis is inhibited at lower concentrations (Lyr *et al.* unpublished results). The EC_{50} of tridemorph is 28 μM for succinate cytochrome C oxydoreductase and 16 μM for NADH-oxidase. For fenpropimorph these values are 33 and 8 μM, respectively.

Tridemorph as Calixin inhibits the chitin synthetase of *Mucor rouxii in vitro* (Lyr and Seyd 1978). At 10^{-4} M of tridemorph the incorporation of UDP-^{14}C-glucosamine in chitin is 59 % of the control. The EC_{50} is reported to be 5×10^{-4} M. The inhibition is ascribed to the lytic activity of tridemorph, due to the alkyl chain and positive charge of the morpholine ring. Phospholipids are believed to be essential for the activity of the enzyme, and lysis leads to inactivation of chitin synthetase.

Recently Mercer *et al.* (unpublished results) have described the effects of morpholines on the sterol Δ^{14} reductase using microsomes of *Saccharomyces cerevisiae*. The yeast S_8 preparation from anaerobically grown yeast has been used in 0.1 M phosphate buffer, pH 6.2. Several concentrations of the fungicides are tested. The substrate is radiolabelled ignosterol and the product ergost-8-en-3β-ol. The radioactivity in ignosterol and ergost-8-en-3β-ol in the control incubation is used to determine the % conversion of substrate to product. Taking the % conversion in the control as 100 %, the degree of inhibition in the fungicide-containing samples is calculated. Determination of the I_{50} values from dose-response curves reveals that fenpropimorph is a much more potent inhibitor of the Δ^{14} reductase than tridemorph (100-fold) and that fenpropidin with the same N-alkyl substituents, but a different N-heterocycle as fenpropimorph, has an almost identical I_{50} value as fenpropimorph. From these observations they conclude that the Δ^{14} reductase inhibitory activity of these three fungicides resides principally in the N-alkyl substituent rather than in the heterocyclic ring.

Discussion

Although various authors have demonstrated that morpholines strongly inhibit sterol biosynthesis, it is still questioned whether this inhibition is the sole basis of toxicity in various organisms. Inhibition of growth in fungi and higher plants may not only result from the effect of the drug on the sterol profile of membranes but also from side effects. NADH-oxidase and succinate cytochrome C reductase are sensitive to inhibition (Lyr *et al.* unpublished results). In chloroplasts a repression of the netto assimilation in wheat leaves is observed which cannot be explained by inhibition of sterol synthesis (Lyr *et al.* unpublished results).

Gram-positive bacteria like *Bacillus subtilis*, *Streptococcus lactis*, *Streptococcus faecalis* and *Mycobacterium phlei*, are sensitive to tridemorph (Kerkenaar *et al.* 1979). Tridemorph as Calixin is also active against the phytopathogenic bacteria *Corynebacterium michiganense*, *Corynebacterium fascians*, and other *Corynebacterium* spp. *in vitro* (Oros and Süle 1983).

Since sterols are of limited occurrence in most bacteria (Ourisson and Rohmer 1982), it is difficult to believe that the antibacterial activity of tridemorph originates from interference with sterol biosynthesis.

The effects, altered morphology and inhibition of dry-weight increase, observed on fungal growth, are preceded by an irregular deposition of polysaccharides, probably chitin. These effects, however, appear much later than the observed effects on sterol biosynthesis. The irregular deposition of probably chitin is contradictory to the reported inhibition of chitin synthetase from *Mucor*. The latter inhibition is suggested to be a consequence of phospholipid breakdown, whereas the former effects are attributed to the altered sterol pattern. These and other effects based on membrane changes could account for a lethal action in filamentous fungi and a highly detrimental action in budding fungi, due to differences in cell wall composition (Kerkenaar and Barug 1984).

In fungi two sites of action have been indicated: the $\Delta^8 \rightarrow \Delta^7$ isomerase and the Δ^{14} reductase. Also in higher plants two sites have been described: the $\Delta^8 \rightarrow \Delta^7$ isomerase and the cycloeucalenol-obtusifoliol isomerase (Fig. 12.3). In the case of tridemorph or fenpropimorph and fungi the inhibition site varies from organism to organism and seems to be dependent on culturing conditions. In plants the inhibition site may vary from organism to organism and organ to organ. If the cycloeucalenol-obtusifoliol isomerase or the Δ^{14} reductase is completely inhibited, inhibition of the $\Delta^8 \rightarrow \Delta^7$ isomerase cannot be detected.

The differences existing between plants and fungi could be explained partly in considering that the cycloeucalenol-obtusifoliol isomerase is present in photosynthetic eukaryotes and has never been found in non-photosynthetic eukaryotes (Benveniste *et al.* 1984; Berg 1983; Berg *et al.* 1983). On the other hand, it is assumed that in

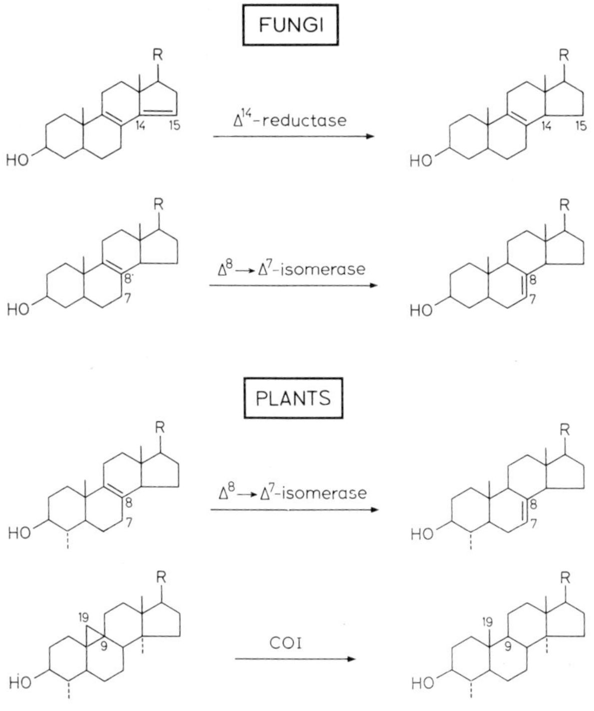

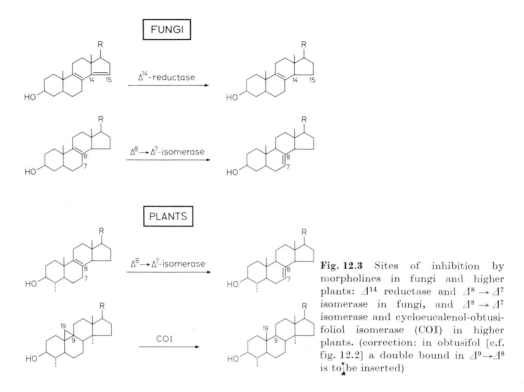

Fig. 12.3 Sites of inhibition by morpholines in fungi and higher plants: Δ^{14} reductase and $\Delta^8 \rightarrow \Delta^7$ isomerase in fungi, and $\Delta^8 \rightarrow \Delta^7$ isomerase and cycloeucalenol-obtusifoliol isomerase (COI) in higher plants. (correction: in obtusifol [c.f. fig. 12.2] a double bound in $\Delta^9 \rightarrow \Delta^8$ is to be inserted)

higher plants the sterol Δ^{14} reductase is also active as in fungi (Bottema and Parks 1978) and mammals (Paik *et al.* 1984). Surprisingly, this enzyme has not been reported to become inhibited, whereas $\Delta^8 \rightarrow \Delta^7$ isomerase becomes inhibited. The enzymes Δ^{14} reductase, cycloeucalenol-obtusifoliol isomerase and $\Delta^8 \rightarrow \Delta^7$ isomerase show some similarities. It has been suggested that $\Delta^8 \rightarrow \Delta^7$ isomerization reaction starts with a Δ^8 hydrogenation (Benveniste *et al.* 1984; Berg 1983; Berg *et al.* 1983). This hydrogenation is assumed to be an NADPH/H+-dependent reduction of the Δ^8 double bond, similar to the NADPH/H+-dependent reduction of the Δ^{14} double bond. Also the ring opening of the 9β,19-cyclopropane ring could depend on such a mechanism. Benveniste *et al.* (1984) propose the formation of a carbonium ion high energy intermediate upon addition of a proton at C-14 for the reductase, C-9 for the cycloeucalenol-obtusifoliol isomerase and C-8 for the $\Delta^8 \rightarrow \Delta^7$ isomerase. These carbonium ions are very close to each other (less than 0.15 nm). It is known that N-alkyl morpholines (pK_a around 7—8) lead to the formation of ammonium (morpholinium) ions at pH 7.4. They suggest that morpholinium ions, having electronic and broad structural similarities with the three carbonium ion high energy intermediates, would have a high affinity for the three enzymes and would be able to inhibit them. The likelihood of this hypothesis has been verified recently by these researchers (Rahier *et al.* manuscript in preparation).

The inhibition of the NADH-oxidase may fit in the concept of inhibition of NADPH-oxydoreductases and probably also the effect on the 2,3,5-triphenyltetrazolinium-chloride (TTC) reaction in the TTC test in plants as indicator of phytotoxic effects (Lyr *et al.* unpublished results).

The difference in effect of tridemorph and fenpropimorph on $\Delta^8 \rightarrow \Delta^7$ isomerase and Δ^{14} reductase is believed to be due to the structural differences of these fungicides. It is hypothesized that the 2,6-dimethyl morpholine ring binds more efficiently to the sterol B ring than other rings, resulting in inhibition of the $\Delta^8 \rightarrow \Delta^7$ isomerase. The N-substituent of fenpropimorph gives a better fit and more efficient binding of the N-substituent to the sterol moiety, consisting of the rings C and D plus side chain in the enzyme system than the N-substituent of tridemorph. The observations that *trans*-1,4-cyclohexyl isomers of fenpropimorph are much more potent inhibitors, and presumably sterol Δ^{14} reductase inhibitors, than the *cis*-1,4-isomers, are in agreement with this hypothesis. The *trans*-1,4-isomers have an essentially planar structure that matches the sterol rings C, D and side chain well (BALOCH *et al.* 1984). The validity of this hypothesis has to be proven.

Much work is in progress to elucidate the effects of morpholines on enzymes in cell-free extracts and on purified enzymes. Hopefully this will be the subject of a next edition.

Acknowledgements: The author wishes to thank Drs. B. C. BALDWIN, P. BENVENISTE, D. BERG, L. R. BERG, D. HOLLOMON, P. LEROUX, H. LYR, E. I. MERCER, A. POLAK-WYSS, and H. D. SISLER for cooperation and for kindly providing unpublished results and/or manuscripts in press. Thanks are also due to my colleagues D. BARUG and F. J. RITTER.

References

BALOCH, R. I., MERCER, E. I., WIGGINS, T. E., and BALDWIN, B. C.: Where do morpholines inhibit sterol biosynthesis? Brit. Crop. Protect. Conf. on Pests Dis. vol. 3 (1984 a): 893—898.

— — — — Inhibition of ergosterol biosynthesis by tridemorph, fenpropimorph, and fenpropidin. Phytochemistry **23** (1984 b): 2219—2226.

BARUG, D., and KERKENAAR, A.: Cross-resistance of UV-induced mutants of *Ustilago maydis* to various fungicides which interfere with ergosterol biosynthesis. Meded. Rijksfac. Landbouwwet. Gent **44** (1979): 421—427.

— — Resistance in mutagenic induced mutants of *Ustilago maydis* to fungicides which inhibit ergosterol biosynthesis. Pesticide Sci. **15** (1984): 78—84.

BENVENISTE, P., BLADOCHA, M., COSTET, M.-F., and EHRHARD, A.: Use of inhibitors of sterol biosynthesis to study plasmalemma structure and functions. In: GONDIA, E. M. (Ed.): Membranes and Compartmentization in the Regulation of Plant Functions. Ann. Proc. Phytochem. Soc. Eur. vol. **24**, 1984.

— Biochemical mode of action of fungicides: Studies on ergosterol biosynthesis inhibitors. BURDICK and JACKSON Symp. 188th ACS National Meeting, Philadelphia, Pennsylvania, USA, August 27 (1984): 1—26.

— KRAEMER, W., REGEL, E., BUECHEL, K.-H., HOLMWOOD, G., PLEMPEL, M., and SCHEINPFLUG, H.: Mode of action of fungicides: studies on ergosterol biosynthesis inhibitors. Brit. Crop. Protect. Conf. on Pests Dis. **3** (1984): 887—892.

BERG, L. R.: The effects of triarimol and tridemorph on the sterol biosynthesis of five species of Oomycetes. Ph.D. Thesis, Univ. of Maryland, College Park, MD (1983): 1—43.

— PATTERSON, G. W., and LUSBY, W. R.: Effects of triarimol and tridemorph on sterol biosynthesis in *Saprolegnia ferax*. Lipids **18** (1983): 448—452.

BERGMANN, H.: Die Wirkung von Tridemorph auf *Torulopsis candida*. In: LYR, H., and POLTER, C. (Eds.): Systemfungizide (Int. Symposium Reinhardsbrunn 1974). Akademie-Verlag, Berlin 1975.

— Wirkung von Tridemorph auf *Torulopsis candida*. Z. Allg. Mikrobiol. **19** (1979): 155—162.

— LYR, H., KLUGE, E., and RITTER, G.: Untersuchungen zur Wirkungsweise von Tridemorph. In: LYR, H., and POLTER, C. (Eds.): Systemfungizide (Int. Symp. Reinhardsbrunn 1974). Akademie-Verlag, Berlin 1975.

— MÜLLER, R., and NEUHAUS, W.: Bestimmung der phytotoxischen Wirkung von Calixin auf Kulturpflanzen. In: LYR, H., and POLTER, C. (Eds.): Systemische Fungizide und antifungale Verbindungen (Int. Symp. Reinhardsbrunn 1982). Akademie-Verlag, Berlin 1983, S. 355—359.

Bladocha, M., and Benveniste, P.: Manipulation by Tridemorph, a systemic fungicide, of the sterol composition of maize leaves and roots. Pl. Physiol., Lancaster **71** (1983): 756—762.

Bottema, C. K., and Parks, L. W.: Δ^{14}-sterol reductase in *Saccharomyces cerevisiae*. Biochim. Biophys. Acta **531** (1978): 301—307.

Buchenauer, H.: Various response of cell membranes of plant and fungal cells to different systemic fungicides. Proc. Int. Plant Prot. Congr. Moscov **3** (1975): 94—110.

Cortés, F., Escalza, P., Moreno, J., and López-Campos, J. L.: Effects of the fungicide tridemorph on mitosis in *Allium cepa*. Cytobios. **34** (1982): 181—190.

Fisher, D. J.: A note on the mode of action of the systemic fungicide tridemorph. Pestic. Sci. **5** (1974): 219—224.

Garg, D. K., and Mehrotra, R. S.: Effect of some fungicides on growth, respiration and enzyme activity of *Fusarium solani* f. sp. *pisi*. Z. Pflanzenschutz **10** (1975): 570—576.

Hollomon, D. W.: The effects of tridemorph on barley powdery mildew: its mode of action and cross sensitivity relationships. Phytopathol. Z. **105** (1982): 279—287.

Hosokawa, G., Patterson, G. W., and Lusby, W. R.: Effects of triarimol, tridemorph and triparanol on sterol biosynthesis in carrot, tobacco and soybean suspension cultures. Lipids **19** (1984): 449—456.

Kaars Sijpesteijn, A.: Effects on fungal pathogens. Chapter 7. In: Marsh, R. W. (Ed.): Systemic Fungicides. 2nd ed., Longman, London 1972, p. 153.

Kato, T.: Biosynthetic processes of ergosterol as the target of fungicides. In: Pesticide Chemistry: Human Welfare and the Environment. Proc. of the 5th Intern. Congr. of Pesticide Chem. Kyoto, Vol. 3. Pergamon Press, Oxford 1983, pp. 33—41.

— Shoami, M., and Kawase, Y.: Comparison of Tridemorph with Buthiobate in antifungal mode of action. J. Pesticide Sci. **5** (1980): 69—79.

Kerkenaar, A.: Mode of action of tridemorph and related compounds. In: Pesticide Chemistry: Human Welfare and the Environment. Proc. of the 5th Intern. Congr. of Pesticide Chem., Kyoto, Vol. 3. Pergamon Press, Oxford 1983, pp. 123—127.

— and Barug, D.: Fluorescence microscopic studies of *Ustilago maydis* and *Penicillium italicum* after treatment with imazalil and fenpropimorph. Pesticide Sci. **15** (1984): 199—205.

— — and Kaars Sijpesteijn, A.: On the antifungal mode of action of tridemorph. Pesticide Biochem. Physiol. **12** (1979): 195—204.

— and Kaars Sijpesteijn, A.: On a difference in the antifungal activity of tridemorph and its formulated product Calixin. Pesticide Biochem. Physiol. **12** (1979): 124—129.

— Uchiyama, M., and Versluis, G. G.: Specific effects of tridemorph on sterol biosynthesis in *Ustilago maydis*. Pesticide Biochem. Physiol. **16** (1981): 97—104.

— van Rossum, J. M., Versluis, G. G., and Marsman, J. W.: Effect of fenpropimorph and imazalil on sterol biosynthesis in *Penicillium italicum*. *Pesticide* Sci. **15** (1984): 177—187.

Leroux, P., and Gredt, M.: Effects de quelques fongicides systémiques sur la biosynthèse de l'ergostérol chez *Botrytis cinerea* Pers., *Penicillium expansum* Link et *Ustilago maydis* (DC) Cda. Ann. Phytopathol. **10** (1978): 45—60.

— — Études sur les inhibiteurs de la biosynthèse des stérols fongiques: I. — Fongicides provoquant l'accumulation des desméthylstérols. Agronomie **3** (1983): 123—130.

Lyr, H., and Seyd, W.: Hemmung der Chitin-Synthetase von *Mucor rouxii in vitro* durch Fungizide und andere Wirkstoffe. Z. Allg. Mikrobiol. **18** (1978): 721—729.

Müller, W., and Schewe, T.: Zur Wirkung von Tridemorph auf Elektronentransportpartikel aus Rinderherz-Mitochondrien. In: Lyr, H., and Polter, C. (Eds.): Systemfungizide (Int. Symp. Reinhardsbrunn 1974). Akademie-Verlag, Berlin 1975, S. 189—196.

— — Das Systemfungizid Tridemorph als Hemmstoff der Atmungskette von Elektronentransportpartikeln aus Rinderherzmitochondrien. Acta Biol. Med. Ger. **35** (1976): 693—707.

Oros, G., and Süle, S.: Antibacterial activity of Calixin on phytopathogenic bacteria. In: Lyr, H., and Polter, C. (Eds.): Systemische Fungizide und antifungale Verbindungen (Int. Symp. Reinhardsbrunn 1982). Akademie-Verlag, Berlin 1983, S. 417—420.

Ourisson, G., and Rohmer, M.: Prokaryotic polyterpenes: Phylogenetic precursors of sterols. In: Bronner, F., and Kleinzeller, A. (Eds.): Current Topics in Membranes and Transport, vol. 17: Membrane Lipids of Prokaryotes. Academic Press, New York 1982, pp. 153—182.

Paik, Y.-K., Trzaskos, J. M., Shafiee, A., and Gaylor, J. L.: Microsomal enzymes of cholesterol biosynthesis from lanosterol. Characterization, solubilization, and partial purification of NADPH-dependent $\Delta^{8,14}$-steroid 14-reductase. J. biol. Chem. **259** (1984): 13413—13423.

Polak-Wyss, A., Lengsfeld, H., and Oesterhelt, G.: Effect of oxiconazole and Ro 14-4767/002 on sterol pattern in *Candida albicans*. Sabouraudia (1985): in press.

Polter, C., and Casperson, G.: Einfluß von Tridemorph auf Ultrastruktur und Lipidmetabolismus von *Botrytis cinerea*. In: Lyr, H., and Polter, C. (Eds.): Systemfungizide (Int. Symp. Reinhardsbrunn 1974). Akademie-Verlag, Berlin 1975, S. 225—241.

Rahier, A., Bouvier, P., Cattel, L., Narula, A., and Benveniste, P.: Inhibition of 2,3-oxidosqualene-β-amyrin cyclase, (*S*)-adenosyl-L-methionine-cycloartenol-C-24-methyltransferase and cycloeucalenol-obtusifoliol isomerase by rationally designed molecules containing a tertiary amine function. Biochem. Soc. Trans. **11** (1983): 537—543.

Schlüter, K., and Weltzein, H. C.: Ein Beitrag zur Wirkungsweise systemischer Fungizide auf *Erysiphe graminis*. Meded. Rijksfac. Landbouwwet. Gent **36** (1971): 1159—1164.

Schmitt, P., Benveniste, P., and Leroux, P.: Accumulation of 9β,19-cyclopropyl sterols in suspension cultures of bramble cells cultured with Tridemorph. Phytochemistry **20** (1981): 2153—2158.

— Rahier, A., and Benveniste, P.: Inhibition of sterol biosynthesis in suspension cultures of bramble cells. Physiol. Veg. **20** (1982): 559—571.

Siddiqui, K. A., and Haahr, V.: Different reactions of wheat mutants to a systemic fungicide. Naturwissenschaften **58** (1971): 415—416.

Lyr, H. (Ed.): Modern Selective Fungicides — Properties, Applications, Mechanisms of Action. Longman Group UK Ltd., London, and VEB Gustav Fischer Verlag, Jena, 1987.

Chapter 13

Sterol biosynthesis inhibiting piperazine, pyridine, pyrimidine and azole fungicides

H. Scheinpflug and K. H. Kuck
Institut für Pflanzenkrankheiten, BAYER Aktiengesellschaft Monheim b. Leverkusen, FRG

Introduction

In the second half of the 1960's, several independent research groups synthesized numerous fungicides from chemically heterogeneous classes. Subsequently these fungicides have been shown to have a common biochemical target in fungal metabolism; that is the biosynthesis of ergosterol.

In 1967, BASF presented the first morpholine derivative, dodemorph. Within the same year, several imidazole and pyrimidine derivatives were patented by various other companies. The imidazoles, clotrimazole, miconazole, econazole and isoconazole were patented by Bayer and Janssen and the pyrimidines triarimol, fenarimol and nuarimol were patented by Eli Lilly.

In 1968 the piperazine derivative triforine from Boehringer and the first triazole compound fluotrimazole from Bayer were patented.

With Sumitomo's presentation of buthiobate, a pyridine derivative, in 1970, one or more representatives of each group of fungicides now classified as ergosterol biosynthesis inhibiting fungicides (EBI's) (also known as "sterol biosynthesis inhibitors" (SBI's) had been synthesized.

Biochemical investigations by the group of Sisler (cited in Sisler *et al.* 1984), Kato (1975), Buchenauer (1977) and others, have shown that with the exception of the morpholines, all other compounds have a common site of action within the fungal sterol biosynthesis pathway. That is an inhibition of demethylation at position 14 of lanosterol or 24-methylen dihydrolanosterol which are precursors of sterols in fungi (chapter 14).

All compounds with this mechanism of action have been grouped together as demethylation inhibitors or DMIs in order to distinguish them from the morpholines. Morpholines are sterol biosynthesis inhibitors but have been shown to inhibit a later step in fungal sterol biosynthesis (see chapter 9).

The imidazole and triazole derivatives have proven to be the most important DMI fungicides when measured by the number of compounds reaching an advanced stage of development or registration.

The imidazole compounds have become an important tool in the medical field as antimycotics as well as in the field of agriculture as fungicides. Some medically active imidazoles will be presented in this chapter.

Triazoles are by far the most rapidly growing class of fungicides. A breakthrough in this field was the introduction of triadimefon in the early seventies. This compound combined a broad spectrum of fungicidal activity against important pathogens of cereals with excellent systemic properties and low application rates.

Spectrum of fungicidal activities

In spite of their chemical heterogenicity *in vitro* as well as *in vivo* the usable spectrum of fungicidal activity of most DMI-fungicides has some similarities. *In vitro* studies have shown that most fungi tested in the Ascomycetes, Basidiomycetes and Fungi Imperfecti are more or less sensitive to DMIs. Oomycetes, which are important in the field of agriculture, are generally not sensitive to DMIs.

Nearly all of these compounds are effective to some extent against powdery mildew and rust fungi, which are obligate parasites that cause yield losses in many economically important crops. A multitude of other leaf spot pathogens such as *Pyrenophora* spp., *Mycosphaerella* spp., *Venturia* spp. and *Septoria* spp. are effectively controlled by several of these compounds.

For a long time not more than suppression of two economically important pathogens, *Botrytis cinerea* and *Pseudocercosporella herpotrichoides*, could be found within the DMI fungicides. Recently introduced compounds, such as prochloraz and DPX H6573, provide an economically acceptable level of control of eye spot caused by *Pseudocercosporella herpotrichoides*.

Substantial progress has been made in the field of cereal seed treatment, since the introduction of new systemic and plant compatible products such as triadimenol and imazalil. In many countries these new products serve as a replacement for mercury-based seed dressings. These new compounds control a broad complex of seed- and soilborne pathogens which previously had only been controlled by mercury compounds. Additionally, the systemicity and the broad fungicidal spectrum of the new compounds provides control of pathogens outside the spectrum of activity of mercury; such as *Ustilago* spp., *Typhula incarnata* and *Pyrenophora teres* and early plant infections by powdery mildew and rust.

Systemic properties

Variation in chemical structure and physico-chemical properties influence the compound's ability to enter the outer barriers of the plant and to be subsequently translocated. All of the DMIs are able to penetrate the plant cuticle and/or the seed coat to some extent. Having penetrated the plant surface, a compound may be further translocated either in the apoplast or symplast (Edgington 1981) (chapter 1).

Symplastic (or basipetal) transport, is transport in the living part of the plant (phloem or cytoplasm). Symplastic movement would allow for the control of root and lower stem diseases following foliar application. Thus far, DMIs have shown only minimal symplastic movement within plants.

What is thought to be basipetal transport of DMIs is sometimes misidentified in those cases where the vapour pressure of a compound is high enough to allow translocation of biologically effective amounts of chemical via the gas phase. In dense plant stands this method of transport has been shown to improve fungicidal efficacy (Schein*flug and Paul 1977), but may cause experimental errors in small plot experiments (Jenkyn *et al.* 1983).

The use of systemic fungicides has considerably facilitated and improved the control of plant diseases. Pathogens which have already become established in the plant can be controlled by a curative treatment. In many cases this application provides control after first infections are already established. The systemic uptake of chemicals into a plant not only prevents them from weathering, but may also result in a more complete control of fungal pathogens, since areas not covered during the initial spray treatment are provided with protection through redistribution. Since all systemic DMI-fungicides are translocated predominately in the apoplast, protection is confined to such cases where sufficient quantities of active ingredient have been applied to the basal parts of a leaf or shoot.

Side effects of applications on plants

Distinct plant growth regulatory effects on monocotyledonous and generally stronger effects on dicotyledonous plants have been reported since the first use of DMI fungicides. Typical plant growth regulatory effects which may be seen following foliar application are: shorter shoots and internodes, and/or smaller, darker green leaves. Non-specific signs of plant incompatibility such as necrosis or leaf drop are quite rare within this group.

Investigations on the biochemical mechanisms causing these growth alterations in plants have revealed two possible targets.

Appreciable evidence indicates that one of the primary targets of plant growth regulating compounds in higher plants are reactions that depend on cytochrom P-450. These reactions assist in the conversion of kaurene to kaurenoic acid in the gibberellin biosynthesis pathway.

This finding is confirmed by the observation that the growth retardant action at low concentrations of the fungicide triarimol, or the related growth-regulant ancymidol, was reversed, either fully or in part, by application of gibberellic acid (Sisler *et al.* 1984).

However, with other DMI-fungicides or at higher concentrations, the reversal of growth-retardant action by gibberellic acid was poor. Secondly, some fungicides such as fenarimol and triadimenol, have been shown to inhibit demethyl-sterol synthesis in plants (Buchenauer 1977; Schmitt and Benveniste 1979). Future research is needed to elucidate which of both possible targets in given plant species is of importance with a given DMI-fungicide.

Resistance of fungi to DMI fungicides

Shortly after the introduction and field use of the first systemic single site inhibitor fungicides (the benzimidazoles and hydroxypyrimidines), resistance of certain fungi had been found. It is therefore understandable, that soon after the introduction of the first DMI-fungicides, research on the risk of resistance was initiated. Since no naturally-occurring resistant organisms were obvious during the first years of agricultural use, laboratory studies were conducted using artificially induced resistant strains.

From a multitude of publications, many of which originated from Wageningen in the Netherlands (for example Fuchs *et al.* 1977; van Tuyl 1977; de Waard and Gieskes 1977), which have been reviewed by Dekker (1985), one can summarize the following:

1) Most artificially induced laboratory strains initially showed a relatively low level of resistance.

2) Studies have shown that resistant strains of phytopathogenic fungi were usually reduced in fitness and virulence and were not as competitive as the wild strains.

3) Genetic studies with *Aspergillus nidulans* were able to detect 10 different loci coding for resistance. From this one can assume that more than one gene is responsible

for resistance (VAN TUYL 1977). This finding was in principle confirmed by HOLLOMON *et al.* (1984) with barley powdery mildew.

4) Although DE WAARD and VAN NISTELROY (1979, 1980) were able to demonstrate that strains, resistant to fenarimol, showed a reduced uptake of the fungicide, the mechanism of resistance is unknown on the molecular level.

5) Strains with resistance to one DMI-fungicide generally were cross-resistant to other DMI-fungicides.

Based on these results, and information available from the use of DMI fungicides under field conditions up to that time, DEKKER (1981) gave the prognosis that with DMIs the risk of resistance was relatively low.

With the increased use of DMIs at the beginning of the 1980's, field studies of powdery mildew in cucumbers and cereals became more frequent. SCHEPERS (1985) recently summarized his experience from several years of work with cucumber powdery mildew *(Sphaerotheca fuliginea)* from commercial greenhouses. According to this author, in 1981, the sensitivity of powdery mildew strains isolated from greenhouses were found to be lower than that of wild strains. In 1982 and 1983, a further decrease in sensitivity had been found. SCHEPERS called these strains "resistant" to DMIs. SCHEPERS demonstrated a clear correlation between reduced sensitivity and frequency of fungicide application. So far the level of resistance to DMIs found in SCHEPERS's isolates seems to be only moderate. In this study triforine was weak in performance and bitertanol, fenarimol and imazalil still provided an acceptable level of control when applied at shorter intervals.

The level of resistance found in *Sphaerotheca fuliginea*-strains originating from the mediterranean countries seems to be distinctly higher (HUGGENBERGER *et al.* 1984).

WOLFE and FLETCHER (1981) reported for the first time an increased frequency of barley powdery mildew strains with a decreased sensitivity to triadimefon and triadimenol. Thereafter, other authors reported similar findings (GILMORE 1984; LIMPERT *et al.* 1983). In addition to the selection pressure from fungicide application, changes within the powdery mildew population seem to be connected with the cultivation of highly susceptible varieties such as the barley cultivar "Golden Promise" and the wheat cultivar "Kanzler".

Even though most of the DMI-fungicides have a broad spectrum of fungicidal activity, insensitivity problems have thus far only been found within the powdery mildew fungi. DMI-fungicides are still fully effective against a multitude of other pathogens against which they have been used.

From the examples discussed above, it may be concluded that in comparison to other groups of fungicides (such as the benzimidazoles and the acylalanines), resistance problems with DMIs should not appear suddenly with large populations having a high level of resistance. If resistance should develop, it would tend to develop stepwise within individual fungal species.

Piperazines, pyridines and pyrimidines

Triforine was introduced in 1969 (SCHICKE and VEEN 1969) and remains the only piperazine derivative which is in agricultural use. Information on the broad spectrum of activity of triforine has been published by FUCHS and DRANDAREVSKI (1973). Today the compound is primarily applied as a foliar treatment against powdery mildew, rust, scab and several other pathogens such as *Colletotrichum* and *Monilinia* spp. Triforine was also introduced as a systemic seed treatment of barley for control of powdery mildew (ROHRBACH 1977). Unlike most other SBI-fungicides, triforine has a distinct effect against red spider mites *(Panonychus ulmi)* (MANTINGER and VIGL 1975).

The systemic properties of triforine are well documented. Following leaf application, triforine penetrates the tissue (DRANDAREVSKI and MAYER 1974) and is then translocated over short distances, thus enabling a curative action against fungal pathogens such as *Venturia inaequalis*. When applied to the roots, triforine is readily taken up and translocated to stems and leaves. VON BRUCHHAUSEN and STIASNI (1973) reported that after soil drench application in sandy soil, barley shoots contained up to 7.5 ppm active ingredient two days after treatment. Soil with a high organic matter content decreased the rate of uptake (FUCHS *et al.* 1976a, b). An apparent lack of accumulation of triforine in the roots (EBENEBE *et al.* 1974) and a rather high turnover of the compound in plants (FUCHS *et al.* 1972) seems to account for a maximum concentration in the shoots 2—8 days after application of the chemical (FUCHS and OST 1976).

In contrast to most imidazole and triazole derivatives triforine does not cause typical growth regulator effects on treated plants. Papers dealing with plant incompatibility of triforine describe only nonspecific symptoms in ornamentals (ATTABHANYA and HOLCOMB 1976) or European larch (BOUDIER 1981). Several authors report that triforine inhibits the germination of apple (CHURCH and WILLIAMS 1978) and blueberry (BRISTOW 1981) pollen.

Surprisingly, triforine has been shown to be very phytotoxic to certain lettuce cultivars. Inheritance studies by GLOBERSON and ELIASI (1979) and SMITH (1979) showed that resistance to leaf damage caused by triforine is recessive and is determined by a single gene.

Buthiobate is the only pyridine-based SBI fungicide to have reached a commercial status thus far. KATO *et al.* (1975) demonstrated through *in vitro* studies that buthiobate inhibited growth of several Ascomycetes and Fungi Imperfecti at low rates, but was ineffective against Basidiomycetes. In practice, this compound is mainly used in Japan against powdery mildew fungi.

Buthiobate's mobility in plants seems to be limited. OHKAWA *et al.* (1976) reported an absorption of the compound by leaves and roots but only minimal translocation into other parts of the plant. Using paper discs, KATO *et al.* (1975) were able to demonstrate a distinct vapour action of buthiobate against *Sphaerotheca fuliginea* on cucumbers.

Pyrifenox (Ro 15–1297) is a new experimental pyridine derivative from Maag Ltd. The compound is being researched as a systemic foliar fungicide in pomefruits, stonefruits and vegetables against a wide range of leaf spot pathogens such as *Cercospora* spp., *Cercosporidium* spp., *Septoria* spp. and blossom blight and brown rot (*Monilinia* spp.).

A series of three substituted pyrimidine-5-ylmethanol fungicides, triarimol, fenarimol and nuarimol, were discovered and developed by Eli Lilly since the late sixties. Triarimol (EL-273), the first of the three products presented to the public (BROWN *et al.* 1970), was an effective fungicide against scab and powdery mildew in fruit but was deleted from further development.

Fenarimol like triarimol has been mainly developed in fruits and vegetables. The compound shows activity against a broad spectrum of powdery mildews, scabs, rusts and a multitude of leaf spot pathogens. According to Brown and HALL (1981), fenarimol has been primarily used to provide protective and curative control of scab (*Venturia inaequalis*) and powdery mildew (*Podosphaera leucotricha*) on apples. On grapes, cucurbits, tomatoes, peppers and peach it has been used against other powdery mildew fungi. *In vitro*, fenarimol controls a wide range of Ascomycetes, Basidiomycetes and Fungi Imperfecti at low rates (BUCHENAUER 1979). As expected, there is a strong similarity in biological activity to the structurally-related compound nuarimol.

Greenhouse studies from BROWN and HALL (1981) demonstrated that fenarimol is rapidly absorbed by leaf tissue. These authors demonstrated an apoplastic movement through the leaf, and a measurable vapour phase effect against powdery mildew using treated paper discs, separated or not, from the leaf with aluminium foil discs.

Root uptake of fenarimol has been reported (BUCHENAUER and RÖHMER 1982), but seems to be insufficient to provide disease control in crops grown under field conditions.

Another pyrimidin-5-ylmethanol derivative, ancymidol, is a potent growth regulant in higher plants (COOLBAUGH *et al.* 1982). Therefore, it is not surprising that a distinct growth retardant side-effect of the fungicidal analogues fenarimol and nuarimol has sometimes been reported. BUCHENAUER (1977) reported an inhibition in shoot elongation of tomato and wheat plants, and ABDEL-RAHMAN (1977) noted a reduction of terminal shoot elongation and a reduction of fruit thinning in apples following application of fenarimol. A detailed analysis on the biochemical interrelationships of fungitoxicity and plant growth regulation within the pyrimidin-5-ylmethanols has been published by SISLER *et al.* (1984).

In contrast to fenarimol, which was developed for use in fruits and vegetables, **nuarimol** was developed mainly for use against cereal diseases.

As a seed treatment in cereals, nuarimol is sometimes used in mixture with imazalil to control several seed- and soilborne pathogens such as *Ustilago* spp. and *Fusarium (Gerlachia) nivale*. In addition, control of the airborne pathogen *Erysiphe graminis* is obtained on young seedlings (CASANOVA *et al.* 1977). Growth retarding effects of nuarimol from seed treatments were mentioned by several authors (CASANOVA *et al.* 1977; DÖHLER and MERTZ 1979; BUCHENAUER 1977) who mostly regarded the effect as being transitory. As a foliar fungicide, nuarimol is used in cereals against powdery mildew and leaf blotch *(Rhynchosporium secalis)*. Similar to fenarimol, nuarimol is further recommended against a broad range of fungal diseases in crops such as stonefruits, bananas, sugar beets and peanuts.

The systemic activity of nuarimol is similar to that of fenarimol. The somewhat higher hydrophilicity of nuarimol, as compared to fenarimol, indicates that the

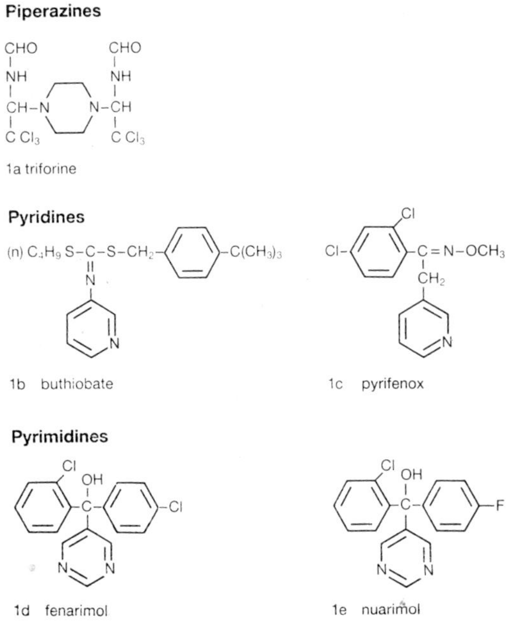

Fig. 13.1 Piperazine-, Pyridine-, and Pyrimidine compounds.

compound is more readily translocated in the apoplast of the plant. CARDOSO *et al.* (1979) and BUCHENAUER and RÖHMER (1982) describe a systemic uptake of the chemical into soybean hypocotyls and barley, respectively, after seed- or soil treatment. This uptake is also obvious from the effect the fungicide has on powdery mildew on cereal seedlings grown from treated seed.

Table 13.1 Piperazine-, Pyridine-, and Pyrimidine-Fungicides

1) Common name {1} [Experim. No.] 2) Chemical name {2} 3) Trade name(s)	1) Originating Company 2) Patent No., Year {3} 3) Developing company 4) Year of presentation	1) Acute toxicity {4} 2) no-effect-level {5} 3) Solubility (water) {6} 4) Vapour pressure {7}
Piperazines		
1) **Triforine** [CELA W 254] 2) N,N′-[1,4-piperazinediylbis (2,2,2-trichloroethylidene) bisformamide] 3) Saprol®, Funginex® Triforine 20®, Prodressan®	1) Ch. Boehringer & Sohn 2) DOS 1901421, 1968 3) Celamerck 4) 1969	1) > 16,000 2) 625 (r;2a) 100 (d;2a) 3) 30 [RT] 4) 2.6×10^{-7} [25 °C]
Pyridines		
1) **Buthiobate** [S-1358] 2) Butyl[4-(1,1-dimethylethyl) phenyl]methyl-3-pyridinyl-carbonimidodithioate 3) Denmert®	1) Sumitomo 2) DOS 21 19 174, 1970 3) Sumitomo 4) 1975	1) 2,700—4,900 2) 3) 0.96 4) 6×10^{7}
1) **Pyrifenox** [Ro 15—1297] 2) 1-(2,4-dichlorophenyl)-2-(3-pyridinyl)-ethanone 0-methyloxime 3)	1) Hoffmann-La Roche 2) EP 49854, 1980 3) Maag 4)	1) 2900 2) 3) 115 [pH 7] 4) 1.9×10^{-5}
Pyrimidines		
1) **Fenarimol** [EL 222] 2) α-(2-chlorophenyl)-α-(4-chlorophenyl)-5-pyrimidine methanol 3) Rubigan®; Bloc®	1) Eli Lilly 2) Fr. 1,569940, 1967 3) Elanco 4) 1975	1) 2500 2) 50 (r;3m) 800 (d;3m) 3) 13.7 [pH 7; 25 C] 4) $<1.3 \times 10^{-7}$ [25 °C]
1) **Nuarimol** [EL 228] 2) α-(2-chlorophenyl)-α-(4-fluorophenyl)-5-pyrimidine-methanol 3) Trimidal®, Trimunol®	1) Eli Lilly 2) FR. 1,569940, 1967 3) Elanco 4) 1975	1) 1250 2) 50 (r;3m) 3) 26 [pH 7, 25 C] 4) $<2 \times 10^{-8}$ [25 °C]

Footnote to tables 13.1, 13.3, 13.5

{1}: approved or proposed common name; {2}: Chemical Abstracts (CA) nomenclature (where available); {3}: year of priority date; {4}: oral acute toxicity in rats (LD_{50} mg/kg); {5}: no-effect-level in subchronic or chronic studies [mg/kg feed] in rats (r) or dogs (d); duration is given in months (m) or years (a); {6}: solubility in water [mg/kg] at 20 °C; variations in temperature or pH are given in brackets []; {7}: vapour pressure in mbar at 20 °C; variation in temperature at which vapour pressure was taken appears in brackets []; (RT = room temperature)

The following table 13.2 and its equivalent presented in the following chapters (Tab. 13.4. and 13.6) provide an overview of many of the important crops, diseases and use notes for several DMI fungicides. This information should be considered as a biological profile and does not constitute a label or recommendation, since many of the uses of compounds are not registered.

Data presented in these tables were obtained from technical data sheets from the respective companies or from a current literature review. Crops and diseases for each organism may not be all inclusive due to space limitations. General or specific literature citations are given beside their respective indication. An asterisk (∗) is used to mark experimental compounds or experimental uses of a registered compound.

Table 13.2 Fungicidal spectrum and application rates of piperazine-, pyridine-, and pyrimidine fungicides. For general comments see page 176 ff.

Compound Crop/Pathogen 1) Foliar Application 2) Seed Treatment 3) Post-harvest Treatment 4) Other Treatments	Rate of application (g a.i./ha or % a.i. for foliar sprays; g a.i./100 kg seed for seed treatments)	Literature
Triforine		
1) Stonefruit: *Monilinia* spp., powdery mildew rust, *Coccomyces hiemalis*	0.019—0.029 %	DRANDAREVSKI & SCHICKE 1976;
Apples: powdery mildew, scab, rust, *Monilinia* spp.	0.024 %	ADLUNG & DRANDAREVSKI 1971
Berry fruits: powdery mildew, rust, *Monilinia* spp.	0.029 %	
Grapes, Hops, Mango: powdery mildew		
Vegetables: powdery mildew, rust, *Colletotrichum* spp.,	190—380 g/ha	
various leaf spot diseases	0.019—0.029 %	
Cereals: *Erysiphe graminis, Puccinia* spp.	285 g/ha	EBENEBE *et al.* 1971
Others: powdery mildew, rust, leaf spot diseases in tobacco, sugar beet, ornamentals		
2) Barley: *Erysiphe graminis*	270 g/100 kg	ROHRBACH 1977
Buthiobate		
1) Vegetables, Fruits, Ornamentals: powdery mildew	0.007—0.02 %	KATO *et al.* 1975;
*Pyrifenox		
1) Pomefruits: *Podosphaera leucotricha, Venturia* spp.	0.0025—0.0075 %	
Stonefruits: *Monilinia* spp.	0.005—0.01 %	
Grapes: *Uncinula necator*	70—210 g/ha	
Peanuts: *Mycosphaerella* spp.	70—210 g/ha	
Others: powdery mildew and leaf spot diseases in pecans, sugar beets, vegetables		

Table 13.2 (continued)

Compound Crop/Pathogen 1) Foliar Application 2) Seed Treatment 3) Post-harvest Treatment 4) Other Treatments	Rate of application (g a.i./ha or % a.i. for foliar sprays; g a.i./100 kg seed for seed treatments)	Literature
Fenarimol		
1) Pomefruits: *Venturia* spp., *Podosphaera leucotricha*	0.0018—0.0036 %	KELLEY & JONES 1981; VERHEYDEN 1981; BERAUD *et al.* 1980
Fruits: powdery mildew, *Monilinia* spec., rust		
Sugar beets: *Cercospora beticola*, *Erysiphe* sp.	90—120 g/ha	
Soybeans: *Diaporthe* sp., *Glomerella* sp., *Cercospora* sp.		
Peanuts: *Mycosphaerella* spp.	90—120 g/ha	
Turf: *Sclerotinia homoecarpa*, *Rhizoctonia* spp., *Ustilago* sp., *Fusarium* spp.	60—6000 g/ha	
Grapes: *Uncinula necator*	0.0018—0.0036 %	VERGNES & PISTRE 1982
Nuarimol		
1) Cereals: *Erysiphe graminis*, *Rhynchosporium secalis*	40—90 g/ha	CASANOVA *et al.* 1977 FRATE *et al.* 1979
Sugar beets: *Cercospora beticola*, *Erysiphe* sp.		
Bananas: *Mycosphaerella* spp.	90 g/ha	
2) Cereals: *Ustilago* spp., *Tilletia* spp., *Pyrenophora* spp. *Fusarium* sp., *Rhynchosporium secalis* *Septoria nodorum*, *Erysiphe graminis*	5—40 g/100 kg	LUZ & VIEIRA 1982 PIENING *et al.* 1983

Imidazoles used in agriculture

Imazalil, the first agricultural imidazole fungicide, is mainly used today as a seed treatment in cereals but has additional applications as a foliar treatment in bananas, vegetables and ornamentals and as a postharvest treatment in citrus fruits. Information on the broad, *in vitro*, activity of imazalil is available from studies of BUCHENAUER and RÖHNER (1979).

Because of imazalil's specific activity against seed- and soilborne pathogens, such as *Pyrenophora* (especially against *P. graminea*), *Fusarium* and *Septoria*, it is most often used in combination with other fungicides. Fungicides such as guazatine, triadimenol and nuarimol, which largely replaced mercury containing seed dressing products, broaden the spectrum of activity of imazalil.

With the aid of tritium-labelled imazalil, VONK and DEKHUIJZEN (1979) demonstrated that imazalil is taken up by plants after seed treatment or root application. In 3 week-old barley plants grown from seed treated with 1.8 g imazalil/100 kg seed, the authors found about 6 % of the labelled material to be present in the leaves. This resulted in an extractable concentration of only 0.07 ppm active ingredient in the leaves. REISDORF *et al.* (1983) confirmed these results and concluded that the excellent activity of imazalil against *Pyrenophora graminea* may not be derived from the relatively low concentrations of the product taken up into the plants, but mainly from its fungicidal activity at the seed surface. REISDORF *et al.* (1983) reported that high concentrations of imazalil caused distinct growth retardation of roots and shoots of barley seedlings. BUCHENAUER and RÖHNER (1979) observed similar symptoms at the rate of 25—50 g a.i./100 kg seed. Physiological studies by these authors showed that imazalil caused both, an inhibition of Gibberellic acid biosynthesis and of C-4-desmethylsterol production in barley seedlings.

Fenapanil, also known as phenapronil or fenapronil, is an experimental broad spectrum imidazole whose *in vitro* activity is similar to that of imazalil, triadimefon and fenarimol (MARTIN and EDGINGTON 1982). Reports on the effectiveness of fenapanil deal mainly with its activity against scab and powdery mildew in pomefruits (SHABI *et al.* 1981; KELLEY and JONES 1981, and as a cereal seed treatment against various seed- and soilborne diseases (HOFFMANN and WALDHER 1981; HANSEN 1981).

According to KELLEY and JONES (1982) about 50 % of fenapanil, when applied to apple leaves, was taken up within 1—3 days. MARTIN and EDGINGTON (1981, 1982) who used barley and soybean plants for studies on the systemicity of fenapanil stated that it exhibited primarily apoplastic transport in barley. In soybeans fenapanil was found to be ambimobile with 13.5 % of the material available for translocation transported basipetally.

Contrary to the usual growth regulating pattern of DMIs, fenapanil inhibited growth of roots and shoots of barley seedlings stronger than those of soybeans. Symptoms described in barley were stunting and chlorosis of the foliage with some necrosis at the leaf tips (MARTIN and EDGINGTON 1982).

Prochloraz, presented in 1977 by the Boots Company, is a broad spectrum fungicide which has a pronounced activity against Ascomycetes and Fungi Imperfecti, and somewhat lesser activity against Basidiomycetes (BIRCHMORE *et al.* 1977). The fungicidal profile of prochloraz in cereals includes activity against *Pseudocercosporella herpotrichoides.* This is unique among the azolyl fungicides which have reached a market stage. In addition, good activity is seen against pathogens such as leaf blotch, net blotch, and *Septoria* spp. Prochloraz has shown no crossresistance to benzimidazole fungicides: This fact permits a wide use especially in cereal stands, where resistance to carbendazim has been detected.

In France and other European countries, combination-formulations with carbendazim are available for improved control of eyespot in wheat. This combination product also showed efficacy in oil seed rape against a complex of pathogens such as *Phoma lingam, Sclerotinia sclerotiorum,* and *Alternaria* spp.

Plant compatibility of the normally used EC-formulation of prochloraz on various broadleaved crops is critical. A prochloraz-manganese complex, formulated as 50 WP, is recommended for use or testing in most dicotyledonous plants (BIRCHMORE *et al.* 1979).

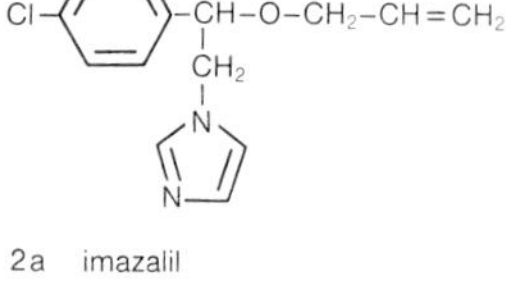

2a imazalil

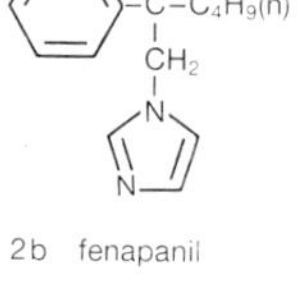

2b fenapanil

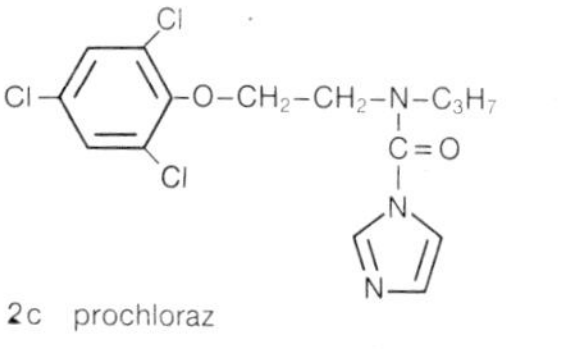

2c prochloraz

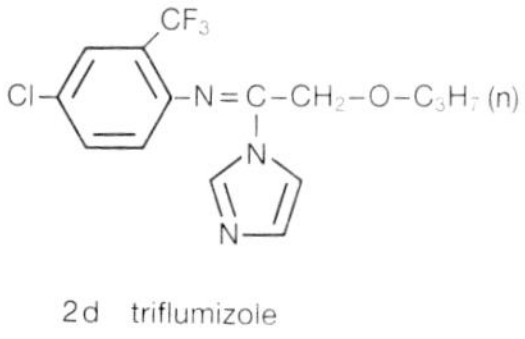

2d triflumizole

Fig. 13.2 Imidazole compounds.

The mobility of prochloraz in plants seems to be the rather limited. According to de SAINT-BLANQUAT and MY (1982) prochloraz is readily absorbed from plant surfaces, but not translocated over longer distances. COOKE *et al.* (1979), who used strawberries as a test plant system confirmed this finding.

Triflumizole, the most recent imidazole compound, is being tested in Japan and the United States mainly in fruits and vegetables against a wide range of pathogens such as powdery mildew, scab and *Monilinia* spp. The compound is described as a preventive and curative fungicide which has translaminar but not systemic activity.

Table 13.3 Imidazole-Fungicides

1) Common name {1} [Experim. No.] 2) Chemical name {2} 3) Trade name(s)	1) Originating Company a) Patent No., Year {3} 3) Developing Company 4) Year of presentation	1) Acute toxicity {4} 2) no-effect-level {5} 3) Solubility (water) {6} 4) 9.3×10^{-8}
1) **Imazalil** [R 23 979 (free base)] 2) 1-[2(2,4-dichlorophenyl)-2-(2-propenyloxy)ethyl]-1H-imidazole 3) Fungaflor®, Fungazil® Fecundal® a.o.	1) Janssen 2) BP 1 244 530, 1969 3) Janssen 4) 1972	1) 227—343 2) 80 [r,2a, mg/kg body wt.] 3) 293 4) 9.3×10^{-8}
1) **Fenapanil** [RH-2161] (Phenapronil) 2) α-butyl-α-phenyl-1H-imidazole-1-propanenitrile 3) Sisthane®	1) Rohm & Haas 2) DOS 2 604 047, 1976 3) Rohm & Haas 4) 1978	1) 1590 2) 3) 4)
1) **Prochloraz** [BTS 40542] 2) N-propyl-N-[2-(2,4,6-trichlorophenoxy)ethyl]-1-imidazole-1-carboxamide 3) Sportak®, Sporgon®	1) Boots Co. Ltd. 2) GB 1 469 772, 1973 3) FBC, Schering 4) 1977	1) c. 1600—2400 2) 3) 55 [25 °C] 4) 7.6×10^{-10}
1) **Triflumizole** [NF-114, A815] 2) 4-chloro-N-[1-(1H-imidazol-1-yl)-2-propoxyethyliden]-2-(trifluoromethyl)-benzenamin 3) Trisosol®, Trifludol®	1) Nippon Soda 2) DOS 2 814 041, 1978 3) Nippon Soda, Uniroyal 4) 1983	1) 1057 m 1780 f 2) 3) 12.5 4)

For footnote see table 13.1

Table 13.4 Fungicidal spectrum and application rates of agricultural imidazole fungicides for general comments see page 180ff.

Compound Crop/Pathogen(s) 1) Foliar Application 2) Seed treatment 3) Post-harvest Treatment 4) Other Treatments	Rate of application (g a.i./ha or % a.i. for foliar sprays; g a.i./100 kg seed for seed treatments)	Literature
Imazalil		
1) Vegetables, Ornamentals: Powdery mildew, various leaf and stem diseases	0.005—0.03 %	Reisdorf *et al.* 1983
Bananas: *Mycosphaerella* spp.		Melin *et al.* 1975
2) Cereals: *Pyrenophora* spp., *Fusarium* spp., *Septoria* spp.	4—5 g/100 kg	
Potatoes: *Phoma exigua*, *Polyscytalum pustulans*	10—30 g/1000 kg	Cayley *et al.* 1981
3) Fruits: post harvest diseases	0.015—0.5 %	
*Fenapanil		
1) Pomefruits: powdery mildew, scab, rust	0.03—0.062 %	Shabi *et al.* 1981
Others: powdery mildew, rust, leaf spots in cereals, fruits, vegetables		
2) Cereals: smuts, bunts, *Pyrenophora graminea*	30—120 g/100 kg	Hoffmann & Waldher 1981
Prochloraz		
1) Cereals: *Pseudocercosporella* sp., *Pyrenophora teres*, *Rhynchosporium secalis*, *Septoria* spp., *Erysiphe graminis*	400—450 g/ha	Tromas & Cornier 1983 Harris & Barnes 1981
Rapes: *Sclerotinia sclerotiorum*, *Phoma lingam*	500 g/ha	Wakerley & Russel 1985
*Bananas: *Mycosphaerella musicola*, *M. fijiensis*	100—200 g/ha	
2) Cereals: *Pyrenophora* spp., *Fusarium* spp., *Leptosphaeria nodorum*	20—50 g/100 kg	
Rice: *Gibberella fujikuroi*, *Helminthosporium oryzae*	12.5 g/hl	
3) *Citrus, Bananas: various moulds and rots	25—300 g/hl	
4) Mushrooms: *Verticillium fungicola*, *Mycogone perniciosa*	3—15 g/10 m²	v. Zaayen & v. Adrichem 1982
*Triflumizole		
1) Apples: *Venturia inaequalis*, *Podosphaera leucotricha* *Gymnosporangium* spp.	0.015—0.03 %	
Stonefruits: *Monilinia* sp.	0.006—0.01 %	
Grapes: *Uncinula necator*	140—280 g/ha	
Vegetables: powdery mildew, *Rhizoctonia* sp.	70—350 g/ha	
Ornamentals: powdery mildew, rust, *Rhizoctonia* sp.	70—280 g/ha	
Cucurbits: powdery mildew		

Triazoles

Fluotrimazole was first mentioned in literature by GREWE and BÜCHEL (1972). Following introduction and testing, it became the first triazole-fungicide to be marketed. This compound is characterized by a very narrow usable spectrum of fungicidal activity, restricted primarily to powdery mildew fungi. *In vitro*, a broader spectrum has been demonstrated (BUCHENAUER 1979). The importance of fluotrimazole is limited due to the introduction of other broader spectrum SBI fungicides such as triadimefon.

Information on the systemic properties of fluotrimazole was given by STEFFENS and WIENEKE (1981). Six weeks after soil-treatment, only 0.3—0.4 % of C-14 labelled active ingredient was taken up by wheat and tomato plants. Local systemic uptake of fluotrimazole into the plant after foliar application is responsible for a distinct curative effect. Phytotoxic side effects of fluotrimazole have seldom been seen probably due to the limited uptake of this compound.

Triadimefon, the first systemic broad spectrum triazole fungicide, was presented to the public a short time after fluotrimazole (GREWE and BÜCHEL 1973; KASPERS *et al.* 1975). *In vitro*, testing of triadimefon demonstrated its effectiveness at 0.1 to 2 ppm against a broad spectrum of Ascomycetes, Basidiomycetes and Fungi imperfecti (BUCHENAUER 1979).

Triadimefon soon became known under its trade name Bayleton®. The advantage of triadimefon over other fungicides was, that it provided simultaneous protective and curative control of several economically important diseases in important crops, such as cereals.

When triadimefon is applied to plants, it is quickly reduced to its corresponding secondary alcohol, triadimenol. In wheat leaves, about 60 % of the active ingredient which penetrates the epidermis is reduced to triadimenol within 2 days (KUCK 1986). This reduction is regarded as an activation since triadimenol is the more active of the two fungicides. The fungicidal activity of the precursor-compound triadimefon is itself difficult to define since most fungi are also able to perform this reduction step (GASZTONYI and JOSEPOVITS 1979).

Triadimefon is a racemic mixture of two optical isomers (— and +), whose fungicidal activities are similar (KRÄMER *et al.* 1983).

The systemic properties of triadimefon contribute to a high degree to its curative efficacy. From a multitude of investigations it is shown that triadimefon (and/or its metabolite triadimenol) rapidly enters root-, stem- and leaf tissues and is translocated acropetally in the apoplast (FÜHR *et al.* 1978; BRANDES *et al.* 1978; KRAUS 1981; SANDERS *et al.* 1978; BUCHENAUER 1976). For example FÜHR *et al.* (1978) found that in barley up to 51 % of the ^{14}C-labelled triadimefon applied to the basal half of leaves had been translocated to the apical half within 12 days. In these experiments basipetal translocation accounted for only 0.3 %. KRAUS (1981) reported that only 40 % of the applied triadimefon was water-removable from grape leaves 45 minutes after application.

A marked vapour phase activity of triadimefon was shown against powdery mildew on cucumbers and barley during greenhouse studies by SCHEINPFLUG and PAUL (1977). Field experiments confirmed that the gas phase may contribute to the performance of the fungicide (SCHEINPFLUG *et al.* 1978).

Several aspects of the plant growth regulatory side effects of triadimefon and triadimenol have been studied by FÖRSTER *et al.* (1980a, b, c) using barley plants. These authors reported that in plants grown from treated seed a transient retardation of roots, shoots and coleoptiles and a disturbed geotropism of the plants could be observed. Thirty days after sowing differences could no longer be found in the fresh and dry weights of shoots. Buchenauer and Röhner (1981) concluded from physiological studies that triadimefon and triadimenol interfere in gibberellin and sterol biosynthesis in barley seedlings by inhibiting oxidative demethylation reactions.

As most other SBI-fungicides, triadimefon causes stronger plant growth regulatory side effects on dicotyledonous than on monocotyledonous plants. In red raspberries earlier ripening and shorter and thinner canes were reported by Pepin *et al.* 1980. On apple trees reduction of leaf area, shortening of leaf petioles (Abdel-Rahman 1977) and increase in fruit set (Strydom *et al.* 1981) have been reported.

More recently, Fletcher and coworkers have investigated possible beneficial side effects of triadimefon. Treated wheat, peas and soybeans showed a reduced transpiration which prevented the leaves of water-stressed plants from wilting and increased yield (Fletcher and Nath 1984). Beans proved to be more resistant to ozone, and cabbage and barley seedlings were shown to be more resistant to chilling after root application of triadimefon (Fletcher and Hofstra 1985; Fletcher 1985).

Triadimenol was initially introduced as the active ingredient of the systemic seed treatment Baytan®. Due to its broad fungicidal spectrum (Buchenauer 1979) and its systemic uptake into the plant, triadimenol proved to be active against a wide range of seed-, soil- and airborne pathogens of cereal seedlings.

Smuts (*Ustilago* spp), bunts (*Tilletia* spp.), Pyrenophora-diseases and early powdery mildew and rust infections are simultaneously controlled by a Baytan seed treatment (Frohberger 1978). For complete control of *Pyrenophora graminea* and *Fusarium (Gerlachia) nivale*, imazalil and fuberidazol have been used as co-fungicides.

More recently, triadimenol has been further developed as a foliar fungicide (trade name: Bayfidan®). Triadimenol has shown advantages over triadimefon in curative and eradicative treatments. Due to its close chemical relationship to triadimefon the application rates and the fungicidal spectrum of triadimenol are essentially the same.

Triadimenol resembles triadimefon in being a highly systemic fungicide. The uptake of ^{14}C-triadimenol into cereal seedlings after seed treatment has been studied by Steffens *et al.* (1982). According to these authors, up to 7.7% of the applied radioactivity was taken up by wheat seedling shoots within 51 days. The largest quantity of radioactivity was found in the first three leaves, with the primary leaf containing 3.9% and leaves two and three containing 1.4 and 0.5% of the applied radioactivity, respectively.

The reduction of triadimefon introduces a second chirality centre into its metabolite triadimenol. Thus triadimenol can exist in the two diastereomeric forms triadimenol A (= I or threo) and triadimenol B (= II or erythro). Each of these consists of two enantiomers. Thus, triadimenol A includes the enantiomeric forms A (—)(= 1S, 2R) and A (+)(= 1R, 2S), while triadimenol B contains the enantiomers B (—)(= 1S, 2S) and B (+)(= 1R, 2R).

The fungicidal properties of the diastereomeric forms, and more recently, of the enantiomers, as well as the reduction of triadimefon by fungi and plants have been studied intensively.

Buchenauer (1979) reported that triadimenol A was more fungitoxic than triadimenol B and remarked that various fungi differed significantly in the capability to reduce triadimefon and in their capability of producing certain diastereomeric mixtures.

According to Krämer *et al.* (1983) the fungicidal activity of individual enantiomers is in the order A (—) > B (+) ≫ A (+), B (—). Efforts to explain the level of sensitivity to triadimefon of a given fungus with its capability to reduce triadimefon (Gasztonyi and Josepovits 1978, 1979) were recently reexamined on the enantiomeric level by Deas *et al.* (1984a, b). These authors demonstrated that the amount and the pattern of reduction of triadimefon to triadimenol gives an explanation for the high sensitivity of some fungi to triadimefon. However, this study fails to explain the low sensitivity of some fungal species such as *Fusarium culmorum*. It was therefore concluded that a simple relationship between sensitivity to triadimefon and preferential metabolic production of the most active A(—) enantiomer does not exist.

Bitertanol, although structurally related to triadimenol and triadimefon, is distinctly different from these compounds in regard to its spectrum of fungicidal activity. *In vitro*, studies show the effects of bitertanol against the Fungi imperfecti to be more pronounced than that of triadimefon (Kraus 1979).

When applied as a protective or curative foliar fungicide, the main uses of bitertanol are for the control of scab (*Venturia* spp.) in pomefruits, leaf diseases (*Puccinia* spp., *Mycosphaerella* spp.) in peanuts, *Mycosphaerella* spp. in bananas, and rusts, powdery mildews and *Monilinia* diseases in vegetables, ornamentals and stonefruits.

As a seed treatment, bitertanol is effective against certain seed- and soilborne pathogens of cereals. Bitertanol controls dwarf bunt of wheat *(Tilletia controversa)*, a pathogen which is difficult to control with other fungicides.

As with triadimenol, bitertanol is a mixture of four enantiomeric forms. According to Büchel (1984) the 1S, 2R enantiomer was the most active form in experiments involving bean rust *(Uromyces phaseoli)*.

Due to its high lipophilicity, bitertanol penetrates plant surfaces readily and is locally redistributed, but is hardly translocated over longer distances in the plant (Scheinpflug and van den Boom 1981). In accordance with its low vapour pressure, redistribution of the compound via a gas-phase is minimal.

Because of the low mobility of bitertanol inside and outside the plant, growth regulator effects are seldom seen (Kelley and Jones 1981; Brandes *et al.* 1979).

Propiconazole, originally discovered by Janssen, was presented in 1979 by Ciba-Geigy (Urech *et al.* 1979), and rapidly became one of the most successful SBI-fungicides of its time.

Chemically speaking, propiconazole is closely related to its ethyl-analogue, etaconazole. Biologically, it is characterized as a very broad spectrum fungicide which is active at low rates. Propiconazole was initially developed for the control of a complex of cereal pathogens, but also showed efficacy in a multitude of other crops such as peanuts, grapes and bananas.

In cereals, propiconazole is used partially in mixture with carbendazime or captafol in order to improve the control of eyespot *(Pseudocercosporella herpotrichoides)* and *Septoria* spp., respectively (Smith and Speich 1981).

Quantitative studies have not been published, but Urech and Speich (1981) described propiconazole as a highly systemic compound which was absorbed by leaves and stems of cereal plants within 24 hours of application and transported acropetally in the plant.

The vapour phase activity of propiconazole was estimated to be biologically similar to that of triadimefon (Rathmell and Skidmore 1982).

Propiconazole is a typical SBI-fungicide in regard to its plant growth regulator activities. Buchenauer *et al.* (1981), who investigated the plant growth regulatory effects of several triazoles, stated that when applied as a barley seed-treatment at 25—50 g a.i./100 kg Propiconazole, caused a growth inhibition of leaves, roots and coleoptiles. Buchenauer *et al.* also described plant alterations which might be considered beneficial, such as a delayed senescence of chlorophyll and an increased tolerance of the seedlings towards dryness, frost and salt stress.

Etaconazole, the second triazole fungicide presented by Ciba Geigy in 1979 (Staub *et al.* 1979), has basically the same biological properties as the related compound propiconazole. Thus, etaconazole has a broad *in vitro* fungicidal spectrum of activity with ED_{50} values often between 0.1 and 1 ppm.

According to URECH *et al.* (1979), this compound is especially suited for the control of deseases of deciduous fruit. Similar to penconazole, which more recently was introduced by Ciba-Geigy, etaconazole provides simultaneous control of scab and powdery mildew on apples and pears. Additional activity has been shown against a broad spectrum of diseases in peanuts, cucumbers, peaches and other agriculturally important crops (STAUB *et al.* 1979).

The plant-growth-regulatory side-effects of etaconazole on barley seedlings were reported by BUCHENAUER *et al.* (1981) to be more pronounced than those of propiconazole. In apples, several reports describe smaller, thicker, darker green leaves and retarded tree growth when compared to the untreated check (KELLEY and JONES 1981; SZKOLNIK 1981).

Etaconazole is a fully systemic fungicide, which is transported acropetally in plants. In apple leaves KELLEY and JONES (1982) reported an uptake of approximately 90% of the compound within 12 hours. SZKOLNIK (1982) studied the pronounced vapour phase activity of etaconazole. When applied to the shading cloth in greenhouses, etaconazole controlled several genera of powdery mildew on crops such as apples, grapes, vegetables and roses.

The presence of two asymmetric carbon atoms in the molecules of propiconazole and etaconazole results in four stereoisomers. For the latter compound, the antifungal activity of the isomeric forms has been described (VOGEL *et al.* 1982; HEERES 1984). The activity of the individual isomers depended strongly on the fungi used as test organisms. For example, *in vitro* tests with *Botrytis cinerea* showed the order 2S, 4R (cis) $>$ 2S, 4S (trans) $\gg$ 2R, 4S (cis), 2R, 4S (trans). Against *Erysiphe graminis* on barley the 2R, 4S-isomer was the most active, followed by the 2S, 4R and the 2R, 4R-forms, and the 2S,4S-isomer was the weakest one in this test system.

Penconazole, introduced by Ciba-Geigy in 1983 (EBERLE *et al.* 1983), is another triazole fungicide patented by Janssen and developed by Ciba-Geigy. According to the authors mentioned above, this compound has a broad *in vitro* spectrum of antifungal activity. It is especially active in controlling diseases such as powdery mildew on grapes and apples, scab on apples and pears, and black rot on grapes. When used alone, the protective activity of penconazole against scab is not always sufficient. Mixtures with a residual fungicide such as captan are recommended.

Very little has been published on the systemicity of penconazole. EBERLE *et al.* (1983) mention that the compound penetrated rapidly into plant tissues and was translocated acropetally. These authors stated that in their trials, symptoms of phytotoxicity such as reduction in leaf size could not be detected.

ICI presented **diclobutrazol** in 1979 (BENT and SKIDMORE 1979) as a foliar fungicide for the control of cereal diseases. Besides control of powdery mildew, *Rhynchosporium secalis* and *Typhula incarnata*, diclobutrazol provides especially good control of *Puccinia* spp. in wheat and barley. Good control of coffee leaf rust has also been reported (JAVED 1981).

According to BENT and SKIDMORE (1979) diclobutrazol is a systemic fungicide which shows translaminar movement and acropetal translocation. These authors reported some vapour activity against cereal powdery mildew in greenhouse trials. This activity was estimated to be weak under field conditions (RATHMELL and SKIDMORE 1982).

Diclobutrazol, like the chemically related compound triadimenol, contains two chiral centres. It should be noted that the common name diclobutrazol applies only to the 2 RS, 3 RS diastereomer which is over 100 times more active as a fungicide than the 2 RS, 3 RS diastereomer. Resolution of the 2 RS, 3 RS-form showed that the 2 R, 3 R enantiomer was responsible for almost all of the fungicidal activity (BALDWIN and WIGGINS 1983).

The plant-growth regulatory side effects of diclobutrazol are rather distinct. BENT and SKIDMORE (1979) reported a reduction in shoot growth and a darkening of foliage in apples. In greenhouse tests using various plant species soil drenches tended to have greater plant growth regulating effects than foliar sprays. Dicotyledonous plants were affected more than monocotyledonous species.

BUCHENAUER *et al.* (1981) mentioned that diclobutrazol caused stronger growth regulating effects then etaconazole and propiconazole in barley.

Flutriafol is a systemic cereal fungicide from ICI, which was presented in 1983 (SKIDMORE *et al.* 1983).

This systemic foliar fungicide, when used alone, controls several important cereal pathogens such as powdery mildew, rust and net blotch *(Pyrenophora teres)*. Mixtures with carbendazim are necessary for good control of eyespot and improved control of *Fusarium* spp. Mixtures with captafol have shown increased activity against *Septoria nodorum* (NORTHWOOD *et al.* 1983).

SKIDMORE *et al.* (1983) tested flutriafol as a cereal seed treatment at 7.5 g/100 kg seed and reported good control of smuts and bunts (*Ustilago* spp., *Tilletia* spp.). The addition of a co-fungicide, such as thiabendazol, was needed to be fully effective against *Pyrenophora* and *Fusarium* spp. Doses higher than 10 g/100 kg seed caused delays in seedling emergence in wheat. This rate also provided systemic control of *Erysiphe graminis* and *Puccinia hordei* in winter barley.

In combination with ethirimol and thiabendazol flutriafol is further marketed as a seed treatment in cereals under the trade name of Ferrax (R). This treatment has shown excellent activity against powdery mildew strains with decreased sensitivity to inhibitors of C-14-demethylation of dihydrolanosterol, such as triadimenol or flutriafol (NORTHWOOD *et al.* 1984).

Flusilazol = **(DPX) H6573** was introduced by Du Pont in 1984 and is the first agrochemical or pharmaceutical azole compound to contain silicon in the active form of the molecule (MOBERG *et al.* 1985).

This broad-spectrum fungicide has shown promising results at low rates against several diseases of cereals including *Pseudocercosporella* foot rot (FORT and MOBERG 1984). Excellent activity has also been shown against several important tree fruit diseases such as *Venturia* spp., *Podosphaera leucotricha* and *Monilinia* spp.

In greenhouse tests, BRUHN *et al.* (1985) showed that DPX H6573 was able to penetrate plant tissue and to move within the plant's transpiration stream allowing for preventative and curative activity. DAVIS *et al.* (1985) reported that at test rates, DPX H6573 caused no phytotoxic effects on tree fruits.

BAS 454 06 F is another new systemic triazole compound which was presented by POMMER and ZEEH (1983). The product is under investigation against a wide range of pathogens on crops which usually are treated with broad spectrum DMIs.

Diniconazol = **(S-3308)** belongs to a new group of triazole derivatives containing potent, broad spectrum fungicides as well as effective plant growth regulators.

The compound is being tested in cereals, grapes, apples and peanuts against powdery mildews, rusts and other leaf spotting diseases. TAKUNO *et al.* (1983) showed control of *Ustilago*, *Tilletia* and *Pyrenophora graminea* when S-3308 was used as a seed treatment. According to FUNAKI *et al.* (1983), the compound exhibited curative properties and was partially systemic.

Stereo-isomeric forms exist within the structure of S-3308, which express different biological properties. The R(—) isomer was the more potent fungicide, while the S(+) isomer was shown to be a better plant growth regulator (FUNAKI *et al.* 1983). The 4-chlorophenyl analogue of S-3308, S-3307, is under development as a plant growth regulator (IZUMI *et al.* 1983, 1984).

PP 969 is an experimental fungicide with uncommon systemic properties. As PP 969 is a highly water soluble compound with a low partition coefficient (log P

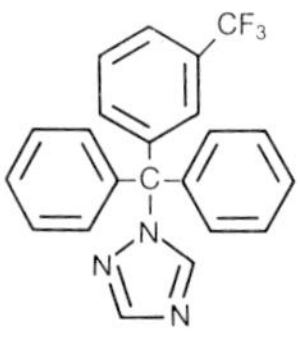

3a fluotrimazole

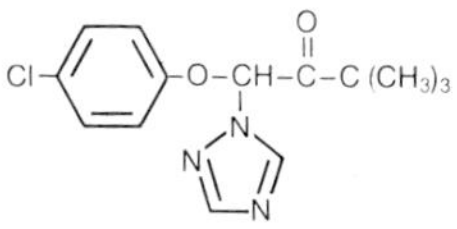

3b triadimefon

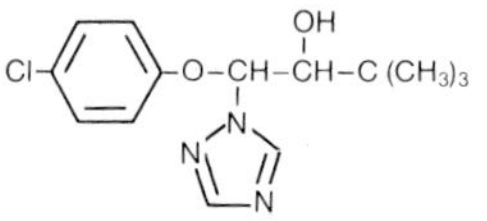

3c triadimenol

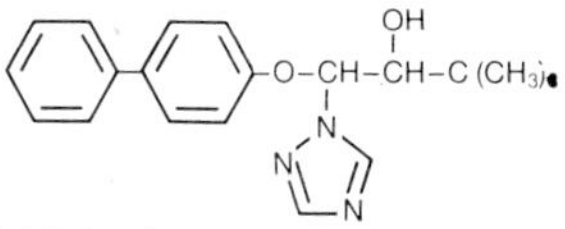

3d bitertanol

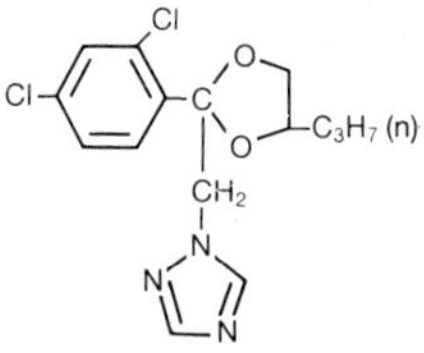

3e propiconazole

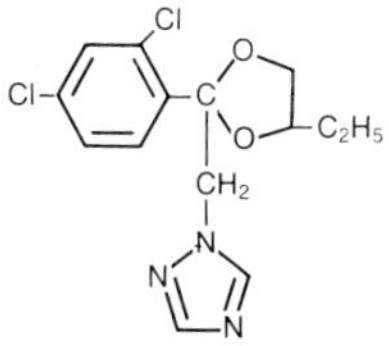

3f etaconazole

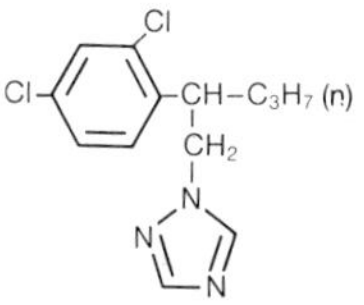

3g penconazole

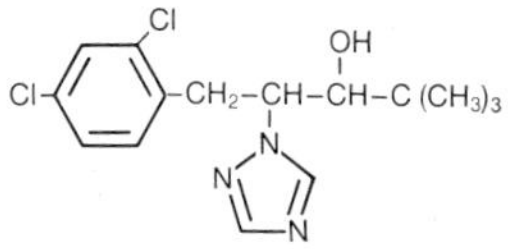

3h diclobutrazol

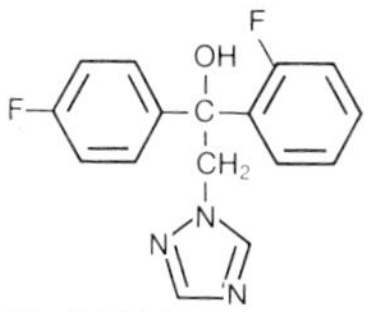

3i flutriafol

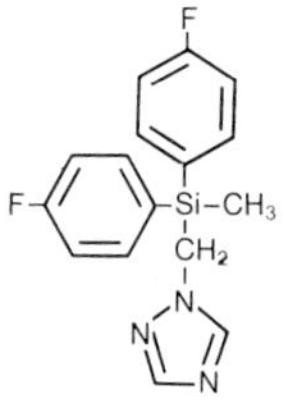

3j DPX H6573

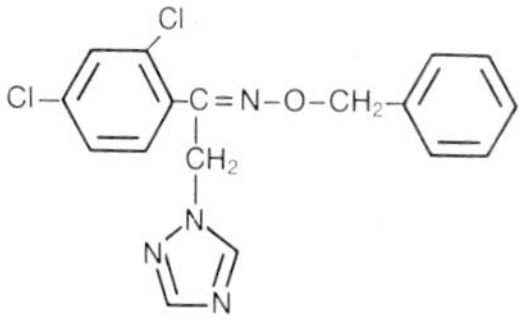

3k BAS 45 406 F

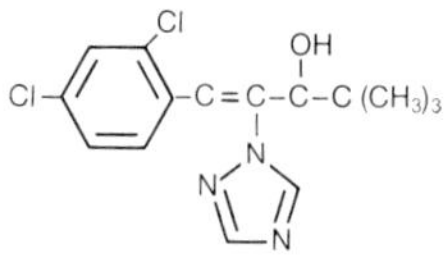

3l S-3308

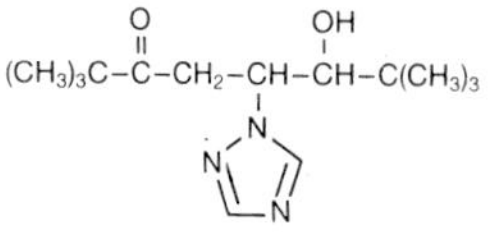

3m PP 969

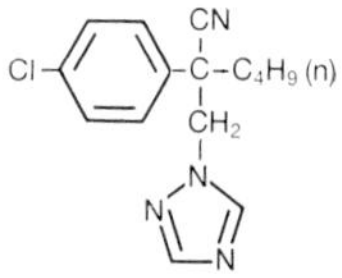

3n RH-3866

Fig. 13.3 Triazole compounds (DPX H 6573 = flusilazol, S-3308 = diniconazol, RH-3866 = myclobutanil) octanol-water = 1.95), this compound seems to be relatively ineffective in penetrating the cutinized surface of leaves but is highly mobile in woody plants when applied as a soil drench or stem injection (SHEPARD and FRENCH 1983; SHEPARD 1985). Single application as a soil drench or injection has provided control of leaf diseases of coffee, bananas and apples for up to 30 weeks.

Myclobutanil (RH-3866) is a new experimental fungicide from Rohm and Haas, which has been tested in the United States. This compound is structurally similar to the older imidazole fungicide fenapil. Myclobutanil is being tested as a foliar fungicide against a broad spectrum of fungal diseases such as rusts, powdery mildews, scab and various leaf spot diseases. It is also being tested on monocotyledonous plants as a seed treatment against a complex of soil- and seed-borne seedling diseases. From the activity of myclobutanil against loose smuts, it can be assumed that it is at least partially systemic in plants.

Table 13.5 Triazole-Fungicides

1) Common name {1} [Experim. No.] 2) Chemical name {2} 3) Trade name(s)	1) Originating Company 2) Patent No., Year {3} 3) Developing Company 4) Year of Presentation	1) Acute toxicity {4} 2) no-effect-level {5} 3) Solubility (water) {6} 4) Vapour pressure {7}
1) **Fluotrimazol** [BAY BUE 0620] 2) diphenyl-(3-fluoromethyl-phenyl)-1,2,4-triazole-1-yl-methan 3) Persulon®	1) Bayer 2) DOS 1795249, 1968 3) Bayer 4) 1973	1) >5000 2) 15 (r,3m) 150 (d,3m) 3) 0.5 4) 3.6×10^{-7}
1) **Triadimefon** [BAY MEB 6447] 2) 1-(4-chlorophenoxy)-3,3-dimethyl-1-(1H-1,2,4-triazole-1-yl)-2-butanone 3) Bayleton®	1) Bayer 2) DOS 2 201 063, 1972 3) Bayer 4) 1973	1) 750—1200 2) 50,500 (r,f,m,2a) 330 (d,2a) 3) 70 4) $<10^{-6}$
1) **Triadimenol** [BAY KWG 0519] 2) β-(4-chlorophenoxy)-α-(1,1-dimethylethyl)-1H-1,2,4-triazole-1-ethanol 3) Baytan®, Bayfidan®, Summit®	1) Bayer 2) DOS 2 324 010, 1973 3) Bayer 4) 1977	1) 700—1200 2) 125 (r,2a) 600 (d,3m) 3) 95 4) $\ll 10^{-6}$
1) **Bitertanol** [BAY KWG 0599] 2) β-([1,1'-biphenyl]-4-yloxy)-α-(1,1-dimethylethyl)-1H-1,2,4-triazole-1-ethanol 3) Baycor®, Sibutol®	1) Bayer 2) DOS 2 324 010, 1973 3) Bayer 4) 1978	1) >5000 2) 100 (r,3m) 3) 5 4) $\ll 10^{-6}$
1) **Propiconazole** [CGA 64 250] 2) 1-[2-(2,4-dichlorophenyl)-4-propyl-1,3-dioxolan-2-yl]-methyl]-1H-1,2,4-triazole 3) Tilt®, Desmel®	1) Janssen 2) DOS 2 551 560, 1974 3) Ciba Geigy 4) 1979	1) 1517 2) 3) 110 4) 1.3×10^{-6}
1) **Etaconazole** [CGA 64 251] 2) 1-[2-(2,4-dichlorophenyl)-4-ethyl-1,3-dioxolan-2-yl]-methyl]-1H-1,2,4-triazole 3) Vangard®, Sonax®	1) Janssen 2) DOS 2 551 560, 1974 3) Ciba-Geigy 4) 1979	1) 1343 2) 3) 80 4)

Table 13.5 (continued)

1) Common name {1} [Experim. No.] 2) Chemical name {2} 3) Trade name(s)	1) Originating Company 2) Patent No., Year {3} 3) Developing Company 4) Year of Presentation	1) Acute toxicity [4] 2) no-effect-level {5} 3) Solubility (water) {6} 4) Vapour pressure {7}
1) **Penconazole** [CGA 71 818] 2) 2-(2,4-dichlorophenyl)- pentyl-1H-1,2,4-triazole 3) Topas®	1) Janssen 2) DOS 2 735 872, 1976 3) Ciba-Geigy 4) 1983	1) 2125 2) 3) 70 4) 2.1×10^{-6}
1) **Diclobutrazol** [PP 296] 2) (2R,3R)- and (2S,3S)-1-[2,4- dichlorophenyl)-4,4-dimethyl- 2-(1,2,4-triazole-1-yl)pentan-3-ol 3) Vigil®	1) ICI 2) DOS 2 737 489, 1976 3) ICI 4) 1979	1) ~4000 2) 50 (r,3m) 3) 9 4) $1.3—2.6 \times 10^{-8}$
1) **Flutriafol** [PP 450] 2) RS-2,4′-difluor-α-(1H-1,2,4- triazol-1-yl-methyl)-benz- hydrylalcohol 3) Impact®, Ferrax®	1) ICI 2) EP 15 756, 1979 3) ICI 4) 1983	1) 1140 m 1480 f 2) 3) 130 4) 4×10^{-9}
1) **[DPX H6573]** = **Flusilazol** 2) bis(4-fluorophenyl)methyl- (1H-1,2,4-triazol[-1-yl]- methyl)silane 3) Nustar®, Punch®	1) Du Pont 2) EP 68 813, 1981 3) Du Pont 4) 1984	1) 1110 m 674 f 2) 3) 54 [pH 7] 4)
1) **[BAS 45 406F]** 2) 1-(2,4-dichlorophenyl)-2- (1H-1,2,4-triazol-1-yl) ethanon- 0-(phenylmethyl)-oxim 3)	1) BASF 2) DOS 3 139 370, 1981 3) BASF 4) 1983	1) 4640 2) 3) 4)
1) [**S-3308**, XE-779] = **Diniconazol** 2) β-[2,4-dichlorophenyl)-methy- len]-α-(1,1-dimethylethyl)- 1H-1,2,4-triazol-1-ethanol 3)	1) Sumitomo 2) DOS 3 010 560, 1979 3) Sumitomo, Chevron 4) 1983	1) 639 m 474 f 2) 3) 4—6 4) $\sim 2.9 \times 10^{-5}$
1) **[PP 969]** 2) (5RS,6RS)-6-hydroxy-2,2-7,7- tetramethyl-5-(1H-1,2,4- triazol-1-yl)octan-3-one 3)	1) ICI 2) EP 27 685, 1979 3) ICI 4) 1983	1) 2) 3) 3600 4)
1) **Myclobutanil** [RH-3866] 2) α-butyl-α′-(4-chlorophenyl)- 1H-1,2,4-triazol-1-propane- nitrile 3)	1) Rohm & Haas 2) US 4 366 165, 1977 3) Rohm & Haas 4)	1) 1600 m 2290 f 2) 100 (r,3m) 10,100 (d,m,f,3m) 3) 142 [25 °C] 4) 2.1×10^{-6} [25 °C]

For footnote see table 13.1

Table 13.6 Fungicidal spectrum and application rates of triazole fungicides.
For general comments see page 185ff.

Compound Crop/Pathogen 1) Foliar Application 2) Seed Treatment 3) Post-harvest Treatment 4) Other Treatments	Rate of application (g/ha or % a.i. for foliar sprays; g/100 kg seed for seed treatments)	Literature
Fluotrimazole		JEFFREY *et al.* 1975
1) Fruits, Ornamentals, Vegetables: powdery mildew		LE BON *et al.* 1978
Triadimefon		
1) Cereals: *Erysiphe graminis, Puccinia* sp., *Rhynchcsporium secalis, Septoria tritici, Typhula incarnata*	125—250 g/ha	FROHBERGER 1975 SCHEINPFLUG *et al.* 1978 MARTIN & MORRIS 1979
Apples: *Podosphaera leucotricha, Gymnosporangium* sp.	0.0025—0.01 %	KOLBE 1982a, b
Grapes: *Uncinula necator*	0.0025—0.005 %	KASPERS 1979
Coffee: *Hemileia vastatrix*	125—500 g/ha	KASPERS & PATEL 1980
Vegetables, Ornamentals: powdery mildew, rust	0.0025—0.0125 %	NOEGEL *et al.* 1977
Turf: various diseases		SANDERS *et al.* 1978
Others: powdery mildew, rust, leaf spot diseases on mango, hops, sugar cane, small fruits etc.		
Triadimenol		
1) Cereals: *Erysiphe graminis, Puccinia* sp., *Rhynchosporium secalis, Septoria tritici, Typhula incarnata*	125—250 g/ha	
Apples: *Podosphaera leucotricha, Gymnosporangium* sp.	0.005—0.01 %	
Grapes: *Uncinula necator Guignardia bildwellii*	0.0025—0.005 %	
Coffee: *Hemileia vastatrix*	125—500 g/ha	
Cucurbits: powdery mildew		
Bananas: *Mycosphaerella* sp.	100—125 g/ha	
2) Cereals: *Ustilago* sp., *Tilletia* sp., *Pyrenophora teres, P. avenae, Leptophaeria nodorum, Typhula incarnata, Rhynchosporium secalis, Gerlachia nivale, Erysiphe graminis, Puccinia* sp.	15—37.5 g/100 kg	FROHBERGER 1978 TRAEGNER-BORN & VAN DEN BOOM 1978
Bitertanol		
1) Pome fruits: *Venturia* spp. *Gymnosporangium* spp.	0.005—0.025 %	BRANDES *et al.* 1979
Stonefruits: *Monilinia* sp., *Stigmina carpophila*	0.0125—0.0375 %	
Bananas: *Mycosphaerella* sp., *Guignardia musae*		SCHEINPFLUG &
Peanuts: *Puccinia arachidis, Phoma arachidicola*		VAN DEN BOOM 1981
Sugar beet: *Cercospora beticola, Erysiphe betae*	0.25—0.6 kg/ha	
Others: various diseases in vegetables ornamentals		
2) Cereals: *Tilletia* sp., *Ustilago* sp., *Gerlachia nivale, Leptosphaeria nodorum, Tilletia controversa*	10—75 g/100 kg	TRÄGNER-BORN & KASPERS 1981

Table 13.6 (continued)

Compound Crop/Pathogen 1) Foliar Application 2) Seed Treatment 3) Post-harvest Treatment 4) Other Treatments	Rate of application (g/ha or % a.i. for foliar sprays; g/100 kg seed for seed treatments)	Literature
Propiconazole		
1) Cereals: *Erysiphe graminis, Puccinia* sp., *Septoria tritici, Pyrenophora teres, Rhynchosporium secalis*	125 g/ha	URECH *et al.* 1979; URECH & SPEICH 1981
Peanuts: *Mycosphaerella arachidicola, M. berkeleyi*	100—150 g/ha	
Grapes: *Uncinula necator, Guignardia bildwellii*	0.0025 %	SCHWINN & URECH 1981
Bananas: *Mycosphaerella* spp.	100 g/ha	MOURICHON & BEUGNON 1982
Others: leaf spot diseases, rust, powdery mildew, scab in pecans, soybeans, rice etc.		
∗Etaconazole		
1) Pome Fruit: *Venturia inaequalis, Podosphaera leucotricha, Gymnosporangium* spp.	0.002 %	SZKOLNIK 1981; KELLEY & JONES 1981
Stonefruits: *Monilinia* sp., *Coccomyces hiemalis*		
3) Citrus: *Geotrichum candidum, Penicillium* sp.		GUTTER 1982; SCHACHNAI 1982
Penconazole		
1) Pomefruits: *Podosphaera leucotricha, Venturia* spp.	0.0025 %	EBERLE *et al.* 1983
Grapes: *Uncinula necator, Guignardia bildwellii*	0.0015—0.005 %	
Others: powdery mildew, rust a.o.	0.0025—0.015 %	
Diclobutrazol		
1) Cereals: *Puccinia* sp., *Erysiphe graminis, Rhynchosporium secalis, Typhula incarnata*	125 g/ha	BENT & SKIDMORE 1979
Coffee: *Hemileia vastatrix*		JAVED 1981
Flutriafol		
1) Cereals: *Erysiphe graminis, Puccinia* sp., *Septoria* sp., *Rhynchosporium secalis, Helminthosporium* spp.	125 g/ha	SKIDMOORE *et al.* 1983 DAWSON *et al.* 1984
2) Cereals: *Tilletia* sp., *Ustilago* sp., *Erysiphe graminis*	7.5—30 g/100 kg	NORTHWOOD *et al.* 1984
∗DPX H6573 = Flusilazol		
1) Apples: *Venturia inaequalis, Podosphaera leucotricha*	70—140 g/ha	DAVIS *et al.* 1985
Grapes: *Uncinula necator, Guignardia bildwellii*	35—140 g/ha	
Cereals: *Puccinia* spp., *Erysiphe graminis, Pseudocercosporella herpotrichoides, Septoria* spp., *Pyrenophora* spp., *Rhynchosporium* sp.	100—250 g/ha	MOBERG *et al.* 1985
Peanuts: *Cercospora* spp.	70—140 g/ha	
Bananas: *Mycosphaerella* spp.		
Sugar beets: *Cercospora beticola, Erysiphe betae*		

Table 13.6 (continued)

Compound Crop/Pathogen 1) Foliar Application 2) Seed Treatment 3) Post-harvest Treatment 4) Other Treatments	Rate of application (g/ha or % a.i. for foliar sprays; g/100 kg seed for seed treatments)	Literature
*BAS 454 06 F		
1) Cereals: *Erysiphe graminis, Leptosphaeria nodorum*	38—1000 g/ha	POMMER & ZEEH 1983
Coffee: *Hemileia vastatrix*		
Apples: *Venturia inaequalis*		
Grapes: *Uncinula necator*		
Peanuts: *Cercospora* spp.		
*S-3308 = Diniconazole		
1) Apples: *Venturia inaequalis, Podosphaera leucotricha* *Gymnosporangium juniperi-virginianae*	28—56 g/ha	FUNAKI *et al.* 1983
Stonefruits: *Monilinia* spp.		
Vegetables: rust, powdery mildew		
Wheat: *Puccinia* spp., *Erysiphe graminis*	28—56 g/ha	
Grapes: *Uncinula necator*	28—84 g/ha	
*PP 969		
4) Coffee: *Hemileia vastatrix*		SHEPARD *et al.* 1983
Bananas: *Mycosphaerella musicola*		
Apples: *Podosphaera leucotricha, Venturia inaequalis*		
*Myclobutanil		
1) Apples: *Venturia inaequalis, Podosphaera leucotricha, Gymnosporangium juniperi-virginianae*	0.006—0.0075 %	
Grapes: *Uncinula necator, Guignardia bildwellii*	56—70 g/ha	
Wheat: *Erysiphe graminis, Puccinia* spp., *Septoria* spp., *Pyrenophora tritici-repentis, Fusarium* spp.	224—280 g/ha	
2) Cereals: *Ustilago* spp., *Tilletia* spp., *Pyrenophora* spp., *Fusarium* spp.	20—60 g/100 kg	

Imidazoles in medical and other uses

The introduction of three antifungal imidazole compounds (clotrimazole, miconazole and econazole) in 1969 by Bayer and Janssen opened a new competitive field of research. Many researchers regarded the introduction of the new highly active imidazole antimycotics as a landmark in the control of human mycoses (SCHWINN 1983).

The medically important imidazole compounds have been shown to have a broader antimycotic spectrum and a higher activity against dermatophytes, pathogenic yeasts, filamentous and dimorphic fungi than previous antimycotics. Some of these imidazol-fungicides are also reported to be active against some of the pathogenic grampositive bacteria.

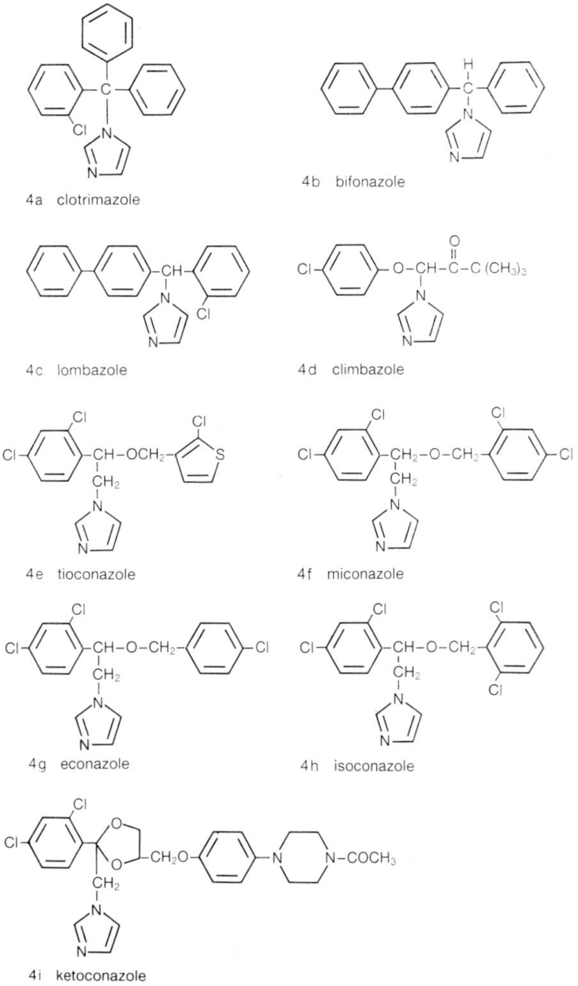

Fig. 13.4 Antifungal imidazoles in medical and other uses.

The most prevalent mycotic infections of humans and animals are cutaneous or superficial in nature and can be treated with topically applied antifungal substances. Undesirable toxicological side effects or insufficient solubility of most imidazoles antimycotics limit the use of this chemical group in treating systemic fungal infections in humans by oral and intravenous applications. The introduction of ketaconazole in 1979 was therefore regarded as promising progress in the chemotherapy of systemic fungal infections.

This article is not intended to give a comprehensive review of medically important antimycotics, but will present an overview on some clinically used compounds in the imidazole group. For additional information, the reader should refer to current reviews on medically important antimycotics (Fromtling 1984; Zirngibl 1983).

Table 13.7 Azoles in medical and other uses

1) Common name {1} 2) Chemical name {2} 3) Trade name(s)	1) Company 2) Patent No., Year {3} 3) Use	Literature
1) **Clotrimazole** 2) 1-[(2-chlorophenyl)-di-phenylmethyl]-1H-imidazol 3) Canesten®, Empecid® Lotrimin®, Mycelex®	1) Bayer 2) DOS 1 617 481, 1967 3) topical antimycotic	BÜCHEL *et al.* 1972 IWATA *et al.* 1973
1) **Bifonazole** 2) 1-(1,1′-biphenyl)-4-yl-phenyl-methyl-1H-imidazol 3) Mycospor®	1) Bayer 2) DOS 2 461 406,1974 3) topical antimycotic	PLEMPEL *et al.* 1983 YAMAGUCHI *et al.* 1983
1) **Lombazole** 2) 1-[(4-biphenyl)-(2-chloro-phenyl)-methyl]-1H-imidazole 3) Twent®	1) Bayer 2) DOS 2 461 406, 1974 3) antiacne compound	BÜCHEL & PLEMPEL 1983
1) **Climbazole** 2) 1-(4-chlorophenoxy)-1-(1H-imidazol-1-yl)-3,3-dimethylbutanon 3) Baysan®, Baypival® Ceox®	1) Bayer 2) DOS 2 105 490, 1971 3) antidandruff compound	BÜCHEL & PLEMPEL 1983
1) **Tioconazole** 2) 1-[2-[(2-chloro-3-thienyl) methoxy]-2-(2,4-dichloro-phenyl)ethyl]-1H-imidazole	1) Pfizer 2) US 4 062 966, 1977 3) topical antimycotic	JEVONS *et al.* 1979
1) **Miconazole** 2) 1-[2-[(2,4-dichlorophenyl) methoxy]-2-(2,4-dichlorophenyl)-ethyl]-1H-imidazol 3) Daktar®, Daktarin® Gyno-Monistat®, Epi-Monistat®	1) Janssen, Cilag 2) DOS 1 940 388, 1967 3) topical and intravenous antimycotic	
1) **Econazole** 2) 1-[2-[(4-chlorophenyl)methoxy]-2-(2,4-dichlorophenyl)ethyl]-1H-imidazol 3) Pevaryl®	1) Janssen 2) DOS 1 940 388, 1967 3) local antimycotic	THIENPONT *et al.* 1975
1) **Isoconazole** 2) 1-[(2-(2,6-dichlorophenyl)-methoxy]-2-(2,4-dichloro-phenyl)ethyl]-1H-imidazol 3) Travogen®, Travocort®	1) Janssen, Schering 2) DOS 1 940 388, 1967 3) local antimycotic	WATANABE & NOMURA 1978 WENDT & KESSLER 1979
1) **Ketoconazole** 2) cis-1-acetyl-4-[4-[2-(2,4-dichlorophenyl)-2-(1H-imidazol-1-yl-methyl)-1,3-dioxolan-4-yl-methoxy]phenyl]piperazin 3) Nizoral®	1) Janssen 2) DOS 2 804 096, 1978 3) oral antimycotic	BORELLI *et al.* 1979 N. N. 1980

For footnote see table 13.1

References

ABDEL-RAHMAN, M.: Morphological effects of fungicides on apple trees. Proc. Amer. Phytopathological Soc. **4** (1977): 213.

ADLUNG, K. G., and DRANDAREVSKI, C. A.: The evaluation of "CELA W 524", a systemic fungicide, for the control of powdery mildew and apple scab. Proc. 6th Brit. Insectic. Fungic. Conf. **2** (1971): 577—586.

ATTABHANYO, A., and HOLCOMB, G. E.: Control of Fusarium wilt of mimosa with systemic fungicides. Plant Disease Rep. **60** (1976): 56—59.

BALDWIN, B. C., and WIGGINS, T. E.: Biochemical studies on the diclobutrazol series of systemic fungicides. Symp. Systemische Fungizide und Antifungale Verbindungen, Reinhardsbrunn 1983, Abstracts p. 4.

BENT, K. J., and SKIDMORE, A. M.: Diclobutrazol: a new systemic fungicide. Proc. 1979 Brit. Crop Protect. Conf. (1979): 477—484.

BERAUD, J. M., GUEGUEN, F., LECA, J. L., TUSSAC, M., and BONQUET, G.: Qu'est-ce que le fenarimol? La Defense des Vegetaux **34** (1980): 17—24.

BIRCHMORE, R. J., BROOKES, R. F., COPPING, L. G., and WELLS, W. H.: BTS 40 542 — a new broad spectrum fungicide. Proc. 1977 Brit. Crop Protect. Conf. Vol. 2 (1977): 593—598.

— WELLS, W. H., and COPPING, L. G.: A new group of fungicidally-active metal co-cordination compounds based on prochloraz. Proc. 1979 Brit. Crop Protect. Conf. Vol. 2 (1979): 583—601.

BON, Y. LE, BOURDIN, J., and BERTHIER, G.: Efficacy of various fungicides against rose mildew *(Sphaerotheca pannosa* var. *rosae)*. Phytiatrie-Phytopharmacie **27** (1978): 199—205.

BORELLI, D., FUENTES, J., LEIDERMANN, R., RESTREPO, M. A., BRAN, J. L., LEGENDRE, R., LEVINE, H. B., and STEVENS, D. A.: Ketoconazole, an oral antifungal: Laboratory and clinical assessment of imidazole drugs. Postgrad. Med. J. **55** (1979): 657—661.

BOTTER, A. A.: Topical treatment of nail and skin infections with miconazole, a new broad-spectrum antimycotic. Mykosen **14** (1971): 187—191.

BOUDIER, B.: Essai de lutte contre le dessèchement de aiguilles du a Meria laricis sur melezes d' Europe (Larix decidua) en pepinieres. Revue Forestière Francaise **33** (1981): 394—399.

BRANDES, W., STEFFENS, W., FÜHR, F., and SCHEINPFLUG, H.: Further studies on translocation of [^{14}C] triadimefon in cucumber plants. Pflanzenschutz-Nachr. Bayer **31** (1978): 132—144. Physiolog. Plant **62** (1984): 422—426.

— KASPERS, H., and KRÄMER, W.: ®Baycor, a new foliar-applied fungicide of the biphenyloxytriazolylmethane group. Pflanzenschutz-Nachr. Bayer **32** (1979): 1—16.

BRISTOW, P. R.: Effect of triforine on pollen germination and fruit set in highbusch blueberry. Plant Dis. **65** (1981): 350—353.

BROWN, I. F., and HALL, H. R.: Certain biological properties of fenarimol applicable to its field use. Proceedings 1981 Brit. Crop Protect. Conf. (1981): 573—578.

— — and MILLER, J. R.: EL-273, a curative fungicide for the control of Venturia inaequalis. Phytopathology **60** (1970): 1013—1014.

BRUCHHAUSEN, V., VON, and STIASNI, M.: Transport of the systemic fungicide CELA W 524 (Triforine) in barley plants. II. Uptake and metabolism. Pesticide Sci. **4** (1973): 767—773.

BRUHN, J. A., FORT, T. M., and DENIS, S. J.: DPX-H 6573: a broad spectrum fungicide for the control of field crop diseases. National Meeting of the Amer. Phytopathological Soc., August 1985.

BÜCHEL, K. H., DRABER, W., REGEL, E., and PLEMPEL, M.: Synthesis and properties of clotrimazole and other antimycotic l-triphenylmethyl imidazoles. Drugs Germ. **15** (1972): 79—94.

— and PLEMPEL, M.: The azole story. In: FINDRA, J. S., and LEDNICER, D. (Eds.): Chronicles of Drug Discovery, Vol. 2. John Wiley & Sons Inc. 1983, pp. 235—269.

BUCHENAUER, H.: Systemisch-fungizide Wirkung und Wirkungsmechanismus von Triadimefon (MEB 6447). Mitteilungen aus der Biol. Bundesanst. Land- und Forstw. Berlin-Dahlem **165** (1975): 154—155.

— Mode of action and selectivity of fungicides which interfere with ergosterol biosynthesis. Proc. 1977 Brit. Crop Protect. Conf. (1977): 699—711.

— Comparative studies on the antifungal activity of triadimefon, triadimenol, fenarimol, nuarimol, imazalil and fluotrimazole *in vitro*. Z. Pflanzenkrankh. Pflanzenschutz **86** (1979a): 341—354.

— Conversion of triadimefon into two diastereomeres, triadimenol I and triadimenol II, by fungi and plants. IX. Intern. Congr. Plant Pathol., Washington USA (1979b): Abstr. no. 939.

— Kohts, T., und Roos, H.: Zur Wirkungsweise von Propiconazol (Desmel®), CGA 64 251 und Dichlobutrazol (Vigil®) in Pilzen und Gerstenkeimlingen. Mitteilungen der Biol. Bundesanst. Land- und Forstw. Berlin-Dahlem **203** (1981): 310—311.

— and Röhner, E.: Zum Wirkungsmechanismus von Imazalil in Pilzen und Pflanzen. Proc. V. Intern. Symp. Systemfungizide Reinhardsbrunn 1977, Akademie-Verlag, Berlin 1979, pp. 175—185.

— — Effect of triadimefon and triadimenol on growth of various plant species as well as on Gibberellin content and sterol metabolism in shoots of barley seedlings. Pesticide Biochem. Physiol. **15** (1981): 58—70.

— — Aufnahme, Translokation und Transformation von Triadimefon in Kulturpflanzen. Z. Pflanzenkrankh. Pflanzenschutz **89** (1982): 385—389.

Büchel, K. H.: History of Azole Chemistry. Burdick and Jackson Symp.; 188th ACS National Meeting, Philadelphia 1984.

Byford, W. J.: Experiments with fungicide sprays in late summer and early autumn on sugar beet root crops. Ann. appl. Biol. **86** (1977): 47—57.

Casanova, A., Döhler, R., Farrant, D. M., and Rathmell, W. G.: The activity of the fungicide nuarimol against diseases of barley and other cereals. Proc. Brit. Crop Protect. Conf. (1977): 1—7.

Cayley, G. R., Hide, G. A., and Tillotson, Y.: The determination of imazalil on potatoes and its use in controlling potato storage diseases. Pesticide Sci. **12** (1981): 103—109.

Church, R. M., and Williams, R. R.: Fungicide toxicity to apple pollen in the anthers. J. Horticult. Sci. **53** (1978): 91—94.

Cooke, B. K., Pappas, A. C., Jordan, V. W. L., and Western, N. M.: Translocation of benomyl, prochloraz and procymidone in relation to control of *Botrytis cinerea* in strawberries. Pesticide Sci. **10** (1979): 467—472.

Davis, A. E., Fort, T. M., Denis, S. J., and Henry, M. J.: DPX-H 6573: a broad spectrum fungicide for the control of tree-fruit diseases. National Meeting Amer. Phytopathol. Soc., August, 1985.

Dawson, M., Torcheux, R., and Horrellou, A.: Qu'est-ce que le "Impact" Sopra®? La Defense des Végétaux **226** (1984): 77—85.

Deas, A. H. B., Clark, T., and Carter, G. A.: The enantiomeric composition of triadimenol produced during metabolism of triadimefon by fungi. Part I: Influence of dose and time of incubation. Pesticide Sci. **15** (1984a): 63—70.

— — — The enantiomeric composition of triadimenol produced during metabolism of triadimefon by fungi. Part II.: Differences between fungal species. Pesticide Sci. **15** (1984b): 71—77.

Dekker, J.: Counter measures for avoiding fungicide resistance. In: Dekker, J., and Georgopoulos, S. G. (Eds.): Fungicide Resistance in Crop Protection. Pudoc, Wageningen 1982, pp. 177—186.

— The development of resistance to fungicides. Progr. Pesticide Biochem. and Toxicol., Vol. 4 (1985): 166—209.

Döhler, R., and Mertz, M. V.: Nuarimol — ein neues systemisches Fungizid zur Bekämpfung von Krankheiten in Gerste. Mitt. Biol. Bundesanst. Land- und Forstw. Berlin-Dahlem **191** (1979): 185.

Drandarevski, C. A., and Mayer, E.: Eine Methode zur Untersuchung der Penetration von systemischen Fungiziden durch Blattkutikula und Epidermis. Meded. Fac. Landbouwwet. Rijksuniv. Gent **39** (1974): 1127—1143.

— and Schicke, P.: Formation and germination of spores of *Venturia inaequalis* and *Podosphaera leucotrica* following curative treatment with triforine. Z. Pflanzenkrankh. Pflanzenschutz **83** (1976): 385—396.

Ebenebe, C., Bruchhausen, V., von, and Grossmann, F.: Dosage-response curve of wheat brown rust to triforine supplied via root treatment. Pesticide Sci. **5** (1974): 17—24.

— Fehrmann, H., and Grossmann, F.: Effects of a new systemic fungicide, piperazin.-1,4-diylbis-(1-[2,2,2-trichloro-ethyl]formamide), against wheat leaf rust. Plant Dis. Rep. **55** (1971): 691—694.

Eberle, J., Ruess, W., and Urech, P. A.: CGA 71818, a novel fungicide for the control of grape and pome fruit diseases. Proc. 10th Intern. Congr. Plant Protect., Vol. 1, Brighton (1983): 376—383.

Edgington, L. V.: Structural requirements of systemic fungicides. Ann. Rev. Phytopathol. **19** (1981): 107—124.

Fletcher, R. A.: Plant growth regulating properties of sterol-inhibiting fungicides. In: Purohit, S. S. (Ed.): Hormonal Regulation of Plant Growth and Development. Vol. 2. Agro Botanical Publ. 1985.

— and Hofstra, G.: Triadimefon a plant multi-protectant. Plant and Cell Physiol. **26** (1985): 775—780.

— and Nath, V.: Triadimefon reduces transpiration and increases yield in water stressed plants. Physiolog. Plant **62** (1984): 422—426.

Förster, H., Buchenauer, H., and Grossmann, F.: Side effects of the systemic fungicides triadimefon and triadimenol on barley plants. I. Influence on growth and yield. Z. Pflanzenkrankh. Pflanzenschutz **87** (1980a): 473—492.

— — — Side-effects of the systemic fungicides triadimefon and triadimenol on barley plants. II. Cytokininlike effects. Z. Pflanzenkrankh. Pflanzenschutz **87** (1980b): 640—653.

— — — Side-effects of the systemic fungicides triadimefon and triadimenol on barley plants. III. Further effects of metabolism. Z. Pflanzenkrankh. Pflanzenschutz **87** (1980c): 717—730.

Fort, T. M., and Moberg, W. K.: DPX-H 6573, a new broad spectrum fungicide candidate. Proc. 1984 Brit. Crop Protect. Conf., Vol. 2 (1984): 413—419.

Frate, C. A., Leach, L. D., and Hills, F. J.: Comparison of fungicide application methods for systemic control of sugar beet powdery mildew. Phytopathology **69** (1979): 1190—1194.

Frohberger, P. E.: New approaches to the control of cereal diseases with triadimefon (MEB 6447). VIII. Intern. Plant Protect. Congr. 1975, Moscow, Section III (1975): 247—258.

— Baytan®, ein neues systemisches Breitband-Fungizid mit besonderer Eignung für die Getreidebeizung. Pflanzenschutz-Nachr. Bayer **31** (1978): 11—24.

Fromtling, R. A.: Imidazoles as medically important antifungal agents: an overview. Drugs of Today **20** (1984): 325—349.

Fuchs, A., and Drandarevski, C. A.: Wirkungsbreite und Wirkungsgrad von Triforine *in vitro* und *in vivo*. Z. Pflanzenkrankh. Pflanzenschutz **80** (1973): 403—417.

— and Ost, W.: Translocation, distribution and metabolism of triforine in plants. Arch. Environment. Contamination and Toxicology **4** (1976): 30—43.

— Viets-Verweij, M., and Vries, F. W. de: Metabolic conversion in plants of the systemic fungicide triforine (N,N′-bis-[1-formamido-2,2,2-trichloroethyl]-piperazine; CELA W 524). Phytopathol. Z. **75** (1972): 111—123.

— Vries, F. W. de, and Aalbers, M. J.: Uptake, distribution and metabolic fate of ^{3}H-triforine in plants. I. Short-term experiments. Pesticide Sci. **7** (1976a): 115—126.

— — — Uptake, distribution and metabolic fate of ^{3}H-triforine in plants. II. Long-term experiments. Pesticide Sci. **7** (1976b): 127—134.

— Ruig de, S. P., Tuyl van, J. M., and Vries de, F. W.: Resistance to triforine: a non existent problem? Netherl. J. Plant Pathol. **83**, Suppl. 1 (1977): 189—205.

Führ, F., Paul, V., Steffens, W., and Scheinpflug, H.: Translokation von ^{14}C Triadimefon nach Applikation auf Sommergerste und seine Wirkung gegen *Erysiphe graminis* var. *hordei*. Pflanzenschutz-Nachr. Bayer **31** (1978): 116—131.

Funaki, Y., Ishiguri, Y., Kato, T., and Tanaka, S.: Structure-activity relationships of a new fungicide S-3308 and its derivatives. Proc. Intern. Congr. Pesticides Chem. 5th, 1982 (publ. 1983) Vol. 1: 309—314.

Gasztonyi, M., and Josepovits, G.: Translocation and metabolism of triadimefon in different plant species. Acta Phytopathologica Academiae Scientiarum Hungaricae **13** (1978): 403—415.

— — The activation of triadimefon and its role in the selectivity of fungicide action. Pesticide Sci. **10** (1979): 57—65.

Gestel, J. van, Cutsem, J. van, and Thienpont, D.: Vapour phase activity of imazalil. Chemotherapy **27** (1981): 270—276.

Gilmour, J.: Comparison of some aspects of mildew fungicides use on spring barley in South-East Scotland in 1982 and 1983. Proc. Brit. Crop Protect. Conf. Brighton 1984, Vol. 1 (1984) 109—114.

Globerson, D., and Eliasi, R.: The response to Saprol® (systemic fungicide) in lettuce species and cultivars and its inheritance. Euphytica **28** (1979): 115—118.

GREWE, F., and BÜCHEL, K. H.: Ein neues Mehltaufungizid aus der Klasse der Trityltriazole. Mitt. Biol. Bundesanst. Land- und Forstw. Berlin-Dahlem **151** (1973): 208—209.

GUTTER, Y.: Comparative effectiveness of Sonax, thiabendazole and sodium orthophenylphenate in controlling green mould of citrus fruits. Z. Pflanzenkrankh. Pflanzenschutz **89** (1982): 332—226.

HAAS, E.: Spritzversuche gegen *Oidium* der Reben. Obstbau Weinbau **21** (1984): 82—83.

HANSEN, K. E.: Experiments with cereal seed dressings: II. Field experiments. Tidsskrift for Plante avl. **85** (1981): 77—92.

HARRIS, R. G., and BARNES, G.: Prochloraz: the control of net blotch and Septoria in winter cereals. Proc. 1981 Brit. Crop Protect. Conf. (1981): 268—274.

HEERES, J.: Structure-activity relationships in a group of azoles, with special reference to 1,3-Dioxolan-2-ylmethyl derivatives. Pesticide Sci. **15** (1984): 268—279.

HOFFMANN, J. A., and WALDHER, J. T.: Chemical seed treatment for controlling seedborne and soilborne common bunt of wheat. Plant Dis. **65** (1981): 256—259.

HOLLOMON, D. W., BUTTERS, J., and CLARCK, J.: Genetic control of triadimenol resistance in barley powdery mildew. Proc. Brit. Crop Protect. Conf. Brifhton 1984, Vol. 2 (1984): 477—482.

HORELLOU, A., and DAWSON, M.: Qu'est-ce que le Vigil et le Vigil K?. La Defense des Vegetaux **207** (1981): 9—21.

HUGGENBERGER, F., COLLINS, M. A., and SKYLAKAKIS, G.: Decreased sensitivity of *Sphaerotheca fuliginea* to fenarimol and other ergosterol-biosynthesis inhibitors. Crop Protect. **3** (1984): 137—149.

IZUMI, K., YAMAGUCHI, I., WADA, A., OSHIO, H., and TAKAHASHI, N.: Effects of a new plant growth retardant (E)-1-(4-Chlorophenyl)-4,4-dimethyl-2-(1,2,4-triazol-1-yl)-1-penten-3-ol (S 3307) on the growth and Gibberellin content of rice plants. Plant and Cell Physiol. **25** (1984): 611—617.

— KAMIYA, Y., SAKURAI, A., OSHIO, H., and TAKANASHI, N.: Studies of site of action of a new plant growth retardant (E)-1-(4-Chlorophenyl)-4,4-dimethyl-2-(1,2,4-triazol-1-yl)-1-penten-3-ol (S 3307) and comparative effects of its stereo-isomers in a cell-free system from *Cucurbita maxima*. Plant and Cell Physiol. **26** (1985): 821—827.

IWATA, K., YAMAGUCHI, H., and HIRATANI, T.: The mode of action of clotrimazole. Sabouraudia **11** (1973): 158—168.

JAVED, Z. U. R.: Field trials with new and recommended fungicides for leaf rust control during 1980. Kenya Coffee **46** (1981): 239.

JEFFREY, R. A., ROWLEY, N. K., and SMAILES, A.: The effect of fluotrimazole on powdery mildew of cereals. Proc. 1975 Brit. Insecticide Fungicide Conf., Vol. 2 (1975): 429—436.

JENKYN, J. F., DYKE, G. V., and TODD, A. D.: Effects of fungicide movement between plants in field experiments Plant Pathol. **32** (1983): 311—324.

JEVONS, S., GYMER, G. E., BRAMMER, K. W., COX, D. A., and LEMMING, M. R. G.: Antifungal activity of tioconazole (UK-20, 349), a new imidazole derivative. Antimicrob. Agents Chemother. **15** (1979): 597—602.

KASPERS, H., and PATEL, N. K.: Versuche zur Bekämpfung des Kaffeerostes (*Hemileia vastatrix* Berk. et Br.) in Kenya. Pflanzenschutz-Nachr. Bayer **33** (1980): 152—164.

KATO, T., TANAKA, S., UEDA, M., and KAWASE, Y.: Effects of the fungicide, S-1358, on general metabolism and lipid biosynthesis in *Monilinia fructigena*. Agricult. Biol. Chemistry **38** (1974): 2377—2384.

— — YAMAMOTO, S., KAWASE, Y., and UEDA, M.: Fungitoxic properties of a N-3-pyridylimidadithiocarbonate derivative. Ann. Phytopathol. Soc. of Japan **41** (1975): 1—8.

KELLEY, R. D., and JONES, A. L.: Evaluation of two triazole fungicides for postinfection control of apple scab. Phytopathology **71** (1981): 737—742.

— — Volatility and systemic properties of etaconazole and fenapanil in apple. Canad. J. Plant Pathol. **4** (1982): 243—246.

KOLBE, W.: Zehn Jahre Versuche mit ®Bayleton zur Mehltaubekämpfung im Getreidebau (1971 bis 1981). Pflanzenschutz-Nachr. Bayer **35** (1982a): 36—71.

— Versuche zur Bekämpfung des Obstbaumkrebses mit ®Bayleton. Pflanzenschutz-Nachr. Bayer **35** (1982b): 152—170.

KRÄMER, W., BÜCHEL, K. H., and DRABER, W.: Structure activity correlation in the azoles. In: MIYAMOTA, J., and KEARNEY, P. C. (Eds.): Pesticide Chemistry: Human Welfare and the Environment. Vol. 1. Pergamon Press, Oxford 1983, pp. 223—232.

Kraus, P.: Studies on the mechanism of action of ®Baycor. Pflanzenschutz-Nachr. Bayer **32** (1979): 17—30.

— Untersuchungen zur Aufnahme und zur Verteilung von Bayleton® in Weinreben. Pflanzenschutz-Nachr. Bayer **34** (1981): 197—212.

Kuck, K. H.: Pflanzenschutz-Nachr. Bayer (1986): in preparation.

— Scheinpflug, H., Tiburzy, R., and Reisener, H. J.: Fluorescence microscopy studies of the effect of ®Bayleton and ®Baytan on growth of stem rust in the wheat plant. Pflanzenschutz-Nachr. Bayer **35** (1982): 209—228.

Leroux, P., and Gredt, M.: Cross resistance between ergosterol biosynthesis inhibiting fungicides in *Aspergillus nidulans, Botrytis cinerea, Penicillium expansum* and *Ustilago maydis.* Netherl. J. Plant Pathol. **87** (1981): 240—241 (abstracts).

Limpert, E., and Fischbeck, G.: Regionale Unterschiede in der Empfindlichkeit des Gerstenmehltaus gegen Triadimenol und ihre Entwicklung im Zeitraum 1980 bis 1982. Phytomedizin **13** (1983): 19 (abstracts).

Luz, W. C., and Vieira, J. C.: Seed treatment with systemic fungicides to control *Cochliobolus sativus* on barley. Plant Dis. **66** (1982): 135—136.

Mantinger, H., and Vigl, J.: Ergebnisse des Mehltauversuchs. Obstbau Weinbau **12** (1975): 341—344.

Martin, R. A., and Edgington, L. V.: Comparative systemic translocation of several xenobiotics and sucrose. Pesticide Biochem. Physiol. **16** (1981): 87—96.

— — Antifungal, phytotoxic and systemic activity of fenapanil and structural analogs. Pesticide Biochem. Physiol. **17** (1982): 1—9.

— and Morris, D. B.: ®Bayleton als systemisches Fungizid zur Bekämpfung von Blattkrankheiten bei Sommer- und Wintergerste in Großbritannien. Pflanzenschutz-Nachr. Bayer **32** (1979): 31—82.

Melin, P., Plaud, G., Dezenas Du Montcel, H., and Laville, E.: Activité comparée de l'imazalil sur la cercosporiose du bananier au Cameroun. Fruits **30** (1975): 301—306.

Moberg, W. K., Basarab, G. S., Cuomo, J., and Liang, P. H.: Biologically active organosilicon componds: Fungicidal Silylmethyltriazoles. 190th National Meeting Am. Chemical Soc., Chicago 1985.

Mourichon, X., and Bengnon, M.: Comparative effectiveness of ®Tilt (CGA 64259) on banana leaf spot in the Ivory Coast. Fruits **37** (1982): 595—597.

N. N.: First International Symposium on Ketaconazole. Ref. Inf. Dis., Vol. 2, No. 4, Ed.: A. Restrepo, D. A., Stevens, J. P. Utz (1980).

Noegel, K. A., *et al.:* ®Bayleton: A potential fungicide for ornamental crops. Hortsci. **12** (1977): 408.

Northwood, P. J., Horellou, A., and Heckele, K. H.: PP 450: Field experience with a new cereal fungicide. Proc. 10th Intern. Congr. Plant Protect. 1983, Vol. 3 (1983): 930.

— Paul, J. A., and Gibbard, M.: FF 4050 seed treatment — a new approach to control barley diseases. Proc. 1984 Brit. Crop Protect. Conf. Vol. 1 (1984): 47—52.

Ohkawa, H., Shibaike, R., Okihara, V., Moridawa, M., and Miyamoto, J.: Degradation of the fungicide Denmert® (s-n-butyl-S'-p-tert-butylbenzyl N-3-pyridyldithiocarbaonimidate, S-1358) by plants, soils and light. Agricult. Biol. Chemistry **40** (1976): 943—951.

Pepin, H. S., MacPherson, E. A., and Clements, S. J.: Effect of triadimefon on the growth of Willamette red raspberry. Canad. J. Plant Sci. **60** (1980): 1203—1208.

Piening, L. J., Duczek, L. J., Atkinson, T. J., and Davidson, J. G. N.: Control of common root rot and loose smut and the phytotoxicity of seed treatment fungicides on Gateway barley. Canad. J. Plant Pathol. **5** (1983): 49—53.

Plempel, M., Regel, E., and Büchel, K. H.: Antimycotic efficacy of Bifonazole *in vitro* and *in vivo.* Drug Res. **33** (1983): 517—524.

Pommer, E. H., and Zeeh, B.: BAS 454 06 F, a new triazole derivative for the control of phytopathogenic fungi. 4th Intern. Congr. Plant Pathol., Melbourne 1983.

Rathmell, W. G., and Skidmore, A. M.: Recent advances in the chemical control of cereal rust diseases. Outlook on Agriculture **11** (1982): 37—43.

Reisdorf, K., Wurzer-Fassnacht, U., and Walther, F. H.: Zur systemischen Wirkung von Imazalil. Z. Pflanzenkrankh. Pflanzenschutz **90** (1983): 641—649.

ROHRBACH, K. U.: Der Einsatz von Triforine als Beizmittel zu Sommergerste. Mitt. Biol. Bundesanst. Land- und Forstw. **178** (1977): 147—148.

SAINT-BLANQUAT DE, A., and MY, J.: Qu'est-ce que le Prochloraz? La Defense des Vegetaux **221** (1983): 121—141.

SANDERS, P. L., BURPEE, L. K., COLE, H. jr., and DUICH, J. M.: Uptake, translocation and efficacy of triadimefon in control of turfgrass pathogens. Phytopathology **68** (1978): 1482—1487.

SCHACHNAI, A.: Evaluation of the fungicides CGA 64 251, Guazatine, Sodium o-Phenylphenate, and imazalil for the control of sour rot on lemon fruits. Plant Dis. **66** (1982): 733—735.

SCHEINPFLUG, H., and VAN DEN BOOM, T.: ®Baycor, a new fungicide for tropical and subtropical crops. Pflanzenschutz-Nachr. Bayer **34** (1981): 8—28.

— and PAUL, V.: On the mode of action of triadimefon. Netherl. J. Plant Pathol. **83** (1977): 105—111.

— — and KRAUS, P.: Studies on the mode of action of Bayleton® against cereal diseases. Pflanzenschutz-Nachr. Bayer **31** (1978): 110—115.

SCHEPERS, H. T. A. M.: Development and persistence of resistance to fungicides in *Sphaerotheca fuliginea* in cucumbers in the Netherlands. Doktorthesis Proefschrift van de Landbouwhogeschool te Wageningen (1985): 1—55.

SCHICKE, P., and VEEN, K. H.: A new systemic, CELA W 524 (N,N'-bis-[1-formamido-2,2,2-trichloroethyl)-piperazine) with action against powdery mildew, rust and apple scab. Proc. 5th Brit. Insecticide Fungicide Conf. (1969): 569—575.

SCHMITT, P., and BENVENISTE, P.: Effect of fenarimol on sterol biosynthesis in suspension cultures of bramble cells. Phytochemistry **18** (1979): 1659—1665.

SCHNEIDER, A., KREMER-SCHILLINGS, W., and DROSIHN, G.: Sportak® Alpha — eine Fungizidkombination gegen parasitären Halmbruch in Weizen. Ges. Pflanzen **37** (1985): 23—29.

SCHWINN, F.: Ergosterol biosynthesis inhibitors. An overview of their history and contribution to medicine and agriculture. Pesticide Sci. **15** (1983): 40—47.

SCHWINN, F. J., and URECH, P. A.: New approaches for chemical disease control in fruit and hops. Proc. 1981 Brit. Crop Protect. Conf. (1981): 819—833.

SHABI, E., ELISHA, S., and ZELIG, Y.: Control of pear and apple diseases in Israel with sterol-inhibiting fungicides. Plant Dis. **65** (1981): 992—994.

SHEPHARD, M. C.: Fungicide behaviour in the plant: systemicity. Proc. Bordeaux Mixture Centenary Meeting. Vol. 1 (BCPC Monograph No. 31) (1985): 99—196.

— FRENCH, P. N., and RATHMELL, W. G.: PP 969: A broad spectrum systemic fungicide for injection or soil application. Proc. 10th Intern. Congr. Plant Protect. Vol. 2 (1983): 521.

— — Biochemical and cellular aspects of the antifungal action of ergosterol biosynthesis inhibition. In: TRINCI, A. P. J., and RYLEY, J. F. (Eds.): Mode of Action of Antifungal Agents. Cambridge Univ. Press, Cambridge 1984, pp. 257—282.

SISLER, H. D., RAGSDALE, N. N., and WATERFIELD, W. W.: Biochemical aspects of the fungitoxic and growth regulatory action of fenarimol and other pyrimidin-5-ylmethanols. Pesticide Sci. **15** (1984): 167—176.

SKIDMORE, A. M., FRENCH, P. N., and RATHMELL, W. G.: PP 450: a new broad-spectrum fungicide for cereals. Proc. 10th Intern. Congr. Plant Protect. 1983, Vol. 1 (1983): 368—375.

SMITH, J. M., and SPEICH, J.: Propiconazole: disease control in cereals in Western Europe. Proc. 1981 Brit. Crop Protect. Conf. (1981): 291—297.

SMITH, J. W. M.: Triforine sensitivity in lettuce *(Lactuca sativa):* A potentially useful genetic marker. Euphytica **28** (1979): 351—360.

STAUB, T., SCHWINN, F., and URECH, P.: CGA-64251, a new broad spectrum fungicide. Phytopathology **69** (1979): 1046.

STEFFENS, W., FÜHR, F., KRAUS, P., and SCHEINPFLUG, H.: Uptake and distribution of ®Baytan in spring barley and spring wheat after seed treatment. Pflanzenschutz-Nachr. Bayer **35** (1982): 171—188.

— and WIENECKE, J.: Behaviour and fate of the ^{14}C-labelled mildew fungicide fluotrimazole in plant and soil. I. Radioactivity distribution in treated spring barley and in the soil of field lysimeters and uptake by untreated rotational crops. Z. Pflanzenkrankh. Pflanzenschutz **88** (1981): 343—354.

STRYDOM, D. K., and HONEYBORNE, G. E.: Increase in fruit set of "Starking Delicious" apple with triadimefon. Hort Sci. **16** (1981): 51.

SZKOLNIK, M.: Physical modes of action of erol-inhibiting fungicides against apple diseases. Plant Dis. **65** (1981): 981—985.

SZKOLNIK, M.: Unique vapour activity by CGA-64251 (Vangard) in the control of powdery mildews roomwide in the greenhouse. Plant Dis. **67** (1983): 360—366.

TAKANO, H., OGURI, Y., and KATO, T.: Mode of action of (E)-1-(2,4-dichlorophenyl)-4,4-dimethyl-2-(1,3,4-triazol-lyl)-1-penten-3-ol (S-3308) in *Ustilago maydis*. J. Pesticide Sci. **8** (1983): 575 to 582.

THIEMPONT, D., VAN CUTSEM, J., VAN NUETEN, J. M., NIEMEGEERS, D. J. E., and MARSBOOM, R.: Biological and toxicological properties of econazole, a broad spectrum antimycotic. Arzneimittel-Forsch. **25** (1975): 224—230.

THOMAS, J., and CORNIER, A.: Étude du prochloraz sur céréales. Défense des Vógétaux **221** (1983): 143—156.

TRÄGNER-BORN, J., and VAN DEN BOOM, T.: Über Ergebnisse von Freilandversuchen mit ®Baytan, einem neuen systemischen Getreidebeizmittel. Pflanzenschutz-Nachr. Bayer **31**/1978): 25—37.

— and KASPERS, H.: Zur Bekämpfung von Zwergsteinbrand an Winterweizen mit ®Sibutol. Pflanzenschutz-Nachr. Bayer **34** (1981): 1—7.

TUYL, VAN, J. M.: Genetics of fungal resistance to systemic fungicides. Meded. Landbouwwet. Wageningen, **77—2** (1977): 137.

URECH, P. A., SCHWINN, F. J., SPEICH, J., and STAUB, T.: The control of airborne diseases of cereals with CGA 64 250. Proc. Brit. Crop Protect. Conf. (1979): 508—515.

URECH, P. A., and SPEICH, J.: Propriétés du CGA 64250 (Tilt®) et activité contre les maladies des cereales. Phytiatrie-Phytopharmacie **30** (1981): 21—26.

VANDEN BOSSCHE, H., LAUWERS, W., WILLEMSENS, Y., MARICHAL, P., CORNELISSEN, F., and COOLS, W.: Molecular basis for the antimycotic and antibacterial activity of N-substituted imidazoles and triazoles: the inhibiton of isoprenoid biosynthesis. Pesticide Sci. **15** (1984): 188—198.

VERGNES, A., and PISTRE, R.: Sur les traitments contre l'oidium. Progres Agricole et Viticole **99** (1982): 528—530.

VERHEYDEN, C.: Curative control of apple scab (Venturia inaequalis). Meded. Fac. Landbouwwet. Rijsuniv. Gent **46** (1981): 955—960.

VOGEL, C., STAUB, T., RIST, G., and STURM, E.: The four isomers of etaconazole (CGA 64251) and their fungicidal activity. In: MIYAMOTO, S., and KEARNEY, P. C. (Eds.): Pesticide Chemistry: Human Welfare and the Environment. Vol. I, Pergamon Press, Oxford 1983, pp. 303—308.

VONK, J. W., and DEKHUIJZEN, H. M.: Transport and metabolism of ^{3}H-Imazalil in barley and cucumber. Meded. Fac. Landbouwwet. Rijsuniv. Gent **44** (1979): 927—934.

WAARD DE, M. A., and GIESKES, S. A.: Characterization of fenarimol-resistant mutants of *Aspergillus nidulans*. Netherl. J. Plant Pathol. **83**, Suppl. 1 (1977): 177—188.

— and NISTELROOY VAN, J. G. M.: Mechanism of resistance to fenarimol in *Aspergillus nidulans*. Pesticide Biochem. Physiol. **10** (1979): 219—229.

— — An energy-dependent efflux mechanism for fenarimol in a wild-type strain and fenarimol-resistant mutants of *Aspergillus nidulans*. Pesticide Biochem. Physiol. **13** (1980): 255—266.

— and FUCHS, A.: Resistance to ergosterol biosynthesis inhibitors. II. Genetic and physiological aspects. In: DEKKER, J., and GEORGOPOULOS, S. G. (Eds.): Fungicide Resistance in Crop Protection. Pudoc, Wageningen 1982, pp. 87—100.

WAKERLEY, S. B., and RUSSEL, P. E.: Prochloraz — a decade of development. Proc. Bordeaux Mixture Centenary Meeting, BCPC Monograph No. 31, Vol. 2 (1985): 257—260.

WATANABE, S., and NOMURA, H.: Clinical effects of Isoconazole on dermatomycoses and its antimycotic activities *in vitro*. Acta Dermatologica **73** (1978): 209.

WENDT, H., and KESSLER, H. J.: Antimikrobielle Wirkung des Breitspektrum-Antimykotikums Isoconazolnitrat im Humanversuch. Drug Res. **29** (1979): 846.

WOLFE, M. S., and FLETCHER, J. T. (1981): Insensitivity of *Erysiphe graminis* f. sp. *hordei* to triadimefon. Netherl. J. Plant Pathol. **87** (1981): 239.

YAMAGUCHI, H., HIRATANI, T., and PLEMPEL, M.: *In vitro* studies of a new imidazole-antimycotic, Bifonazole, in comparison with Clotrimazole and Miconazole. Drug Res. **33** (1983): 546—551.

ZAAYEN, A. VAN, and ADRICHEM, J. C. J. VAN: Prochloraz for control of fungal pathogens of cultivated mushrooms. Netherl. J. Plant Pathol. **88** (1982): 203—213.

ZIRNGIBL, L.: Fifteen years of structural modifications in the field of antifungal monocyclic 1-substituted 1H-azoles. Progr. Drug Res. **27** (1983): 253—383.

LYR, H. (Ed.): Modern Selective Fungicides — Properties, Applications, Mechanisms of Action. Longman Group UK Ltd., London, and VEB Gustav Fischer Verlag, Jena, 1987.

Chapter 14

Mechanism of action of triazolyl fungicides and related compounds

H. BUCHENAUER
Institute of Plant Disease and Plant Protection University Hannover, FRG

Introduction

The sterol C^{14}-demethylation inhibiting fungicides (SDIs) comprise a large number of economically highly important substances for controlling a broad spectrum of fungal diseases in plants and animals (cf. chapter 13). The structural variety in both the heterocyclic rings and the lipophilic substituents may account for their selectivity and variation of the antifungal spectrum.

Pyrimidine-fungicides include triarimol, fenarimol, and nuarimol. Furthermore, closely related derivatives of the pyrimidine methanol fungicides have been synthesized and developed as plant growth retardants (e.g. ancymidol) (SHIVE and SISLER 1976; COOLBAUGH and HAMILTON 1976) (Fig. 14.1).

Imidazole antifungal compounds have been developed for control of fungal diseases in plants as well as in humans and animals.

Triazole fungicides comprise the largest group of SDIs. Many have already been or are being, developed for use as agricultural fungicides.

Pyridine-derivatives contain buthiobate (KATO *et al.* 1974, 1975) and EL-241 (SHERALD *et al.* 1973).

The **piperazine**-compound triforine (SHERALD and SISLER 1975) differs from the aforementioned fungicide groups with respect to the N-heterocyclic ring.

Mode of action in fungi

Effect on ergosterol synthesis

The antifungal compounds belonging to the N-substituted triazole, imidazole, pyrimidine and pyridine as well as the piperazine derivatives interfere at low concentrations in sterol synthesis of fungi. In the presence of these fungicides ergosterol synthesis was early and profoundly inhibited while other metabolic processes in treated fungi were almost unimpaired.

Incorporation studies with [^{14}C]-acetate revealed that inhibition of ergosterol synthesis became evident already after short term incubation (e.g. 0.5—1 h) (RAGSDALE 1975; KATO *et al.* 1975). While ergosterol synthesis was markedly inhibited simultaneously C-4,4 dimethyl-(eburicol), C-4 methyl-(obtusifoliol) and C-14 methyl (14-methyl fecosterol) sterols accumulated in a number of fungi tested, e.g. in sporidia of *Ustilago maydis* (RAGSDALE and SISLER 1972; RAGSDALE 1975; HENRY and SISLER 1979; EBERT 1983), mycelium of *Aspergillus fumigatus* (SHERALD and SISLER 1975), *Monilinia fructigena* (KATO et al. 1975), *Botrytis cinerea* (KATO 1980), *Penicillium expansum* (LEROUX and GREDT 1978) and *Ustilago avenae* sporidia (BUCHENAUER 1975, 1976, 1977a, b, c, 1978a, b; BUCHENAUER and KEMPER 1981; GIRARDET and BUCHENAUER 1980; KRAUS 1979) (Fig. 14.2).

After increasing incubation periods (e.g. 13 h) a novel sterol metabolite, 14α-methylergosta 8,24(28)-diene-3β,6α-diol accumulated in etaconazole treated *U. maydis* sporidia (EBERT *et al.* 1983) and small amounts of the unusual sterol ergosta-5,7-dien-3β-ol were detected in triarimol treated *A. fumigatus* (SHERALD and SISLER 1975).

While in most fungi the principal functional sterol is considered to be ergosterol, there are certain groups of fungi producing instead of ergosterol other 4,14-demethyl sterols. For example, in *Taphrinales* species brassicasterol (ergosta-5,22-dienol) was identified as the principal functional sterol. In propiconazole treated *T. deformans* methyl sterols such as lanosterol and 24-methylene dihydrolanosterol (eburicol) accumulated (WEETE *et al.* 1983).

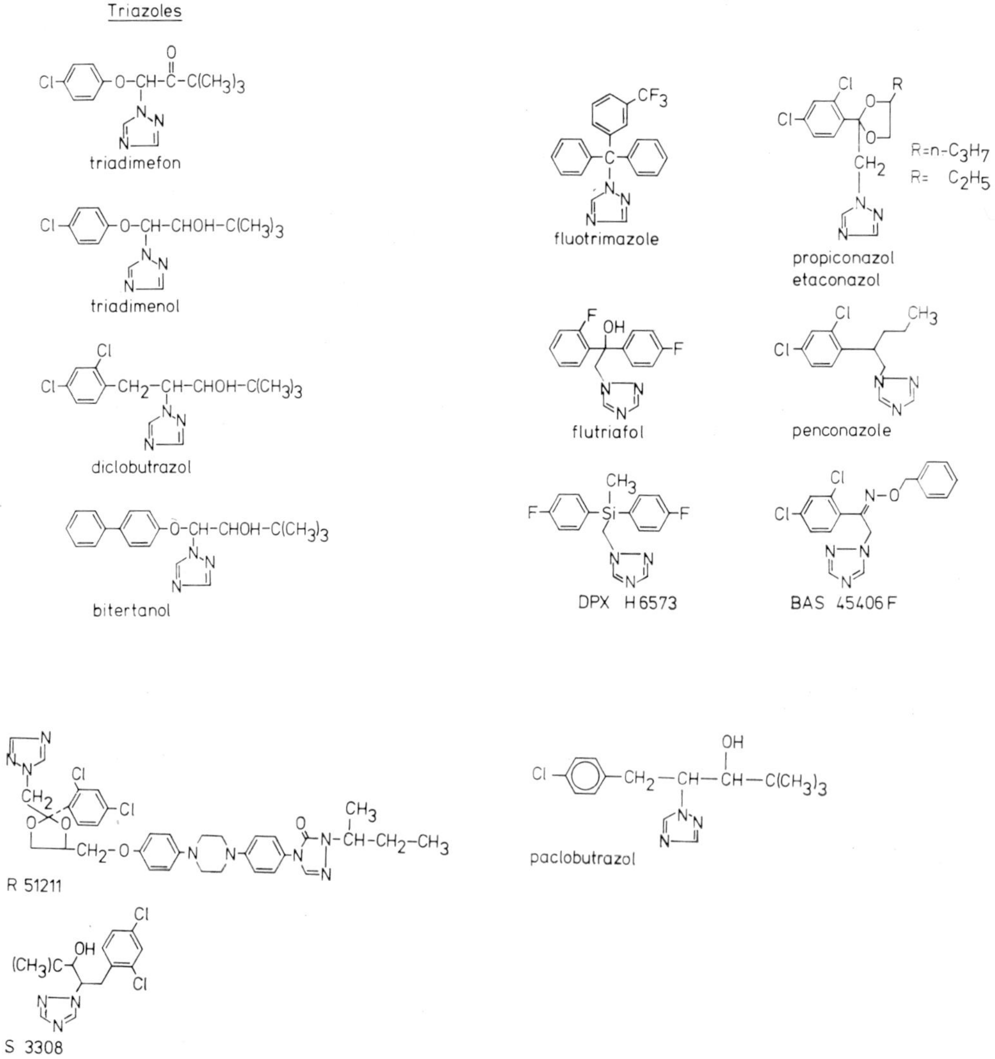

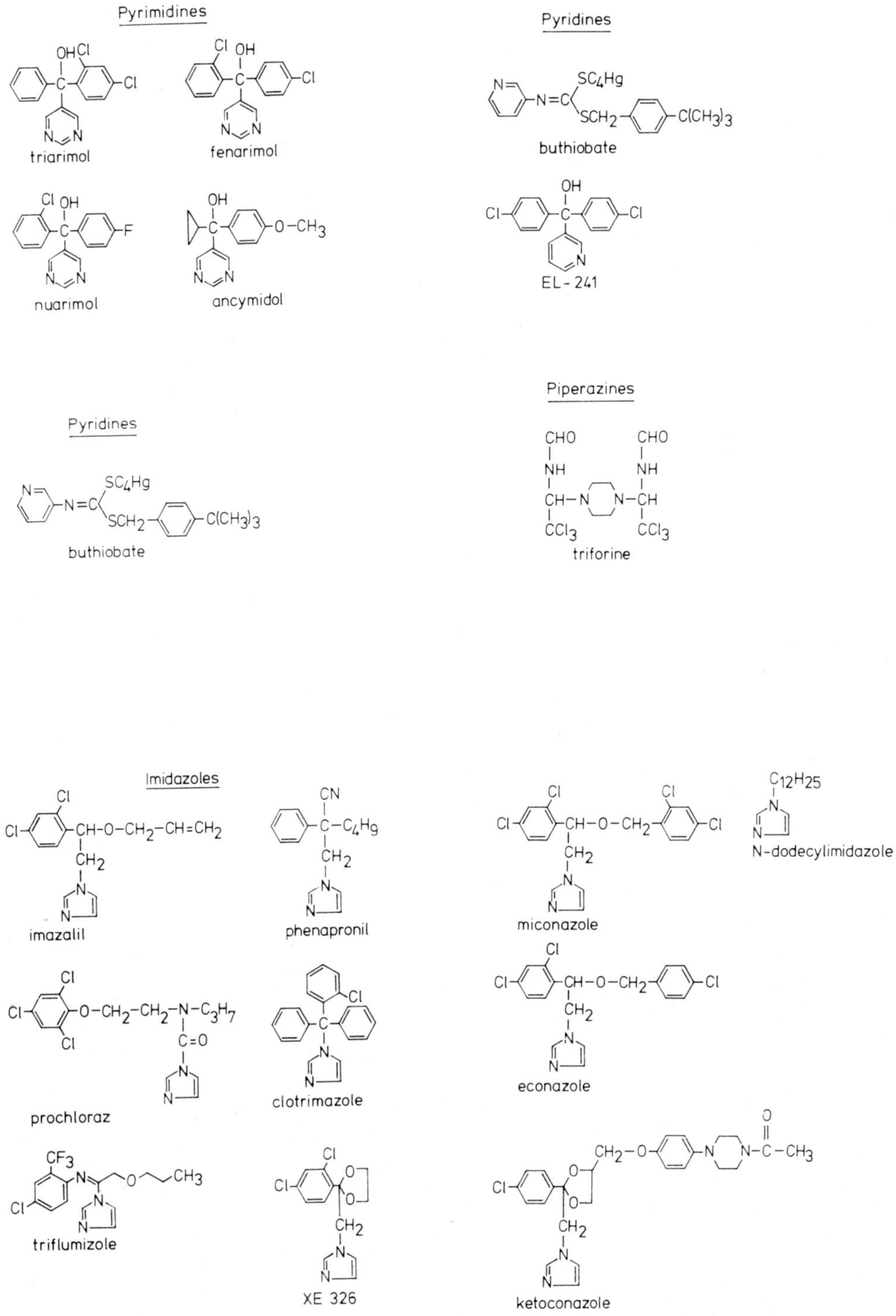

Fig. 14.1 Structural formulae of pyrimidine, pyridine, piperazine, imidazole and triazole derivatives.

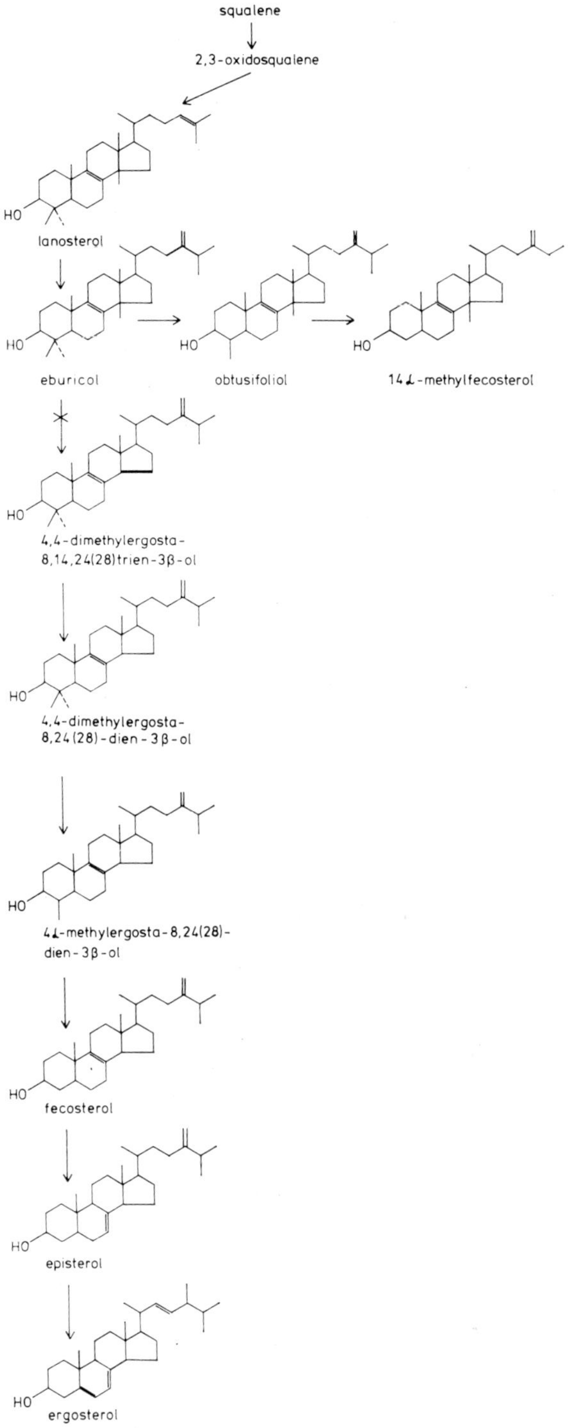

Fig. 14.2 Scheme of ergosterol pathway showing site of action of C-14-demethylation inhibitors (→).

The medicinally used antimycotic imidazole (e.g. clotrimazole, miconazole, ketoconazole, econazole, parconazole, terconazole, bifonazole) and the triazole-derivatives (e.g. R 51211) are also potent inhibitors of ergosterol synthesis in yeasts and dermatophytes (e.g. *Candida albicans, Microsporum canis, Trichophyton mentagrophytes* and *Epidermophyton floccosum*) (Van den Bossche *et al.* 1978, 1980, 1984a; Berg *et al.* 1983; Mariott 1980). Treatment of *C. albicans* with some of these compounds resulted in accumulation of sterols containing a 14α-methyl group: lanosterol, 24-methylenedihydrolanosterol, 4,14-dimethylzymosterol, obtusifoliol and 14-methyl-24-methyleneergosterol (Berg *et al.* 1983; Van den Bossche *et al.* 1978, 1980, 1982a, 1984a).

Oomycetes species (e.g. *Saprolegnia ferax*) which do not synthesise sterols accumulate in the presence of triarimol lanosterol while in the nontreated mycelium predominantly desmethylsterols were detected (Berg *et al.* 1983).

It has been shown in numerous studies that fungi treated with SDIs generally contain higher sterol contents than untreated ones (Sherald and Sisler 1975; Ragsdale and Sisler 1973; Ragsdale 1975; Buchenauer 1977a, b, 1978b; Leroux and Gredt 1978; Weete *et al.* 1983; Baldwin and Wiggins 1984). The accumulation of the methyl sterol intermediates, especially of eburicol, obtusifoliol and 14α-methylfecosterol may probably be reduced to a deranged feedback mechanism caused by low levels of the sterol end products (Sisler and Ragsdale 1984).

Since in the presence of these compounds no sterol intermediates were found beyond the C-14 demethylation reactions of the sterol pathway it is generally considered that the inhibition of C-14 demethylation is the most sensitive site of action of these antifungal agents (Fig. 14.2).

In the removal of the 14α-methyl group three NADPH-dependent oxygenase reactions are involved. In the first step the 14α-methyl group (CH_3) is oxidized to a 14α-hydroxymethyl group (CH_2OH), in the second step the hydroxymethyl group is oxidized to a formoyl group (CHO) and in the third step the latter is oxidized and lost as formate (HCOOH) along with the 15α-hydrogen (Mitropoulos *et al.* 1976; Gibbons *et al.* 1979; Mercer 1984). Of the mono-oxygenases catalysing the first three steps, the first step is catalysed by a cytochrome P-450 while the other two are not (Gibbons *et al.* 1979). In a further step, the resulting 14, 15-double bond is saturated by an NADPH-requiring reductase. The absence of oxygenated intermediates during the C-14 demethylation reactions indicates that the first oxygenation step which is catalysed by a cytochrome P-450 enzyme is the primary site of action of the SDIs.

It is suggested that the SDIs may interact with a specific form of the microsomal cytochrome P-450 that is involved in the biosynthesis of cholesterol or ergosterol. Microsomes are known to contain various cytochromes of the P-450 type participating in mixed function oxidase reactions. All cytochromes have in common the same haem system, and their different activities may be attributed to the protein portions of the enzymes (Mitropoulos *et al.* 1976). It is likely therefore, that a specific cytochrome exists for the specific oxidation of lanosterol and lanosterol-like compounds (Fig. 14.3).

The pyrimidine, pyridine, imidazole and triazole derivatives with the exception of triforine have in common an unsubstituted nitrogen atom in position 3 of the heterocyclic ring and this nitrogen atom is also not hindered by substituents at the adjacent ring positions.

It has been shown that imidazole derivatives interact with iron haem proteins and possess a high affinity to cytochrome P-450 (Wilkinson *et al.* 1972) producing a typical type II difference spectrum (Hajek *et al.* 1982; Hajek and Novak 1982). Type II difference spectra were also obtained in studies with triazole derivatives (e.g. diclobutrazole) (Gadher *et al.* 1983; Wiggins and Baldwin 1984) and imidazole derivatives (e.g. miconazole and ketoconazole) (Van den Bossche *et al.* 1984a)

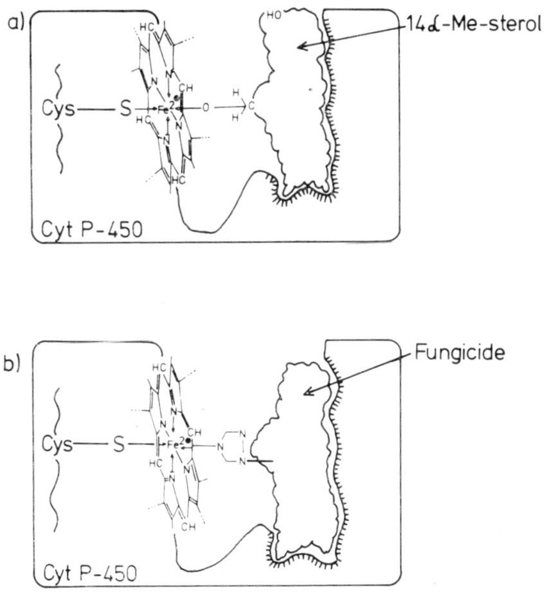

Fig. 14.3 Assumed structures of cytochrome P-450-lanosterol 14-demethylation transition state complex (a) and the cytochrome fungicide complex (b) (GADHER *et al.* 1983).

suggesting a binding of the unsubstituted meta nitrogen atom in the heterocyclic ring of the SDI-fungicides to the haem iron of cytochrome P-450 (Fig. 14.3).

This binding excludes the natural dioxygen reaction leading to the formation of a superoxide-Fe^{3+} complex (Fe^{3+}—O_2^-) necessary for hydroxylation of the 14α-methyl group of sterol intermediates (e.g. lanosterol, 24-dihydrolanosterol and other 14α-methyl sterols) resulting in accumulation of 14α-methylated sterols rather than 14α-hydroxymethyl or 14α-formyl sterols (AOYAMA and YOSHIDA 1978a, b; GADHER *et al.* 1983; VAN DEN BOSSCHE 1984a).

The binding of sterol C-14 demethylation inhibitors to cytochrome P-450 not only depends on interference of the fungicide to the haem iron atom but also on the affinity to the adjacent regions of the enzyme. Studies on the interaction of diclobutrazol and its isomers with a highly purified preparation of cytochrome P-450 from yeast microsomes revealed a correlation between fungicide binding and activity. The more toxic isomer (2 R, 3 R) exhibited a higher binding than the less active isomer (2 S, 3 S) (WIGGINS and BALDWIN 1984). It has been shown that azole fungicides not only bind to cytochrome P-450 of yeast microsomes but also of rat liver microsomes. MITROPOULOS *et al.* (1976) already demonstrated that triarimol also inhibited lanosterol 14α-demethylase of rat liver microsomal fraction, indicating a close similarity between the enzymes participating in 14α-demethylation from fungal and animal cells. The selective toxicity of the antimycotic drugs has been studied by VAN DEN BOSSCHE *et al.* (1984b), WILLEMSENS *et al.* (1980), and HEERES *et al.* (1981) using subcellular fractions of *C. albicans*, *S. cerevisiae*, and rat liver. Cholesterol synthesis in rat liver subcellular fractions used to be 20—70 times less sensitive to ketoconazole or miconazole than ergosterol synthesis in microsomal fractions from *C. albicans* and *S. cerevisiae*. Studies with other azole derivatives (e.g. imazalil, propiconazole, terconazole, itroconazole and R 51211) also indicated that the cytochrome P-450 preparations from rat liver were much less sensitive to these compounds than cytochrome P-450 suspensions from yeast microsomes (VAN DEN BOSSCHE *et al.* 1984a, b). Selective toxicity of ketokonazole on biosynthesis of ergosterol in fungi and cholesterol in mammalian cells was further studied by oral treatment of *C. albicans*-infected rats. On the basis of ED_{50}-values at least a six-fold higher concentration of the substance was required to inhibit cholesterol synthesis in the liver of the rat than to reduce ergosterol synthesis in *C. albicans* (VAN DEN BOSSCHE *et al.* 1984b). These inhibitors not only interfered in C-14 demethylation of fungi and animal cells but also in various species of algae (FRASINEL *et al.* 1978) and higher plants (SCHMITT and BENVENISTE 1979).

While these compounds are highly potent inhibitors of C-14 demethylation they do not interfere with oxidative demethylations at C-4 positions. The C-14 demethylation is carbonmonoxide-sensitive and cyanide-insensitive, the C-4 demethylation is cyanide sensitive and carbonmonoxide-insensitive (GIBBONS *et al.* 1979).

The enzymes of fungi are able to remove the 4α-methyl groups of sterols still containing the 14α-methyl group, whereas enzymes of animals do not possess this capacity. Thus, in the presence of C-14 demethylation inhibitors 4,4,14α-trimethyl sterols usually do not accumulate to appreciable amounts in fungi (SHERALD and SISLER 1975; BUCHENAUER 1977a, b, 1978a, b; RAGSDALE 1975; LEROUX and GREDT 1978; DE WAARD and RAGSDALE 1979).

In animals, however, C-4 demethylation takes place only after removal of the 14α-methyl group and 4,4,14α-trimethyl sterols (lanosterol and 24, 25 dehydrolanosterol) accumulate when 14α-demethylation inhibitors are present. Enzymes of green plants show similar capacities like those from fungi (GIBBONS and MITROPOULOS 1973; MITROPOULOS *et al.* 1976).

The SDIs may differ in their ability to affect other enzymes of sterol metabolism. Triarimol had no effect on HMG-CoA reductase and only slightly interfered in introduction of the sterol C-5(6) double bond at a concentration that inhibited C-14 demethylase activity more than 90%. On the other hand, BERG *et al.* (1984) showed that the antimycotic compound bifonazole, a powerful inhibitor of C-14 demethylase at low concentrations, additionally inhibited at higher concentrations HMG-CoA reductase in the dermatophyt *Trichophyton mentagrophytes*. Clotrimazol did not display this property. Whereas the primary site of action of bifonazole was regarded as fungistatic, its secondary mode of action was considered as fungitoxic (PLEMBEL 1983; BERG *et al.* 1984). 1-Dodecylimidazole inhibiting at low concentrations sterol C-14 demethylation interfered at higher doses in 2,3-oxidosqualene cyclization and subsequent transmethylation (ATKIN *et al.* 1972; HENRY and SISLER 1979).

There are indications that triarimol inhibits the introduction of C-22 double bond in *U. maydis sporidia* (RAGSDALE 1975), *A. fumigatus* mycelium (SHERALD *et al.* 1975), and algae (FRASINEL *et al.* 1978).

It has been shown recently that a monoxygenase which contains cytochrome P-450, participates in C-22 desaturase activity of *S. cerevisiae* (HATA *et al.* 1981).

The SDIs hardly interferred in other reactions during synthesis of ergosterol, such as introduction of the C-14 methyl group and the shift of the double bond to C-24(28) and that from C-8(9) to C-7 as well as in the introduction of a double bond at C-5(6) (SHERALD and SISLER 1975; RAGSDALE 1975).

Sterols interact particularly with the acyl groups of phospholipids to modulate the degree of membrane fluidity (BUTTKE and BLOCH 1980; NES *et al.* 1978; YEAGLE *et al.* 1980; DEMEL *et al.* 1972).

Qualitative changes of sterol composition (e.g. presence of C-14 methyl sterols) in plasma membranes containing high levels of sterols will probably result in altered membrane fluidity. Because of the planarity of the predominant sterols such as ergosterol and cholesterol, extra methyl groups on the α-face increase the thickness of the molecule and prevent the correct packing into the lipid bilayer of the membrane resulting in disorder of the membrane structure (BLOCH 1982).

Changes of the functional properties of the fungal plasma membrane after treatment with C-14 demethylation inhibitors have been reported. Plasma membranes of *T. deformans* treated with propiconazole contained C-14 methyl sterols and these sterols were not functionally equivalent to 4,14-demethylsterols (WEETE *et al.* 1983). Polarisation studies indicated that membranes of *S. cerevisiae* and *U. maydis* cells treated with diclobutrazol were more disordered than those of control cells (BALDWIN and WIGGINS 1984).

Differences in interactions between ergosterol and lanosterol with phospholipids in artificial membranes have been obtained. Incorporation of ergosterol into unilamellar vesicles composed of phosphatidylcholine, -ethanolamine and -glycerol (phospholipid: ergosterol ratio 2: 1) reduced the release of entrapped glucose by 57 %. On the other hand, lanosterol did not retain the release of entrapped glucose from the vesicles (VAN DEN BOSSCHE *et al.* 1982a; VAN DEN BOSSCHE *et al.* 1984b). Different interactions of ergosterol and lanosterol with phospholipids in membranes were also obtained by using differential calorimetry. Lanosterol affected the enthalpy of melting to a smaller degree than ergosterol indicating that lanosterol had a lower effect on membrane fluidity than ergosterol (BERG *et al.* 1984; VAN DEN BOSSCHE *et al.* 1984b). Lanosterol affects the orientation of phospholipids not only in artificial membranes but also in those of intact cells. *S. cerevisiae* cells grow under aerobic conditions when ergosterol is supplied, whereas lanosterol did not allow growth under these conditions (VAN DEN BOSSCHE *et al.* 1980; NES 1974; BUTTKE and BLOCH 1980; TAYLOR and PARKS 1980).

Mutants of *S. cerevisiae* lacking the sterol C-14 demethylation were viable with 14-methylfecosterol as the major sterol (PIERCE *et al.* 1978; TROCHA *et al.* 1977).

A strain of *U. maydis* deficient in sterol C-14 demethylation showed a similar growth rate and morphology as the fenarimol-treated wild-type strain (WALSH and SISLER 1982), and a sterol deficient *S. cerevisiae* strain grew in the presence of cholesterol and 14α-methylcholest-7-en-3β-ol, but not when C-4 methyl sterols were added (BUTTKE and BLOCH 1981). The SDIs proved to be fungicidal in *M. fructicola* and *C. lunata*, since they are unable to remove the two C-4 methyl groups in the presence of the C-14 methyl groups (SISLER *et al.* 1984).

These studies show that SDIs act fungistatically when fungi are able to remove C-4 methyl groups in presence of the C-14 methyl group and display fungicidal activity when fungi do not possess this capacity. This differential effectiveness is obviously organism specific.

Changes in membrane structure also resulted in increase of membrane permeability, differences in activities of membrane bound enzymes and differential uptake of nutrients. On increased membrane permeability and membrane damage in fungi after treatment with SDIs has been reported. Furthermore, at higher concentrations miconazole, ketoconazole and clotrimazole exhibited a direct effect on fungal plasma membrane causing leakage of intracellular constituents (fungicidal activity), whereas at lower dosages of the compounds fungistasis was observed correlating with the inhibition of C-14 demethylase (IWATA *et al.* 1973; DE NOLLIN and BORGERS 1975; SUD and FEINGOLD 1981; UNO *et al.* 1982). The structural changes of membranes caused by the SDIs may affect specific activities of membrane bound enzymes. Differences in activity have been found in plasma membrane ATPase of *S. cerevisiae* (DUFOUR *et al.* 1980). The alterations in the membrane structure induce a general activation of chitin synthetase. It is assumed that under normal growth conditions chitin synthetase which has been found distributed in the membrane in a latent stage, is probably activated by a protease at the specific sites of chitin synthesis (ROBERTS *et al.* 1983). The SDIs may increase the liberation of the protease which results in an uncontrolled chitin synthetase activation. The integrity of the plasma membrane is considered as a prerequisite to concentrate amino acids from the medium into the cells. In the presence of diclobutrazole the ability of *S. cerevisiae* and *U. maydis* to accumulate f.e. histidine, leucine and arginine was significantly reduced (THOMAS 1983; see BALDWIN and WIGGINS 1984).

Effect on synthesis of other lipid components

The specific inhibition of C-14 demethylation during sterol synthesis also resulted in alterations of other major lipid fractions. The accumulation of free fatty acids in cells of *U. maydis* (RAGSDALE and SISLER 1973; SHERALD and SISLER 1975; RAGSDALE 1975; BALDWIN and WIGGINS 1984), *U. avenae* (BUCHENAUER 1976b, 1977a, b, c, d, 1978a, b; KRAUS 1979) and *T. deformans* (WEETE *et al.* 1983) was delayed compared to the inhibition of ergosterol synthesis. In *U. maydis sporidia* treated with triarimol the portion of 18: 2 acids was increased whereas the percentage of 18: 0 and 18: 1 acids decreased in the free fatty acid, di- and triglyceride fractions (RAGSDALE 1975). Similar changes in fatty acids were obtained in triadimefon and triadimenol treated *U. avenae* sporidia (BUCHENAUER 1977a, 1978b) and in propiconazole treated *T. deformans* cells (WEETE *et al.* 1983). The increase in the free fatty acid content probably contributes to the toxicity, since it is known that high concentrations of free fatty acids are toxic to fungal cells (RIETH 1977; SUMRELL *et al.* 1978). The increased degree of unsaturation of fatty acids also occurred in the phospholipid fractions. Studies with propiconazole treated *T. deformans* cells showed that especially phosphatidylcholine (PC), phosphatidylethanolamine (PE), phosphatidylinositol (PI), and cardiolipin contained a higher degree of free fatty acids (WEETE *et al.* 1983). This effect was probably necessary to maintain a certain degree of membrane fluidity. Furthermore, a reduction in polar lipid fraction has been observed in various fungi treated with SDIs (SHERALD and SISLER 1975; BUCHENAUER 1978a, b; SISLER and RAGSDALE 1977; KATO *et al.* 1974; WEETE *et al.* 1983). This effect has been attributed to the breakdown rather than to the impaired synthesis of the polar lipids (SISLER and RAGSDALE 1977).

In propiconazole treated *T. deformans* cells the most significant quantitative reductions were found in the phospholipid fractions PC and PI and it is assumed that PI might be involved in cell wall formation (WEETE *et al.* 1983). SMITH and BERRY (1974) demonstrated that *S. cerevisiae* forms in an inositol deficient medium buds but these are unable to separate from mother cells. These morphological effects correspond to that induced by SDIs in budding fungi (SANCHOLLE *et al.* 1983; BUCHENAUER 1976).

Effect on respiration

It has been reported in numerous papers that SDIs did not appreciably affect respiration (glucose or acetate oxidation) in fungi at concentrations that severely inhibited ergosterol synthesis. The degree of respiration inhibition paralleled that of dry weight increase (RAGSDALE and SISLER 1973; KATO *et al.* 1974; SHERALD and SISLER 1975; BUCHENAUER 1976a, 1977a, b, c, 1978a, b; SIEGEL and RAGSDALE 1978; KRAUS 1979; DE WAARD and VAN NISTELROOY 1980; SISLER *et al.* 1984).

On the other hand, SHIGEMATSU *et al.* (1982) and UNO *et al.* (1982) assumed that ketoconazole would specifically interfere in cytochrome C oxidase in isolated mitochondria (concentrations $>0.4\,\mu M$) and in intact cells of *C. albicans* (concentrations >0.1 mM). However, this effect may be secondary compared to the sensitivity of ergosterol synthesis, which was inhibited by 50% at 5 nM ($= 5\times10^{-9}$ M) of the compound (VAN DEN BOSSCHE *et al.* 1980). Studies of SISLER *et al.* (1984) and DE WAARD and NISTELROOY (1980) indicated that the cytochrome-mediated respiration of *S. cerevisiae*, *U. maydis* and *A. nidulans* proved not to be very sensitive to fenarimol,

Effect on nucleic acid and protein synthesis

Synthesis of nucleic acids (DNA and RNA) and proteins were also only slightly affected by the sterol C-14 demethylation inhibitors in fungi tested, the retardation of these metabolic pathways corresponded approximately to that of dry weight increase (RAGSDALE and SISLER 1973; KATO *et al.* 1974; SHERALD and RAGSDALE 1975; BUCHENAUER 1976a, 1977a, b, c, 1978a, b; SIEGEL and RAGSDALE 1978; KRAUS 1979; DE WAARD and VAN NISTELROOY 1980; KATO *et al.* 1980). With increasing exposure, however, the uptake of exogenous nucleotides (e.g. uridine) and amino acids (e.g. arginine, leucine, histidine) gradually declined which possibly resulted from changes in membrane properties (BALDWIN and WIGGINS 1984).

Effects on growth, morphology, and cytology

The SDIs induce distinct morphological alterations in fungi, but their ultimate effects on multiplication of sporidia, hyphal growth and spore germination depend on the individual fungal species tested. One of the characteristics of the sporidia of *U. maydis* and *U. avenae* as well as of the bud cells of *T. deformans* in the presence of SDIs is that multiplication of sporidia and buds continued to form in liquid culture, however, daughter cells did not separate from mother cells. Compared to control cells treated sporidia appeared multicellular, often branched and swollen (RAGSDALE and SISLER 1973; BUCHENAUER 1975, 1976a, b, 1977a, b, c; KRAUS 1979; KERKENAAR and BARUG 1984; WEETE *et al.* 1983). The effects on reduction of dry weight became evident already at low concentrations of the inhibitors but growth was not completely arrested even when higher concentrations were applied. Similar effects on growth pattern and morphological alterations have been observed in *C. albicans* in presence of various antimycotic drugs (DE NOLLIN and BORGERS 1975). In filamentous fungi the compounds caused distorted as well as exaggeratedly branched and swollen germ tubes and hyphae, e.g. in *C. cucumerinum* (SHERALD *et al.* 1973; DE WAARD and GIESKES 1977; BUCHENAUER and RÖHNER 1977); *M. fructigena* (KATO *et al.* 1974), and *B. cinerea* (KATO *et al.* 1980). While spore germination of *C. cucumerinum* was hardly affected by various SDIs (SHERALD *et al.* 1973; BUCHENAUER 1977e), germination of spores in other fungal species, e.g. *P. italicum*, *P. expansum*, is significantly inhibited already at low concentrations of imazalil or etaconazole and conidia were swollen (SIEGEL *et al.* 1977; TEPPER and YODER 1982; KERKENAAR and BARUG 1984).

It has been mentioned already that the accumulation of 14α-methyl sterols and the decreased ergosterol content induced by the SDIs results in an activation of uncoordinated chitin synthetase. Chitin is an essential component of the primary septum in budding fungi necessary for separating daughter and mother cells (CABIB *et al.* 1982).

It also plays an important role in septum and primary wall formation in filamentous fungi (GOODAY 1978; ROGERS *et al.* 1980).

Imazalil and fenpropimorph caused an irregular deposition of β-1,3- and β-1,4-polysaccharides (probably chitin) in sporidia of *U. maydis* and hyphae of *P. italicum* observed by fluorescence microscopy using an optical brightener (e.g. "Diethanol") which binds preferentially to the chitin component β-1,4 polymer of N-acetylglucosamine (KERKENAAR and BARUG 1984). In *C. albicans* cells treated with ketoconazole and R 51211 chitin-like material was irregularly distributed while in control cells chitin-like material was observed only at the site of the septum and below of the apex of hyphae (VAN DEN BOSSCHE *et al.* 1983). It is assumed that the increased and irregular deposition of chitin affects in budding fungi the normal processes of cell separation leading to chains and clusters of inter-connected cells and in filamentous fungi to excessive branched, distorted and thickened hyphae and in some cases to bursting of cells (Sisler *et al.* 1983; Kerkenaar and Barug 1984; Fuchs *et al.* 1977).

Ultrastructural studies with *C. albicans* (Voigt and Plembel 1974; Voigt and Schnell 1974; Iwata *et al.* 1973; De Nollin and Borgers 1974, 1976; Borgers 1980), *Trichophyton rubrum* (Preusser 1976), *Histoplasma capsulatum* and *Paracoccidioides brasiliensis* (Negroni de Bonvehi *et al.* 1981), *Botrytis allii* (Richmond and Pring 1979) and *Ustilago avenae* (Hippe 1982) revealed early changes in the plasmalemma after treatment with SDIs suggesting a direct relationship between the biochemical and structural alterations. The plasmalemma was characterised by invaginations and accumulation of extra cytoplasmic vesicles indicating an excessive exocytosis. Scanning electron microscopy studies after freeze fracture showed that triadimefon, nuarimol and imazalil caused hemispherical pits, protrusions, invaginations and aggregation of intramembrane particles on plasmalemma of *U. avenae* sporidia (Hippe 1984). These alterations may refer to structural and functional damages of cellular membranes.

With extended incubation period the membraneous structures of the plasmalemma and the inner mitochondria became fragmentary followed by lysis of the intracytoplasmic membrane systems.

Treatment of *C. albicans, U. avenae* and other fungi also resulted in increase in number and size of vacuoles and lipid droplets. In vacuoles membraneous like material and vesicles accumulated that could be extruded by exocytosis processes and which were thought to cause the cell wall abnormalities (De Nollin and Borgers 1976; Borgers 1980; Hippe *et al.* 1980). In treated *U. avenae* sporidia endoplasmic reticulum near the cell wall accumulated, the number and size of mitochondria and lipid bodies increased and the mitochondria showed an irregular shape. The mitochondrial alterations may also be caused by the interference of the fungicides in ergosterol synthesis. For normal development and possibly also function mitochondria seem to require ergosterol (Taylor and Parks 1980; Thompson and Parks 1971; Mandal *et al.* 1978). The relative slow effect of the C-14 demethylation inhibitors on respiration may be due to the lower ergosterol content in mitochondrial membranes or to slower turnover of sterol in mitochondria.

The ultrastructural studies confirmed an increase in thickness of cell walls and septa as well as an incomplete septa formation in treated *C. albicans* (Voigt and Plembel 1974; De Nollin and Borgers 1974), *B. allii* (Richmond and Pring 1979), *U. avenae* (Hippe *et al.* 1980) and *T. deformans* (Weete *et al.* 1983).

Resistance

Fungal strains resistant to SDIs can easily be obtained in laboratory experiments by selection of untreated or mutagen-treated (either following irradiation with ultraviolet light or treatment with N-methyl-N′-nitro-N-nitrosoguanidine) spores on agar media containing lethal concentrations of the SDIs: *C. cucumerinum* (Sherald *et al.* 1973; Fuchs *et al.* 1977; Buchenauer and Röhner 1977), *A. nidulans* (De Waard and Sisler 1976; De Waard and Gieskes 1977), *U. maydis* (Barug and Kerkenaar 1979), *B. cinerea* (Leroux and Gredt 1984). Generally, the resistant mutants are cross resistant to most of the SDIs (Sherald *et al.* 1973; De Waard and Sisler 1976; Buchenauer and Röhner 1977; Barug and Kerkenaar 1979). However, some mutants in different fungal species did not exhibit cross resistance. For instance, imazalil resistant strains of *C. cucumerinum* (Fuchs *et al.* 1977; van Tuyl 1977), *Phialophora cinerescens* (van Tuyl 1977) and *A. nidulans* (De Waard and Gieskes

1977) proved to be sensitive to fenarimol. Similar results have been reported in a number of mutants of *U. maydis* (Barug and Kerkenaar 1979, 1984).

Studies of Fuchs and De Vries (1984) revealed that the diastereomers of some triazole fungicides displayed a differential toxicity against wild-type and resistant strains of *C. cucumerinum*. While the activity ratio of the diastereomers of bitertanol, diclobutrazol and etaconazol was almost 1 for the wild type strain, this ratio increased as the degree of resistance of the mutants increased. On the other hand, the activity ratio of the diastereomers of triadimenol towards the wild-type was 1 : 36 and the ratio became progressively smaller with increasing resistance of the fungal strains.

Resistance to SDIs in fungi was often associated with reduction in fitness and pathogenicity. The degree of resistance of triforine and triarimol resistant strains of *C. cucumerinum* appeared to be negatively correlated with their pathogenicity (Fuchs and Viets-Verweij 1975; Fuchs *et al.* 1977). Reduced fitness of resistant isolates has also been observed with mutants of *A. nidulans*. The degree of resistance was found to be inversely proportional to germ tube elongation and mycelium growth. Mutant strains of *C. cucumerinum* (Sherald and Sisler 1975; Fuchs and Drandarevski 1976; Buchenauer and Röhner 1977) and *A. nidulans* (De Waard and Sisler 1976; De Waard and Gieskes 1977) showed a diminished capacity in spore production, and these spores were often not viable (Sherald and Sisler 1975; Fuchs and Drandarevski 1976).

However, resistance to SDIs was not alwys accompanied with reduced pathogenicity. Mutants of *P. expansum* resistant to imazalil (van Tuyl 1977) and *P. italicum* resistant to fenarimol (De Waard and van Nistelrooy 1984) were not inferior in their virulence compared to the wild-type strains.

Mutations in fungi for resistance to SDIs are sometimes associated to pleiotropic effects, which are defined as a single-gene mutation affecting besides resistance to a toxicant also other characteristics of the cell (Georgopoulos 1977). Amongst imazalil resistant mutants of *A. nidulans* some strains showed a higher sensitivity to unrelated fungitoxic agents, e.g. acriflavine, cycloheximide, chloramphenicol and neomycin as well as to lower temperatures (van Tuyl 1977). Furthermore, several imazalil and fenarimol resistant strains of *A. nidulans* were more sensitive to 8-azaguanidine, p-fluorophenylalanine, D-serine and thiourea (De Waard and van Nistelrooy 1979) and to pimaricin (De Waard and Sisler 1976). Triarimol-resistant mutants of *B. cinerea* showed a higher sensitivity to cycloheximide than the wild-type strain (Leroux *et al.* 1976). Some of the *U. maydis* mutants (Barug and Kerkenaar 1984) and *P. italicum* (De Waard and van Nistelrooy 1984) showed an increased sensitivity to fenpropimorph.

Van Tuyl (1977) studied the genetics of resistance in *A. nidulans* to imazalil. Eight loci allocated to six different linkage groups were identified in an analysis of 21 imazalil resistant mutants. Two other genes for resistance to cycloheximide impart also resistance to imazalil. Mutations at particular loci occurred more frequently than at other sites. The large number of at least 10 different genes involved in imazalil resistance indicates that the loci confering resistance are distributed over the genome of *A. nidulans*. Allelic mutations in a single locus often resulted in different degrees of resistance, indicating that mutations in a single gene may differ. Generally, the levels of resistance to imazalil were low (van Tuyl 1977). These findings imply that several alterations at a certain site in the fungal cell may occur.

Investigations with heterozygous diploid imazalil-resistant strains of *A. nidulans* showed that the degree of resistance of such strains was intermediate between sensitive diploid and resistant haploid strains. By combining different single gene mutations for imazalil resistance the recombinant strain showed a higher level of resistance, indicating additive interaction of the single gene mutations (van Tuyl 1977).

The association between decreased fitness or pathogenicity and resistance in fungi to SDIs might also be regarded as pleiotropic effects. This assumption may be supported by the mechanisms of resistance of fungal mutants to SDIs (De Waard and van Nistelrooy 1979, 1980). Resistance in *A. nidulans* to fenarimol was attributed to differences in uptake of the toxicant by the wild-type and resistant strains. The wild-type strain rapidly accumulated fenarimol during the initial 10 min of incubation followed by a slow release of the compound and with extended incubation an equilibrium was reached. It was suggested that the amount of fenarimol accumulating during the initial phase is sufficient for saturation and thus for inhibition of the target site. While the influx proved to be a passive process, efflux which is inducible was regarded as an energy-dependent process.

On the other hand, the resistant strains of *A. nidulans* took up the fungicide only in a constant low level during incubation. It was assumed that the low concentrations of the toxicants in the fungal cells from the beginning of exposure is the result of a high energy-dependent efflux activity. Because of the high efflux rate of fenarimol in resistant strains inhibitory concentrations are not reached at the target site. Efflux may be inhibited by low temperature, anaerobiosis and respiratory inhibitors resulting in a considerably increased accumulation (De Waard and van Nistelrooy 1979, 1980, 1982). It was suggested that the decreased pathogenicity or fitness of the resistant strains that are characterized by a high energy-dependent efflux activity may be attributed to higher constitutive membrane transport activities that will have impaired effects on other cell processes.

If the energy-dependent efflux mechanism is involved in resistance of phytopathogenic fungi to SDIs it is of interest to modify the uptake of these inhibitors by substances interfering in efflux activity (De Waard and Fuchs 1982).

The findings that fungal strains resistant to SDIs are inferior to wild-type strains in their fitness and virulence might be of great significance with respect to development of resistance in practice, since in absence of selection pressure the competitive ability of resistant strains would be severely reduced. In addition, the resistant strains generally possess a low to medium degree of resistance. It was concluded that development of resistance to SDIs in practice would be retarded since the majority of resistant strains can be controlled by recommended fungicide dosages (Fuchs and Drandarevsky 1976; Brown and Hall 1979). However, there are indications that strains of *P. expansum* and *P. italicum* resistant to imazalil and fenarimol, respectively, did not differ appreciably from the wild-type strains in their pathogenicity (van Tuyl 1977; De Waard and van Nistelrooy 1984).

Decreased sensitivity of cucumber powdery mildew *(Sphaerotheca fuliginea)* as well as barley *(Erysiphe graminis* f. sp. *hordei)* and wheat *(Erysiphe graminis* f. sp. *tritici)* powdery mildew has been described. In the Netherlands, for control of cucumber powdery mildew in the greenhouses triforine as the first SDI has been applied from 1972 to 1977. After introduction of imazalil and fenarimol in 1977 and 1981, respectively, the use of SDIs increased rapidly. Monitoring experiments in 1981 revealed that the sensitivity of *S. fuliginea* was already lower than that of control isolates and during 1982 and 1983 the number of isolates with decreased sensitivity to SDIs further increased. Under high disease pressure triforine failed to control cucumber powdery mildew, the efficacy of imazalil was limited and normal application rates of fenarimol still controlled the resistant isolates (Schepers 1983, 1984, 1985b). Resistance of *S. fuliginea* to SDIs had also been reported by Huggenberger (1984). Positively correlated cross resistance of *S. fuliginea* isolates to other SDIs (e.g. bitertanol, buthiolate) was observed (Schepers 1983; Huggenberger 1984). Resistant greenhouse isolates of *S. fuliginea* showed no reduced competitive ability compared to the reference isolates (Schepers 1985a).

With the continuing use of SDIs to control powdery mildew of barley and wheat in various western European countries the population of powdery mildews became increasingly less sensitive during 1981 und 1984 (FLETCHER and WOLFE 1981; WOLFE *et al.* 1982, 1983; LIMPERT and FISCHBECK 1983; BENNETT and VAN KINTS 1983; BUTTERS *et al.* 1984; BUCHENAUER 1983; BUCHENAUER *et al.* 1984a; BUCHENAUER and HELLWALD 1985; DE WAARD *et al.* 1984). Surveys of barley mildew indicated that the relative number of colonies developing on detached leaves of barley seedlings grown from seed treated with 0,025 g a.i. of triadimenol/kg seed was 21 % in 1981 and increased to 100 % in 1984; the relative colony number on leaves from seed treated with 0.075 g a.i. of the fungicide/kg seed increased from 23 % in 1982 to almost 70 % in 1984 (WOLFE *et al.* 1984).

Isolates of *E. graminis* f. sp. *hordei* and *E. graminis* f. sp. *tritici* with decreased sensitivity to triadimefon showed cross-resistance to other SDIs, such as triadimenol, propiconazole, diclobutrazol, prochloraz and nuarimol. The lack of cross-resistance to the morpholine derivatives (tridemorph and fenpropimorph) agreed with the different site of interference of these chemicals in the sterol pathway (KATO *et al.* 1980; KERKENAAR *et al.* 1981) (chapter 12). No cross-sensitivity to pyrazophos was detected (BUCHENAUER 1984; BUCHENAUER et al. 1984a; BUCHENAUER and HELLWALD 1985). The decrease in sensitivity of barley powdery mildew isolates to SDIs was associated with an increased sensitivity to ethirimol (HOLLOMON 1982; BUCHENAUER and HELLWALD 1985).

The portion of isolates with higher level of resistance decreased as the selection pressure decreased during winter and spring and the powdery mildew population tended to an intermedian sensitivity (WOLFE *et al.* 1984).

The development of resistance of powdery mildew of cucumber and barley against SDIs resembles observations in build-up of resistance in cucumber powdery mildew to dimethirimol and in barley powdery mildew to ethirimol during 1970 and 1975 (BRENT 1982).

It is assumed that resistance in powdery mildew fungi is not determined by one major gene but rather by several different genes, similar to imazalil resistance in *A. nidulans* (VAN TUYL 1977). This possibly may be the cause for low development of high levels of resistance in powdery mildew fungi.

Antagonism and synergism

According to the primary mode of action of SDIs it should be expected that exogeneously applied ergosterol would annul the toxicity of these compounds. However, addition of ergosterol to the culture medium only partially alleviated growth inhibition of some fungi by the fungicides but did not restore growth (SHERALD *et al.* 1973; RAGSDALE and SISLER 1972, 1973; LEROUX *et al.* 1976; BUCHENAUER 1979), and in various cases ergosterol did not exhibit any antagonistic activity. Partial reversal of toxicity of SDIs in some fungi has been described (e.g. *C. cucumerinum*, *B. cinerea*, *U. maydis*, *U. avenae*, *A. nidulans*, *C. albicans*) by different lipophilic compounds, e.g. squalene, farnesol, progesterone, testosterone, β-carotene and vitamin A, phospholipids (e.g. phosphatidyl inosit, phosphatidylglycerol, phosphatidylserin, phosphatidylcholin, phosphatidylethanolamin), acylglycerides (α-mono-olein, 1,2-diolein, triolein), free fatty acids (e.g. oleic acid, linoleic acid) (SHERALD *et al.* 1973; LEROUX and GREDT 1978; KERKENAAR *et al.* 1979; BUCHENAUER 1979; DE WAARD and VAN NISTELROOY 1982; YAMAGUCHI 1977, 1978; KURODA *et al.* 1978). Toxicity of a number of SDIs was reduced by non-ionic detergents (e.g. Tween 20 and 40) (SHERALD *et al.* 1973; BUCHENAUER 1979; DE WAARD and VAN NISTELROOY 1982) calcium and magnesium chlorides as well as by fungicides (e.g. carboxin and dialkyldithiocarbamates) (DE WAARD and VAN NISTELROOY 1982).

Apparently, different factors operate in alleviation of the toxicity of SDIs. Antagonism may be due to complex formation between the fungicide and the antagonist (e.g. non-ionic detergents) in the medium; other possible explanations of antagonism are effects on fungal membranes that result in reduced uptake or a decreased affinity of the fungicide to the membrane-bound target enzyme. The variable results with certain antagonists may be due to the different activity of these compounds in the different fungi tested.

Numerous different compounds displayed synergistic activity; these included hydrochloric acid, sodium hydroxide, cationic and anionic surfactants, sodium orthovanadate, respiratory inhibitors (e.g. oligomycin, dicyclohexylcarbodiimide) and conventional fungicides (e.g. folpet, chlorothalonil) (DE WAARD and VAN NISTELROOY 1982, 1984a, b). It is assumed that an increased solubility of the toxicants by some compounds (e.g. HCl, NaOH) in the medium might be a possible mechanism of synergism. Other synergism mechanisms may include increased uptake of the fungicides which may be reduced either to membrane alterations (e.g. cationic and anionic agents) or to inhibition of energy-dependent efflux mechanism of the fungicide (e.g. respiratory inhibitors, sodium orthovanadate) (DE WAARD and VAN NISTELROOY 1984a, b; DE WAARD and DEKKER 1983) (cf. chapter 24).

Mode of action in bacteria

Various SDIs (e.g. miconazole and ketoconazole) are also active against Gram-positive bacteria (e.g. *Staphylococcus aureus, Clostridium multifermentans, Propionibacterium acnes*) (VAN DEN BOSSCHE *et al.* 1982). Studies on the mechanism of action suggest that miconazole may interfere in biosynthesis of polyisoprenoids of *Staphylococcus aureus* (VAN DEN BOSSCHE *et al.* 1982b). Incorporation experiments with [^{14}C]-mevalonate revealed that miconazole (at concentrations $> 1\,\mu$M) inhibited biosynthesis of C-55 isoprenoid alcohol and 2-methyl-3-prenyl-1,4-naphthoquinone derivatives (e.g. menaquinone-7, menaquinone-8, vitamins of K-type), concomitantly C-40 isoprenoid alcohol accumulated (VAN DEN BOSSCHE *et al.* 1982b, 1984b). The azole derivatives R-39519 and ketoconazole, possessing a lower antibacterial activity than miconazole, also inhibited at higher concentrations (R-39519 $\geq 10\,\mu$M, ketoconazole $\geq 30\,\mu$M) synthesis of C-55 isoprenoid alcohol and vitamin K. The phosphorylated derivative of C-55 isoprenoid alcohol plays a role as carrier lipid in the synthesis of polymers of bacterial cell walls (e.g. peptidoglycan, lipopolysaccharides, teichoic acids) and membranes (ROGERS *et al.* 1980).

Side effects in higher plants

Effects on growth

Ancymidol is an effective inhibitor of plant growth in monocotyledons and dicotyledons (TSCHABOLD *et al.* 1970; LEOPOLD and WRIGHT 1970; LEOPOLD 1971; COOLBAUGH and HAMILTON 1976; SHIVE and SISLER 1976; COOLBAUGH *et al.* 1982a). The structurally related fungicides triarimol (Shive and SISLER 1976; COOLBAUGH and HAMILTON 1976; COOLBAUGH *et al.* 1982a), nuarimol (SHIVE and SISLER 1976; BUCHENAUER 1979; COOLBAUGH *et al.* 1982a; KONSTANTINIDOU-DOLTSINI *et al.* 1986a) and fenarimol (SHIVE and SISLER 1976; COOLBAUGH *et al.* 1982a) show also growth retarding activity.

Likewise, on plant growth retarding effects of various triazole fungicides in growth chamber or greenhouse experiments have been reported. After seed treatment (0.25 g a.i./kg seed) with some triazole derivatives retardation of elongation growth of primary leaves of barley, wheat, oats, and rye seedlings increased in the following order: triadimenol $<$ propiconazol $<$ etaconazol $<$ diclobutrazol $<$ paclobutrazol. Primary leaves of barley and wheat seedlings were more affected by the triazole fungicides than those of oats and rye (BUCHENAUER *et al.* 1984b). Growth retardation in seedlings of barley by triadimefon and triadimenol (FÖRSTER *et al.* 1980a), diclobutrazol, etaconazol and propiconazol (BENT and SKIDMORE 1979; BUCHENAUER *et al.* 1981), paclobutrazol (FROGATT *et al.* 1982) and wheat by nuarimol and imazalil (KONSTANTINIDOU-DOLTSINI *et al.* 1986a) has been described. Root growth of cereal seedlings was less severely retarded than that of primary leaves by the triazole fungicides after seed treatment (BUCHENAUER and GROSSMANN 1977; FÖRSTER *et al.* 1980a; BUCHENAUER and RÖHNER 1981; BUCHENAUER *et al.* 1984b). On the other hand, root length of bean seedlings was more affected than that of stems by triarimol (SHIVE and SISLER 1976). In field experiments, growth of barley and wheat seedlings was affected to a much lesser degree than in greenhouse and growth chamber tests by triadimefon, triadimenol (FÖRSTER *et al.* 1980a) and nuarimol (KONSTANTINIDOU-DOLTSINI *et al.* 1986a).

Following root treatment, the various triazole fungicides retarded elongation growth of tomato shoots as follows: etaconazol $>$ paclobutrazol $>$ triadimenol $>$ diclobutrazol $>$ bitertanol (BUCHENAUER *et al.* 1984b). Soil drench applications of paclobutrazol retarded growth of apple (SWIETLIK and MILLER 1983; CURRY and WILLIAMS 1983) chrysanthemums (MENHENETT 1981; McDANIEL 1983), poinsettias (MENHENETT 1981; WILFRET 1981), tulips and lilies (MENHENETT and HANKS 1982/83; MENHENETT *et al.* 1983) and sunflowers (WAMPLE and CULVER 1983).

Because of their asymmetric carbon atoms a great number of SDIs possess diastereomeric forms. The diastereomers or their enantiomers may differ in their plant regulatory and fungicidal activity. While the diastereomers of triadimenol I exhibited a markedly higher fungitoxic activity than the diastereomeric form II, with respect to their growth regulatory activity both diastereomeric forms behaved contrarily (BUCHENAUER 1979). Similar results have been reported for the enantiomers of paclobutrazol. The (+) enantiomeric form (2 R, 3 R) showed high activity against powdery mildews and rusts of cereals and low plant regulatory activity in apple seedlings, whereas the (—) enantiomeric form (2 S, 3 S) displayed the opposite effects (SUGAVANAM 1984).

Effects others than growth retardation

Besides retardation of elongation growth the pyrimidine and triazole fungicides caused various morphological and physiological effects in cultural plants. Growth retardation of primary leaves of barley seedlings was attributed to a decreased cell extension; the number of cells per leaf of treated seedlings did not differ significantly from that of the control (FÖRSTER *et al.* 1980a). Ancymidol caused lateral expansion of stem segments of oats (MONTAGUE 1975) and stems of bean seedlings (SHIVE and SISLER 1976). The diameter of epidermal cells in oats stem segments were increased (MONTAGUE 1975). Primary leaves of cereal seedlings were thicker and contained a higher water content than untreated leaves (FÖRSTER *et al.* 1980a; BUCHENAUER *et al.* 1984b; KONSTANTINIDOU-DOLTSINI *et al.* 1986a). At seed treatment with higher concentrations of the fungicides (e.g. 0.5—1.0 g a.i./kg seed) geotropism of cereal seedlings was disturbed (FÖRSTER *et al.* 1980a, KONSTANTINIDOU-DOLTSINI *et al.* 1986a). Ancymidol and triarimol induced at high concentrations ethylene-like responses in bean seedlings (SHIVE and SISLER 1976).

The chlorophyll content of primary leaves of barley was reduced after seed treatment with triadimefon and triadimenol when based on fresh weight (FÖRSTER *et al.* 1980b). At higher concentrations paclobutrazol reduced the photosynthetic rate in sunflowers (WAMPLE and CULVER 1983). While xanthophyll, carotin, and RNA contents were decreased in treated primary barley and wheat leaves, DNA and protein contents were unchanged. The contents of free amino acids were increased in barley and wheat seedlings (FÖRSTER *et al.* 1980b; KONSTANTINIDOU-DOLTSINI *et al.* 1986b). Also nitrate reductase activity was increased in nuarimol treated wheat seedlings (KONSTANTINIDOU-DOLTSINI *et al.* 1986a). In ageing treated primary leaves degradation of pigments and nucleic acids was delayed. The retarded diminution of the RNA contents could partly be attributed to higher synthetic processes maintained over a longer period of time in the leaf tissue of treated seedlings (FÖRSTER *et al.* 1980b). In field tests, the influence of the pigments and nucleic acids contents could be demonstrated to a lesser extent only in the primary leaves after seed treatment (FÖRSTER *et al.* 1980b; KONSTANTINIDOU-DOLTSINI 1981).

Peroxidase activity was reduced in barley and wheat seedlings (FÖRSTER *et al.* 1980c; KONSTANTINIDOU-DOLTSINI 1981). Likewise, PAL and TAL activity was lower in barley and wheat seedlings (BUCHENAUER 1979; KONSTANTINIDOU-DOLTSINI 1981).

Seed treatment of propiconazol, etaconazol, and diclobutrazol induced in barley seedlings an increased tolerance to frost, drought and salt (BUCHENAUER *et al.* 1981). PENNYPACKER *et al.* (1982) found that triadimefon improved survival of annual bluegrass *(Poa annua)* during periods of summer heat and drought stress.

Interaction between gibberellic acid

Growth retardation (e.g. shoot height, leaf area) and other parameters induced by low concentrations of the pyrimidine and triazole derivatives were fully reversed by application of gibberellic acid (GA), suggesting that growth retardation of plants by these compounds is attributed to interactions with GA dependent processes (LEOPOLD and WRIGHT 1970; TSCHABOLD *et al.* 1970; LEOPOLD 1971; COOLBAUGH and HAMILTON 1976; SHIVE and SISLER 1976; BUCHENAUER and GROSSMANN 1977; BUCHENAUER and RÖHNER 1977; BUCHENAUER and RÖHNER 1981; COOLBAUGH *et al.* 1982a; WAMPLE and CULVER 1983; CURRY and WILLIAMS 1983). At higher concentrations of the inhibitors, however, growth retardation was only partly annulled by simultaneous GA application in bean (SHIVE and SISLER 1976), pea (COOLBAUGH *et al.* 1982), and barley and wheat seedlings (BUCHENAUER and GROSSMANN 1977; BUCHENAUER 1979; BUCHENAUER and RÖHNER 1981). This suggests that growth inhibition may result from interaction of the inhibitor at sites other than those affecting GA biosynthesis directly.

Effects on gibberellin synthesis in plants

It has been shown that at higher dosages ancymidol and triarimol reduced the extractable GA-like activity in bean plants (SHIVE and SISLER 1976). The triazole derivatives triadimenol, triadimefon, paclobutrazol, diclobutrazol, etaconazol and propiconazol, the pyrimidine derivative nuarimol and the imidazole compound imazalil diminished the activity of GA-like compounds in primary leaves of barley and wheat seedlings (Buchenauer and Grossmann 1977; Buchenauer and Röhner 1977; Buchenauer and Röhner 1981; Buchenauer *et al.* 1984b).

Studies of Coolbaugh and Hamilton (1976) and Coolbaugh *et al.* (1978) indicated that ancymidol interfered in three reactions of GA biosynthetic pathway by specific inhibition of ent-kaurene, ent-kaurenol and ent-kaurenal oxidation in microsomes of *Marah macrocarpus* and *Marah oreganus.* These three cytochrome P-450 dependent oxidative reactions showed a similar degree of sensitivity to ancymidol (Coolbaugh *et al.* 1978).

Coolbaugh *et al.* (1982a) extended their studies by comparing several pyrimidine derivatives (including ancymidol as plant growth retardant and triarimol, fenarimol and nuarimol as fungicides) on their activity to retard growth of pea plants and to inhibit ent-kaurene oxidations of microsome preparations of *M. macrocarpus.* They found a good correlation between the effectiveness of the substances as inhibitors of ent-kaurene oxidation and as plant growth retardants. The compounds tested could be grouped into two categories. Ancymidol, pyrimidol (EL-509), EL-93807 and EL-75253 proved to be highly active on both processes, whereas the fungicides nuarimol, triarimol, fenarimol and EL-72303 showed a markedly reduced effectiveness both as inhibitors of elongation growth of pea plants and microsomal ent-kaurene oxidation in the cell free system. Of the fungicides tested only nuarimol exhibited growth regulatory activity. The I_{50} of ancymidol and its isopropyl analogues for microsomal kaurene oxidation ranged between 1 and 100 nM, whereas that for triarimol, fenarimol and nuarimol varied between 1 and 100 μM. No corresponding studies have been carried out with the triazole derivatives.

Effects on growth and gibberellin synthesis in Gibberella fujikuroi

In comparative studies the effects of the fungitoxic compounds triarimol, fenarimol and nuarimol and of the plant growth regulators ancymidol and EL-509 were tested on growth and gibberellin synthesis in *Gibberella fujikuroi (Fusarium moniliforme).* The effectiveness of the different analogues was opposite to that observed in higher plants. While the three fungitoxic agents proved to be more effective in retardation of mycelium growth than ancymidol and EL-509, all five substances inhibited gibberellin production of the fungus in culture filtrates. The activity of GA-compounds closely paralleled that of growth retardation of *G. fujikuroi.* At subinhibitory concentrations of EL-509 and ancymidol GA_3 synthesis in *G. fujikuroi* was stimulated. All compounds interfered in ent-kaurene oxidation of microsomal preparations from the mycelium of the fungus (Coolbaugh *et al.* 1982b). It is assumed that in oxidation of ent-kaurene in both the fungus *G. fujikuroi* and the plant *M. macrocarpus* a cytochrome P-450 component of the microsomal mixed function oxygenases is involved (Coolbaugh *et al.* 1978). However, the structure-activity relationships suggest that the enzymes from the higher plant and the fungus may differ in some subtle properties. The fungicides may bind more strongly to those of fungal cytochrome P-450 and the growth retardant ancymidol as well as its closely related structural derivatives may interfere more strongly with the plant cytochrome P-450.

The effects of the triazole derivatives on growth, GA-like activity and sterol synthesis in *G. fujikuroi* was studied (Kutzner and Buchenauer 1986). The compounds (at each 10^{-4} M) retarded the increase of mycelium dry weight of the fungus: etaconazol (95 %), paclobutrazol (70 %), diclobutrazol (60 %), bitertanol (30 %) and triadimenol (0 %). All five compounds diminished the production of gibberellin-like substances of the fungus in the culture filtrates to a smaller extent than dry weight increase. Furthermore, at 10^{-4} M the triazole derivatives interfered with ergosterol synthesis in *G. fujikuroi*. While ergosterol synthesis was inhibited, sterols containing C-4 and C-14 methyl groups accumulated (KUTZNER and BUCHENAUER 1986).

Effects on sterol synthesis in plants

It has been shown recently that sterol C-14 demethylation inhibitors (SDIs) in fungi also interfere at higher concentrations in sterol synthesis of higher plants. Fenarimol inhibited 14α-demethylase and Δ24(28)-reductase in suspension cultures of bramble *(Rubus fructicosus)* cells (SCHMITT and BENVENISTE 1979). Of the 14α-methyl sterols, obtusifoliol particularly accumulated. Furthermore, the following sterols, which were absent in control cells, were present in treated cells: 14α-methyl-5α-ergosta-8,24(28)-dien-3β-ol, 14α-methyl-5α-stigmasta-8,Z-24(28)dien-3β-ol, 4α,14α-dimethyl-5α-stigmasta-8Z-24(28)-dien-3β-ol. The second alkylation reaction at C-24 was inhibited by the presence of the 14α-methyl group. Similar to the findings in fungi (e.g. RAGSDALE 1975) bramble cells grown in the presence of fenarimol contained a higher total sterol content than control cells (SCHMITT and BENVENISTE 1979).

Triarimol treatment of carrot, tobacco, and soybean suspension cultures resulted in inhibition of sterol C-14 demethylation and second alkylation of the side chain at C-24 (HOSAKAWA *et al.* 1984). All three cultures produced 14α-methyl-5α-ergost-8-en-3β-ol and obtusifoliol. In treated carrot cultures small amounts of cycloeucalenol and 5α-stigmast-7-en-3β-ol accumulated. Higher plant cells grown in suspension cultures appear somewhat less sensitive to SDIs than fungi.

In the three *Chlorella species (C. emersonii, C. ellipsoidea, C. sorokiniana)* triarimol blocked 14α-demethylation leading to accumulation of 14α-methyl sterols. In two *Chlorella* species the second alkylation at C-24 and in all three species the introduction of the C-22 double bond was inhibited in the presence of the fungicide. Triarimol treatment increased the quantity of unsaturated fatty acids in the three *Chlorella* species (FRASINEL *et al.* 1978).

The triazole fungicides triadimefon, triadimenol, propiconazol, etaconazol, diclobutrazol and paclobutrazol inhibited in detached leaves of barley seedlings the synthesis of C-4-desmethyl sterols, simultaneously sterols containing C-4 and C-14 methyl groups were accumulated (BUCHENAUER 1977d; BUCHENAUER and RÖHNER 1981; BUCHENAUER *et al.* 1984b). By comparing the relative sensitivity it was assumed that GA synthesis was more affected than sterol synthesis in primary barley leaves by the triazole fungicides. On the other hand, at growth retarding concentrations triarimol did not influence sterol synthesis in bean seedlings (Shive and SISLER 1976).

It is concluded that plant growth retardations induced by high inhibitor concentrations which could not be annulled by GA applications might be due to interference in sterol synthesis in addition to GA synthesis. Plant growth retardants and fungicides interfering in GA and sterol synthesis may affect membrane structure and function since both sterols and gibberellins are involved in membrane phenomena.

References

AOYAMA, Y., and YOSHIDA, Y.: Interaction of lanosterol to cytochrome P-450 purified from yeast microsomes: evidence for contribution of cytochrome P-450 to lanosterol metabolism. Biochemical and Biophysical Research Communications **82** (1978a): 33—38.

— — The 14α-demethylation of lanosterol by a reconstituted cytochrome P-450 system from yeast microsomes. Biochemical and Biophysical Research Communications **85** (1978b): 28—34.

Atkin, S. D., Morgan, B., Baggaley, K.-H., and Green, J.: The isolation of 2,3-oxidosqualene from the liver of rats treated with 1-dodecylimidazole a novel hypocholesterolamic agent. Biochem. J. **130** (1972): 153—157.

Baldwin, B. C., and Wiggins, T. E.: Action of fungicidal triazoles of the diclobutrazol series of *Ustilago maydis*. Pesticide Sci. **15** (1984): 156—166.

Barug, D., and Kerkenaar, A.: Cross resistance of UV-induced mutants of *Ustilago maydis* to various fungicides which interfere with ergosterol-biosynthesis. Meded. Fac. Landbouwwet. Rijksuniv. Gent **41** (1979): 421—427.

Bennet, F. G. A., and van Kints, T. M. C.: Powdery mildew of wheat. Ann. Rep. Pl. Breed. Inst. (1983): 85—87.

Bent, K. J., and Skidmore, A. M.: Diclobutrazol: a new systemic fungicide. Brit. Crop Prot. Conf. — Pests and Diseases, Vol. 2 (1979): 477—484.

Berg, D., Kraemer, W., Regel, E., Buechel, K.-H., Holmwood, G., Plembel, M., and Scheinpflug, H.: Mode of action of fungicides: Studies on ergosterol biosynthesis inhibitors. British Crop. Protection Conf. — Pests and Diseases, Vol. 3 (1984): 887—892.

Berg, L. R., Patterson, G. W., and Lusby, W. R.: Effect of triarimol and tridemorph on sterol biosynthesis in *Saprolegnia ferax*. Lipids **18** (1983): 448—452.

Bloch, K.: Sterols and membranes. In: Mertonosi, A. N. (Ed.): Membranes and Transport. Plenum 1982, pp. 25—35.

Borgers, M.: Mechanism of action of antifungal drugs with special reference to the imidazole derivatives. Rev. Infect. Dis. **2** (1980): 520—534.

Brent, K. J.: Case study 4: Powdery mildews of barley and cucumber. In: Dekker, J., and Georgopoulos, S. G. (Eds.): Fungicide Resistance in Crop Protection. Pudoc, Wageningen 1982, pp. 219—230.

Brown, I. F., and Hall, H. R.: Induced and natural tolerance of fenarimol (EL-222) in *Cladosporium cucumerinum* and *Venturia inaequalis*. Phytophatology **69** (1979): 914.

Buchenauer, H.: Systemisch-fungizide Wirkung und Wirkungsmechanismus von Triadimefon (MEB 6447). Mitt. Biol. Bundesanst. Land.-Forstw. Berlin-Dahlem, H. 165 (1975): 154—155.

— Wirkungsmechanismus von ®Bayleton (Triadimefon) in *Ustilago avenae*. Pfl. Schutz.-Nachr. (Bayer) **29** (1976a): 281—302.

— Hemmung der Ergosterolbiosynthese in *Ustilago avenae* durch Triadimefon und Fluotrimazol. Z. Pfl. Krankh. Pfl. Schutz **83** (1976b): 363—367.

— Mode of action of triadimefon in *Ustilago avenae*. Pesticide Biochem. Physiol. **7** (1977a): 309—320.

— Biochemical effects of the systemic fungicides fenarimol (EL-222) and nuarimol (EL-228) in *Ustilago avenae*. Z. Pfl. Krankh. Pfl. Schutz **84** (1977b): 286—299.

— Studies on the mode of action of imazalil in *Ustilago avenae*. Z. Pfl. Krankh. Pfl. Schutz **84** (1977c): 440—450.

— Mode of action and selectivity of fungicides which interfere with ergosterol biosynthesis. Proc. Brit. Crop Prot. Conf. — Pests and Diseases Vol. 3 (1977d): 699—711.

— Analogy in the mode of action of fluotrimazole and clotrimazole in *Ustilago avenae*. Pesticide Biochem. Physiol. **8** (1978a): 15—25.

— Inhibition of ergosterol biosynthesis by triadimenol in *Ustilago avenae*. Pesticide Sci. **9** (1978b): 507—512.

— Untersuchungen zur Wirkungsweise und zum Verhalten verschiedener Fungizide in Pilzen und Kulturpflanzen. Habilitationsschrift, Bonn 1979.

— Interaction of different lipid components with various fungicides. Z. Pfl. Krankh. Pfl. Schutz **87** (1980): 423—426.

— Wirkungsweise moderner Fungizide in Pilzen und Kulturpflanzen. Ber. Deutsch, Bot. Ges. **96** (1983): 427—457.

— Stand der Fungizidresistenz bei Getreidekrankheiten am Beispiel der Halmbruchkrankheit und des Echten Mehltaus. Gesunde Pflanzen **36** (1984): 132—142.

— Budde, K., Hellwald, K. H., Taube, E., and Kirchner, R.: Decreased sensitivity of barley powdery mildew isolates to triazole and related fungicides. Proc. Brit. Crop Protec. Conf. — Pests and Diseases, Vol. 2 (1984a): 483—488.

— and Grossmann, F.: Triadimefon: mode of action in fungi and plants. Netherl. J. Plant Path. **83**, Suppl. 1 (1977): 93—103.

— and Hellwald, K.-H.: Resistance of *Erysiphe graminis* on barley and wheat to sterol C-14-demethylation inhibitors. EPPO Bull. **15** (1985): 459—466.

— and KEMPER, K.: Wirkungsweise von Propiconazol (CGA 64250) in verschiedenen Pilzarten. Meded. Fac. Landbouwwet. Rijksuniv. Gent **46** (1981): 909—921.

— KOHTS, T., and ROOS, H.: Zur Wirkungsweise von Propiconazol (Desmel®), CGA 64251 und Diclobutrazol (Vigil®) in Pilzen und Gerstenkeimlingen. Mitt. Biol. Bundesanst. Land.- u. Forstwirtsch., Berlin-Dahlem, H. 203 (1981): 310—311.

— KUTZNER, B., and KOHTS, T.: Wirkung verschiedener Triazol-Fungizide auf das Wachstum von Getreidekeimlingen und Tomatenpflanzen sowie auf die Gibberellin-Gehalte und den Lipidstoffwechsel von Gerstenkeimlingen. Z. Pfl. Krankh. Pfl. Schutz **91** (1984b): 506—524.

— and RÖHNER, E.: Einfluß von Nuarimol, Fenarimol, Triadimefon und Imazalil auf den Gibberellin- und Lipidstoffwechsel in jungen Gersten- und Weizenpflanzen. Mitt. Biol. Bundesanst. Land.- u. Forstwirtsch., Berlin-Dahlem, H. 178 (1977): 158.

— — Effect of triadimefon and triadimenol on growth of various plant species as well as on Gibberellin content and sterol metabolism in shoots of barley seedlings. Pesticide Biochem. Physiol. **15** (1981): 58—70.

BUTTERS, J., CLARK, J., and HOLLOMON, D. W.: Resistance to inhibitors of sterol biosynthesis in barley powdery mildew. Meded. Fac. Landbouww. Rijksuniv. Gent **49** (1984): 143—151.

BUTTKE, T. M., and BLOCH, K.: Comparative responses of the yeast mutant strain GL 7 to lanosterol, cycloartenol, and cyclolandenol. Biochem. Biophys. Acta **531** (1980): 301—307.

— — Utilization and metabolism of methyl-sterol derivatives in the yeast mutant strain GL 7. Biochemistry **20**, 3267—3272.

CABIB, E., ROBERTS, R., and BOWERS, B.: Ann. Rev. Biochem. **51** (1982): 783—793.

COOLBAUGH, R. C., and HAMILTON, R.: Inhibition of entkaurene oxidation and growth by α-cyclopropyl-α-(p-methoxyphenyl)-5-pyrimidine methyl alcohol. Plant Physiol. **57** (1976): 245—248.

— HEIL, D. R., and WEST, C. A.: Comparative effects of substituted pyrimidines on growth and Gibberellin biosynthesis in *Gibberella fujikuroi*. Plant Physiol. **69** (1982b): 712—716.

— HIRANO, S. S., and WEST, C. A.: Studies on the specificity and site of action of ancymidol, a plant growth regulator. Plant Physiol. **62** (1978): 571—576.

— SWANSON, D. J., and WEST, C. A.: Comparative effects of ancymidol and its analogs on growth of peas and ent-kaurene oxidation in cell-free extracts of immature *Marah macrocarpus* endosperm. Plant Physiol. **69** (1982a): 707—711.

CURRY, E. A., and WILLIAMS, M. W.: Promalin or GA_3 increase pedicel and fruit length and leaf size of 'Delicious' apples treated with paclobutrazol. Hort Sci. **18** (1983): 214—215.

DEMEL, R. A., BRUCKDORFER, K. R., and VAN DEENEN, L. L. M.: Structural requirements of sterols for the interaction with lecithin at the air-water interface. Biochem. Biophys. Acta **255** (1972): 311—320.

DE NOLLIN, S., and BORGERS, M.: The ultrastructure of *Candida albicans* after *in vitro* treatment with miconazole. Sabouraudia **12** (1974): 341—351.

— — Scanning electron microscopy of *Candida albicans* after *in vitro* treatment with miconazole. Antimicrob. Agents Chemother. **7** (1975): 704—711.

— — An ultrastructural and cytochemical study of *Candida albicans* after *in vitro* treatment with imidazoles. Mykosen **19** (1976): 317—328.

DE WAARD, M. A., and DEKKER, J.: Resistance to pyrimidine fungicides which inhibit ergosterol biosynthesis. In: MIYAMOTO, J., and KEARNY, P. C. (Eds.): IUPAC Pesticide Chemistry: Human Welfare and the Environment. Vol. 3. Pergamon Press, Oxford 1983, pp. 43—49.

— and FUCHS, A.: Resistance to ergosterol-biosynthesis inhibitors II. Genetic and physiological aspects. In: DEKKER, J., and GEORGOPOULOS, S. G. (Eds.): Fungicide resistance in crop protection. Pudoc, Wageningen 1982, pp. 87—100.

— and GIESKES, S. A.: Characterization of fenarimol-resistant mutants of *Aspergillus nidulans*. Netherl. J. Plant Path. **83** (1977): 177—188.

— KIPP, E. C. M., and VAN NISTELROOY, J. G. M.: Variatie in gevoeligheid van tarwemeeldauw voor fungiciden die die ersterolbiosynthese remmen. Gewasbescherning **15**, 8 (Abstr.) 1984.

— and RAGSDALE, N. N.: Fenarimol, a new systemic fungicide. In: LYR, H., and POLTER, C. (Eds.): Systemfungizide. Akademie-Verlag, Berlin 1979, pp. 187—194.

— and SISLER, H. D.: Resistance to fenarimol in *Aspergillus nidulans*. Meded. Fac. Landbouwwet. Rijksuniv. Gent **41** (1976): 571—578.

— and VAN NISTELROOY, J. G. M.: Mechanism of resistance to fenarimol in *Aspergillus nidulans*, Pesticide Biochem. Physiol. **10** (1979): 219—229.

De Waard, M. A., and van Nistelrooy, J. G. M.: An energy dependent efflux mechanism for fenarimol in a wild-type strain and fenarimol resistant mutants of *Aspergillus nidulans*. Pesticide Biochem. Physiol. **13** (1980): 255—266.

— — Antagonistic and synergistic activities of various chemicals on the toxicity of fenarimol to *Aspergillus nidulans*. Pesticide Sci. **13** (1982): 279—286.

— — Differential accumulation of fenarimol by a wild-type isolate and fenarimol-resistant isolates of *Penicillium italicum*. Netherl. J. Plant Pathol. **90** (1984a): 143—153.

— — Effects of phthalimide fungicides on the accumulation of fenarimol by *Aspergillus nidulans*. Pesticide Sci. **15** (1984b): 56—62.

Dufour, J. P., Boutry, M., and Goffeau, A.: Plasma membrane ATPase of yeast. J. Biol. Chem. **255** (1980): 5735—5741.

Ebert, E., Gaudin, J., Muecke, W., Ramsteiner, K., Christian, V., and Fuhrer, H.: Inhibition of ergosterol biosynthesis by etaconazole in *Ustilago maydis*. Z. Naturforschung **38** (1983): 28—34.

Fletcher, J. T., and Wolfe, M. S.: Insensitivity of *Erysiphe graminis* f. sp. *hordei* to triadimefon, triadimenol and other fungicides. Proc. Brit. Crop Prot. Conf. — Pests and Diseases Vol. 2 (1981): 633—640.

Förster, H., Buchenauer, H., and Grossmann, F.: Nebenwirkungen der systemischen Fungizide Triadimefon und Triadimenol auf Gerstenpflanzen. I. Beeinflussung von Wachstum und Ertrag. Z. Pfl. Krankh. Pflanzenschutz **87** (1980a): 473—492.

— — — Nebenwirkungen der systemischen Fungizide Triadimefon und Triadimenol auf Gerstenpflanzen. II. Cytokininartige Effekte. Z. Pfl. Krankh. Pflanzenschutz **87** (1980b): 640—653.

— — — Nebenwirkungen der systemischen Fungizide Triadimefon und Triadimenol auf Gerstenpflanzen. III. Weitere Beeinflussungen des Stoffwechsels. Z. Pfl. Krankh. Pflanzenschutz **87** (1980c): 717—730.

Fort, T. M., and Moberg, W. K.: DPX H 6573, a new broad spectrum fungizide candidate. Proc. Brit. Crop Prot. Conf. — Pests and Diseases Vol. 2 (1984): 413—420.

Frasinel, C., Patterson, C. W., and Dutky, S. R.: Effect of triarimol on sterol and fatty acid composition of three species of *Chlorella*. Phytochem. **17** (1978): 1567—1570.

Frogatt, P. J., Thomas, W. D., and Batch, J. J.: The value of logging control in winter wheat as exemplified by the growth regulator PP 333. In: Hawkins, H. F., and Jeffcoat, B. (Eds.): Opportunities for Manipulation of Cereal Productivity. BPGRG Monograph **7** (1982): 71—87.

Fuchs, A., de Ruig, S. P., van Tuyl, J. M., and de Vries, F. W.: Resistance to triforine: a nonexistant problem? Netherl. J. Plant Pathol. **83** (Suppl. 1) (1977): 189—205.

— and de Vries, F. W.: Diastereomer-selective resistance in *Cladosporium cucumerinum* to triazole-type fungicides. Pesticide Sci. **15** (1984): 90—96.

— and Drandarevski, Ch. A.: The likelihood of development of resistance to systemic fungicides which inhibit ergosterol biosynthesis. Netherl. J. Plant Path. **82** (1976): 85—87.

— and Viets-Verweij, M.: Permanent and transient resistance to triarimol and triforine in some phytopathogenic fungi. Meded. Fac. Landbouwwet. Rijksuniv. Gent **40** (1975): 699—706.

Gadher, P., Mercer, E. J., Baldwin, B. C., and Wiggins, T. E.: A comparison of the potency of some fungicides as inhibitors of sterol 14-demethylation. Pesticide Biochem. Physiol. **19** (1983): 1—10.

Georgopoulos, S. G.: Development of fungal resistance to fungicides. In: Siegel, M. R., and Sisler, H. R. (Eds.): Antifungal Compounds. Vol. 2. Marcel Dekker Inc., New York 1977, pp. 439—495.

Gibbons, G. F., and Mitropoulos, K. A.: The effect of carbon monoxide on the nature of the accumulated 4,4-dimethyl sterol precursors of cholesterol during its biosynthesis from [2-^{14}C] mevalonic acid in vitro. Biochem. J. **132** (1973): 439—448.

— Pullinger, C. R., and Mitropoulos, K. A.: Studies on the mechanism of lanosterol 14α-demethylation; a requirement for two distinct types of mixed function oxidase systems. Biochemical J. **183** (1979): 309—315.

Girardet, F., and Buchenauer, H.: Wirkungsweise von Phenapronil (Sisthane®) gegenüber *Ustilago avenae* und *Erysiphe graminis* f. sp. *hordei*. Meded Fac. Landbouwwet. Rijksuniv. Gent **45** (1980): 435—443.

Gooday, G. W.: The enzymology of hyphal growth. In: Smith, J. E., and Berry, D. R. (Eds.): The Filamentous Fungi. 3. Development Biology. Edward Arnold, London 1978, pp. 51—77.

Hajek, K., Cook, N. J., and Novak, R. F.: Mechanism of inhibition of microsomal drug metabolism by imidazole. J. Pharmacol. Experim. Therapeutics **223** (1982): 97—104.

— and NOVAK, R. F.: Spectral and metabolic properties of liver microsomes from imidazole-pretreated rabbits. Biochem. Biophys. Res. Commun. **108** (1982): 664—672.

HASHIMOTO, S., SANO, S., and MORISHIMA, Y.: Ergosterol biosynthesis inhibition and morphological changes in the sporidia of *Ustilago maydis* treated with triflumizole. Fungicides for Crop Protection. BCPC Monograph No. 31, Vol. 2 (1985): 277—280.

HATA, S., T. NISHINO, T., KOMORI, M., and KATSUKI, H.: Involvement of cytochrome P-450 in Δ^{22}-desaturation in ergosterol biosynthesis of yeast. Biochem. Biophys. Commun. **103** (1981): 272—277.

HEERES, J., DE BRABANDER, M., and VANDEN BOSSCHE, H.: Ketoconazole: Chemistry and basis for selectivity. In: PERITI, P., and GRASSI, G. (Eds.): Curr. Chemother. Immunoth. The American Society for Microbiology, Washington, D.C. 1981: 1007—1009.

HENRY, M. J., and SISLER, H. D.: Effects of miconazole and dodecylimidazole on sterol biosynthesis in *Ustilago maydis*. Antimicrob. Agents Chemother. **15** (1979): 603—607.

— — Inhibition of ergosterol biosynthesis in *Ustilago maydis* by the fungicide 1-[2-(2,4-dichlorophenyl)-4-ethyl-1,3-dioxolan-2-ylmethyl]-1H-1,2,4-triazole. Pesticide Sci. **12** (1981): 98—102.

HOSAKAWA, G., PATTERSON, G. W., and LUSBY, W. R.: Effects of triarimol, tridemorph and triparanol on sterol biosynthesis in carrot, tobacco and soybean suspension cultures. Lipids **19** (1984): 449—456.

HIPPE, S.: Ultrastrukturelle Veränderungen in Sporidien von *Ustilago avenae* nach Behandlung mit systemischen Fungiziden. Diss. Univ. Stuttgart-Hohenheim 1982.

— Ultrastructural changes induced by the systemic fungicides triadimefon, nuarimol, and imazalil nitrate in sporidia of *Ustilago avenae*. Pesticide Sci. **15** (1984): 210—214.

— BUCHENAUER, H., and GROSSMANN, F.: Einfluß von Triadimefon auf die Feinstruktur der Sporidien von *Ustilago avenae*. Z. Pfl. Krankh. Pflanzenschutz **87** (1980): 423—426.

HOLLOMON, D. W.: The effects of tridemorph on barley powdery mildew: its mode of action and cross sensitivity relationships. Phytopath. Z. **105** (1982): 279—287.

HUGGENBERGER, F., COLLINS, M. A., and SKYLAKAKIS, G.: Decreased sensitivity of *Sphaerotheca fuliginea* to fenarimol and other ergosterol biosynthesis inhibitors. Crop Protect. **3** (1984): 137—149.

IWATA, K., YAMAGUCHI, H., and HIRATANI, T.: Mode of action of clotrimazole. Sabouraudia **11** (1973): 158—166.

KATO, T., SHOAMI, M., and KAWASE, Y.: Comparison of tridemorph with buthioate in antifungal mode of action. J. Pesticide Sci. **5** (1980): 69—79.

TANAKA, S., UEDA, M., and KAWASE, Y.: Inhibition of sterol biosynthesis in *Monilinia fructigena* by the fungicide S-1358. Agr. Biol. Chem. **38** (1974): 2377—2384.

— — — Inhibition of sterol biosynthesis in *Monilinia fructigena* by the fungicide S-1358. Agr. Biol. Chem. **39** (1975): 169—174.

KERKENAAR, A., and BARUG, D.: Fluorescence microscope studies of *Ustilago maydis* and *Penicillium italicum* after treatment with imazalil or fenpropimorph. Pesticide Sci. **15** (1984): 199 to 205.

— — and KAARS SIJPESTEIJN, A.: On the antifungal mode of action of tridemorph. Pesticide Biochem. Physiol. **12** (1979): 195—204.

— UCHIYAMA, M., and VERSLUIS, G. G.: Specific effects of tridemorph on sterol biosynthesis in *Ustilago maydis*. Pesticide Biochem. Physiol. **16** (1981): 97—104.

KONSTANTINIDOU-DOLTSINI, S.: Einfluß der systemischen Fungizide Nuarimol und Imazalil auf den Stoffwechsel von Weizenpflanzen. Diss. Univ. Stuttgart-Hohenheim 1981.

— BUCHENAUER, H., and GROSSMANN, F.: Einfluß der systemischen Fungizide Nuarimol und Imazalil auf Wachstum und Stoffwechsel von Weizenpflanzen. I. Wachstum. Z. Pfl. Krankh. Pflanzenschutz **93** (1986a): (in press).

— — — Einfluß der systemischen Fungizide Nuarimol und Imazalil auf Wachstum und Stoffwechsel von Weizenpflanzen. II. N-Metabolismus. Z. Pfl. Krankh. Pflanzenschutz **93** (1986b).

KRAUS, P.: Untersuchungen zum Wirkungsmechanismus von Baycor. Pflanzenschutz-Nachrichten (Bayer) **32** (1979): 17—30.

KURODA, S., UNO, J., and ARAI, T.: Target substances of some antifungal agents in the cell membrane. Antimicrob. Agents Chemother. **13** (1978): 454—459.

KUTZNER, B., and BUCHENAUER, H.: Effect of varoius triazole fungicides on growth and lipid metabolism of *Fusarium moniliforme* as well as on gibberellin contents in fungus filtrates. Z. Pfl. Krankh. Pflanzenschutz **93** (1986) (in press).

Leopold, A. C.: Antagonism of some Gibberellin actions by a substituted pyrimidine. Plant Physiol. **48** (1971): 537—540.

— and Wright, W. L.: An apparent antagonist of Gibberellin. Plant Physiol. **46** (1970): 19.

Leroux, P., and Gredt, M.: Effet de l'imazalil sur la biosynthèse de l'ergosterol chez *Penicillium expansum* link. Comptes Rendus Académie des Sci. Paris Ser. D, **286** (1978 a): 427—429.

— — Effets de quelques fongicides systemiques sur la biosynthèse de l'ergosterol chez *Botrytis cinerea* Pers., *Penicillium expansum* Link et *Ustilago maydis* (DC.) (da.) Ann. Phytopath. **10** (1978 b): 45—60.

— — Resistance to fungicides which inhibit ergosterol biosynthesis in laboratory strains of *Botrytis cinerea* and *Ustilago maydis*. Pesticide Sci. **15** (1984): 85—89.

Limpert, E., and Fischbeck, G.: Regionale Unterschiede in der Empfindlichkeit des Gerstenmehltaus gegen Triadimenol und ihre Entwicklung im Zeitraum 1980 bis 1982. Phytomedizin **13**, 19 (Abstr.) (1983).

Mandal, S. B., Sen, P. S., and Chartkrabarte, P.: Effect of respiratory deficiency and temperature on the mitochondrial lipid metabolism of *Aspergillus niger*. Can J. Microbiol. **24** (1978): 586—592.

Mariott, M. S.: Inhibition of sterol biosynthesis in *Candida albicans* by imidazole containing antifungals. J. Gen. Microbiol. **117** (1980): 253—255.

Martin, R. A., and Edgington, L. V.: Antifungal, phytotoxic, and systemic activity of fenapanil and structural analogs. Pesticide Biochem. Physiol. **17** (1982): 1—9.

McDaniel, G. L.: Growth retardation activity of paclobutrazol on chrysanthemum. Hort Sci. **18** (1983): 199—200.

Menhenett, R.: Studies with plant growth regulators. Rep. Glasshouse Crops Res. Inst. (1980): 76—77.

— and Hanks, G. R.: Comparisons of a new triazole retardant PP 333 with ancymidol and other compounds on pot-grown tulips. Plant Growth Regulation **1** (1982/83): 173—181.

Mercer, E. I.: The biosynthesis of ergosterol. Pesticide Sci. **15** (1984): 133—155.

Mitropoulos, K. A., Gibbons, G. F., Connell, G. M., and Woods, R. A.: Effect of triarimol on cholesterol biosynthesis in rat liver subcellular fractions. Biochem. Biophys. Res. Commun. **71** (1976): 892—900.

Montague, M. J.: Inhibition of gibberellic acid-induced elongation in *Avena* stem segments by a substituted pyrimidine. Plant Physiol. **56** (1975): 167—170.

Negroni de Bonvehi, M. B., Borgers, M., and Negroni, R.: Ultrastructural changes produced by ketoconazole in the yeast-like phase of *Paracoccidioides brasiliensis* and *Histoplasma capsulatum*. Mycopathologia **74** (1981): 113—118.

Nes, W. R.: Role of sterols in membranes. Lipids **9** (1974): 596—612.

— Sekula, B. C., Nes, W. D., and Adler, J. D.: The functional importance of structural features of ergosterol in yeast. J. Biol. Chem. **253**, 6218—6225.

Pappas, A. C., and Fisher, D. J.: A comparison of the mechanism of action of vinclozolin, procymidone, iprodione and prochloraz against *Botrytis cicerea*. Pesticide Sci. **10** (1979): 239—246.

Pennypacker, B. W., Sanders, P. L., Gregory, L. V., Gilbride, E. P., and Cole, Jr., H.: Influence of triadimefon on the foliar growth and flowering of annual bluegrass. Can. J. Plant Pathol. **4** (1982): 259—262.

Pierce, A. M., Pierce, H. D., Unrau, A. M., and Oehlschläger, A. C.: Lipid composition and polyene antibiotic resistance of *Candida albicans* mutants. Can. J. Biochem. **56** (1978): 135 to 142.

Plembel, M., Regel, E., and Büchel, K. H.: Antimycotic efficacy of bifonazole *in vitro* and *in vivo*. Arzneim.-Forsch. **33** (I) (1983): 517—524.

Pommer, E. H., and Zeeh, B.: BAS 45406 F: a new triazole derivative for the control of phytopathogenic fungi. 4th Intern. Congr. Plant Pathol., Melbourne 231 (Abstr.) 1983.

Preusser, H. J.: Effects of *in vitro* treatment with econazole on the ultrastructure of *Candida albicans*. Mykosen **19** (1976): 304—316.

Ragsdale, N. N.: Specific effects of triarimol on sterol biosynthesis in *Ustilago maydis*. Biochim. Biophys. Acta **380** (1975): 81—96.

— and Sisler, H. D.: Inhibition of ergosterol biosynthesis in *Ustilago maydis* by the fungicide triarimol. Biochem. Biophys. Res. Commun. **46** (1972): 2048—2053.

— — Mode of action of triarimol in *Ustilago maydis*. Pesticide Biochem. Physiol. **3** (1973): 20—29.

RICHMOND, D. V., and PRING, R. J.: Some morphological and cytological effects of fungicides. In: LYR, H., and POLTER, C. (Eds.): Systemic Fungicides. Akademie-Verlag, Berlin 1979, pp. 293—298.

RIETH, H.: Die unterschiedliche Bedeutung der Fettsäuren in ihrer fungistatischen Wirkung. Fette-Seifen-Anstrichmittel **79** (1977): 120—121.

ROBERTS, R. L., BOWERS, B., SLATER, M. L., and CABIB, E.: Chitin synthesis and localization in cell division cycle mutants of *Saccharomyces cerevisiae*. Molecular and Cellular Biology **3** (1983): 922—930.

ROGERS, H. J., PERKINS, H. R., and WARD, J. B.: Biosynthesis of wall components in yeast and filamentous fungi. In: Microbial Cell Walls and Membranes, Chapman and Hall, London 1980, pp. 478—50.

SCHEPERS, H. T. A. M.: Decreased sensitivity of *Sphaerotheca fuliginea* to fungicides which inhibit ergosterol biosynthesis. Netherl. J. Plant Path. **89** (1983): 185—187.

— Persistence of resistance to fungicides in *Sphaerotheca fuliginea*. Netherl. J. Plant Path. **90** (1984): 165—171.

— Fitness of isolates of *Sphaerotheca fuliginea* resistant or sensitive to fungicides which inhibit ergosterol biosynthesis. Netherl. J. Plant Path. **91** (1985a): 65—76.

— Changes over a three-year period in the sensitivity to ergosterol biosynthesis inhibitors of *Sphaerotheca fuliginea* in the Netherlands. Netherl. J. Plant Pathol. **89** (1985b): 105—118.

SCHMITT, P., and BENVENISTE, P.: Effect of fenarimol on sterol biosynthesis in suspension cultures of bramble cells. Phytochemistry **18** (1979): 1659—1665.

SHEPHARD, M. C., FRENCH, P. N., and RATHMELL, W. G.: PP 969; a broad spectrum systemic fungicide for injection or soil application. Proc. 10th Intern. Congr. Plant Protect. Vol. **2**, (1983): 521.

SHERALD, J. L., RAGSDALE, N. N., and SISLER, H. D.: Similarities between the systemic fungicides triforine and triarimol. Pesticide Sci. **4** (1973): 719—727.

— and SISLER, H. D.: Antifungal mode of action of triforine. Pesticide Biochem. Physiol. **5** (1975): 477—488.

SHIGEMATSU, M. L., UNO, J., and ARAI, T.: Effect of ketokonazole on isolated mitochondria from *Candida albicans*. Antimicrobiol. Agents Chemother. **21** (1982): 919—924.

SHIVE, J. B., and SISLER, H. D.: Effects of ancymidol (a growth retardant) and triarimol (a fungicide) on the growth, sterols and gibberellins of *Phaseolus vulgaris* (L.). Plant Physiol. **57** (1976): 640—644.

SIEGEL, M. R., and RAGSDALE, N. N.: Antifungal mode of action of imazalil. Pesticide Biochem. Physiol. **9** (1978): 48—56.

SISLER, H. D., and RAGSDALE, N. N.: Biochemical and subcellular aspects of the antifungal action of ergosterol biosynthesis inhibitors. In: TRINCI, H. P. J., and RYLEY, J. F. (Eds.): Mode of Action of Antifungal Agents. Cambridge Univ. Press 1984, pp. 257—282.

— — Fungitoxicity and growth regulation involving aspects of lipid biosynthesis. Netherl. J. Plant Path. **83** (Suppl. 1) (1977): 81—91.

— — and WATERFIELD, W. F.: Biochemical aspects of fungitoxic and growth regulatory action of fenarimol and other pyrimidine methanols. Pesticide Sci. **15** (1984): 167—176.

— WALSH, R. C., and ZIOGAS, B. N.: Ergosterol biosynthesis: A target of fungitoxic action. In: MATSUNAKA, S., HUTSON, D. H., and MURPHY, S. D. (Eds.): Pestic. Chem.: Human Welfare and the Environment. Vol. **3**, Mode of Action, Metabolism and Toxicology. Pergamon Press, New York 1983, pp. 129—134.

SKIDMORE, A. M., FRENCH, P. N., and RATHMELL, W. G.: PP 450: a new broad-spectrum fungicide for cereals. 10th Intern. Congr. Plant Protect. Vol. 1 (1976): 368—375.

SUD, I. J., and FEINGOLD, D. S.: Heterogenity of action mechanisms among antimycotic imidazoles. Antimicrob. Agent Chemother. **20** (1981): 71—74.

SUGAVANAM, B.: Diastereoisomers and enantiomers of paclobutrazol: their preparation and biological activity. Pesticide Sci. **15** (1984): 296—302.

SUMRELL, G., MOD, R. R., and MAGUE, F. C.: Antimicrobial activity of some fatty acid derivatives. J. Amer. Oil Chem. Soc. **55** (1978): 395—397.

SWIETLIK, D., and MILLER, S.: The effect of paclobutrazol on growth and response to water stress of apple seedlings. J. Amer. Soc. Hort. Sci. **108** (1983): 1076—1080.

TAYLOR, F. R., and PARKS, W. L.: Adaptation of *Saccharomyces cerevisiae* to growth on cholesterol: selection of mutants defective in the formation of lanosterol. Biochem. Biophys. Res. Commun. **95** (1980): 1437—1445.

Taylor, F. R., Rodriquez, R. J., and Parks, L. W.: Relationship between antifungal activity and inhibition of sterol biosynthesis in miconazole, clotrimazole, and 15-Azasterol. Antimicrobiol. Agents Chemother. **23** (1983): 515—521.

Tepper, B. L., and Yoder, K. S.: Postharvest chemical control of blue mold of apple. Plant Dis. **66** (1982): 829—831.

Thompson, E. D., and Parks, L. W.: The effect of altered sterol composition on cytochrome oxidase and S-adenosyl-methionine: Δ^{24} sterol methyltransferase enzymes of yeast mitochondria. Biochem. Biophys. Res. Commun. **57** (1974): 1207—1213.

Trocha, P. J., Jasne, S. J., and Sprinson, D. B.: Yeast mutants blocked in removing the methyl group of lanosterol at C-14. Separation of sterols by high-pressure liquid chromatography. Biochemistry **16** (1977): 4721—4726.

Tschabold, E. E., Taylor, H. M., Davenport, J. D., Hackler, R. E., Krunkalns, E. V., and Meredith, W. S.: A new plant growth regulator. Plant Physiol. **46** (1970): 19.

Uno, J., Shigematsu, M. L., and Arai, T.: Primary site of action of ketokonazole on *Candida albicans*. Antimicrob. Agents Chemother. **21** (1982): 912—918.

Vanden Bossche, H., Lauwers, W. F., and Cornelissen, F.: The antimycotic miconazole: an inhibitor of the biosynthesis of polyisoprenoids in *Staphylococcus aureus* B. 180. Arch. Intern. de Physiologie et Biochemie **90**, B 78 (1982b).

— — Willemsens, G., Marichal, F., Cornelissen, F., and Cools, W.: Molecular basis for the antimycotic and antibacterial activity of N-substituted imidazoles and triazoles: inhibition of isoprenoid biosynthesis. Pesticide Sci. **15** (1984b): 188—198.

— Marichal, P., Lauwers, W. F., and Willemsens, G.: Biochemical differences between yeast and mycelia. Do they determine the antimycotic activity of ketoconazole? In: Spitzy, K. H., and Karrer, K. (Eds.): Proc. 13th Intern. Congr. Chemother. PS 4.8/3—9.: Verl. H. Egermann, Vienna 1983.

— Ruysschaert, J. M., Defrise-Quertain, F., Willemsens, G., Cornelissen, F., Marichal, P., Cools, W., and van Cutsem, J.: The interaction of miconazole and ketoconazole with lipids. Biochem. Pharmadol. **31** (1982a): 2609—2617.

— Willemsens, G., Cools, W., Cornelissen, F., Lauwers, W. F. J., and van Cutsem, J. M.: *In vitro* and *in vivo* effects of the antimycotic drug ketoconazole on sterol synthesis. Antimicrob. Agents Chemother. **17** (1980): 922—928.

— — — Lauwers, W. F. J., and Le Jeune, L.: Biochemical effects of miconazole on fungi. II. Inhibition of ergosterol biosynthesis in *Candida albicans*. Chem.-Biol. Interactions **21** (1978): 59—78.

— — Marichal, P., Cools, W., and Lauwers, W.: The molecular basis for the antifungal activities of N-substituted azole derivatives. Focus on R 51 211. In: Trinci, A. P. J., and Ryley, J. F. (Eds.): Mode of Action of Antifungal Agents. Cambridge Univ. Press. 1984a, pp. 321—341.

van Tuyl, J. M.: Genetics of fungal resistance to systemic fungicides. Meded. Landbouwhogeschool Wageningen **77—2** (1977): 1—136.

Voigt, W.-H., and Plembel, M.: Elektronenmikroskopische Untersuchungen an human-pathogenen Pilzen. I. Mitteilung: Ultrastrukturelle Veränderungen von *Candida albicans*-Zellen durch Clotrimazol im Tierexperiment. Arzneim.-Forsch. (Drug Res.) **24** (1974): 508—515.

— and Schnell, J. D.: Elektronenmikroskopische Untersuchungen an human-pathogenen Pilzen. II. Mitteilung: Ultrastrukturelle Veränderungen von *Candida albicans*-Zellen am menschlichen Vaginalepithel unter der Behandlung mit Clotrimazol. Arzneim.-Forsch. (Drug Res.) **24** (1974): 516—521.

Walsh, R. C., and Sisler, H. D.: A mutant of *Ustilago maydis* deficient in sterol C-14 demethylation: characteristics and sensitivity to inhibitors of ergosterol biosynthesis. Pesticide Biochem. Physiol. **18** (1982): 122—131.

Wample, R. L., and Culver, E. B.: The influence of paclobutrazol, a new growth regulator on sunflowers. J. Amer. Soc. Hort. Sci. **108** (1983): 122—125.

Weete, J. D., Sancholle, M. S., and Montant, C.: Effects of triazoles on fungi. II. Lipid composition of *Taphrina deformans*. Biochim. Biophys. Acta **752** (1983): 19—29.

Wiggins, T. E., and Baldwin, B. C.: Binding of azole fungicides related to diclobutrazol to cytochrome P-450. Pesticide Sci. **15** (1984): 206—209.

Wilfret, G. J.: Height retardation of poinsettia with ICI-PP 333. Hort. Sci. **16** (1981): 443.

Wilkinson, C. F., Hetnarski, K., and Yellin, T. O.: Imidazole derivatives — a new class of microsomal enzyme inhibitors. Biochem. Pharmacol. **21** (1972): 3187—3192.

WILLEMSENS, G., COOLS, W., and VANDEN BOSSCHE, H.: Effects of miconazole and ketoconazole on sterol synthesis in a subcellular fraction of yeast and mammalian cells. In: VANDEN BOSSCHE, H. (Ed.): The Host Invader Interplay. Elsevier Biomedical Press, Amsterdam 1980, pp. 691 to 694.

WOLFE, M. S., MINCHIN, P. N., and SLATER, S. E.: Powdery mildew of barley. Ann. Rep. Pl. Breed, Inst. (1982): 92—96.

— — — Dynamics of triazole sensitivity in barley mildew, nationally and locally. Proc. Brit. Crop. Prot Conf. — Pests and Diseases, Vol. 2 (1984): 465—470.

— SLATER, S. E., and MINCHIN, P. N.: Fungicide insensitivity and host pathogenicity in barley mildew. Proc. 10th Intern. Congr. Pl. Protect. Vol. 2 (1983): 645.

WORTHING, C. R.: The Pesticide Manual. 6th Brit. Crop Protect. Council, London.

YAMAGUCHI, H.: Antagonistic action of lipid components of membranes from *Candida albicans* and various other lipids on two imidazole antimycotics, clotrimazole and miconazole. Antimicrob. Agents Chemother. **12** (1977): 16—25.

— Protection by unsaturated lecithin against the imidazole antimycotics, clotrimazole and miconazole. Antimicrob. Agents Chemother. **13** (1978): 423—426.

YEAGLE, P. L., MARTIN, R. B., LALA, A. K., LIN, H. K., and BLOCH, K.: Differential effects of cholesterol and lanosterol on artificial membranes. Proc. Nat. Acad. Sci. USA **74** (1977): 4924—4926.

LYR, H. (Ed.): Modern Selective Fungicides — Properties, Applications, Mechanisms of Action. Longman Group UK Ltd., London, and VEB Gustav Fischer Verlag, Jena, 1987.

Chapter 15

Benzimidazole and related fungicides

C. J. Delp

Agricultural Consultant, Washington, D.C., (formerly Du Pont Co. Wilmington, DE, USA)

Introduction

Benzimidazole fungicides and the thiophanates which are transformed to benzimidazoles represented a new era in fungicide use when they were introduced in the late 1960's. They are effective at relatively low doses for the inhibition of a broad range of fungi. Even more important is the property of systemic action in the host plant, which has led to extensive study and to the control of pathogens even after infection. The enthusiastic acceptance of benzimidazoles by plant pathologists throughout the world has resulted in countless reports on new and improved disease control. A specific site of action contributes to their high selective activity, and unfortunately also to the risk for resistance problems. Since the benzimidazoles have been used widely as the most effective control measures of so many destructive pathogens, resistance can be a major problem. Reports of new benzimidazole resistance problems and indepth studies have pioneered this field of fungicide research. This chapter covers the major benzimidazole and related products and their characteristics. It is not a review of the voluminous literature on the subject. Among the many general articles on this subject are Marsh 1972; Nene and Thapliyal 1979; Siegel and Sisler 1977. The genetics of fungal sensitivity and mode of action of benzimidazoles are covered in the following chapter 16 by L. Davidse.

Chemicals

Since the introduction of benomyl (Delp and Klopping 1968), other benzimidazole compounds with similar properties have been developed for practical plant disease control. Carbendazim, the degradation product of benomyl (Clemons and Sisler 1969) and thiophanate-methyl (Vonk and Sijpesteijn 1971), appears to be the chemical active in fungi for these fungicides. The names and structures (Fig. 15.1) of the six most important benzimidazole fungicides are tabulated below (Table 15.1). Although the thiophanates are not benzimidazoles until transformed, all of the compounds will be referred to as benzimidazoles in this chapter.

Spectrum of bio-activity

Although the relative toxicity and effectiveness for disease control differs among these related fungicides, they have similar patterns of selective action (Bollen and Fuchs 1970; Edgington *et al.* 1971). Most Ascomycetes, some of the Basidiomycetes and Deuteromycetes, and none of the Phycomycetes are sensitive.

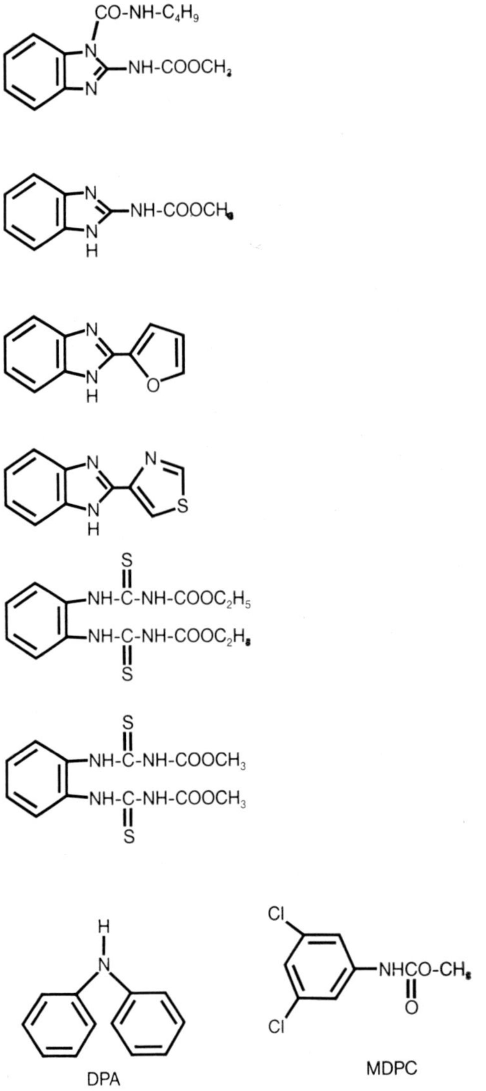

Fig. 15.1 Chemical structures of benzimidazole and related fungicides and of compounds exhibiting a negative cross resistance

Compound name	Patent No.	Filing Date	Company
a) benomyl	USP 3,541,213	1966	Du Pont
b) carbendazim	USP 3,852,460	1967	Du Pont
c) fuberidazole	DAS 1,209,799	1964	Bayer
d) thiabendazole	USP 3,017,415	1960	Merck
e) thiophanate	DOS 1,806,123	1967	Nippon Soda
f) thiophanate-methyl	DOS 1,806,123	1967	Nippon Soda
g) diphenylamine			
h) N-(3,5 dichlorophenyl)-carbamate			

Table 15.1 Major benzimidazole fungicides

Common Name	*Chemical Name*	*Manufacturer, Trade Name*	*Structure*
Benomyl	*methyl 1-(butylcarbamoyl)-2-benzimidazole-carbamate*	*Du Pont* *Benlate®* *Tersan® 1991* *Chinoin* *Fundazol®*	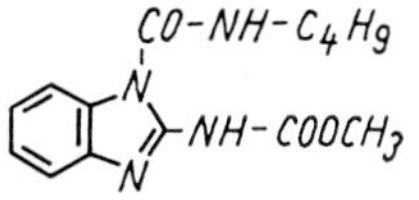
Carbendazim	*methyl benzimidazole-2-yl carbamate (MBC)*	*Du Pont* *Delsene®* *BASF* *Bavistin®* *Hoechst* *Derosal®*	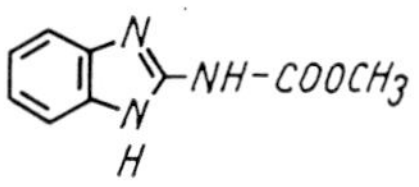
Fuberidazole	*2-(2'-furyl)-1H-benzimidazole*	*Bayer* *Neo-Voronit®*	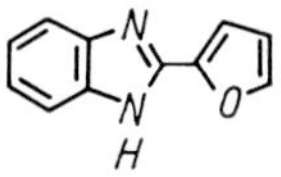
Thiabendazole	*2-(4'-thiazolyl)-benzimidazole (TBZ)*	*Merck* *Mertect®* *Tecto®*	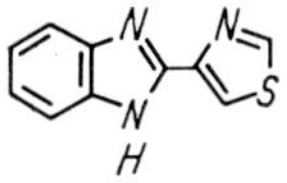
Thiophanate	*diethyl 4,4'-o-phenylene-bis (3-thioallophanate)*	*Nippon Soda* *Cercobin®* *Topsin®*	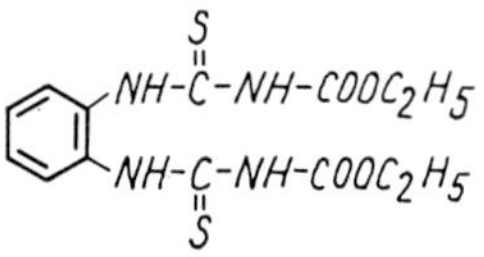
Thiophanate-methyl	*dimethyl 4,4'-o-phenylene-bis (3-thioallophanate)*	*Nippon Soda* *Cercobin-M®* *Topsin-M®*	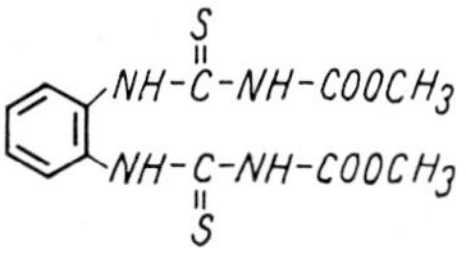

Sensitivity in the cabbage club root pathogen *Plasmodiophora brassica* appears to be an exception. Their selective toxicity provides safety to animals, bees and plants with a corresponding high degree of activity against certain fungi, insects, mites and worms.

Because of differences in activity and marketing strategies, the benzimidazoles have been developed for different uses. Thiabendazole, first sold primarily as an anthelmintic, was then developed on crops for post-harvest fruit treatments, and later a special formulation was developed for treatment of elm trees infected with *Ceratocystis ulmi*. Fuberidazole has been used primarily in Europe for cereal seed treatment. Benomyl, carbendazim and thiophanates have been developed for the uses listed in Table 15.2. The effects on systemic pathogens like *Verticillium*, *Fusarium*, and *Ceratocystis* were promising under controlled conditions, but distribution in soil and host is frequently insufficient for desired results.

An important non-plant pathogenic fungus controlled by benzimidazoles is *Pithomyces chartarum*, the causal agent of facial eczema in sheep and cattle (Campbell and Sinclair 1968).

Table 15.2 Major practical applications

Crops	Pathogens
Some of the diseases controlled by benzimidazole fungicides:	
Almond *(Prunus)*	— *Monilinia* spp.
Asparagus	— *Phoma asparagi*
Avocado *(Persea)*	— *Cercospora perseae, Colletotrichum gloeosporioides, Fusarium* spp., and *Verticillium albo-atrum*
Banana *(Musa)*	— *Ceratocystis paradoxa, Colletotrichum musae, Fusarium* spp., *Mycosphaerella musicola, M. fijiensis* var. *difformis, Nigrospora,* and *Penicillium* spp.
Beans *(Phaseolus)*	— *Botrytis cinerea, Colletotrichum* spp., *Sclerotinia sclerotiorum* and *Cercospora* spp.
Blueberry *(Vaccinium)*	— *Botrytis cinerea, Gloeosporium* sp., and *Monilinia vaccinii-carombosi*
Bushberry *(Rubus)*	— *Botrytis cinerea, Penicillium* spp., and *Sphaerotheca humuli*
Carrot *(Daucus)*	— *Cercospora carota*
Cassava *(Manihot)*	— *Cercospora* spp.
Celery *(Apium)*	— *Cercospora apii,* and *Septoria apii*
Chestnut *(Castanea)*	— *Colletotrichum castanea*
Citrus	— *Botrytis cinerea, Colletotrichum gloeosporiodes, Diplodia natalensis, Elsinoe fawcetti, Guignardia citricarpa, Mycosphaerella citri, Penicillium* spp., *Phomopsis citri,* and *Oidium tingitaninum*
Clove *(Eugenia)*	— *Cylindrocladium quinqueseptatum*
Coffee	— *Cercospora coffeicola, Colletotrichum coffeanum,* and *Pellicularia koleroga*
Cole Crops, Canola or Rape *(Brassica)*	— *Fusarium oxysporum, Mycosphaerella brassicola, Phoma lingam,* and *Sclerotinia sclerotiorum*
Cucurbits & Melons *(Cucumis)*	— *Cladosporium cucumerinum, Colletotrichum gossypii, Colletotrichum lagenarium, Erysiphe cichoracearum, Fusarium* spp., *Mycosphaerella citrullina, Oidium* spp., *Rhizoctonia solani,* and *Sclerotinia sclerotiorum*
Currant *(Ribes)*	— *Drepanopeziza* sp., *Sphaerotheca humuli,* and *Cronartium* sp.
Eggplant *(Solanum)*	— *Colletotrichum capsici, Colletotrichum gloeosporiodes,* and *Corynespora melongenae*
Garlic *(Allium)*	— *Penicillium* sp., and *Sclerotinia sclerotiorum*
Grape *(Vitis)*	— *Botrytis cinerea, Gloeosporium ampelophagum, Guignardia bidwellii, Melanconium fuligenum, Pseudopeziza* sp., and *Uncinula necator*
Lettuce *(Lactuca)*	— *Botrytis cinerea,* and *Sclerotinia sclerotiorum*
Mango *(Mangifera)*	— *Colletotrichum gloeosporiodes,* and *Oidium*
Mulberry *(Morus)*	— *Phyllactinia* sp.
Mushroom	— *Dactylium* sp., *Mycogone perniciosa,* and *Verticillium malthousi*
Olive *(Olea)*	— *Cycloconium oleaginum*
Onion *(Allium)*	— *Botrytis allii, Colletotrichum, gloeosporiodes,* and *Fusarium oxysporum*
Ornamentals and Trees	— *Ascochyta, Botrytis cinerea, Ceratocystis* spp., *Cercospora* spp., *Colletotrichum* spp., *Corynespora* spp., *Cylindrocladium, Didymellina* spp., *Diplocarpon rosae, Entomosporium* sp., *Fabrae maculata, Fusarium* spp., *Gnomonia leptostyla, Oidium* spp., *Ovulinia azaleae, Penicillium* spp., *Phomopsis* spp., *Phyllostictina* spp., *Ramularia* spp., *Rhizoctonia* spp., *Sclerotinia* spp., *Sphaerotheca pannosa,* and *Thielaviopsis* spp.

Table 15.2 (continued)

Some of the diseases controlled by benzimidazole fungicides:

Crops	Pathogens
Papaya *(Carica)*	— *Oidium caricae*
Peanut *(Arachis)*	— *Ascochyta* spp., *Aspergillus* spp., *Cercospora arachidicola, Cercosporidium personatum,* and *Mycosphaerella arachidicola*
Pea *(Pisum)*	— *Mycosphaerella pinoides,* and *Oidium* spp.
Pecan *(Carya)*	— *Cercospora fusca, Cladosporium effusum, Cristulariella pyramidalis, Leptothyrium caryae, Microsphaera alni* and *Mycosphaerella caryigena*
Pepper *(Capsicum)*	— *Botrytis cinerea, Fusarium piperi,* and *Fusarium solani*
Persimmon *(Diospyros)*	— *Cercospora kaki,* and *Phyllactinia kaki*
Pineapple *(Ananas)*	— *Fusarium moniliforme, Rhizoctonia* sp., and *Thielaviopsis paradoxa*
Plantain *(Plantago)*	— *Colletotrichum musae, Fusarium* spp., and *Penicillium* spp.
Pome Fruit *(Malus* and *Pyrus)*	— *Botrytis cinerea, Cladosporium* spp., *Gloeodes pomigena, Gymnosporangium* spp., *Marssonina mali, Microthyriella rubi, Mycosphaerella pomi, Penicillium* spp., *Phyllactinia pyri, Physalospora* spp., *Podosphaera leucotricha, Rosellinia necatrix, Valsa ceratosperma,* and *Venturia* spp.
Potato *(Solanum)*	— *Fusarium solani, Oospora* sp., *Phoma* sp., and *Rhizoctonia solani*
Rice *(Oryza)*	— *Acrocylindrium oryzae, Cercospora oryzae, Gibberella fujikuroi, Piricularia oryzae, Thanatephorus cucumeris,* and *Rhizoctonia solani* (stem rot)
Rubber *(Hevea)*	— *Ceratocystis fimbriata, Microcyclus ulei,* and *Mycosphaerella* spp.
Soybean *(Glycine)*	— *Cercospora kikuchii, C. sojina, Colletotrichum truncatum, Diaporthe* spp., *Phomopsis sojae,* and *Septoria glycinea*
Stone Fruit *(Prunus)*	— *Cladosporium carpophilum, Coccomyces hiemalis, Cytospora leucostoma, Fusarium* spp., *Monilinia* spp., *Oidium* spp., *Phomospsis persicae,* and *Sphaerotheca pannosa*
Strawberry *(Fragaria)*	— *Botrytis cinerea, Cercospora* sp., *Fusarium oxysporum, Mycosphaerella fragariae, Odium* spp., *Sphaerotheca humali,* and *Verticillium* spp.
Sugar Beet *(Beta)*	— *Cercospora beticola*
Sugar Cane *(Saccharum)*	— *Ceratocystis paradoxa,* and *Cercospora* spp.
Sweet Potato and Yam *(Ipomoea* and *Dioscores)*	— *Ceratocystis fimbriata,* and *Monilochaetes infuscans*
Tea *(Thea)*	— *Gloeosporium theae-sinensis, Elsinoe leucospila,* and *Rosellinia necatrix*
Tobacco *(Nicotiana)*	— *Ascochyta nicotianae, Cereospora nicotianae, Helicobasidium mompa,* and *Pellicularia filamentosa*
Tomato *(Lycopersicon)*	— *Botrytis cinerea, Cercospora* spp., *Cladosporium* sp., *Colletotrichum phomoides, Corynespora melongenae, Fusarium oxysporum, Odium* spp., *Phoma destricitiva, Sclerotinia sclerotiorum,* and *Septoria lycopersici*
Turf (*Poa, Agrostis,* etc.)	— *Fusarium* spp., *Gloeotinia temulenta, Pellicularia filamentosa,* and *Sclerotinia homeocarpa*
Wheat *(Triticum)*	— *Erysiphe graminis, Fusarium* spp., *Pseudocercosporella herpotrichoides, Rhynchosporium* sp., *Septoria avenae,* and *Ustilago tritici*

Non-target effects and undeveloped uses

The target organisms discussed above are so responsive to benzimidazole treatment that it is natural for some sensitive, non-target organisms to be effected. Of course, contact with the benzimidazole is most critical to any non-target effects. For example, mites in a treated apple tree or earthworms feeding on treated foliage under an apple tree may be exposed to active rates of a benzimidazole during the season of treatment for apple disease control. But earthworms and microorganisms below the soil surface would contact very little benzimidazole, and most of the chemical present would be so tightly bound to soil that it would not be available for reaction.

The mite ovicide activity, reported in 1968 by DELP and KLOPPING, has not been developed commerically, but the population suppression effects are significant for the spider mites *Panonychus ulmi* and *Tetranychus urticae* and their predators *Amblyseius fallacis* (NAKASHIMA and CROFT 1974) and *Agistemus fleschneri* (CHILDERS and ENNS 1975). These and other non-target mites also develop resistance to benzimidazoles, and population suppressant effects can be only temporary.

There is also evidence for the control of some insects and diseases of cultured insects and bees. Examples include reductions in the population of aphids (BINNS 1970; DELORME 1976; ENGELHARD and POE 1971; and BAILISS *et al.* 1978) and cabbage maggots *Hylemya brassicae* (REYES and STEVENSON 1975). HARVEY and GAUDET (1977) found that 75 ppm of benomyl in the artificial diet of spruce budworm reduced the microsporidian levels and also reduced worm growth and fertility. The protozoan-microsporida *Nosema*, a parasite which constitutes a major problem in laboratory culturing of many insects, is controlled by adding benomyl to the artificial diet of alfalfa weevils *Hyper postica* (HSIAO and HSIAO 1973). Bees of the genera *Aphis, Megachile* and *Nomia* may be affected by the chalkbrood disease caused by the fungus, *Ascosphaera* spp. Benomyl treatments can reduce chalkbrood and at the same time are safe to bees (MOELLER and WILLIAMS 1976).

Thiabendazole and thiophanate are commercial anthelmintics, and the other benzimidazoles have varying degrees of animal worm activity. Their effects on nematodes and earthworms cover a broad number of genera if sufficient chemical is contacted. There are reports of decreased root penetration by nematodes, elimination of the ability of the dagger nematode, *Xiphinema*, to transfer tobacco ring spot virus, and reduction in populations of the rice nematode, *Aphelenchoides*, which causes white tip disease.

The non-target effects on earthworms have been the subject of considerable concern. Under some conditions where earthworms feed on benzimidazole treated residues, populations are temporarily reduced, for example, earthworms which feed on apple foliage on the orchard floor in the UK (STRINGER and LYONS 1974). Excessive use where the presence of earthworms is critical should be avoided. On the other hand, some airport managers have used high rates to reduce the danger of earthworms on runways (TOMLIN *et al.* 1981).

Behaviour on plants

Among the most exciting aspects of these fungicides are their characteristics on and in plants. They are effective protectants, and also penetrate the host plant to inhibit post-infection or move in the apoplast to untreated portions of the plant for preventive or curative effects. In practice, most disease control is accomplished by direct contact or protectant action. They are tightly bound to plant surfaces and degrade slowly, thus they have good residual activity. Excessive deposits serve for surface redistribution. Penetration into plants varies greatly among compounds and also varies with adjuvants applied with the compound, the kind of plant and the location or maturity of the plant. For example Upham and Delp (1973), found 20 times more active compound in herbaceous plants when applying benomyl as compared to applying carbendazim. Thiophanate and benomyl penetrate apple cuticle more rapidly than carbendazim while thiabendazole has the slowest penetration rate (Solel and Edgington 1973).

In plants where penetration is sufficient, translaminar movement will stop the infection process into untreated leaf surfaces opposite the site of application. Although there is some evidence of detectible movement out of a treated leaf, the quantities are too small for practical disease control. Movement in the leaf is in the transpiration water, and carbendazim accumulates at the margins and tip.

Uptake by roots is more efficient and results in more complete distribution throughout a plant than from foliar treatment. Areas of low transpiration, for example some fruit, tend to accumulate less benzimidazole from systemic movement.

The use of benzimidazoles at recommended rates for plant disease control has resulted in very few undesirable plant responses. It is remarkable that these chemicals which are highly toxic to a broad number of pathogenic fungi are so safe even when taken systemically into host plants. Massive doses, especially when applied on seed and soil treatments, may delay emergence or cause stunting or chlorosis of some plants. When used on certain apple varieties, benomyl (and to a lesser extent the other benzimidazoles) may amplify the naturally occurring fruit-finish problems of russeting or opalescence.

Most plants have no injurious responses, but there are other plant responses which are noteworthy. Under some conditions, treatments have resulted in antisenescent, cytokinin and yield boosting effects, and masking of virus symptoms and ozone injury. The cytokinin-like properties of benzimidazoles (Skene 1972; Thomas 1974) may contribute to the observations of delayed senescence, cereal tillering, and symptom masking. Although the yield boosting effects on soybeans and peanuts appear to be directly related to the control of diseases, this is not the case with cereals. Both increased tillering and antisenescence have been correlated with increased yields of cereals treated with a benzimidazole in the absence of traditional disease symptoms (Peat and Shipp 1981; Fehrmann *et al.* 1978; Tripathi *et al.* 1982).

Ozone injury on some crop foliage can be reduced by soil drench systemic applications of benomyl or thiophanate (Pellissier *et al.* 1972; Moyer *et al.* 1974) and also by foliar sprays (Manning *et al.* 1974).

Among the plant responses attributed to carbendazim treatment are the masking of the mosaic and yellows symptoms in tobacco inoculated with tobacco mosaic virus and in lettuce infected with beet western yellows virus (Tomlinson *et al.* 1976).

Behaviour in soil

Practical control of soil-borne, root rot, vascular and foliage pathogens is attained when relatively high rates of benzimidazoles are mixed in the soil of the root zone. This is possible with container-grown plants, incorporation into plant beds and in-furrow applications with small, high-value crops with confined root systems like strawberries and some vegetables and ornamentals. Under other field conditions, disease control with soil applications has been disappointing because the compounds are so tightly adsorbed to soil colloids and organic matter that they are virtually immobile and in some cases unavailable to affect sensitive organisms (Peeples 1974). The penetration into soil can be increased by addition of some surfactants (Pitblado and Edgington 1972). Carbendazim is relatively stable in soil, but studies with uneconomically high (100 ppm) rates of application may lead to unrealistic conclusions about the sensitivity of non-target soil organisms (Baude *et al.* 1974).

The temporary effects on sensitive organisms such as *Fusarium*, *Penicillium*, and *Trichoderma* appear to be compensated for by the increased growth of other less sensitive organisms. This compensation can maintain a balanced soil biological activity as measured by CO_2 release.

Resistance

Benzimidazole resistance represented the beginning of serious fungicide resistance problems. This happened in the 1970's because benzimidazoles were used widely, alone and intensively for crop protection, because they are specific-site inhibitors, and also because most fungi contain resistant strains in their natural populations. Most resistant fungal populations are fit for survival, and benzimidazoles may be ineffective for several seasons after resistant strains dominate large crop production areas. There are some exceptions where these fungicides can be used after a period of abstinence. For example, powdery mildew of cucurbits in New York state was the first to be reported as resistant (Schroeder and Provuidenti 1969), but other powdery mildew populations have developed resistance more slowly, and in some locations, rapidly revert to sensitive populations when the selective pressure from benzimidazoles is removed.

Because benomyl provided excellent control of sugar beet, peanut and celery *Cercospora* leaf spots; of apple and pear *Venturia* scab and *Botrytis* of grapes and other crops, it was frequently used exclusively. Under such conditions, resistance developed within 2 to 4 seasons. On the other hand, where mixtures of benomyl and unrelated fungicides have been used from the beginning, resistance has not become a problem or was delayed for several seasons (Delp 1980).

Under the organization of the Benzimidazole Working Group of industry's Fungicide Resistance Action Committee FRAC, the basic benzimidazole manufacturers launched a cooperative study of potential resistance in *Pseudocercosporella herpotrichoides*, the cereal eyespot pathogen (Delp 1984). The group brought together key European investigators and coordinated studies and recommendations in France, Germany and the UK. The impact of resistance was often difficult to interpret since treatment of fields with resistant populations might fail to control eyespot but still result in yield benefits. High risk situations were identified, and alternative programs, which were generally more expensive, were recommended. Although many of the major uses of benzimidazoles have been lost, there are situations where strategies have been used successfully to delay resistance problems. In general the strategies involved the reduced exposure to selective fungicide pressure by use of fewer treatments, lower rates and mixtures with unrelated companion fungicides. Where mixtures were used from the beginning, some *Cercospora* and *Botrytis* populations retained sensitivity. Also, controlled use of a benzimidazole on the banana Sigatoka *Mycosphaerella* has prolonged effectiveness and even made reentry possible in areas where resistance was emerging. Fifteen years after resistance problems became serious these fungicides are still effective for the control of many pathogens.

Cross-resistance, multiple-resistance and negative cross-resistance

Most fungal strains resistant to one benzimidazole also have reduced sensitivity to the other benzimidazoles (cross-resistance) (chapter 3). Therefore, field resistance problems for one compound are also problems for the others. This precludes switching or combining among benzimidazoles to avoid or solve resistance problems. Although there is no evidence of cross-resistance to non-benzimidazole fungicides, it is possible for a benzimidazole-resistant strain to also develop resistance to an unrelated fungicide (multiple-resistance). Multiple-resistance is common in some citrus fruit packing houses as reported by J. ECKERT (DEKKER and GEORGOPOULOS 1982).

There have been reports of strains which appear resistant to one benzimidazole and more sensitive to another (negative cross-resistance), but this phenomenon, as discussed by DAVIDSE, has been of no practical value. On the other hand, there is a group of carbamate compounds active only against certain strains of fungi resistant to benzimidazoles and dicarboxamides. This negative cross-resistance was reported for N-phenyl carbamate herbicides and benzimidazole fungicides (LEROUX and GREDT 1979). More recently it has been demonstrated that methyl N-(3,5 dichlorophenyl)-carbamate (MDPC) (Fig. 15.1h) represents a class of chemicals which are effective for plant disease control and can be combined with benzimidazoles to overcome resistance (KATO *et al.* 1984). Although in some situations strains resistant to both benzimidazole and MDPC have developed, there are highly effective analogs in field evaluations which could help overcome resistance problems.

The concept may have been in practical use before it was understood that diphenylamine (DPA) (Fig. 15.1g) in combination with a benzimidazole controls both benzimidazole-sensitive and resistant *Penicillium expansum*, the cause of post-harvest decay of apples. ROSENBERGER and MEYER (1985) showed that DPA combined with a benzimidazole is more effective for the control of some resistant isolates than either used alone. DPA (Fig. 15.1g), used as a post-harvest treatment to control the physiological disorder storage scald, is most effective for the control of strains highly resistant to benzimidazoles. Some strains of *P. expansum* are resistant to both DPA and benzimidazole, but this combination may be the first commercial application of negative cross resistance for the solution of a fungicide resistance problem.

Residue, environmental and toxicology issues

Residue deposits are most important in relation to environmental and toxicology issues. From analysis of treated produce, residue tolerances are established well below the levels which have a toxic effect on test animals (FAO 1983).

Residue analysis of crops treated for disease control taken throughout the world have been used to establish tolerances (the allowable and safe residues of benzimidazoles on raw agricultural commodities). Typical residue tolerances are in Table 15.3. Analytical methods for benzimidazoles can be found in ZWEIG and SHERMA 1982.

Carbendazim is moderately stable to photodegradation (FLEEKER and LACY 1977). In plants and soil there is relatively slow decomposition to 2-aminobenzimidazole. HELWEG 1977, demonstrated biodegradation in soil. RHODES and LONG 1974, showed that carbendazim and 2-aminobenzimidazole are immobile in soil and do not leach or move significantly from the site of application. Adsorption and stability in soil is directly influenced by pH and organic matter. Thiabendazole appears to be even more strongly adsorbed than carbendazim.

The conversion of benomyl and thiophanates to carbendazim varies greatly under different conditions. Some laboratory evidence for rapid conversion of benomyl in dilute solution may be misleading because under field use (BAUDE *et al.* 1973) a major portion of benomyl stays intact on plants. The residual benzimidazole on foliage 28 days after treatment of cucumber, banana, orange and grape is more than 60 % benomyl. CHIBA and VERES 1981, confirmed these results on apple foliage. This probably has little influence on the contact or protectant activity because benomye

Table 15.3 Typical* residue tolerances

Commodity	parts per million (mg/kg)
Meat, milk and eggs	0.1
Nuts including peanut and soybean	0.2
Sugarbeet roots (tops 15 ppm)	0.2
Cereal grains	0.5
Cucurbits, banana and avocado	1
Beans and most crops in the Netherlands	2
Celery & mango	3
Rice, strawberry & tomato	5
Pome fruits & berries	7
Citrus, grape and mushroom	10
Stone fruits	15
Pineapple	35

* tolerances selected from different products and countries

and carbendazim have similar protectant efficiency. But the presence of benomyl could have a pronounced effect on systemic and post-infection action since benomyl treatments accumulate higher systemic concentrations of active ingredient in the host than carbendazim treatments (Upham and Delp 1973). On the other hand, the slow transformation of thiophanates to carbendazim (Vonk and Sijpesteijn 1971) has different practical implications since carbendazim is so much more active than intact thiophanates. It is possible to lose some effectiveness because of a transformation delay. There is little or no hydrolysis to carbendazim from pH 1 to 7 (25 °C to 27 °C), and the half-life at pH 9 is about 17 hours.

Compatibility of benzimidazole wettable powder formulations with most other agricultural chemicals in tank mixtures has been good except with highly alkaline pesticides such as Bordeaux mixture or lime sulfur. Of course, it is necessary to determine the compatibility of each mixture by reference to the product labels and by small-scale, "jar" tests.

Toxicology studies for benzimidazoles show safe acute values of greater than 7,500 mg/kg in rats. Reproductive, teratogenic and mutagenic studies also assure that these products can be used safely. Thiabendazole and thiophanates are administered to animals as anthelminthics. The United States Environmental Protection Agency (EPA) issued a "Rebuttable Presumption Against Registration (RPAR)" against benomyl based on potential teratogenic and sperm effects. However, after careful scientific review, the EPA determined in 1983 that benomyl is safe for recommended uses. Instructions for aerial application loaders to wear dust masks to avoid excessive inhalation was added to the Benlate® label as required by the EPA.

Conclusions

Benzimidazoles represent a group of highly effective, broad-spectrum, systemic fungicides which are widely used for efficient plant disease control. Their mild cytokinin effects on some plants tend to retain chlorophyll and in some cases increase yields, delay maturity, but otherwise there are few significant plant responses. Effects on non-target organisms are minimal because of selective toxicity and strong adsorption to plants and soil. Progressive limitations on use have occurred because of resistance in major pathogens where these products were used intensively and exclusively. With increased awareness of effective ways to cope with resistance, users should be less likely to develop resistance problems, and these fungicides will be among the favoured disease-control agents for many years to come.

References

Bailiss, K. W., Partis, G. A., Hodgson, C. J., and Stone, E. V.: Some effects of benomyl and carbendazim on *Aphis fabae* and *Acyrthosphon pisum* on field bean *(Vica faba)*. Ann. appl. Biol. **89** (1978): 443—449.

Baude, F. J., Gardiner, J. A., and Han, J. C-Y, jr.: Characterization of residues on plants following foliar spray applications of benomyl. J. Agricult. Food Chem. **21** (1973): 1084—1090.

— Pease, H. L., and Holt, J.: Fate of benomyl on field soil and turf. J. Agricult. Food Chem. **22** (1974): 413—418.

Binns, E. S.: Aphicidal activity of benomyl. Glasshouse Crops Res. Inst. Ann. Rep. 1969 (1970): 113.

Bollen, G. J., and Fuchs, A.: On the specificity of the *in vitro* and *in vivo* antifungal activity of benomyl. Netherl. J. Plant Pathol. **76** (1970): 299—312.

Campbell, A. G., and Sinclair, D. P.: Control of facial eczema in lambs by use of fungicides. Proc. Ruakura Farmers' Conf. Week. N. Z. (1968): 3—12.

Chiba, M., and Veres, D. F.: Fate of benomyl and its degradation compound methyl 2-benzimidazolecarbamate on apple foliage. J. Agricicult. Food Chem. **29** (1981): 588—590.

Childers, C. C., and Enns, W. R.: Field evaluation of early season fungicide substitutions on Tetranychid mites and the predators Neoseinlus follacis and Agistemus flescbneri in two Missouri apple orchards. J. Econ. Entomol. **68** (1975): 719—724.

Clemons, G. P., and Sisler, H. D.: Formation of a fungitoxic derivative from Benlate. Phytopathology **59** (1969): 705—706.

Dekker, J., and Georgopoulos, S. G.: Fungicide Resistance in Crop Protection. Centre for agricultural publishing and documentation, Wageneningen, Netherl. 1982.

Delorme, R.: Evaluation en laboratoire de la toxicité pour *Diaeretiella* rapae (Hym. Aphidaidae) des pesticides utilises en traitement des parties aeriennes des plantes. Entomophaya **21** (1976): 19—29.

Delp, C. J.: Coping with resistance to plant disease control agents. Plant Disease **64** (1980): 651—657.

— Industry's response to fungicide resistance. Crop Protect. **3** (1984): 3—8.

— and Klopping, H. L.: Performance attributes of a new fungicide and mite ovicide candidate. Plant Disease Rep. **52** (1968): 95—99.

Edgington, L. V., Khew, K. L., and Barron, G. L.: Fungitoxic spectrum of benzimidazole compounds. Phytopathology **61** (1971): 42—44.

Engelhard, A. W., and Poe, S. L.: Combinations of fungicides and insecticides for control of disease, insects and mites on chrysanthemums. Proc. Florida State Horticult. Soc. **84** (1971): 435—441.

FAO Plant Production and Protection Paper 56. Pesticide Residues in Food. Rep. joint meeting on pesticide residues held in Geneva, December 5—14, Rome **4.4** (1983): 12—20.

Fehrmann, H., Reinecke, P., and Weihofen, U.: Yield increase in winter wheat by unknown effects of MBC-fungicides and captafol. Phytopath. Z. **93** (1978): 359—362.

Fleeker, J. R., and Lacy, H. M.: Photolysis of methyl 2-benzimidazolecarbamate. J. Agricult. Food Chem. **25** (1977): 51—55.

Harvey, G. T., and Gaudet, P. M.: The effects of benomyl on the incidence of microsporidia and the developmental performance of eastern spruce budworm (Lepidoptera: Tortricidae). The Canadian Entomologist **109** (1977): 987—993.

Helweg, A.: Degredation and absorption of carbendazim and 2-aminobenzimidazole in soil. Pesticide Sci. **8** (1977): 71—78.

Hsiao, T. H., and Hsiao, C.: Benomyl: A novel drug for controlling a microsporidan disease of the alfalfa weevil. J. Inventebrate Pathol. **22** (1973): 303—304.

Kato, T., Suzuki, D., Takahasbi, J., and Kamoshita, K.: Negatively correlated cross-resistance between benzimidazole fungicides and methyl N-(3,4-dichlorophenyl)-carbamate. J. Pesticide Sci. **9** (1984): 485—495.

Leroux, P., and Gredt, M.: Phenomenes de resistance croisee negative chez *Botrytis cinerea* Pers. entre les fongicides benzimidazoles et des herbicides carbamates. Phytiatr. Phytopharm, **28** (1979): 79—86.

Manning, W. J., Feder, W. A., and Vardaro, P. M.: Suppression of oxidant injury by benomyl. J. Environment. Quality **3** (1974): 1—3.

Marsh, R. W. (Ed.): Systemic Fungicides. Longman, London 1972.

Moeller, F. E., and Williams, P. H.: Chalkbrood research at Madison, Wisconsin. Am. Bee J. **116** (1976): 484—486.

Moyer, J., Cole, H., and La Casse, N. L.: Reduction of ozone injury on poa annua by benomyl and thiophanate. Plant Disease Rep. **58** (1974): 41—44.

Nakashima, M. J., and Croft, B. A.: Toxicity of benomyl to the life stages of *Amblyseius follacis.* J. Econ. Entomol. **67** (1974): 675—677.

Nene, Y. L., and Thapliyal, P. N.: Fungicides in Plant Disease Control. Oxford and IBH, New Delhi 1979.

Peat, W. E., and Shipp, D. M.: The effects of benomyl on the growth and development of wheat. EPPO Bull. **11** (1981): 287—293.

Peeples, J. L.: Microbial activity in benomyl-treated soils. Phytopathology **64** (1974): 857—860.

Pellissier, M., Lacasse, N. L., and Cole, H.: Effectiveness of benzimidazole, benomyl and thiabendazole in reducing ozone injury to pinto beans. Phytopathology **62** (1972): 580—582.

Pitblado, R. E., and Edgington, L. V.: Movement of benomyl in field soils as influenced by acid surfactants. Phytopathology **62** (1972): 513—516.

Reyes, A. A., and Stevensen, A. B.: Toxicity of benomyl to the cabbage maggot *Hyhemya brassicae* in greenhouse tests. Can. Entomologist **107** (1975): 685—687.

Rhodes, R. C., and Long, J. D.: Run-off and mobility studies on benomyl in soils and turf. Bull. Environment. Contamination and Toxicology **12** (1974): 385—393.

Rosenberger, D. A., and Meyer, F. W.: Negatively correlated cross-resistance to diphenylamine in benomyl-resistant *Penicillium* expansum. Phytopathology **75** (1985): 74—79.

Schroeder, W. T., and Provvidenti, R.: Resistance to benomyl in powdery mildew of cucurbits. Plant Disease Rep. **53** (1969): 271—275.

Siegel, M. R., and Sisler, H. D.: Antifungal Compounds, Vol. 2, M. Dekker Inc., New York 1977.

Skene, K. G. M.: Cytokinin-like properties of the systemic fungicide benomyl. J. Horticult. Sci. **47** (1972): 1979—1982.

Stringer, A., and Lyons, C. H.: The effect of benomyl and thiophanate-methyl on earthworm populations in apple orchards. Pesticide Sci. **5** (1974): 189—196.

Thomas, T. H.: Investigations into the cytokinin-like properties of benzimidazole-derived fungicides. Ann. appl. Biol. **76** (1974): 237—241.

Tomlin, A. D., Tolman, J. H., and Thorn, G. D.: Suppression of earthworm *(Lumbricus terrestris)* populations around an airport by soil applications of the fungicide, benomyl. Protect. Ecol. **2** (1981): 319—323.

Tomlinson, J. A., Faithfull, E. M., and Ward, C. M.: Chemical suppression of the symptoms of two virus diseases. Ann. appl. Biol. **84** (1976): 31—41.

Tripathi, R. K., Kommal, K., Schlösser, E., and Hess, W. M.: Effect of fungicides on the physiology of plants. Pesticide Sci. **13** (1982) 395—400.

Upham, P. M., and Delp, C. J.: Role of Benomyl in the Systemic Control of Fungi and Mites on Herbaceous Plants. Phytopathology **63** (1973): 814—820.

Vonk, J. W., and Sijpesteijn, A. K.: Methyl benzimidazol-2-ylcarbamate, the fungitoxic principle of thiophanate-methyl. Pesticide Sci. **2** (1971): 160—164.

Zweig, G., and Sherma, J.: Analytical methods for pesticides and plant growth regulators. Vol. 12, Chap. 2, Academic Press, New York 1982.

Lyr, H. (Ed.): Modern Selective Fungicides — Properties. Applications, Mechanisms of Action. Longman Group UK Ltd., London, and VEB Gustav Fischer Verlag, Jena, 1987.

Chapter 16

Biochemical aspects of benzimidazole fungicides — action and resistance

L. C. DAVIDSE
Laboratory of Phytopathology, Agricultural University, Wageningen, The Netherlands

Introduction

The wide application of benzimidazoles in agriculture and veterinary medicine as fungicides and antihelminthic drugs (chapter 15) and their experimental use in cancer chemotherapy have led to intensive research to elucidate their mode of action in detail. Results of these studies have been intensively reviewed (BURLAND and GULL 1984; CORBETT *et al.* 1984; DAVIDSE 1982; DAVIDSE and DE WAARD 1984; DEKKER 1985 and LANGCAKE *et al.* 1983). In this chapter an extensive review of the complete literature will not be given but rather those experiments will be discussed that made essential contributions to the development of the research area. It appears that much of our knowledge has been obtained from research in cell biology where benzimidazoles have been used as tools to study the organization and function of microtubules. Although the latter subject is beyond the scope of this chapter, some aspects that are crucial for a general understanding of the antifungal activity of the benzimidazoles, are mentioned. More details can be found in a recent review (DAVIDSE 1986), in which the impact of benzimidazoles on fungal cell biology and the molecular genetics of tubulin is discussed.

Mode of action

Inspired by the observations of HASTIE (1970) that benomyl induced nuclear instability in *Aspergillus nidulans* diploids initial studies on the mechanism of action of the benzimidazoles were focused on DNA and RNA synthesis. Soon it became clear that the observed inhibition of DNA synthesis, that in some fungi becomes evident after an initial lag period, was a secondary effect and that blockage of nuclear division was primarily responsible for this effect (CLEMONS and SISLER 1971; DAVIDSE 1973; HAMMERSCHLAG and SISLER 1973).

In arresting nuclear division of fungi the benzimidazoles show a striking resemblance with the secondary plant metabolite colchicine. This compound disrupts mitosis and meiosis in animal and plant cells by inactivation of the spindle. The spindle is composed of microtubules and colchicine inhibits microtubule assembly by binding to tubulin, the major component of microtubules. Tubulin is a heterodimer; both its subunits, usually designated with α- and β-tubulin, have a molecular weight of circa 50,000. Detailed information about microtubules and their functions can be obtained from the books of DUSTIN (1984) and ROBERTS and HYAMS (1979). The molecular biology and genetics of tubulin have recently been discussed by BORISY *et. al.* (1984) and CLEVELAND and SULLIVAN (1985).

Inhibition of microtubule assembly by colchicine disturbs a great number of cellular processes in which microtubules are involved such as mitosis and meiosis, intracellular

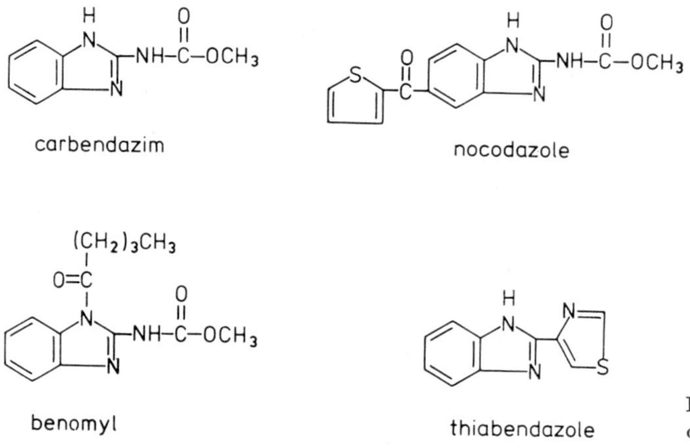

Fig. 16.1 Chemical structures of antifungal benzimidazoles.

transport of molecules, particles and organelles, maintenance of cell shape and cellular mobility through ciliar and flagellar action. Contrary to its effect on cellular functions of a large number of organisms, colchicine is not very active on fungi and its inability to disrupt microtubules in these organisms has hampered research on fungal microtubules and their functioning (Heath 1978).

Binding experiments using ^{14}C-carbendazim and crude mycelial extract of three strains of *Aspergillus nidulans* indicated that binding to a cellular protein, with characteristics typical for tubulin, was involved in the mechanism of action of benzimidazole fungicides (Davidse and Flach 1977). Binding affinity correlated with the sensitivity of the strain involved (Tab. 16.1). Additional experiments with cell free extracts of a number of fungi being either sensitive or resistant to carbendazim indicated that binding activity was only present in extracts of sensitive fungi or strains. Competitive inhibition of ^{14}C-carbendazim-binding by colchicine and nocodazole supported the idea that the carbendazim-binding protein was fungal tubulin. Nocodazole (initially named oncodazole) is structurally related to carbendazim (Fig. 16.1). It is an effective inhibitor of mitosis in mammalian cells in culture and was initially introduced as an experimental antitumoral drug (De Brabander *et al.* 1975, 1976). Nocodazole exerts its action by disrupting microtubules in a manner similar to that of colchicine and competitively inhibits ^{3}H-colchicine binding to mammalian tubulin (Hoebeke *et al.* 1976). Nocodazole was even more active against mycelial growth of *A. nidulans* than carbendazim and the three strains studied responded to a similar manner to both compounds (Tab. 16.1).

The potency of presumed antimicrotubular drugs to inhibit microtubule assembly can be determined in in vitro assays using partially purified tubulin. Initially the high

Table 16.1 Sensitivity to carbendazim and the dissociation constants of the carbendazim-tubulin complex in strains of *Aspergillus nidulans*

Strain	mutation	EC_{50}-value against growth (μM)		Dissociation constant of the carbendazim-tubulin complex (μM)
		carbendazim	nocodazole	
003	—	4.5	0.50	2.2
186	benA16	1.5	0.23	0.6
R	benA15	95	20	27

Table 16.2 Effect of carbendazim, nocodazole and colchicine on yeast and pig brain tubulin (KILMARTIN 1981)

Inhibitor	EC_{50}-value against tubulin assembly (μM) pig brain	yeast	EC_{50}-value against growth (μM)
carbendazim	>1,300	4	30
nocodazole	7	1	4
colchicine	3	2,000	>1,000

concentration tubulin needed, allowed only mammalian brain to be used as the tubulin source. Since then, however, substantial progress has been made in the development of assembly systems for tubulin from other sources.

Colchicine and nocodazole are effective inhibitors of microtubule assembly of tubulin from mammalian sources, whereas carbendazim only slightly affects this process (FRIEDMAN and PLATZER 1978; HOEBEKE *et al.* 1976). Assembly of yeast tubulin, however, is sensitive to both nocodazole and carbendazim. Its sensitivity to colchicine is much lower (Tab. 16.2; KILMARTIN 1981). Similar results were obtained with tubulin from myxamoebae of *Physarum polycephalum* (QUINLAN *et al.* 1981). These in vitro studies strongly supported the idea that the cytological effects of benzimidazoles in fungi are caused by an interference with the normal functioning of microtubules by binding to tubulin.

Interference with microtubule assembly in vivo has convincingly been demonstrated by HOWARD and AIST (1977, 1980), who studied the various effects of carbendazim on hyphal tip cells of *Fusarium acuminatum* by light and electron microscopy. A variety of effects could be ascribed to disappearance of microtubules. Effects included displacement of mitochondria from hyphal apices, disappearance of Spitzenkörpers which are presumed to function in hyphal linear elongation, reduction of linear growth rate, and metaphase arrest of all mitosis. D_2O, a known stabilizer of microtubules antagonized the action of carbendazim, additionally proving that carbendazim destabilizes microtubules. Ultrastructural aspects of carbendazim inhibition of mitosis in *A. nidulans* have been studied by KÜNKEL and HÄDRICH (1977) and KÜNKEL (1980). In the presence of carbendazim spindle formation did not take place, although the spindle pole bodies duplicated and in some cases even quadrupled indicating that both processes are independently regulated.

Inhibition of microtubuli formation apparently is a common property of all biocidal benzimidazoles and may account for most effects observed in a wide range of eukaryotes. Differential affinity of the various benzimidazoles to tubulins of different organisms is one of the factors that determine their selectivity within the eukaryotes as has been illustrated above for carbendazim and nocodazole with respect to animals and fungi. In addition differential metabolism may also play a role. Whether host-parasite selectivity with respect to plants and their pathogenic fungi is also based on differential affinity is not known yet. Final evidence should come from studies on in vitro assembly of plant microtubules, which is feasible now (MOREJOHN and FOSKET 1984).

Mechanism of resistance and properties of resistant strains

Studies on the mechanism of resistance to benzimidazole fungicides closely paralleled those on the mode of action and were greatly facilitated by the availability of genetically characterized resistant mutants of *A. nidulans*. Using standard UV- or chemical mutagenesis techniques resistant mutants of this fungi can be easily obtained (HASTIE and GEORGOPOULOS 1971). In this fungus three loci *benA*, *benB* and *benC* are involved in resistance (VAN TUYL 1977). Allelic *benA* mutants carrying the mutations *benA15* and *benA16* together with a wild-type strain have been used to determine binding affinity of carbendazim to tubulin. Previous work had ruled out that resistance to carbendazim was due to a reduced uptake or increased metabolic conversion (DAVIDSE 1976). *BenA15* governs resistance to both carbendazim and thiabendazole, whereas *benA16* governs supersensitivity to carbendazim and resistance to thiabendazole. The strain carrying the latter mutation has been isolated among several others on thiabendazole-containing medium (VAN TUYL *et al.* 1974). The affinity of carbendazim to *benA15* tubulin was lower and that to *benA16* tubulin was higher than to wild-type tubulin (Tab. 16.1; DAVIDSE and FLACH 1977). Binding affinities of thiabendazole to *A. nidulans* tubulin could not be determined with ^{14}C-thiabendazole in the routinely used binding assay because of high aspecific binding of thiabendazole to components of the crude extracts and the rather low specific binding (DAVIDSE and FLACH 1978). Thiabendazole, however, inhibited ^{14}C-carbendazim-binding to *benA16* tubulin significantly less than to wild-type tubulin. This latter effect could be quantified using a modified binding assay with partially purified tubulin from a wild-type *Penicillium expansum* strain and one that was resistant to thiabendazole but disp'ayed supersensitivity to carbendazim. The binding affinity of the mutant tubulin to each of the benzimidazoles had been changed in opposite directions. Binding affinity to thiabendazole of mutant tubulin was lower than that of wild-type tubulin, whereas binding affinity to carbendazim of mutant tubulin was higher than that of wild-type tubulin (Table 16.3).

Additional binding studies with wild-type and resistant mutants of other benzimidazole sensitive species such as *Botrytis cinerea*, *Fusarium oxysporum* f. sp. *lycopersici*, *P. brevicompactum* and *P. corymbiferum* and species with natural resistance to benzimidazoles such as *Alternaria brassicae* and *Pythium irregulare* (DAVIDSE and FLACH 1977; GESSLER *et al.* 1980) showed that only extracts of the sensitive types showed binding activity. The affinity of the target site to benzimidazoles apparently solely determines whether a benzimidazole has antifungal activity or not.

Affinity changes of tubulin from the *benA* class of mutants of *A. nidulans* are associated with changes in the primary structure of β-tubulin (SHEIR-NEISS *et al.* 1978). Two dimensional gel electrophoresis of tubulin of 26 *benA* mutants, among which *benA15* and *benA16*, showed that 18 of them had altered β-tubulins, with respect to isoelectric point, electrophoretic mobility in the SDS gel dimension or relative amount present. The *benA15* allele produced a β-tubulin with an isoelectric point of pH 5.0 which is 0.1 pH unit lower than that of wild-type β-tubulin. *BenA16* tubulin was electrophoretically identical with the wild-type tubulin. α-tubulins from all mutants appeared to be normal. This indicates that the structure of the β-tubulin is the major determinant for binding of benzimidazoles. This study also revealed the heterogeneity of the tubulins of *A. nidulans*. Three major species of α-tubulin, α_1, α_2 and α_3 and at least two major species of β-tubulin, β_1 and β_2 were distinguishable

Table 16.3 Sensitivity to carbendazim and thiabendazole and the dissociation constants of the carbendazim- and thiabendazole-tubulin complex in *Penicillium expansum*

	EC_{50} value against growth (μM)		Dissociation constant of the complex (μM)	
Strain	carbendazim	thiabendazole	carbendazim	thiabendazole
S	0.4	7	0.9	34
SS	0.07	85	0.2	68
R	>2,500	225	10	N. D.

on the gels. When a *benA* mutant had an altered β-tubulin, a co-cordinated shift occurred with both β-tubulins indicating that both species, designated β_1 and β_2, are products of the single *benA* gene.

The *benA* mutants of *A. nidulans* are until now the most intensively studied benzimidazole resistant mutants. Our knowledge on the molecular aspects of benzimidazole resistance is for the greater part based on studies with this type of mutants.

A number of the *benA* mutants have also proved their usefulness in tubulin research. The temperature sensitivity (ts^-) for growth of *benA11* allowed the selection of a revertant (ts^+) that carried an extragenic suppressor mutation. Both α_1 and α_3 tubulin of this revertant proved to be electrophoretically altered (MORRIS *et al.* 1979; WEATHERBEE and MORRIS 1984). This mutation designated as tubA1 thus identified a structural gene for α-tubulin, that apparently encodes two types of α-tubulin. The *benA33* mutation is of particular interest because at the restrictive temperature, strains carrying this mutation are blocked in mitosis although spindles are formed. This could be ascribed to a hyperstability of microtubules (OAKLEY and MORRIS 1981). In contrast the other *benA* mutations affected the assembly of microtubules because no spindles were formed at the restrictive temperature. Benomyl, nocodazole and thiabendazole partially suppressed the heat sensitivity conferred by the *benA33* mutation. A cold sensitive revertant of *benA33* carried the extragenic suppressor *tubA4* in the α-tubulin gene (OAKLEY 1983). In recombinants with the wild-type the *tubA4* mutation as well as the *tubA1* mutation that also suppresses *benA33*, cause supersensitivity to nocodazole, benomyl, and griseofulvin as well as partial cold sensitivity. The latter phenomenon could be attributed to a decreased stability of microtubules (GAMBINO *et al.* 1984). Hyperstability of microtubules conferred by *benA33* and caused by a structural alteration in β-tubulin apparently is counteracted by a structural alteration in α-tubulin conferred by *tubA* mutations. As a result normal stability is restored. Since the benzimidazoles also restore normal stability of the microtubules, because they suppress the heat sensitivity conferred by *benA33*, they obviously interact with tubulin, destabilizing the hyperstable microtubules. This would imply that *benA33* tubulin displays wild-type binding affinity for the benzimidazoles. The fact that *benA33* also confers resistance to griseofulvin and p-fluorphenylalanine, that are antimicrotubule agents which do not show structural resemblance to the benzimidazoles and that both these compounds suppress heat sensitivity supports the idea that hyperstability of microtubules rather than decreased affinity of tubulin to benzimidazoles is causing benzimidazole resistance in mutants carrying the *benA33* mutation.

The biochemistry of benzimidazole resistance has also been studied in *S. cerevisiae*, *Schizosaccharomyces pombe* and *Physarum polycephalum*. In *S. cerevisiae* a mutation leading to resistance could be localized in the gene encoding β-tubulin (NEFF *et al.* 1983). The available data for *S. pombe* are consistent with the idea that structural changes of tubulin are involved (YAMAMOTO 1980; ROY and FANTES 1982; TODA *et al.* 1983, 1984; UMESONO *et al.* 1983). In *P. polycephalum* two of four unlinked loci in which mutations conferring resistance to benzimidazoles occurred could be assigned to β-tubulin DNA sequences (BURLAND and GULL 1984). So it seems that changes in tubulin structure may be a general principle upon which benzimidazole resistance can be based.

Studies on mechanisms of action of benzimidazoles and mechanisms of resistance in fungi pathogenic to plants are very limited. By contrast numerous data are available on development of resistance and characteristics of resistant strains. In addition the genetics of resistance are known in a number of target fungi.

The available data fit well in the general concept of benzimidazoles as inhibitors of microtubule assembly and benzimidazole resistance as due to altered microtubule structure or functioning, as has been derived from the model studies with *A. nidulans*, *P. polycephalum*, *S. cerevisiae* and *S. pombe*.

Benzimidazole resistance inherited as a single Mendelian gene in *Ceratocystis ulmi* (Nishijima and Smalley 1979), *Neurospora crassa* (Borck and Braymer 1974), *Venturia inaequalis* (Katan *et al.* 1983; Kiebacher and Hoffmann 1981), *V. nashicola* (Ishii and Yanase 1983) and *V. pirina* (Shabi and Katan 1979). Although single gene based, a wide variation was observed in levels of resistance of the various mutants. This is reconcilable with the idea that different mutations within the tubulin genes did occur, leading to different levels of resistance. Unfortunately biochemical studies supporting this idea have not been done with fungal plant pathogens, except for *V. nashicola* (Ishii, private communication). Crude mycelial extracts of resistant strains of this fungus displayed less carbendazim-binding activity than those of wild-type strains. Genetical studies on thiabendazole resistance in *P. italicum* indicated that two or three linked genes might be involved (Beraha and Garber 1980). Again the biochemistry of resistance is not known.

The genetics of benzimidazole resistance indicate that mutations can occur in several loci. A number of these may be tubulin DNA sequences, but coding regions for other components of the microtubular system or other cellular components may be involved as well.

Decreased uptake has been proposed as a mechanism of resistance in *Sporobolomyces roseus* (Nachmias and Barash 1976) and *P. polycephalum* (Burland and Gull 1984). In *Verticillium malthousei* increased acid production is associated with benzimidazole resistance (Lambert and Wuest 1975). Both processes, however, may be related to changes of the microtubular system via impairment of membrane functions.

Of particular interest is the phenomenon of negatively correlated cross resistance of benzimidazole resistant strains to N-phenylcarbamates. These compounds, being developed as herbicides, interfere with cellular and nuclear division of plant cells by affecting microtubule functioning. For a recent review see Corbett *et al.* (1984). N-phenylcarbamates affect fungal mitosis as well (Gull and Trinci 1973; White *et al.* 1979, 1981).

Leroux *et al.* (1979a, b) first noticed that among a number of benzimidazole resistant strains of *B. cinerea* and *P. expansum* some strains showed an increased sensitivity to N-phenylcarbamates. Later on this phenomenon has also been found among resistant strains of *Pseudocercosporella herpotrichoides* (Leroux and Cavalier 1983a, b) and *V. nashicola* (Ishii *et al.* 1984). Only highly resistant strains of the latter fungus showed an increased sensitivity to N-phenylcarbamates, whereas intermediately and weakly-resistant strains did not.

The observations initiated a search among the N-phenylcarbamates to compounds that did not show phytotoxicity but were still inhibitory to fungi (Kato *et al.* 1984). One such compound, methyl N-(3,5-dichlorophenyl)carbamate (MDPC) inhibited growth of benzimidazole-resistant field isolates of *B. cinerea*, *Cercospora beticola*, *F. nivale* and *Mycosphaerella melonis* on nutrient medium and controlled disease incited by these strains on their various hosts. Growth of benzimidazole sensitive isolates was not inhibited, nor could disease control be achieved with MDPC, when the latter isolates were involved. MDPC induced similar morphological changes in germ tubes of benzimidazole resistant strains of *B. cinerea* as did carbendazim in that of sensitive strains. Furthermore MDPC arrested mitosis in a similar manner (Suzuki *et al.* 1984) as carbendazim. Increased sensitivity to N-phenylcarbamates, however, is not necessarily associated with resistance to benzimidazoles. For instance a mutation in the *nda2* gene of *S. pombe* that codes for α-tubulin, confers supersensitivity to benzimidazoles as well as to N-phenylcarbamates (Umesono *et al.* 1983). Supersensitivity to thiabendazole and CIPC was also conferred by mutations in the *ben4* gene of this organism that causes resistance to benomyl (Roy and Fantes

1982). Since both *ben4* and *nda2* mutations result in a cold-sensitive phenotype as well, it is tempting to suggest that the *nda2* gene and the *ben4* gene are identical and thus encode α-tubulin but as yet results of decisive genetical experiments have to be awaited.

Resistance to CIPC in *Dictyostelium discoideum* the growth of which is rather sensitive to N-phenylcarbamates was associated with resistance to thiabendazole and nocodazole, temperature sensitivity and defects in development (WHITE *et al.* 1981). CIPC resistance was not always expressed at all developmental stages. A number of resistant mutants with respect to growth even displayed supersensitivity to CIPC with respect to development. Temperature sensitivity of one of the mutants could be suppressed by CIPC (WHITE *et al.* 1979). In this respect this mutant resembles the *benA33* mutants of *A. nidulans* that displayed hyperstability of microtubules due to a change in β-tubulin structure.

The N-phenylcarbamates obviously interfere in some way with microtubule functioning and at least in *S. pombe* interaction with α-tubulin seems to be involved. If this also applies to other fungi and to plant pathogens in particular is still unknown. It implies that research on the biochemistry and the molecular genetics of tubulin from the latter fungi is urgently needed. Results will not only increase our knowledge on the fundamental aspects of microtubule functioning but also may be helpful in designing new molecules that interact with tubulin and that are potential new fungicides.

The temperature dependent effect of diphenylamine (DPA) on benzimidazole activity against benzimidazole resistant strains of *P. expansum* is another intriguing phenomenon that has been observed in crop protection practice (ROSENBERGER and MEYER 1985). At 2 °C (but not at ambient temperature) DPA proved to have an inhibitory effect on blue mold decay of apples caused by benomyl resistant strains. In combination with benomyl this effect was even more enhanced and resulted in acceptable decay control. An inhibitory effect of DPA was not noticed with benzimidazole sensitive strains.

Growth in vitro of benzimidazole resistant strains was more sensitive to DPA than that of benzimidazole sensitive strains although exceptions occurred. Actual colony growth rates of resistant strains on unamended media were generally lower for benomyl resistant strains than for sensitive ones. The highly-resistant isolates had the lowest growth rate. Growth rates were enhanced by subinhibitory levels of both DPA and benomyl.

These observations are compatible with the idea that hyperstability of *P. expansum* microtubules is causing resistance to benzimidazoles and that DPA has a destabilizing effect on the mutant microtubules. At lower temperature the latter effect apparently becomes more pronounced. When benzimidazoles are additionally present reassembly is inhibited, as a consequence of which microtubules do not form. It would be of great value to determine the effect of known microtubule destabilizers such as griseofulvin, p-fluorophenylalanine, and stabilizers such as taxol (SCHIFF and HORWITZ 1980) and D_2O on these particular strains to see if this hypothesis is correct, and to prove the character of DPA as an antimicrotubule agent.

The fungal cytoskeleton, a target for new fungicides

Benzimidazoles developed as agricultural fungicides additionally have evolved into useful tools to study the formation and functioning of microtubules and in the molecular biology and genetics of fungal tubulin. Microtubules and actin filaments are major components of the cytoskeleton, a unique feature of the eukaryotic cell. It should be clear that each interference with the functioning or structural integrity of the cytoskeleton will have important consequences for cell growth and development. The availability of interfering agents will increase our knowledge of the cytoskeleton, which ín its turn might be helpful in designing new molecules that interact with its various components.

Together with microtubules actin filaments make up the principal constituents of the cytoskeleton. Therefore interference with the formation and functioning of actin filaments would also lead to cellular desintegration. Actin filaments are composed of actin monomers. Like microtubules they are dynamic structures. Actin filaments can lengthen at one end by incorporation of monomers while shortening at the other end by release of subunits. The rate at which this occurs determines whether an actin filament grows or shortens.

The cytochalasins are a group of fungal metabolites that inhibit a variety of cellular functions by interference with actin filament formation. They act by inhibition of the addition of actin monomers to the growing end of the filaments. Since release of monomers continues the actin filaments disappear.

The effect of cytochalasins on fungal growth (Grove and Sweigard 1980) and differentiation (Staples and Hoch 1982) and enzyme secretion (Thomas *et al.* 1974) are a few examples that indicate that also in fungi actin filaments are involved in a number of cellular processes.

The relation between actin and tubulin distribution to bud growth has been studied in wild-type and morphogenetic-mutant *S. cerevisiae* using immunofluorescence microscopy with fluorescent phallotoxin probes, which bind specific to actin, and anti-tubulin antibodies (Kilmartin and Adams 1984; Adams and Pringle 1984). The changing arrangements of actin during the cell cycle suggest that actin may be involved in the localized deposition of cell wall materials, and the selective growth of the bud.

Interference with actin filament formation or functioning by any of the fungicides used in crop protection has not been definitely proven. Kato (1983), however, has proposed that dicarboximides, aromatic hydrocarbons and the organophosphorus compound, tolclofos-methyl act via interference with actin filament functioning. Compounds belonging to these groups induce various morphological changes, recently discussed by Leroux and Fritz (1984), similar to those induced by the cytochalasins. Interference with actin filament functioning might also explain the ability of these compounds to induce somatic segregation (Georgopoulos *et al.* 1976, 1979). This, however, would imply either a function of actin filaments in the spindle or their involvement in the separation of daughter nuclei evidence for which has not been presented yet (Kilmartin and Adams 1984). (For alternative explanations see chapters 6 and 8.)

The functioning of actin filaments may also be impaired by agents that directly or indirectly interfere with their anchorage to the cell membrane. Anchorage is assumed to be essential for the localized deposition of cell wall materials. In budding cells of *S. cerevisiae*, the deposition of a chitin ring preceding the actual emergence of the bud is associated with the presence of actin. This actin is presumably anchored in the cell membrane and orientates the actin filaments of the cytoskeleton to the site of bud initiation. The actin filaments might then be involved in producing a flow of material into the bud by cytoplasmic streaming. Interfering with such anchorage of the actin cytoskeleton to the cell membrane should lead to either absence of bud initiation or irregular deposition of cell wall material. Treatment of *Candida albicans* cells or *P. italicum* hyphae with inhibitors of sterol biosynthesis results in irregular deposition of chitin (Barug *et al.* 1983; Kerkenaar and Barug 1984; Kerkenaar *et al.* 1984; Van den Bossche *et al.* 1984). It suggests that membranes lacking ergosterol are not a suitable substrate for actin anchorage. Therefore some of the ultrastructural changes induced by inhibitors of sterol biosynthesis in fungi (HIPPE 1984) may result from impaired functioning of the cytoskeleton.

It should be clear that our knowledge of the structure and functioning of the cytoskeleton is rapidly expanding. Site specific mutagenesis of genes encoding its various constituents, as is possible now with *S. pombe* and *S. cerevisiae* and will be in the near future with filamentous fungi when suitable integration vectors have been developed, will further clarify the role of each individual component. Once the structure of the many domains at which the components of the cytoskeleton interact is known, it would be possible to design small molecules that would bind to these sites. Such molecules are likely to interfere with the functioning of the cytoskeleton and thus would be potential fungicides. This approach would make innovative design of new fungicides a reality. The reliability of the cytoskeleton as a target for fungicides action have been unjustly discredited by the rapid development of resistance to the benzimidazoles. Therefore, the cytoskeleton should not be considered to be a less suitable target than for instance sterol biosynthesis. Benzimidazoles are just one group of fungicides that interact with cytoskeleton component. Other groups that do not necessarily encounter resistance problems still await their discovery.

Concluding remarks

Elucidation of the mechanism of action of a fungicide is a long process and requires the dedicated efforts of interested biologists.

Getting to know how a compound acts is not a sufficient motive to start a research project in this area. As yet the mechanisms of action of a number of fungicides are still poorly understood, but in itself this does not seem to attract research interest. It clearly needs to be triggered. The rapid development of resistance to benzimidazoles clearly has triggered the earlier research on the mode of action of the benzimidazoles and mechanisms of resistance. Initially these compounds were thought to resolve a number of problems in crop protection, but the development of resistant strains in target fungi was a serious drawback that obviously needed to be investigated. Resistance problems initiated a renewed interest in both epidemiological as well as biochemical aspects of resistance.

When the site of action of the benzimidazoles was localized and resistance could be ascribed to a change at the target site, interest of plant pathologists declined. Cell biologists, however, became excited about the value of these compounds as tools to study microtubule structure and functioning. This, indisputably, has led to a thorough understanding of the mode of action of the benzimidazoles, details of which are still being investigated.

Although this knowledge has not solved resistance problems in crop protection, it greatly facilitates the characterization of resistant strains. The observed negatively correlated cross-resistance of benzimidazole resistant strains to N-phenylcarbamates and diphenylamine are phenomena that can more easily be understood with knowledge obtained from model systems. Temperature sensitivity is frequently encountered among benzimidazole resistant mutants of model organisms. Whether this also occurs with resistant field isolates is still largely unknown. If present it might be of significance for strategies aimed at preventing or retarding the development of resistance.

The benzimidazoles have contributed to a better understanding of the structure and functioning of microtubules, that are just one constituent of the cytoskeleton. Our increasing knowledge in this area will help us to understand the fundamental aspects of cell structure and functioning, but it might well lead to a new era in which specific design of new fungicides has become a reality.

References

ADAMS, A. E. M., and PRINGLE, J. R.: Relationship of actin and tubulin distribution to bud growth in wild-type and morphogenetic-mutant *Saccharomyces cerevisiae*. J. Cell Biol. **98** (1984): 934—945.

BARUG, D., SAMSON, R. A., and KERKENAAR, A.: Microscopic studies of *Candida albicans* and *Torulopsis glabrata* after in vitro treatment with bifonazole. Arzneimittel-Forschung/Drug Research **33** (1983): 528—537.

BERAHA, L., and GARBER, E. D.: A genetic study of resistance to thiabendazole and sodium o-phenyl phenate in *Penicillium italicum* by the parasexual cycle. Botanic. Gazette **141** (1980): 204—209.

BORCK, K., and BRAYMER, H. D.: The genetic analysis of resistance to benomyl in *Neurospora crassa*. J. General Microbiol. **85** (1974): 51—56.

BORISY, G. G., CLEVELAND, D. W., and MURPHY, D. G.: Molecular biology of the cytoskeleton. Cold Spring Harbor Press, New York 1984, 512 pp..

BURLAND, T. G., and GULL, K.: Molecular and cellular aspects of the interaction of benzimidazole fungicides with tubulin and microtubules. In: TRINCI, A. P. J., and RYLEY, J. F. (Eds.): Mode of action of antifungal agents. Brit. Mycologic. Soc. Symposia Series 8, Cambridge Univ. Press, Cambridge 1984, pp. 299—320.

CLEMONS, G. P., and SISLER, H. D.: Localization of the site of action of a fungitoxic benomyl derivative. Pesticide Biochem. Physiol. **1** (1971): 32—43.

CLEVELAND, D. W., and SULLIVAN, K. F.: Molecular biology and genetics of tubulin. Ann. Rev. Biochem. **54** (1985): 331—365.

CORBETT, J. R., WRIGHT, K., and BAILLIE, A. C.: The biochemical mode of action of pesticides. Academic Press 1984, 382 pp..

DAVIDSE, L. C.: Antimitotic activity of methyl benzimidazole-2-ylcarbamate (MBC) in *Aspergillus nidulans*. Pesticide Biochem. Physiol. **3** (1973): 317—325.

— Metabolic conversion of methyl-benzimidazol-2-ylcarbamate (MBC) in *Aspergillus nidulans*. Pesticide Biochem. Physiol. **6** (1976): 538—546.

— Benzimidazole compounds; selectivity and resistance. In: DEKKER, J., and GEORGOPOULOS. S. G. (Eds.): Fungicide Resistance in Crop Protection. Pudoc, Wageningen 1982, pp. 60—70.

— Benzimidazole fungicides: Mechanism of action and biological impact. Ann. Rev. Phytopathol, **24** (1986): 43—65.

— and DE WAARD, M. A.: Systemic Fungicides. In: INGRAM, D. S., and WILLIAMS, P. H. (Eds.): Advances in Plant Pathology, Vol. 2. Academic Press, London 1984, pp. 191—257.

— and FLACH, W.: Differential binding of methyl benzimidazol-2-yl carbamate to fungal tubulin as a mechanism of resistance to this antimitotic agent in mutant strains of *Aspergillus nidulans*. J. Cell Biol. **72** (1977): 174—193.

— — Interaction of thiabendazole with fungal tubulin. Biochimica et Biophysica Acta **543** (1978): 82—90.

DE BRABANDER, M., VAN DE VEIRE, R., AERTS, F., GEUENS, G., BORGERS, M., DESPLENTER, L., and DE CRÉE, J.: Oncodazole (R 17934): a new anti-cancer drug interfering with microtubules. Effects on neoplastic cells cultured in vitro and in vivo. In: BORGERS, M., and DE BRABANDER, M. (Eds.): Microtubules and Microtubule Inhibitors. North Holland/Elsevier, Amsterdam-New York 1975, pp. 509—521.

— — — BORGERS, M., and JANSSEN, P. A. J.: The effects of methyl[5-(2-thienylcarbonyl)-1H-benzimidazol-2-yl]-carbamate (R 17934; NSC 238159), a new synthetic antitumoral drug interfering with microtubules, on mammalian cells cultured in vitro. Cancer Res. **36** (1976): 1011—1018.

DEKKER, J.: The development of resistance to fungicides. In: HUTSON, D. H., and ROBERTS, T. R. (Eds.): Progress in Pesticide Biochemistry and Toxicology, Vol. 4. Wiley and Sons Ltd., New York 1985, pp. 165—218.

DUSTIN, P.: Microtubules. 2nd edit. Springer Verlag, Berlin-Heidelberg 1984, 482 pp..

FRIEDMAN, P. A., and PLATZER, E. G.: Interaction of anthelmintic benzimidazoles and benzimidazole derivatives with bovine brain tubulin. Biochimica and Biophysica Acta **544** (1978): 605—614.

GAMBINO, J., BERGEN, L. C., and MORRIS, N. R.: Effects of mitotic and tubulin mutations on microtubule architecture in actively growing protoplasts of *Aspergillus nidulans*. J. Cell Biol. **99** (1984): 830—838.

GEORGOPOULOS, S. G., KAPPAS, A., and HASTIE, A. C.: Induced sectoring in diploid *Aspergillus nidulans* as a criterion of fungitoxicity by interference with hereditary processes. Phytopathology **66** (1976): 217—220.

— SARRES, M., and ZIOGAS, B. N.: Mitotic instability in *Aspergillus nidulans* caused by the fungicides iprodione, procymidone and vinclozolin. Pesticide Sci. **10** (1979): 389—392.

GESSLER, C., SOZZI, D., and KERN, H.: Benzimidazol-fungicide: Wirkungsweise und Probleme. Ber. Schweizer. Bot. Ges. **90** (1980): 45—54.

GROVE, S. N., and SWEIGARD, J. A.: Cytochalasin A inhibits spore germination and hyphal growth in *Gilbertiella persicaria*. Experiment. Mycol. **4** (1980): 239—250.

GULL, K., and TRINCI, A. P. J.: Griseofulvin inhibits fungal mitosis. Nature **244** (1973): 292—293.

HAMMERSCHLAG, R. S., and SISLER, H. D.: Benomyl and methyl-2-benzimidazole carbamate (MBC): Biochemical, cytological and chemical aspects of toxicity to *Ustilago maydis* and *Saccharomyces cerevisiae*. Pesticide Biochem. Physiol. **3** (1973): 42—54.

HASTIE, A. C.: Benlate-induced instability of *Aspergillus diploids*. Nature **226** (1970): 771.

— and GEORGOPOULOS, S. G.: Mutational resistance to fungitoxic benzimidazole derivatives in *Aspergillus nidulans*. J. General Microbiol. **67** (1971): 371—373.

HEATH, I. B.: Experimental studies of fungal mitotic systems. In: HEATH, I. B. (Ed.): Nuclear Division in the Fungi. Academic Press Inc., New York 1978, pp. 89—176.

HIPPE, S.: Ultrastructural changes induced by the systemic fungicides triadimefon, nuarimol and imazalil nitrate in sporidia of *Ustilago avenae*. Pesticide Sci. **15** (1984): 210—214.

HOEBEKE, J., VAN NYEN, G., and DE BRABANDER, M.: Interaction of oncodazole (R 17934), a new antitumoral drug, with rat brain tubulin. Biochem. Biophys. Res. Comm. **69** (1976): 319—324.

HOWARD, R. J., and AIST, J. R.: Effects of MBC on hyphal tip organization, growth, and mitosis of *Fusarium acuminatum*, and their antagonism by D_2O. Protoplasma **92** (1977): 195—210.

— — Cytoplasmic microtubules and fungal morphogenesis: Ultrastructural effects of methyl benzimidazol-2-yl carbamate determined by freeze-substitution of hyphal tip cells. J. Cell Biol. **87** (1980): 55—64.

ISHII, H., and YANASE, H.: Resistance of *Venturia nashicola* to thiophanate-methyl and benomyl. Formation of the perfect state in culture and its application to genetic analysis of the resistance. Ann. Phytopatholog. Soc. Japan **49** (1983): 153—159.

— — and DEKKER, J.: Resistance of *Venturia nashicola* to benzimidazole fungicides. Meded. Fac. Landbouwwet. Rijksuniv. Gent **49** (1984): 163—172.

KATAN, T., SHABI, E., and GILPATRICK, J. D.: Genetics of resistance to benomyl in *Venturia inaequalis* isolates from Israel and New York. Phytopathology **73** (1983): 600—603.

KATO, T.: Mode of antifungal action of a new fungicide, tolclofos-methyl. In: MIYAMOTO, J., KEARNY, P. C., DOYLE, P., and FUJITA, T. (Eds.): Pesticide Chemistry, Human Welfare and the Environment. Vol. 1. Pergamon Press, Oxford 1983, pp. 153—157.

— SUZUKI, K., TAKAHASHI, J., and KAMOSHITA, K.: Negatively correlated cross-resistance between benzimidazole fungicides and methyl N-(3,5-dichlorophenyl)carbamate. J. Pesticide Sci. **9** (1984): 489—495.

KERKENAAR, A., and BARUG, D.: Fluorescence microscope studies of *Ustilago maydis* and *Penicillium italicum* after treatment with imazalil or fenpropimorph. Pesticide Sci. **15** (1984): 99—205.

— VAN ROSSUM, J. M., VERSLUIS, G. G., and MARSMAN, J. W.: Effect of fenpropimorph and imazalil on sterol biosynthesis in *Penicillium italicum*. Pesticide Sci. **15** (1984): 177—187.

KIEBACHER, J., and HOFFMANN, G. M.: Genetics of benzimidazole resistance in *Venturia inaequalis*. Z. Pflanzenkrankh. u. Pflanzenschutz **88** (1981): 189—205.

KILMARTIN, J. V.: Purification of yeast tubulin by self-assembly in vitro. Biochemistry **20** (1981): 3629—3633.

— and ADAMS, A. E. M.: Structural rearrangements of tubulin and actin during the cell cycle of the yeast *Saccharomyces*. J. Cell Biol. **98** (1984): 922—933.

KÜNKEL, W.: Antimitotische Aktivität von Methylbenzimidazol-2-yl carbamat (MBC). I. Licht-, elektronenmikroskopische und physiologische Untersuchungen an keimenden Konidien von *Aspergillus nidulans*. Z. Allgem. Mikrobiol. **20** (1980): 113—120.

— and HÄDRICH, H.: Ultrastrukturelle Untersuchungen zur antimitotischen Aktivität von Methylbenzimidazol-2-yl carbamat (MBC) und seinen Einfluß auf der Replikation des Kernassoziierten Organells ("centriolar plaque", "MTOC", "KCE") bei *Aspergillus nidulans.* Protoplasma **92** (1977): 311—323.

Lambert, D. H., and Wuest, P. J.: Acid production, a possible basis for benomyl tolerance in *Verticillium malthousei* Ware. Phytopathology **65** (1975): 637—638.

Langcake, P., Kuhn, P. J., and Wade, M.: The mode of action of systemic fungicides. In: Hutson, D. H., and Roberts, T. R. (Eds.): Progress in Pesticide Biochem. Toxicology. Vol. **3.** John Wiley and Sons Ltd., New York 1983, pp. 1—109.

Leroux, P., and Cavelier, N.: Caractéristiques des souches de *Pseudo-cercosporella herpotrichoides* (Agent du piétin-verse des cerealis) résistantes aux fongicides benzimidazoles et thiophanates. La Défense des Végétaux **222** (1983a): 231—238.

— — Phénomènes de résistance du piétin-verse aux benzimidazoles et aux thiophanates. Phytoma-Defense des cultures **353** (1983b): 40—47.

— and Fritz, R.: Antifungal activity of dicarboximides and aromatic hydrocarbons and resistance to these fungicides. In: Trinci, A. P. J., and Ryley, J. F. (Eds.): Mode of action of antifungal agents. Brit. Mycolog. Soc. Symposia Ser. 8, Cambridge Univ. Press, Cambridge 1984, pp. 207—237.

— and Gredt, M.: Effets du barbane, du chlorbufame, du chlorprophame et du prophame sur diverses souches de *Botrytis cinerea* Pers. et de *Penicillium expansum* Link sensibles ou résistantes au carbendazim et au thiabendazole. Comptes Rendues Academie des Sciences Paris, Serie D **289** (1979a): 691—693.

— — Phénomènes de résistance croisée négative chez *Botrytis cinerea* Pers. entre les fongicides benzimidazoles et des herbicides carbamates. Phytiatrie-Phytopharmacie **28** (1979b): 79—86.

Morejohn, L. C., and Fosket, D. E.: Taxol-induced rose microtubule polymerization in vitro and its inhibition by colchicine. J. Cell Biol. **99** (1984): 141—147.

Morris, N. R., Lai, M. H., and Oakley, C. E.: Identification of a gene for α-tubulin in *Aspergillus nidulans.* Cell **16** (1979): 437—442.

Nachmias, A., and Barash, I.: Decreased permeability as a mechanism of resistance to methyl benzimidazol-2-yl carbamate (MBC) in *Sporobolomyces roseus.* J. General Microbiology **94** (1976): 167—172.

Neff, N. F., Thomas, J. H., Grisafi, P., and Botstein, D.: Isolation of the β-tubulin gene from yeast and demonstration of its essential function in vivo. Cell **33** (1983): 211—219.

Nishijima, W. T., and Smalley, E. B.: *Ceratocystis ulmi* tolerance to methyl-2-benzimidazole carbamate and other related fungicides. Phytopathology **69** (1979): 69—73.

Oakley, B. R.: Conditionally lethal mutations in genes that encode α and β-tubulin and other proteins essential to microtubule function in *Aspergillus nidulans.* J. Cell Biol. **97** (1983): 217a.

— and Morris, N. R.: A β-tubulin mutation in *Aspergillus nidulans* that blocks microtubule function without blocking assembly. Cell **24** (1981): 837—845.

Quinlan, R. A., Roobol, A., Pogson, C. I., and Gull, K.: A correlation between in vivo and in vitro effects of the microtubule inhibitors colchicine, parbendazole, and nocodazole on myxamoebae of *Physarum polycephalum.* J. General Microbiology **122** (1981): 1—6.

Roberts, K., and Hyams, J. S.: Microtubules. Academic Press, London 1979, 595 pp.

Rosenberger, D. A., and Meyer, F. W.: Negatively correlated cross-resistance to diphenylamine in benomyl-resistant *Penicillium expansum.* Phytopathology **75** (1985): 74—79.

Roy, D., and Fantes, P. A.: Benomyl resistant mutants of *Schizosaccharomyces pombe* cold-sensitive for mitosis. Current Genetics **6** (1982): 195—201.

Schiff, P. B., and Horwitz, S. B.: Taxol stabilizes microtubules in mouse fibroplast cells. Proc. National Academy Sci. USA **77** (1980): 1561—1565.

Shabi, E., and Katan, T.: Genetics, pathogenicity and stability of carbendazim resistant isolates of *Venturia pirina.* Phytopathology **69** (1979): 267—270.

Sheir-Neiss, G., Lai, M. H., and Morris, N. R.: Identification of a gene for β-tubulin in *Aspergillus nidulans.* Cell **15** (1978): 639—649.

Staples, R. C., and Hoch, H. C.: A possible role for microtubules and microfilaments in the induction of nuclear division in bean rust uredospore germlings. Experimental Mycology **6** (1982): 293—302.

Suzuki, K., Kato, T., Takahashi, J., and Kamoshila, K.: Mode of action of methyl N-(3,5-dichlorophenyl)carbamate in the benzimidazole-resistant isolate of *Botrytis cinerea.* J. Pesticide Sci. **9** (1984): 497—501.

Thomas, D. S., Lutzac, M., and Manavathu, E.: Cytochalasin selectivity inhibits synthesis of a secretory protein, cellulase, in *Achlya.* Nature **249** (1974): 140—141.

Toda, T., Adachi, Y., Hiraoka, Y., and Yanagida, M.: Identification of the pleiotropic cell division cycle gene NDA2 as one of two different α-tubulin genes in *Schizosaccharomyces pombe*. Cell **37** (1984): 233—242.

— Umesono, K., Hirata, A., and Yanagida, M.: Cold-sensitive nuclear division arrest mutants of the fission yeast *Schizosaccharomyces pombe*. J. Molecular Biol. **168** (1983): 251—270.

Umesono, K., Toda, T., Hayashi, S., and Yanagida, M.: Two cell division cycle genes NDA2 and NDA3 of the fission yeast *Schizosaccharomyces pombe* control microtubular organization and sensitivity to antimitotic benzimidazole compounds. J. Molecular Biol. **168** (1983): 271 to 284.

Van den Bossche, H., Willemsens, G., Marechal, P., Cools, W., and Lauwers, W.: The molecular basis for the antifungal activities of N-substituted azole derivatives. Focus on R 51211. In: Trinci, A. J. P., and Ryley, J. F. (Eds.): Mode of action of antifungal agents. Brit. Mycolog. Soc. Symposia Ser. 8, Cambridge Univ. Press, Cambridge 1984, pp. 321—341.

Van Tuyl, J. M.: Genetics of fungal resistance to systemic fungicides. Meded. Landbouwhogeschool **77—2** (1977): 1—137.

— Davidse, L. C., and Dekker, J.: Lack of cross-resistance to benomyl and thiabendazole in some strains of *Aspergillus nidulans*. Netherl. J. Plant Pathol. **80** (1974): 165—168.

Weatherbee, J. A., and Morris, N. R.: *Aspergillus* contains multiple tubulin genes. J. Biol. Chemistry **259** (1984): 15452—15459.

White, E., Scandella, D., and Katz, E. R.: CIPC-resistant mutants of *Dictyostelium discoideum*. J. Cell Biol. **83** (1979): 341 a.

Lyr, H. (Ed.): Modern Selective Fungicides — Properties, Applications, Mechanisms of Action Longman Group UK Ltd., London, and VEB Gustav Fischer Verlag, Jena, 1987.

Chapter 17

Phenylamides and other fungicides against Oomycetes

F. J. SCHWINN and T. STAUB
Ciba-Geigy Ltd., Agricultural Division, Basle, Switzerland

Introduction

In modern taxonomy, the *Mastigomycotina*, i.e. the subdivision of fungal organisms having motile stages (= zoospores) in their life cycle, are considered as being phylogenetically different from the true (= higher) fungi (KREISEL 1969; WHITTACKER 1969; VON ARX 1974). One of the classes in this subdivision is that of the *Oomycetes*, which comprises some 70 genera with 550 species, living partly in aquatic, partly in terrestrial biotopes as saprophytes or plant parasites. Despite their ecological diversity, they are a well defined unit of high physiological and biochemical uniformity, well separated from all other taxa. Their classification is shown in table 17.1.

Among the Oomycetes, the *Peronosporales* comprise most of the plant parasitic genera and species, causing foliar, root and crown diseases on a wide range of annual and perennial crops in temperate and tropical climates. The wide distribution and the high potential of the foliar *Peronosporales*, i.e. the downy mildews and late blight, causing heavy epidemics within very short periods of favourable weather makes them a devastating group of plant pathogens. Their control had high priority since the beginning of modern plant protection.

The first effective foliar fungicides for practical use against late blight and downy mildews were copper compounds, particularly copper sulfate in combination with limestone, the famous Bordeaux mixture (introduced 1885) and cuprous oxide (introduced 1932). They were followed by the ethylene bis-dithiocarbamates (introduced from 1931—1962), the phthalimides captan, folpet and captafol (introduced 1949—1965), the triphenyl tin compounds (introduced since 1954) and chlorothalonil, introduced in 1963 (STAUB and HUBELE 1981). Despite being non-selective biocides

Table 17.1 Taxonomy of the Oomycetes

Subdivision:		*Mastigomycotina*	
Class:		*Oomycetes*	
Order:		*Peronosporales*	
Family:	*Pythiaceae* (non-obligate)	*Peronosporaceae* (obligate) Downy Mildews	*Albuginaceae* (obligate)
Genus:	*Pythium* *Phytophthora*	*Bremia* *Peronospora* *Peronosclerospora* *Plasmopara* *Pseudoperonospora* *Sclerophthora* *Sclerospora*	*Albugo*

they can be used in plant protection because they do not penetrate into the plant tissue and thus do not affect it. Apart from the organo tin compounds which have some curative effect, they are purely residual and protective fungicides, i.e. they protect only those plant parts against diseases which were treated. Since they stay on the plant surface, they are exposed to rainfall and weathering, a fact which requires repeated applications. The level of inherent fungitoxicity, residual behaviour and exposure to wash-off result in fairly high dose rates in the order of 1 to 2.5 kg of active ingredient/ha.

For the control of soil-borne root- or crown-infecting Oomycetes the only available fungicides were until recently non-specific biocidal soil sterilants like vapam or methylbromide, the use of which is limited to plant-free periods due to their high inherent biocidal activity. In addition, some of the foliar fungicides described above are used as soil drenches or root dips, such as captan, zineb, or mancozeb.

One group of Oomycetes which until the introduction of modern, systemic compounds was not amenable to chemical control, are the systemic downy mildews. They occur on a large range of host plants (table 17.2) with a wide geographical distribution, and cause considerable loss in tropical staple food like sorghum, maize and millet.

The control of diseases caused by Oomycetes has ever been a major element in chemical plant protection. The significance of this sector is illustrated by the fact that in 1982 about 25 % of the worldwide expenses for chemical disease control were devoted to the control of these diseases (Schwinn and Urech 1986). The relative economic importance of major diseases in this class is shown in figure 17.1, based on expenditures for their control. The dominating role of downy mildew of grapes and late blight in potatoes is as evident as the low potential of soil-borne diseases.

Table 17.2 Major systemic diseases, caused by Oomycetes, *Peronosporales*

Name	Host plant	Distribution
Peronospora parasitica	brassicas	Europe, Africa, North America
Peronospora tabacina	tobacco	North and Central America, Australia, Europe
Plasmopara halstedii	sunflower	Europe, America, Africa
Pseudoperonospora humuli	hops	Europe, N-America
Peronosclerospora maydis	maize	Asia, Australia
Peronosclerospora sorghi	sorghum	Asia, Africa, America
Sclerospora graminicola	pearl millet	Asia, Africa, N-America

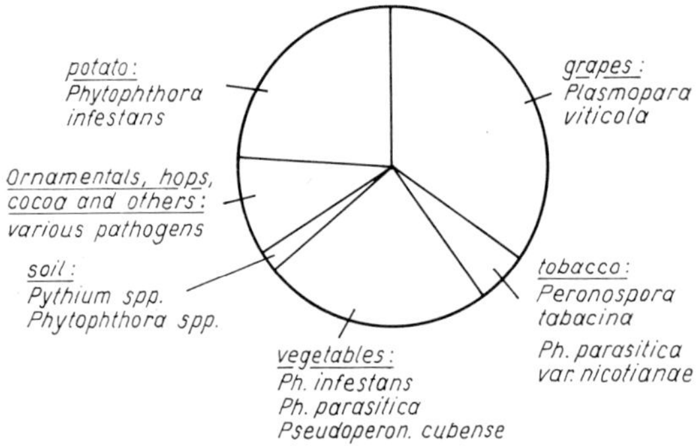

Fig. 17.1 Relative importance of major diseases caused by Oomycetes as reflected in expenses for their chemical control. Source: Ciba-Geigy 1983.

Chemistry

The past decade has witnessed for the introduction of five classes or new types of fungicides controlling diseases caused by Oomycetes:

- the carbamates
- the isoxazoles
- the cyanoacetamide oximes
- the ethyl phosphonates
- the phenylamides.

The term phenylamides, introduced by GISI and WIEDMER (1983), covers three groups of compounds: the acylalanines, the butyrolactones and the oxazolidinones. Of them, the acylalanines were discovered and introduced first. Since their name does not cover the compounds described more recently (see Table 3, 5.2 and 5.3, and Table 4), a broader name had to be found.

Additional new chemical groups, out of which, however, no representatives ever reached the commercial level, like the pyridines and the dinitroanilines (STAUB and

Table 17.3 Chemistry of new fungicides against Oomycetes (common names, chemical names and patent numbers after WORTHING and WALKER 1983, except ofurace, oxadixyl)

1. Carbamates	
prothiocarb (Schering, SN 41703)	*Bastiaansen et al. (1974)*
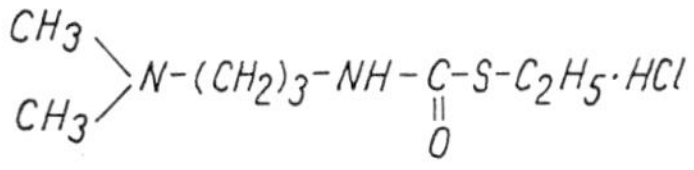	*S-ethyl (3-dimethylaminopropyl) thiocarbamate (DEP 1'567'169)*
propamocarb (Schering, SN66752)	*Pieroh et al. (1978)*
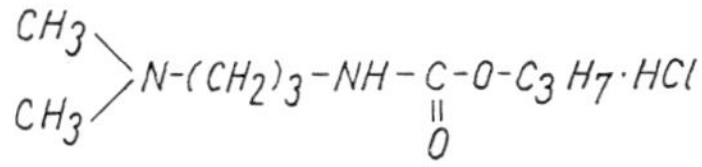	*propyl 3-(dimethylamino) propyl carbamate (DEP 1'567' 169; 1'643'040)*
2. Isoxazoles	
hymexazol (Sankyo, F319, SF-6505)	*Takahi et al. (1974a,b)*
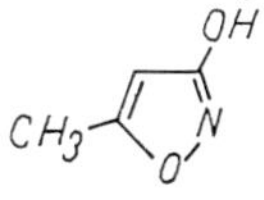	*5-methylisoxazol -3-ol (JPP 518' 249; 532' 202)*
3. Cyanoacetamide-oximes	
cymoxanil (Du Pont, DPX 3217)	*Serres et al. (1976)*
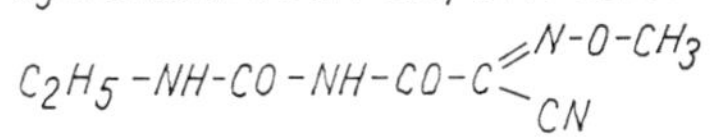	*1-(2-cyano-2-methoxyiminoacetyl) -3-ethylurea (USP 3'957'847)*
4. Ethyl phosphonates	
fosetyl (Rhône-Poulenc, LS 74-783)	*Bertrand et al. (1977), Williams et al. (1977)*
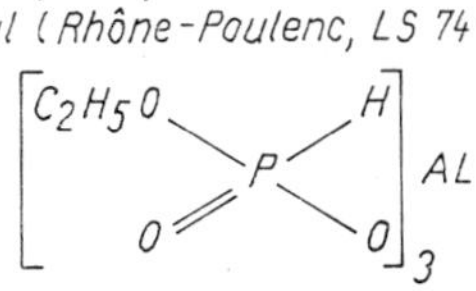	*ethylhydrogen phosphonate, aluminium salt (FP 2'254' 276)*

Table 17.3 (continued)

5. Phenylamides

5.1. Acylalanines

furalaxyl (Giba-Geigy, CGA 38140)

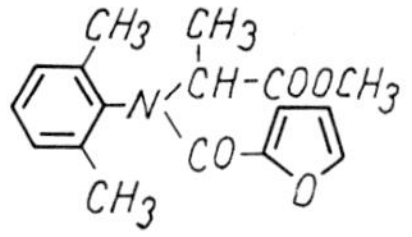

Schwinn et al. (1977a,b)

methyl N-(2-furoyl)-N-(2,6-xylyl)-DL-alaninate (BEP 827' 419; GBP 1' 448' 810)

metalaxyl (Giba-Geigy, CGA 48988)

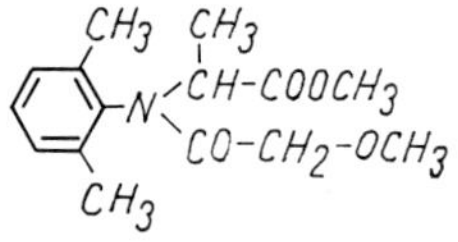

Urech et al. (1977)

methyl N-(2-methoxyacetyl)-N-(2,6-xylyl)-DL-alaninate (BEP 827' 671; GBP 1' 500' 581)

benalaxyl (Montedison, M9834)

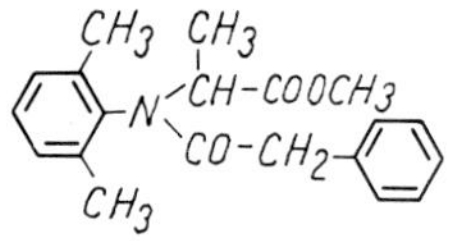

Bergamaschiet al. (1981)

methyl N-phenylacetyl-N-2,6-xylyl-DL-alaninate (BEP 873' 908; DEP 2' 903' 612; ITPA 19' 896-1978)

5.2. Acylamino-Butyrolactones

ofurace (Chevron, RE 20615)*

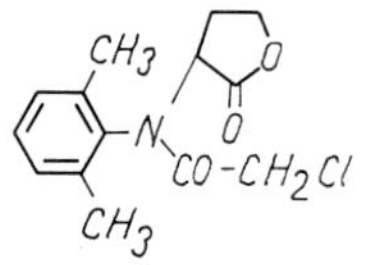

Lukens et al. (1978)

DL-3[N-chloroacetyl-N-(2,6-dimethylphenyl)-amino] γ-butyrolactone (US 3' 933' 860)

**ISO draft proposal*

5.3. Acylamino-Oxazolidinones

cyprofuram (Schering SN 78314)

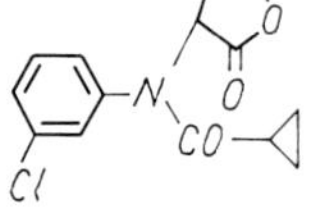

Baumert and Buschhaus (1982)

(±)-α-[N-(3-chlorophenyl)cyclopropane carboxamido] γ-butyrolactone (BP 1' 603' 730)

oxadixyl (Sandoz, SAN 371 F)

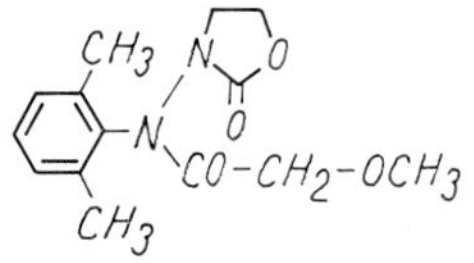

Gisi et al. (1983)

2-methoxy-N-(2-oxo-1,3-oxazolidin-3yl) acet-2', 6'-xylidide (BE 884' 661; GBP 2058-059

Hubele 1981), are not discussed here. Likewise older fungicides with specific activity against Oomycetes, but little commercial impact, such as fenaminosulf, etridiazole and chloroneb, are excluded. They were reviewed by Scheinpflug *et al.* (1977) and Davidse and de Waard (1984) (cf. chapters 5 and 21). (Correction: insert 5.3. Acylamino-Oxazolidinones after cyprofuram, before oxadixyl.)

Table 17.4 New chemicals for control of Oomycetes (in order of their introduction)

Common name	Trade name(s)	First report Formulation(s)
prothiocarb	PREVICUR S 70; SCW	Bastiaansen *et al.* (1974)
hymexazol	TACHIGAREN; EC, SD, D	Takahi *et al.* (1974a, b)
cymoxanil	CURZATE; WP	Serres *et al.* (1976)
furalaxyl	FONGARID; WP, G	Schwinn *et al.* (1977a, b)
fosetyl Al	ALIETTE; WP	Bertrand *et al.* (1977) Williams *et al.* (1977)
metalaxyl	RIDOMIL; WP, G ACYLON; WP APRON; SD	Urech *et al.* (1977)
propamocarb	PREVICUR N; SCW	Pieroh *et al.* (1978)
ofurace	PATAFOL; CALTAN; WP	Lukens *et al.* (1978)
benalaxyl	GALBEN; WP, G	Bergamaschi *et al.* (1981)
cyprofuram	VINICUR; WP	Baumert and Buschhaus (1982)
oxadixyl	SANDOFAN WP	Gisi *et al.* (1983)

D = dust; EC = emulsifiable concentrate; G = granules; SCW = soluble concentrate on water basis; SD = seed treatment; WP = wettable powder

The chemical structure of those compounds of the above five classes that have become commercial products are listed in table 17.3. Whereas in groups 2, 3 and 4 there is only one per group, there are two in the carbamates and six in the phenylamides. The first sales products in this latter group were the acylalanines furalaxyl and metalaxyl, followed by the acylamino butyrolactone ofurace (Staub and Hubele 1981) and the other compounds of this class (table 17.3). The history of discovery of the acylalanines was described by Schwinn and Urech (1986), structure-activity relations were studied by Hubele *et al.* (1983) and Gozzo *et al.* (1985), methods of preparation of enantiomers and their biological activity were investigated by Moser and Vogel (1978), Schwinn and Staub (1982) and Hubele *et al.* (1983). An interesting aspect is the comparatively high water solubility of most of the Oomycetes fungicides, as discussed by Bruin and Edgington (1983). More research is needed in order to find out the role of this factor in fungitoxicity against Oomycetes and the inactivity against pathogens other than these.

Major trade names and formulations as well as first literature quotations are summarized in table 17.4. It is remarkable that the majority of the chemicals mentioned were reported within a very short period of time, i.e. five years only. This reflects the intensity of research in different chemical classes during the years before.

In general, wettable powder formulations are prevailing. Metalaxyl shows the broadest spectrum of formulations, making it the most versatile product in the table in terms of ways of applications.

Biological performance

The biological activity of the new Oomycetes fungicides has been described in several reviews (Schwinn 1979, 1981, 1983; Bruin and Edgington 1983; Davidse and de Waard 1984). Therefore, it is discussed here only briefly, mainly under the aspect

Table 17.5 Spectrum of activity of new Oomycetes fungicides

Common name	Pathogens on root/stem	Foliar pathogens	Additional activity against pathogens outside class Oomycetes
hymexazol	*Aphanomyces* *Pythium*	—	*Corticium sasaki* *Fusarium* spp.
prothiocarb/ propamocarb	*Aphanomyces* *Pythium* *Phytophthora*	*Bremia* *Peronospora* *Pseudoperonospora*	—
cymoxanil	—	*Plasmopara* *Peronospora* *Phytophthora* *Pseudoperonospora*	
osetyl	*Phytophthora* *(Pythium)*	*Bremia* *Plasmopara* *Pseudoperonospora*	*Phomopsis viticola* *Guignardia bidwelli* *Pseudopeziza trachei-phila*
phenylamides	*Peronosclerospora* *Phytophthora* *Pythium* *Sclerospora* *Sclerophthora*	*Albugo* *Bremia* *Peronospora* *Peronosclerospora* *Phytophthora* *Plasmopara* *Pseudoperonospora* *Sclerospora* *Sclerophthora*	—

of practical usefulness. An overview is given in table 17.5. The wide range of variation is obvious; hymexazol on the one end of the scale controls only *Aphanomyces* spp. and *Pythium* spp., the phenylamides on the other end are active against all pathogens in the order of *Peronosporales*. Within the group of phenylamides, however, there are strong differences in the relative activity of the individual molecules. So far, metalaxyl is the most active representative with the broadest use spectrum.

The activity of hymexazol and fosetyl against pathogens other than Oomycetes (table 17.5) is a surprising fact which — on the level of biochemical mode of action — is not yet understood.

Based on spectrum of biological activity, crop tolerance and systemicity, the crop spectrum can be summarized as shown in table 17.6. It shows two main groups of products: in the upper part those which are specialities for particular crops, and in the lower part those with a more or less broad use in agricultural and horticultural crops.

Hymexazol is focussed on just two crops: rice and sugarbeets, mainly as a seed dressing application or as a drench to the rice seedling box. Besides its fungicidal effect, it has a direct growth-promoting activity in rice seedlings (Ota 1975) by stimulation of lateral root and root hair development.

Prothiocarb and furalaxyl are special products for use on ornamentals, a use segment which requires an extremely good crop tolerance against large numbers of cultivars of high economic value. Propamocarb shows similar use profile with additional uses in vegetable crops.

Cymoxanil has made strong inroads into grape vines and potatoes as a foliar fungicide. In contrast, it has no useful activity against soil-borne pathogens. For use against foliar diseases, it has to be mixed with protective fungicides because of the

Table 17.6 Main practical usage of the new Oomycetes fungicides

Compound	Main usage against	Main crops	Application method
prothiocarb/ propamocarb	diseases of roots and stems	ornamentals vegetables	drench
hymexazol	diseases of roots and stems in seedling stage	rice, sugarbeets	drench, seed dressing, dust
furalaxyl	diseases of roots and stems	ornamentals	drench
cymoxanil	foliar diseases	grapes, potatoes	spray
fosetyl	foliar stem and root diseases	grapes, avocado, pineapple, citrus, ornamentals	spray, drench, dip, injection
metalaxyl and related compounds	foliar, stem and root diseases	grapes, potatoes, avocado, pineapples, citrus, tobacco, hops, maize, sorghum, millet	spray, drench, dip, granules, seed dressing

rapid loss of its activity. In combinations at low rates (0.1 kg a.i./ha); 10—15 g a.i./hl), it improves strongly the performance of traditional products like Mancozeb or Folpet. This effect is mainly based on its strong curative action.

Fosetyl has a very broad use spectrum, reaching from foliar application against *Plasmopara viticola* in grapes to trunk injections in avocado trees against root rot caused by *Phytophthora cinnamomi* or dip treatment of pineapple tops against *Phytophthora parasitica* var. *parasitica*. Its strong basipetal systemicity in green and woody tissue (cf. table 17.8) is an outstanding feature of this compound which makes it an extraordinarily flexible tool in the control of a range of diseases on many crops. Its main characteristics are:

- active against foliar downy mildews (not late blight) and root-infecting Oomycetes
- both acropetal and basipetal translocation allows protection of new growth and protection of roots by foliar application
- weak curative activity.

The phenylamides have the most complete range of activity: basically, they control all diseases caused by pathogens of the order *Peronosporales*, including the *Albuginaceae* (table 17.5). However, there are strong differences in performance against various target pathogens between the individual chemicals of this class. So far, metalaxyl has the most complete spectrum of activity: it shows excellent field performance against all pathogens listed in the table. Of all the new compounds discussed here the phenylamides are those with the highest inherent fungitoxic activity with *in vitro* ED_{50} values of 0.01 to 3 ppm for the inhibition of mycelial growth (Bruin 1980, Schwinn and Staub 1982) (cf. chapter 2).

So far, metalaxyl is the most active, versatile and broadly used compound of this class. Its main biological features have been described by several authors (for references see Schwinn 1983; Schwinn and Urech 1986). They can be summarized as follows:

- high inherent fungitoxicity
- protective and curative activity against all *Peronosporales*
- rapid uptake, high acropetal systemicity
- protection of new growth

- good persistence in plant tissue (extended spray intervals)
- control of systemic seed- and soil-borne diseases
- weak on old plant tissue (senescence).

In conclusion, the new fungicides against Oomycetes, above all cymoxanil, fosetyl and the phenylamides have contributed to substantial progress in the practical control of this important group of pathogens. This holds particularly true for root and crown diseases which are well controlled by fosetyl and the phenylamides as well as for the control of systemic diseases by seed dressing application of the phenylamides which thereby offer a new dimension of control. It is worth mentioning that apart from the speciality products propamocarb, prothiocarb and furalaxyl, all new fungicides are used in practice in admixtures with other protectants as illustrated in table 17.7. It shows that all the new fungicides do need mixture partners, however for different reasons.

Table 17.7 Mixture concepts for the new Oomycetes fungicides

New fungicide	Mixture partner	Rationale
hymexazol	metalaxyl	enhancing activity
cymoxanil	protectants, oxadixyl	improving performance, broadening of spectrum
fosetyl	protectants	stabilizing performance, broadening of spectrum
phenylamides	protectants, cymoxanil	anti-resistance strategy, broadening of spectrum, improved end of season performance

Uptake and transport in plants

From a practical point of view, perhaps the most important feature of the mode of action of the new Oomycetes fungicides is their systemicity. The varying degree of apoplastic and symplastic transport of these compounds offers characteristic possibilities for each group to protect plant parts away from the point of application. The range of systemicity is illustrated by cymoxanil with only locally systemic acitivity, by the phenylamides with excellent apoplastic transport and by fosetyl with excellent symplastic and good apoplastic transport (see table 17.8).

With cymoxanil, hymexazole and propamocarb the lack of useful long distance transport appears to be based on different limiting factors. While cymoxanil and hymexazole are taken up rapidly by plant roots, prothiocarb is taken up less readily (Kluge 1978). All three compounds are metabolized rapidly inside the plants which prevents them from showing lasting systemic effects. The short duration of the local protection by cymoxanil (Serres and Carraro 1976) and its rapid metabolism to glycine (Belasco *et al.* 1981) are clear evidence for the limiting role of in vivo stability in the duration of protection. With hymexazole two major glycosidic metabolites are formed which exhibit different biological activities. The O-glycoside is as fungitoxic as hymexazole itself while the N-glycoside is not fungitoxic but has plant growth promoting properties (Kamimura *et al.* 1974).

The phenylamides studied are taken up easily by roots, green stems and leaves and transported apoplastically with the transpiration stream. For metalaxyl it could be

Table 17.8 Systemicity of the new Oomycetes fungicides

Chemical	Characteristics of translocation		
	local (penetration)	apoplastic	symplastic
prothiocarb/propamocarb	++	+	—
hymexazole	++	+	—
cymoxanil	+++	+	—
fosetyl	+++	++	+++
phenylamides	+++	+++	+

+++ = rapid, major factor for performance
++ = intermediate, contributing factor for performance
+ = weak, slow transport, effective only in special circumstances
— = no transport in effective quantities

shown that limited basipetal transport does occur in tomato (Staub *et al.* 1978) and avocado (Zaki *et al.* 1981). In potatoes Rowe (1982) showed that symplastic transport of the experimental phenylamide RE 26745 (Chevron) may be more pronounced than that of metalaxyl. On this crop the sink effect of tubers appears to lead to an accumulation of small but sufficient quantities of metalaxyl (0.02—0.04 ppm) for protection against tuber rot by *Phytophthora infestans* (Bruin *et al.* 1982; Stewart and McClamont 1982) independent of foliar disease control. Transport in the plant tissue is sometimes confounded with distribution by vapour phase.

Based on performance data excellent symplastic transport of fosethyl-Al is much faster than that of phenylamides (Munnecke 1982). However, while the parent compounds of the latter group are translocated apoplastically, the symplastic translocation of the activity of fosetyl appears to be based on the transport of its metabolite H_3PO_3 (Fenn and Coffey 1984). Precise quantitative data on metabolism and transport of fosetyl and H_3PO_3 is lacking.

Mode of action in target fungi and in host-pathogen interaction

Physiological level

An interesting aspect on the mode of action of the modern Oomycete fungicides is their specific inhibition of certain stages in the biology of their target fungi. *In vitro* they are, as a rule, more inhibitory to mycelial growth and sporulation than to spores and spore germination (table 17.9). Their action is fungistatic rather than fungicidal.

The biological site of action of phenylamides in the infection cycle has been studied for several airborne and soilborne Oomycetes. With *Plasmopara viticola* metalaxyl did not inhibit spore germination and initial penetration of the fungus into grape leaves. It inhibited strongly any further fungal growth inside the leaf after the formation of the first haustorium (Staub *et al.* 1980). This is in strong contrast to the site of action of protective multi-site fungicides which only affect spores and germtubes on the leaf surface but do not affect fungal growth inside the leaves (Fig. 17.2). In mixtures of phenylamides with residual compounds the complementary modes of biological action often lead to a synergism in the field situation. However, the exposure time of the pathogen to the systemic compounds is longer than to the residual ones.

With soil-borne Oomycetes the sites of action of metalaxyl appear to be analogous. In the infection cycle of *Phytophthora parasitica* var. *nicotianae* on tobacco seedlings

Table 17.9 Aspects of the mode of action of fungicides against Oomycetes

Compound	Sensitive stages of target fungi	Biochemical targets	Type of inhibition	Reversal of fungi-toxicity by	References
prothiocarb/ propamocarb	mycelium, sporulation	?	fungistatic	sterols, L-methio-nine	KERKENAAR and KAARS SIPJESTEIJN (1977)
hymexazole	mycelium, sporulation	RNA syn-thesis	?	?	NAKANISHI and SISLER (1983)
cymoxanil	mycelium, sporulation	RNA syn-thesis (in B. cinerea)	fungistatic	amino acids	DESPREAUX *et al.* (1981)
fosetyl	mycelium, sporulation	?	fungistatic	H_3PO_4	FENN and COFFEY (1984)
phenyl-amides	mycelium, sporulation	RNA synthesis	fungistatic	?	KERKENAAR and KAARS SIPJESTEIJN (1981) DAVIDSE (chapter 18)

no inhibition of zoospore release from sporangia nor of their migration to and encystment on root tips was found. Inhibition was observed only after penetration of root tissue had occurred. In addition, sporangia formation from infected root tissue and from chlamydospores was also strongly inhibited (STAUB and YOUNG 1980).

Another aspect of the mode of action of systemic Oomycetes fungicides is the question of indirect action via activation of latent host plant resistance mechanisms. For metalaxyl, prothiocarb and especially for fosetyl an indirect effect via the activation of the host's defence reaction has been claimed.

Based on lacking *in vitro* activity of fosetyl and the type of response of fosetyl-treated sensitive grape and tomato plants, respectively, which is similar to that of naturally resistant ones including the accumulation of phytoalexins, RAYNAL *et al.* (1980) and BOMPEIX *et al.* (1980/1981) suggested these indirect effects to be the primary ones. However, resistant type responses and phytoalexin accumulation were also found on metalaxyl treated lettuce (CRUTE 1979), potatoes (BRUCK *et al.* 1980) and soybean (WARD *et al.* 1980), and some isolates also show a discrepancy between *in vivo* and *in vitro* sensitivity (STAUB *et al.* 1979). The latter is also a general characteristic of prothiocarb (KAARS SIJPESTEIJN *et al.* 1974). For metalaxyl and prothiocarb, therefore, effects on host responses have also been implicated, but as secondary ones. The suppression of both phytoalexin production and protection on treated plants by inhibitors of the shikimic pathway (e.g. the herbicide glyphosate) provided additional evidence for some sort of involvement of host responses in the protection effects by fosetyl (FETTOUCHE *et al.* 1981) and metalaxyl (WARD 1984). Recent work by BOWER and COFFEY (1985) with fosetyl resistant mutants of *Phytophthora capsici* showed that *in vitro* resistance to this fungicide is paralleled by *in vivo* resistance on green pepper plants. Similar observations with metalaxyl are common. They suggest that the hypothesis of triggered host resistance effects being primary events in the mode of action of fosetyl, which had been lauded by many reviewers as the new generation of disease control agent, so far has no solid experimental basis. It appears rather that fosetyl or its degradation product H_3PO_4 acts directly on the target fungi leading to growth inhibition. This allows the defence reactions of normally susceptible plants to be triggered. Therefore, we suggest that as with many other systemic fungicides, the protection by fosetyl is the result of combined actions by the fungicide on the

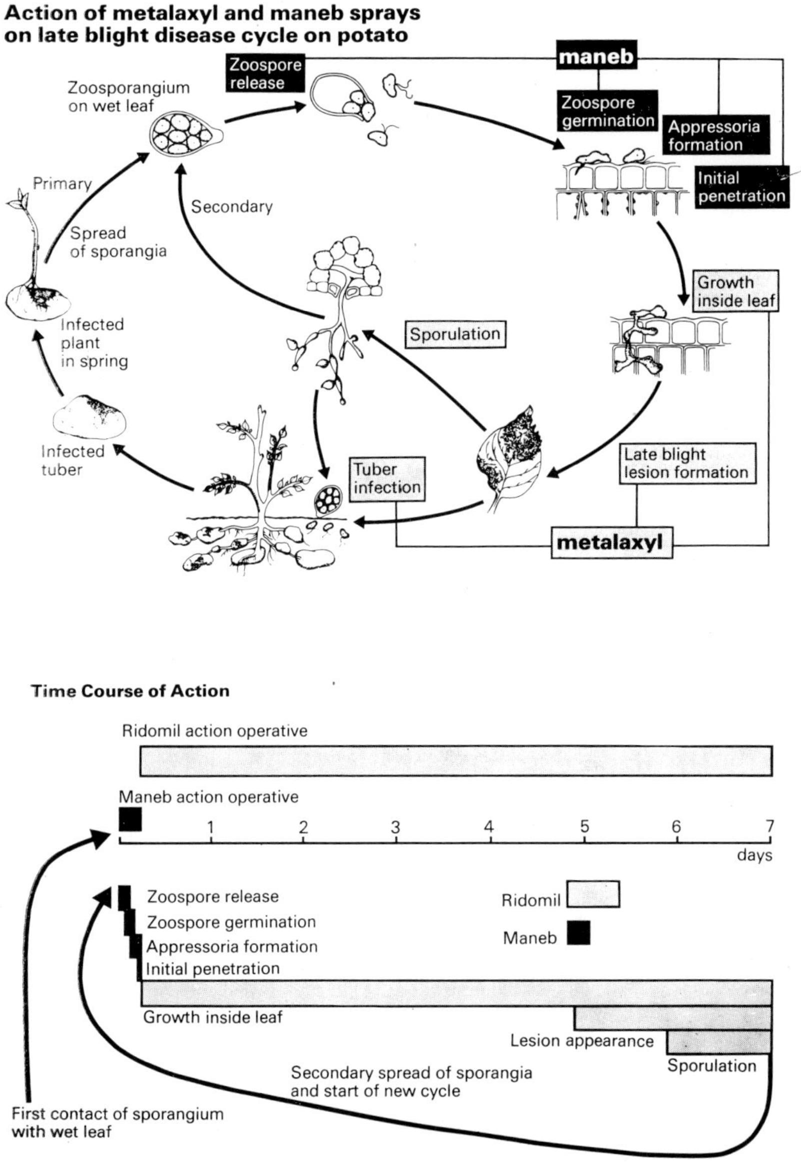

Fig. 17.2 Action of metalaxyl and maneb sprays on late blight disease cycle on potato (upper part) and time course of action in relation to the biological cycle (lower part).

fungus (primary) and the plant on the fungus (secondary). Additional support for direct fungitoxicity as the primary mode of action of fosetyl was provided by FENN and COFFEY (1984) who showed that fosetyl and H_3PO_4 are highly fungitoxic *in vitro* if tested on media with a low phosphate content.

Biochemical level

The biochemical mode of action of the fungicides treated in this chapter is summarized in table 17.9. For the phenylamides details are discussed in chapter 18 by DAVIDSE.

Therefore they are not treated here. Phenylamides interfere with RNA synthesis of their target fungi. Interstingly, RNA synthesis is also implicated as the biochemical mode of action of hymexazole in *Pythium aphanidermatum* (Nakanishi and Sisler 1983) and of cymoxanil in a sensitive *Botrytis cinerea* mutant (Despareaux *et al.* 1981). *Botrytis* is not ordinarily a target for cymoxanil and it remains to be seen whether the mechanism is the same in sensitive Oomycetes.

The biochemical mode of action of prothiocarb and propamocarb is still not known other than that their action *in vitro* can be reversed by the addition of sterols to the medium. However, for certain *Saprolegniales* which are sensitive to prothiocarb only, fungitoxicity is attributed to the degradation product ethylmercaptan and it can be reversed by the addition of L-methionine (Kerkenaar and Kaars Sijpesteijn 1977).

Status of resistance

Of the fungicides discussed in this chapter only the phenylamides have met with resistance problems in the field. The other Oomycete fungicides have so far not produced field resistance. This is not necessarily due to their inherent lower proneness on the biochemical level, as was shown by the easy induction of resistant mutants to fosetyl in *Phytophthora capsici* (Bower and Coffey 1985). They may rather have escaped resistance problems because of their lower fungitoxic efficacy. This in turn made it necessary, especially for the foliar compounds, to use them from the start of introduction in mixtures with residual compounds. Without a mixture partner, neither cymoxanil nor fosetyl appear to achieve reliable protection against foliar Oomycetes under field conditions.

Field resistance to phenylamides has developed shortly after their market introduction in cases of intensive, continuous and exclusive use of the straight product, mainly in *Phytophthora infestans* and *Plasmopara viticola*. Meanwhile, it has also shown up in certain cases against *Peronospora tabacina*. History, parameters and consequences of this development were reviewed by Bruin and Edgington (1983), Schwinn (1983) and Davidse and de Waard (1984).

Detailed studies have been done on cross-resistance patterns in phenylamides. For all practical purposes cross-resistance exists among all phenylamides even though the resistance factors can vary in some strains (Clerjeau *et al.* 1985; Davidse in chapter 18). However, no cross-resistance exists between phenylamides and the other new Oomycetes fungicides or residual products (Leroux and Gredt 1981; Kendall and Carter 1984; Bower and Coffey 1985; Davidse in chapter 18). The only conflicting report indicating cross-resistance between phenylamides and unrelated fungicides is by Cohen *et al.* 1984. However, his results could not be reproduced by others unless they allowed the vapour phase activity of phenylamides to interfere and create artefacts which mimic cross-resistance (Clerjeau *et al.* 1984).

A key question in the context of fungicide resistance is how to prevent the build-up of resistant pathogen populations. Experimental data on this point is very sparse and badly needed. This need and the phenomenon of cross-resistance (as described above for phenylamides), which threatens the effective life of related compounds from different manufactures, has lead industry to react by the formation of a **F**ungicide **R**esistance **A**ction **C**ommittee **(FRAC)** (Delp 1984). It consists of working groups for the major clases of resistance prone modern fungicides: e.g. for the phenylamides. Their main goal is to design and establish use strategies which help to safeguard the availability and effectiveness of the endangered fungicides. For the phenylamides the working group decided on the use of pre-pack mixtures with residual compounds and on a limited number of applications per season as the basic anti-resistance strategy

(Urech 1985). In areas with substantial proportions of resistance in the pathogen population, the number of applications should not exceed two.

Observational evidence suggests that this anti-resistance strategy works in slowing down the build-up of resistance and in preventing crop losses even in cases where resistance is detectable in the fungus population. In addition, model tests with *Phytophthora infestans* populations, starting with a known proportion of phenylamide resistant spores, confirmed the slow-down of the selection process when mixtures with mancozeb were used (Staub and Sozzi 1983). Mixtures are favoured over the use of single products in alternation for practical purposes because their use can be enforced whereas alternations cannot. They also provide a safeguard against crop losses where resistance levels would lead to complete failures of the single product.

References

Bastiaansen, M. G., Pieroh, E. A., and Aelbers, E.: Prothiocarb, a new fungicide to control *Phytophthora fragariae* in strawberries and *Pythium ultimum* in flower bulbs. Meded. Fac. Landbouwwet. Rijksuniv. Gent **39** (1974): 1019—1025.

Baumert, D., and Buschhaus, H.: Cyprofuram, a new fungicide for the control of Phycomycetes. Meded. Fac. Landbouwwet. Rijksuniv. Gent **47** (1982): 979—983.

Belasco, I. J., Han, J. C. Y., Chrzanowsky, R. L., and Baude, F. J.: Metabolism of (14C) cymoxanil in grapes, potatoes and tomatoes. Pesticide Science **12**, 4 (1981): 355—364.

Bergamaschi, P., Borsari, T., Garaveglia, C., and Mirenna, L.: Methyl N-phenyl-acetyl-N-2,6-xylyl-dl-alaninate (M 98834), a new systemic fungicide controlling downy mildew and other diseases caused by *Peronosporales*. Proc. 11. Brit. Crop Prot. Conf. (1981): 11—18.

Bertrand, A., Ducret, J., Debourge, J.-C., and Horriére, D.: Etude des propriétés d'une nouvelle famille de fongicides. Les monoéthylphosphites métalliques. Charactéristiques physicochimiques et propriétés biologiques. Phytiatr. Phytopharm. **26** (1977): 3—18.

Bompeix, G., Fettouche, F., and Saurdressan, P.: Mode d'action du phoséthyl Al. Phytiatr. Phytopharm. **30** (1981): 257—272.

— Ravisé, A., Raynal, G., Fettouche, F., and Durand, M. C.: Modelités de l'obtention des nécroses bloquantes sur feuilles détachées de tomates par l'action du tris-0-éthyl phosphonate d'aluminium (phoséthyl d'aluminium), hypothèse sur son mode d'action *in vivo*. Ann. Phytopathol. **12**, No. 4 (1980): 337—351.

Bower, L. A., and Coffey. M. D.: Development of laboratory tolerance to phosphorons acid, fosetyl-Al and metalaxyl in *Phytophthora capsici*. Can. J. Plant Pathol. **1** (1985): 1—6.

Bruin, G. C. A.: Resistance in Peronosporales to acylalanine-type fungicides. Ph.D. thesis Univ. Guelph, Ontario, Canada 1980, 110 pp.

— and Edgington, L. V.: The chemical control of diseases caused by zoosporic fungi. In: Buczacki, S. T. (Ed.): Zoosporic Plant Pathogens. A Modern Perspective. Academic Press, London 1983, pp. 193—233.

— — and Ripley, B. D.: Bioactivity of the fungicide metalaxyl in potato tubers after foliar sprays. Canad. J. Pl. Path. **4** (1982): 353—356.

Bruck, R. I., Fre, W. E., and Apple, A. E.: Effect of metalaxyl an acylalanine fungicide, on developmental stages of *Phytophthora infestans*. Phytopathology **70** (1980): 597—601.

Clerjeau, M., Irhir, H., Moreau, C., Piganeau, B., Staub, T., and Diriwaechter, G.: Etude de la résistance croisée au métalaxyl et au cyprofurame chez *Plasmopara viticola:* Evidence de plusieurs mécanismes de résistance independants. Proceedings of the Centenary Meeting Bordeaux, Vol. 2 (1985): 303—306.

— Moreau, C., Piganeau, B., Bompeix, G., and Malfatti, P.: Effectiveness of Fosetyl-Al against strains of *Plasmopara viticola* and *Phytophthora infestans* that have developed resistance to anilide fungicides. Proc. Brit. Crop Prot. Conf. Pests Dis. **2** (1984): 497—502.

Cohen, Y., and Samoucha, Y.: Cross-resistance to four systemic fungicides in metalaxyl-resistant strains of *Phytophthora infestans* and *Pseudoperonospora cubensis*. Plant Dis. **68** (1984): 137 to 139.

Crute, I. R.: Lettuce mildew — Destroyer of quality. Agric. Res. Counc. (U.K.) Res. Rev. **5** (1979): 9—12.

Davidse, L. C., and de Waard, M. A.: Systemic fungicides. In: Ingram, D. S., and Williams, P. M. (Eds.): Advances in Plant Pathology **2** (1984): 191—257.

Delp, C. J.: Industry's response to fungicide resistance. Crop Protection **3**, 1 (1984): 3—8.

Despreaux, D., Fritz, R., Leroux, P.: Mode d'Action biochimique du Cymoxanil. Phytiatr. Phytopharm. **30**, No. 4 (1981): 245—255.

Fenn, M. E., and Coffey, M. D.: Studies on the *in vitro* and *in vivo* antifungal activity of fosetyl-Al and phosphorous acid. Phytopathology **74** (1984): 606—611.

Fettouche, F., Ravisé, A., Bompeix, G.: Suppression de la résistance induite — phoséthyl-Al — chez la tomate à *Phytophthora capsici* avec deux inhibiteurs — glyphosate et acide amino-oxyacétique. Agronomie **1**, No. 9 (1981): 826.

Gisi, U., Harr, J., Sandmeier, R., and Wiedmer, H.: A new systemic oxazolidinone fungicide (SAN 371) against diseases caused by *Peronosporales*. Meded. Fac. Landbouwwet. Rijksuniv. Gent **48** (1983): 541—549.

— and Wiedmer, H.: Fungicidal activity of SAN 371F and its combinations against Peronosporales. Proc. Brit. Crop Prot. Conf. Pests Dis. **3** (1983): 1193.

Gozzo, F., Garlaschelli, L., Boschi, P. M., Zagni, A., Overeem, J. C., and de Vries, L.: Recent progress in the field of N-acylalanines as systemic fungicides. Pestic. Sci. **16** (1985): 277—286.

Hubele, A., Kunz, W., Eckhardt, W., and Sturm, H.: The fungicidal activity of acylanilines. In: Miyamoto, J., and Kearney, P. C. (Eds.): IUPAC Pesticide Chemistry. Human Welfare and the Environment. Pergamon Press, Oxford New York 1983, pp. 233—242.

Kaars, Sipjesteijn, A., Kerkenaar, A., and Overeem, J. C.: Observations on selectivity and mode of action of Prothiocarb. Meded. Fac. Landbouwwetensch. **39**, No. 2 (1974): 1027—1034.

Kamimura, S., Nishikawa, M., Saeki, H., and Takahi, Y.: Absorption and metabolism of 3-hydroxy-5-methylisoxazole in plants and the biological activities of its metabolites. Phytopathology **64**, 10 (1974): 1273—1281.

Kendall, S. J., and Carter, G. A.: Resistance of isolates of *Phytophthora infestans* to fungicides. Proc. Brit. Crop Prot. Conf. Pests Dis. **2** (1984): 503—508.

Kerkenaar, A., and Kaars Sipjesteijn, A.: On the Mode of action of prothiocarb. Netherl. J. Plant Pathol. **83**, Suppl. 1 (1977): 145—152.

— — Antifungal activity of metalaxyl and furalaxyl. Pestic. Biochem. Physiol. **15** (1981): 71—78.

Kluge, E.: Vergleichende Untersuchungen über die Wirksamkeit von Systemfungiziden gegen Oomyzeten. Arch. Phytophatol. Pflanzenschutz **14**, No. 2 (1978): 115—122.

Kreisel, H.: Grundzüge eines natürlichen Systems der Pilze. Cramer, Lehre 1969, 245 S.

Leroux, P., and Gredt, M.: Phénomènes de résistance aux fongicides anti-mildious: Quelques résultats de laboratoire. Phytiatr. phytopharm. **30**, No. 4 (1981): 273—282.

Lukens, R. J., Cham, D. C. K., and Etter, G.: Ortho 20615, a new systemic for the control of plant diseases caused by Oomycetes. Phytopath. News **12** (1978): 142.

Moser, H., and Vogel, C.: Preparation and biological activity of the enantiomers of CGA 48988, a new systemic fungicide. 4th Intern. IUPAC Congr. Zürich (1978): Abstracts II-310.

Munnecke, D. E.: Apparant movement of Aliette and Ridomil in *Persea indica* and its effect on root rot. Phytopathology **72** (1982): 970.

Nakanishi, T., and Sisler, H. D.: Mode of action of Hymexazol in *Pythium aphanidermatum*. J. Pestic. Sci. **8** (1983): 173—181.

Ota, Y.: Plant growth promoting activities of 3-hydroxy-5-methylisoxazole. Jap. Agric. Res. Quarterly **9**, No. 1 (1975): 1—7.

Pieroh, E. A., Krass, W., and Hemmen, C.: Propamocarb, ein neues Fungizid zur Abwehr von Oomyceten im Zierpflanzen- und Gemüsebau. Meded. Fac. Landbouwwet. Rijksuniv. Gent **43** (1978): 933—942.

Raynal, G., Ravisé, A., and Bompeix, G.: Action du tris-0-éthylphosphonate d'aluminium (phoséthyl d'aluminium) sur la pathogénie de *Plasmopara viticola* et sur la stimulation des réactions de défense de la vigne. Ann. Phytopathol. **12** (1980): 163—175.

Rowe, R.: Translocation of metalaxyl and RE 26745 in potato and comparison of foliar and soil application for control of *Phytophthora infestans*. Plant Disease **66** (1982): 989—993.

Scheinpflug, H., Schlier, H., and Widdig, A.: Chemie der Fungizide. In: Weyler, R.: Chemie der Pflanzenschutz- und Schädlingsbekämpfungsmittel **4** (1977): 120—238.

Schwinn, F. J.: Control of Phycomycetes; a changing scene. Proc. Brit. Crop Prot. Conf. 10th **3** (1979): 791—802.

— Chemical control of downy mildews. In: SPENCER, D. M. (Ed.): The Downy Mildews. Academic Press, London-New York 1981.

— New developments in chemical control of *Phytophthora*. In: ERWIN, D. C., BARTNICKI-GARCIA, S., and TSAO, P. H. (Eds.): *Phytophthora:* its Biology, Taxonomy, Ecology and Pathology. Amer. Phytopath. Soc., St. Paul, USA 1983.

— Progress in chemical control of diseases caused by Oomycetes. ACS Symposium Series (1986), **304**, 89—106.

— STAUB, T., and URECH, P. A.: A new type of fungicide against diseases caused by Oomycetes. Meded. Fac. Landbouwwet. Rijksuniv. Gent **42** (1977a): 1181—1188.

— — — Die Bekämpfung falscher Mehltaukrankheiten mit einem Wirkstoff aus der Gruppe der Acylalanine. Mitt. Biol. Bundesanstalt Land Forstwirtsch. Berlin-Dahlem **178** (1977b): 145—146.

SERRES, J. M., and CARRARO, G. A.: DPX-3217, a new fungicide for the control of grape downy mildew, potato blight and other *Peronosporales*. Meded. Fac. Landbouwwet. Rijksuniv. Gent **42** (1976): 645—650.

STAUB, T., DAHMEN, H., and SCHWINN, F. J.: Biological characterization of uptake and translocation of fungicidal acylalanines in grape and tomato plants. Z. Pflanzenkr. Pflanzenschutz **85** (1978): 162—168.

— — — Effects of Ridomil on the development of *Plasmopara viticola* and *Phytophthora infestans* on their host plants. Z. Pflanzenkr. Pflanzenschutz **87** (1980): 83—91.

— — URECH, P., and SCHWINN, F.: Failure to select for *in vivo* resistance in *Phytophthora infestans* to acylalanine fungicides. Plant Dis. Rep. **63** (1979): 385—389.

— and HUBELE, A.: Recent advances in the chemical control of Oomycetes. In: WEGLER, R. (Ed.): Chemie der Pflanzenschutz- und Schädlingsbekämpfungsmittel. Vol. 6. Springer-Verlag, Heidelberg 1981, pp. 389—422.

— and SOZZI, D.: Recent practical experiences with fungicide resistance. 10. Internat. Congr. Plant Path. Brighton, U.K. **2** (1983): 591—598.

— and YOUNG, T. R.: Fungitoxicity of metalaxyl against *Phytophthora parasitica* var. *nicotianae*. Phytopathology **70** (1980): 797—801.

STEWART, H. E., and MCCLAMONT, D. C.: The effect of metalaxyl and mancozeb on the susceptibility of potato tubers to late blight. Ann. Appl. Biol. **100**, Suppl. (1982): 54—55.

TAKAHI, Y., NAKANISHI, T., and KAMINURA, S.: Characteristics of hymexazol as a soil fungicide. Am. Phytopath. Soc. Japan **40** (1974a): 362—367.

— — TOMITA, K., and KAMINURA, S.: Effects of 3-hydroxy isoxazoles as soil fungicides in relation to their chemical structure. Am. Phytopath. Soc. Japan **40** (1974b): 354—361.

URECH, P. A.: Management of fungicide resistance in practice. Bulletin EPPO, **15** (1985), in press.

— SCHWINN, F. J., and STAUB, T.: CGA 48988, a novel fungicide for the control of late blight, downy mildews and related soil-borne diseases. Proc. Brit. Crop Prot. Conf., 9th **2** (1977): 623—6731.

— and STAUB, T.: Biological properties of metalaxyl. In: LYR, H., and POLTER, C. (Eds.): Systemic Fungicides and Antifungal Compounds. Akademie-Verlag, Berlin 1982. 123—133.

VON ARX: The genera of fungi sporulating in pure culture. Cramer, Vaduz 1974, 315 pp.

WARD, E. W. B.: Suppression of metalaxyl activity by glyphosate: Evidence that host defence mechanisms contribute to metalaxyl inhibition of *Phytophthora megasperma* f. sp. *glycinae* in soybeans. Physiol. Pl. Pathol. **25** (1984): 381—386.

— LAZAROVITS, G., STÖSSEL, P., BARRIE, S. D., and UNWIN, C. H.: Glyceollin production associated with control of *Phytophthora* rot of soybeans by the systemic fungicide metalaxyl. Phytopathology **70** (1980): 738—740.

WHITTACKER, R. H.: New concepts of kingdoms of organisms. Science **163** (1969): 150—160.

WILLIAMS, D. J., BEACH, B. G. W., HORRIÈRE, D., and MARÉCHAL, G.: LS 74—783, a new systemic fungicide with activity against Phycomycete diseases. Proc. Brit. Crop Prot. Conf., 9th **2** (1977): 565—573.

WORTHING, C. R., and WALKER, S. B.: The pesticide manual. 7. Edition. Brit. Crop. Prot. Council (1983), 693 p.

ZAKI, A. I., ZENTMYER, G. A., and LE BARON, H. M.: Systemic translocation of C-labeled metalaxyl in tomato, avocado, and *Persea indica*. Phytopathology **71** (1981): 509—514.

LYR, H. (Ed.): Modern Selective Fungicides — Properties, Applications, Mechanisms of Action. Longman Group UK Ltd., London, and VEB Gustav Fischer Verlag, Jena, 1987.

Chapter 18

Biochemical aspects of phenylamide fungicides - action and resistance

L. C. Davidse
Laboratory of Phytopathology, Agricultural University, Wageningen, The Netherlands

Introduction

The group of phenylamide fungicides includes a number of structurally related chemicals with high activity against fungi belonging to the Peronosporales and a limited number of other fungi (Fuller and Gisi 1985). The group can be subdivided in the acylalanines metalaxyl, furaxyl and benalaxyl, the butyrolactones, ofurace and cyprofuram, and the oxazolidinones, oxadixyl (Fig. 18.1). The phenylamide fungicides are structurally related to the chloroacetanilide herbicides, that in fact provided lead compounds in screening programs for fungicidal phenylamides. The selective activity of the phenylamides against almost exclusively Peronosporales makes these compounds excellent tools to study unique features of these fungi. This

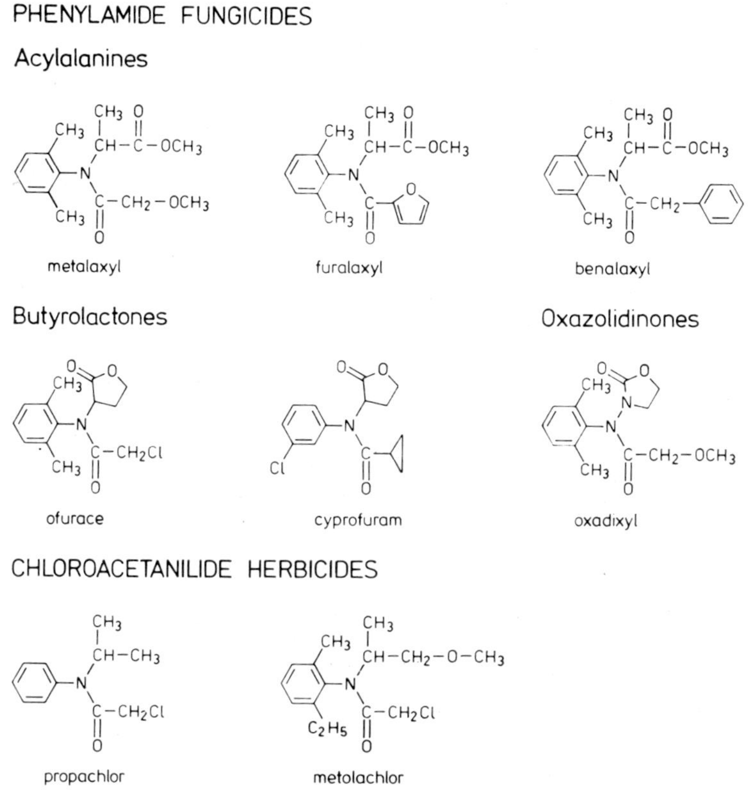

Fig. 18.1 Structures of phenylamide fungicides and related chloroacetanilide herbicides.

together with the ability of sensitive fungi to develop a high level of resistance to the phenylamides stimulated research on their mode of action and the mechanisms of resistance.

Knowledge of the molecular interaction between phenylamides and their target site within the fungal cell facilitates comparative studies on the intrinsic antifungal activity of the various phenylamides and may lead to the design of new compounds with even higher activity. Understanding the mechanism(s) of resistance might be helpful in designing ways to cope with resistance. In addition these studies usually contribute to our knowledge on the physiology of fungi, many details of which are still largely unknown.

The antifungal activity of the phenylamides has been most intensively studied with metalaxyl. Details of the mechanism of action of metalaxyl will be discussed in the next section. A mechanism of resistance to metalaxyl that has been found in resistant strains of *Phytophthora megasperma* f. sp. *medicaginis* and *Phytophthora infestans* will be described in the following section. Subsequently the antifungal mode of action of various phenylamides including related herbicidal compounds is compared with that of metalaxyl.

Mode of action of metalaxyl

Among the biosynthetic processes studied in various fungi incorporation of radiolabelled uridine into RNA proved to be the most sensitive one to metalaxyl (ARP and BUCHENAUER 1981; DAVIDSE *et al.* 1981a, 1983b; FISHER and HAYES 1982; KERKENAAR 1981; WOLLGIEHN *et al.* 1984). Incorporation of precursors into DNA, proteins and lipids is effected less rapidly or to a lesser extent. Respiration is not inhibited by metalaxyl. Since metalaxyl does not inhibit the uptake of uridine nor its conversion into UTP, inhibition of RNA synthesis must be responsible for the observed effects on uridine incorporation. However, even at concentrations that are fully inhibitory to growth, a complete inhibition of uridine incorporation does not occur. Depending on the fungal species used in the experiments incorporation is reduced by metalaxyl to 20—60% of the control value. It indicates that only part of the cellular RNA synthesis is sensitive to metalaxyl.

In eukaryotes RNA is synthesized by three different RNA polymerases each mediating the synthesis of a distinct product. RNA polymerase I (or A) synthesizes r(ibosomal) RNA, that accounts for the majority of the cellular RNA. RNA polymerase II (or B) produces m(essenger) RNA and RNA polymerase III (or C) gives rise to t(ransfer) RNA and the 5S RNA of the ribosomes. Each enzyme has its own characteristics that can be used to purify the individual enzymes. Sensitivity to α-amanitin is a useful property to characterize the polymerases. RNA polymerase II is highly sensitive to this toxin whereas RNA polymerase I is insensitive. Polymerase III of different organisms varies in sensitivity to α-amanitin but in general it is much less sensitive than polymerase II.

A similar differential sensitivity of the RNA polymerases to metalaxyl would explain the partial inhibition of RNA synthesis. Therefore, the effects of metalaxyl on the synthesis of the different classes of RNA were studied. In mycelium of *P. megasperma* f. sp. *medicaginis* synthesis of poly(A)-containing RNA, that represents most of the mRNA appeared to be less affected by metalaxyl than that of total RNA, the majority of which is rRNA (DAVIDSE *et al.* 1983b). rRNA appeared to be selectively inhibited (DAVIDSE and FLEUREN, unpublished results). Similar differential effects were observed in *P. nicotianae* (WOLLGIEHN *et al.* 1984). Inhibition of rRNA synthesis, therefore, can be considered to be the primary mode of action of metalaxyl.

Inhibition of rRNA synthesis would ultimately lead to inhibition of fungal growth because turnover of rRNA will deprive the cell of its ribosomes causing protein synthesis to decrease. It explains the inability of metalaxyl to inhibit germination of encysted zoospores of various fungi. Apparently the reproductive structures possess enough ribosomes to support germ tube formation. Several studies have shown that metalaxyl exerts its inhibiting effect against fungi on plant surfaces after penetration (BRUCK *et al.* 1980; HICKEY and COFFEY 1980; STAUB and YOUNG 1980; STAUB *et al.* 1980). Neither direct nor indirect germination of sporangia, nor the mobility, encystment and germination of zoospores is affected. Inhibition becomes apparent not until after formation of the primary haustorium. Spores apparently carry enough ribosomes along to sustain mycelial growth through critical stages of the life cycle of a fungus.

Inhibition of rRNA synthesis will also lead to accumulation of its precursors because rRNA is the major end product the nucleoside triphosphates are synthesized for. Nucleoside triphosphates are known to stimulate β(1—3) glucan synthetase from various fungi (SZANISLO *et al.* 1985) and are involved in regulation of cell wall synthesis. Therefore, the observed thickening of cell walls of metalaxyl-treated hyphae (GROHMANN and HOFFMANN 1982; MÜLLER and LYR 1983) might be caused by elevated levels of nucleoside triphosphates enhancing cell wall synthesis.

The exact way metalaxyl interferes with rRNA synthesis is not known yet. In *in vitro* experiments metalaxyl does not inhibit the activity of partial purified polymerase I from *P. megasperma* (DAVIDSE *et al.* 1983b) and *P. nicotiana* (WOLLGIEHN *et al.* 1984). Endogenous RNA polymerase activity of nuclei isolated from these fungi, however, is sensitive to metalaxyl, indicating that metalaxyl only interferes with the intact polymerase-template complex. Circa 40 % of the endogenous RNA polymerase activity of isolated nuclei of *P. megasperma* is metalaxyl sensitive and circa 30 % α-amanitin sensitive (DAVIDSE *et al.* 1983b). The effects of metalaxyl and α-amanitin are additive indicating interference of metalaxyl with an RNA polymerase activity different from polymerase II. Synthesis of mRNA being highly sensitive to α-amanitin is apparently not affected by metalaxyl. It confirms the lack of any effect of metalaxyl on the synthesis of poly(A)-containing RNA in the intact organism.

It should be clear from these studies that metalaxyl is a unique inhibitor of rRNA synthesis in those fungi belonging to the Peronosporales. Many details of its action remain to be resolved. Comparative studies with naturally phenylamide-resistant fungi will indicate how rRNA synthesis evolved and may reveal the molecular bases of selectivity of the phenylamides. A similar target may exist in other fungi having a structure that does not allow any interaction with the phenylamides. Characterization of such structures may enable the design of new molecules that will interact and are candidate fungicides.

Mechanism of resistance to metalaxyl

The mechanism of resistance to metalaxyl has been studied in strains of *P. megasperma* f. sp. *medicaginis* in which metalaxyl resistance was induced by nitrosoguanidine mutagenesis (DAVIDSE 1981) and in resistant strains of *P. infestans* obtained from potato fields where metalaxyl failed to control late blight (DAVIDSE *et al.* 1981b, 1983a). In the resistant strains uridine incorporation into RNA appeared to be completely insensitive to metalaxyl at concentrations which maximally inhibited that of the sensitive strains (DAVIDSE *et al.* 1984; DAVIDSE *et al.* unpublished results). Likewise endogenous nuclear RNA polymerase activity of resistant strains of both species was significantly less sensitive to metalaxyl than that of sensitive strains. Apparently, in both species a change in the target site of metalaxyl is responsible for resistance. The fact that both chemically induced resistance and resistance after natural selection have a similar basis proves once more the validity of laboratory studies aimed at evaluating the potential of a fungus to develop resistance to a fungicide.

Mode of action of other phenylamides in comparison with that of metalaxyl

The phenylamide fungicides originate from a screening program in which the antifungal activity displayed by a number of phenylamide herbicides has been optimized with the concurrent elimination of herbicidal activity (HUBELE *et al.* 1983). This development has been possible because herbicidal and antifungal activity require different structural features of the phenylamides. The presence of the chloroacetyl group seems important for herbicidal activity but not for antifungal activity. On the other hand, high antifungal activity requires the presence of an alanine methylester moiety, or an equivalent structure, which does not seem to be essential for herbicidal activity. Chirality dependency of herbicidal activity is also different from that of antifungal activity. R-enantiomers display considerable higher antifungal activity than S-enantiomers. In contrast herbicidal activity is higher with the S-enantiomers (MOSER *et al.* 1982). In view of this the primary mechanism of action of the phenylamides in fungi will likely be different from that in plants (DAVIDSE 1984).

The herbicidal phenylamides have been included in a number of studies on the antifungal mode of action of the phenylamides (DAVIDSE *et al.* 1984). Such comparative studies may yield valuable information about structure-activity relationships and about the relation, if any, between the antifungal and the herbicidal activity of the phenylamides.

Tab. 18.1 and 2 summarize the results of a study in which the activities of four phenylamide fungicides and two antifungal phenylamide herbicides on growth, uridine uptake and uridine incorporation of a wild-type strain and a metalaxyl resistant strain of *P. megasperma* f. sp. *medicaginis* are compared. The existence of cross resistance of the resistant strain to the various phenylamides is evident when the respective EC_{50}-values against growth of the two strains are compared (Tab. 18.1). Endogenous nuclear RNA polymerase activity of the resistant strain is hardly affected by any of the compounds whereas that of the sensitive strain is sensitive to all phenylamides except propachlor (Tab. 18.2). The activity of propachlor might be too low in order to be detected in this assay. Interference with rRNA synthesis evidently is a common basis of the antifungal activity of phenylamides and resistance appears to be due to a change at the target site. Results obtained with wild-type and metalaxyl-resistant strains of *P. infestans* in similar studies support this idea. In this species the endogenous RNA-po'ymerase activity of isolated nuclei from resistant strains was also significantly less sensitive to inhibition by the phenylamides than that of wild-type strains (DAVIDSE *et al.* unpublished).

As is evident from Tab. 18.1 the resistance level of the resistant strain is distinct for various phenylamides. Propachlor, benalaxyl and metolachlor display the highest inhibitory effect on growth of the resistant strain. It might either be due to a residual specific effect on RNA synthesis of the resistant strain or to a second mechanism of action, the effect of which becomes evident at higher concentrations. The herbicides and in particular benalaxyl also inhibit the uptake of uridine by both the sensitive and the resistant strain as a consequence of which uridine incorporation is also affected. Metalaxyl and oxadixyl do not inhibit uridine uptake; cyprofuram shows some effect but it is certainly less pronounced than that of benalaxyl and the two phenyl-

Table 18.1 Effects of phenylamides on radial growth on agar, uptake of uridine and incorporation of uridine into RNA of a wild-type (S) and a metalaxyl-resistant (R) strain of *Phytopthora megasperma* f. sp. *medicaginis*

Phenylamide	EC_{50} (μM) Radial growth[2]		Resistance factor	EC_{50} (μM) uridine uptake[3]		EC_{50} (μM) uridine incorporation[3]	
	S	R	(EC_{50} R)/ (EC_{50} S)	S	R	S	R
metalaxyl	0.03	1,900	63,300	>1,000	>1,000	0.06	>1,000
benalaxyl	0.25	250	1,000	140	100	0.06	60
cyprofuram	10	900	90	1,000	>1,000	1.6	70
oxadixyl	1.2	1,800	1,500	>1,000	>1,000	1.1	>1,000
metolachlor	13	450	35	1,000	700	5.6	220
propachlor	75	200	3	420	420	130	130

[1] Concentration of phenylamide at which the process indicated is inhibited for 50 %.

[2] Determined by measuring colony diameter on a synthetic agar medium (Erwin and Katznelson 1961), amended with the different phenylamides at various concentrations.

[3] 10 ml liquid cultures containing 20—30 mg dry weight of mycelium in a synthetic liquid medium (Erwin and Katznelson 1961) were pulse labelled for 15 min with 0.5 μCi [^{3}H] uridine with or without (control cultures) preincubation with the different phenylamides at various concentrations for 45 min. Uptake of [^{3}H] uridine and its incorporation into RNA were determined using standard procedures (Davidse *et al.* 1983b).

Table 18.2 Effects of phenylamides at concentrations of 10 μg/ml on endogenous RNA polymerase activity of nuclei isolated from a wild-type (S) and a metalaxyl-resistant (R) strain of *Phytophthora megasperma* f. sp. *medicaginis*

Phenylamides	Activities as a percentage of control[1]	
	S	R
metalaxyl	58	91
benalaxyl	51	97
cyprofuram	70	96
oxadixyl	63	92
metolachlor	72	92
propachlor	93	91

[1] Isolated nuclei were incubated in a reaction mixture containing in a total of 100 μl: 50 mM Tris-HCl (pH 7.9), 1 mM dithiothreitol, 1 mM ATP, GTP and CTP, 0.01 mM UTP, 2 μCi [5,6-^{3}H] UTP, 4 mM $MnCl_2$, 50 mM ammonium sulphate, 10 μl of nuclear suspensions and the different phenylamides or solvent (control). Reaction mixtures were incubated at 25 °C and after 30 min acid-precipitable radioactivity was determined using standard procedures (Davidse *et al.* 1983b).

amide herbicides. Apparently the latter three compounds have in addition to their effect on RNA synthesis a second mechanism of action that contributes to the inhibition of growth of the sensitive strain at higher phenylamide concentrations and that is solely responsible for growth inhibition of the resistant strain. Since cyprofuram is much less inhibitory to uridine uptake than benalaxyl but inhibits incorporation of the resistant strain to the same extent, this compound still specifically interferes with RNA synthesis. Apparently the change at the target site of the phenylamides that has led to almost complete resistance to metalaxyl does not completely prevent the interaction of cyprofuram with RNA synthesis at this site.

Similar results were obtained with a wild-type and a metalaxyl-resistant strain of *P. infestans* (Davidse *et al.* unpublished results). Benalaxyl inhibited uridine uptake of both *P. infestans* strains and cyprofuram displayed significant activity on uridine incorporation of the resistant strain, whereas it did not inhibit uridine uptake. Uridine uptake by *P. infestans* was also not affected by metalaxyl and oxadixyl. Both compounds were also inactive at high concentrations on uridine incorporation of the resistant strain but were highly active on this process in the sensitive strain.

Inhibition of uridine uptake as displayed by benalaxyl and the phenylamide herbicides probably will not lead to a reduction of fungal growth because uridine is synthesized by the fungus itself. Inhibition of this process evidently is just one feature of the second mechanism of action of these phenylamides and indicates a general disturbance of metabolism as a consequence of which growth is affected. Suggestions about the nature of the second inhibitory mechanism of action can only be speculative. Interference with membrane functioning may be involved. Metolachlor and to a lesser degree cyprofuram cause lysis of protoplasts of *P. megasperma* f. sp. *medicaginis* (Fisher and Hayes 1985), whereas metalaxyl is almost inactive. Benalaxyl lyses zoospores of *Plasmopara viticola* at concentrations as low as 10 μg/ml (Gozzo *et al.* 1984). These observations and the effect on uridine uptake support the idea that interference with membrane functioning is a second mechanism of action of propachlor, metolachlor and benalaxyl.

The second mechanism of action of benalaxyl and the activity of cyprofuram to interfere specifically with RNA synthesis of the resistant strain could theoretically lead to a better performance of these compounds in disease control as compared with the other phenylamides, when phenylamide resistant strains are involved. Benalaxyl and cyprofuram, however, could not prevent damping-off of lucerne seedlings caused by a metalaxyl-resistant strain of *P. megasperma* f. sp. *medicaginis* in a laboratory test, even at concentrations that were almost phytotoxic (Davidse unpublished results). In greenhouse experiments both fungicides were unable to control disease development incited by metalaxyl-resistant strains of *P. infestans* and *Pseudoperonospora cubensis* on potato and cucumber plants, respectively (Cohen and Samoucha 1984; Katan 1982). In a detached-leaf assay and a leaf disc assay, however, cyprofuram inhibited sporulation of resistant strains of *P. infestans* at lower concentrations than metalaxyl but still at levels at least 10-fold higher than required for inhibition of sporulation of sensitive strains. Similarly the level of phenylamide resistance of *Plasmopara viticola* to cyprofuram was lower than that to metalaxyl (Leroux and Clerjeau 1985).

Field observations on late blight development in experimental plots treated with cyprofuram and benalaxyl, however, did not reveal any disease-controlling effect of these compounds when phenylamide-resistant strains were present at the experimental site. These data indicate that the second mechanism of action of benalaxyl and the ability of cyprofuram to inhibit RNA polymerase activity of phenylamide-resistant strains are only of limited value, if any, under practical conditions.

As yet it is unknown which structural features of the phenylamides are required for maximal expression of the second mechanism of antifungal activity. The low phytotoxicity of benalaxyl indicates that this mechanism is not related to the primary mechanism of action of the phenylamides in plants, as has been suggested previously (Davidse 1984). It would be interesting to know if antifungal activity based on the second mechanism of action could be optimized further. Antifungal activity, however, would have to be increased 100—1,000-fold before a compound would be a promising candidate fungicide.

Knowledge of how the phenylamide target site actually has been changed in phenylamide resistant strains would be valuable in interpreting the interaction of cyprofuram with the altered site. It may help in redesigning phenylamide structures or even in the design of completely new structures that show increased affinity to the altered site. In view of this but also from a more fundamental point of view the site of interaction of the phenylamides in the synthesis of rRNA should be identified further.

Concluding remarks

Studies on the mode of action of the phenylamides identified these compounds as highly specific inhibitors of rRNA synthesis. This property makes the phenylamides excellent tools to study details of this complex process. Research along this line will lead to characterization of the component of the RNA polymerase-template complex that interacts with the phenylamides and its role in rRNA synthesis. When this component, presumably a protein, is characterized isolation of its structural gene would be feasible. Once this has been accomplished an extensive research area will be opened. A search for similar components in other fungi that are naturally resistant to the phenylamides would be possible, indicating how rRNA synthesis in fungi belonging to different taxonomic classes has evolved. From a practical point of view it may even lead to new fungicides that are specifically designed to interact with these components.

From a fundamental point of view phenylamide resistance that threatens the usefulness of phenylamides as commercial fungicides has also its positive aspects. Phenylamide resistance would be a valuable marker in studying the genetics of Peronosporales. The availability of a gene, carrying a mutation governing phenylamide resistance, would provide the molecular geneticist with a selectable marker that could be incorporated in yet to be constructed transformation vectors of fungi belonging to the Peronosporales. In future research on host-pathogen relations the latter approach will be instrumental in the identification of fungal genes specifying pathogenicity and/or avirulence. It is difficult to foresee how research in the areas mentioned above will evolve. The perception, however, that mode of action studies ultimately will contribute to the development of better disease control measures either chemically or using the plants own defence mechanisms, is highly motivating to continue research on the molecular aspects of fungicide action and resistance.

References

Arp, U., and Buchenauer, H.: Untersuchungen zum Wirkungsmechanismus von RE 20615 und metalaxyl in *Phytophthora cactorum* und zur Resistenzentwicklung des Pilzes gegenüber diesen Fungiziden. Mitt. Biol. Bundesanst. (1981): 236—237.

Bruck, R. I., Fry, W. E., and Apple, A. E.: Effect of metalaxyl, an acylalanine fungicide on developmental stages of *Phytophthora infestans*. Phytopathology **70** (1980): 597—601.

Cohen, Y., and Samoucha, Y.: Cross resistance to four systemic fungicides in metalaxyl-resistant strans of *Phytophthora infestans* and *Pseudoperonospora cubensis*. Plant Disease **68** (1984): 137—139.

Davidse, L. C.: Resistance to acylalanine fungicides in *Phytophthora megasperma* f. sp. *medicaginis*. Netherl. J. Plant Pathol. **87** (1981b): 11—24.

— Antifungal activity of acylalanine fungicides and related chloroacetanilide herbicides. In: Trinci, A. P. J., and Ryley, J. F. (Eds.): Mode of Action of Antifungal Agents. Brit. Mycolog. Soc. Symp. Ser. 8, Cambridge Univ. Press, Cambridge 1984, pp. 239—255.

— Danial, D. L., and van Westen, C. J.: Resistance to metalaxyl in *Phytophthora infestans* in The Netherlands. Netherl. J. Plant Pathol. **89** (1983a): 1—20.

DAVIDSE, L. C., GERRITSMA, O. C. M., and HOFMAN, A. E.: Mode d'action du metalaxyl. Phytiatrie-Phytopharmacie **30** (1981 a): 235—244.
— — and VELTHUIS, G. C. M.: A differential basis of antifungal activity of acylalanine fungicides and structurally related chloroacetanilide herbicides in *Phytophthora megasperma* f. sp. *medicaginis*. Pesticide Biochem. Physiol. **21** (1984): 301—308.
— HOFMAN, A. E., and VELTHUIS, G. C. M.: Specific interference of metalaxyl with endogenous RNA polymerase activity in isolated nuclei from *Phytophthora megasperma* f. sp. *medicaginis*. Experiment. Mycology **7** (1983 b): 344—361.
— LOOYEN, D., TURKENSTEEN, L. J., and VAN DER WAL, D.: Occurrence of metalaxyl-resistant strains of *Phytophthora infestans* in Dutch potato fields. Netherl. J. Plant Pathol. **87** (1981 b): 65—68.
ERWIN, D. C., and KATZNELSON, H.: Studies on the nutrition of *Phytophthora cryptogea*. Canad. J. Microbiol. **7** (1961): 15—25.
FISHER, D. J., and HAYES, A. L.: Mode of action of the systemic fungicides furalaxyl, metalaxyl and ofurace. Pesticide Sci. **13** (1982): 330—339.
— — A comparison of the biochemical and physiological effects of the systemic fungicide cyprofuram with those of the related compounds metalaxyl and metolachlor. Crop Protect. **4** (1985): 501—510.
FULLER, M. S., and GISI, U.: Comparative studies of the *in vitro* activity of the fungicides oxadixyl and metalaxyl. Mycologia **77** (1985): 424—432.
GOZZO, F., GARAVAGLIA, C., and ZAGNI, A.: Structure-activity relationships and mode of action of acylalanines and related structures. Proc. of the 1984 Brit. Crop Protect. Conf. Pest and Diseases (1984): 923—928.
GROHMANN, U., and HOFFMANN, G. M.: Licht- und elektronenoptische Untersuchungen zur Wirkung von metalaxyl bei *Pythium*- und *Phytophthora*-Arten. Z. Pflanzenkrankh. u. Pflanzenschutz **89** (1982): 435—446.
HICKEY, E. L., and COFFEY, M. D.: The effects of Ridomil on *Peronospora pisi* parasitising *Pisum sativum*: an ultrastructural investigation. Physiologic. Plant Pathol. **17** (1980): 199—204.
HUBELE, A., KUNZ, W., ECKHARDT, W., and STURM, E.: The fungicidal activity of acylalanines. In: DOYLE, P., and FUJITA, T. (Eds.): Pesticide Chemistry, Human Welfare and the Environment, Vol. 1, Synthesis and Structure-activity Relationships. Pergamon Press, Oxford 1983, pp. 233—242.
KATAN, T.: Cross-resistance of metalaxyl-resistant *Pseudoperonospora cubensis* to other acylalanine fungicides. Canad. J. Plant Pathol. **4** (1982): 387—388.
KERKENAAR, A.: On the antifungal mode of action of metalaxyl, an inhibitor of nucleic acid synthesis in *Pythium splendens*. Pesticide Biochem. Physiol. **16** (1981): 1—13.
LEROUX, P., and CLERJEAU, M.: Resistance of *Botrytis cinerea* (Pers.) and *Plasmopara viticola* (Berk. & Curt.) Berl. and de Toni to fungicides in French vineyards. Crop Protect. **4** (1985): 137—160.
MOSER, H., RIKS, G., and SANTER, H.: Der Einfluß von Atropisomerie und chiralem Zentrum auf die biologische Aktivität des metolachor. Z. f. Naturforsch. **87b** (1982): 451—462.
MÜLLER, H. M., and LYR, H.: Morphologische und zytologische Veränderungen bei *Phytophthora infestans* (Mont.) de Bary und *Phytophthora cactorum* (Leb. et Cohn) Schroet. unter dem Einfluß von metalaxyl. In: LYR, H., and POLTER, C. (Eds.): Systemische Fungizide und Antifungale Verbindungen 1982 IN. Akademie-Verlag, Berlin 1983, pp. 403—410.
STAUB, T. H., DAHMEN, H., and SCHWINN, F. J.: Effects of Ridomil on the development of *Plasmopara viticola* and *Phytophthora infestans* on their host plants. Z. Pflanzenkrankh. u. Pflanzenschutz **87** (1980): 83—91.
— and YOUNG, T. R.: Fungitoxicity of metalaxyl against *Phytophthora parasitica* var. *nicotianae*. Phytopathology **70** (1980): 797—801.
SZANISZLO, P. J., KANG, M. S., and CABIB, E.: Stimulation of $\beta(1 \rightarrow 3)$ glucan synthetase of various fungi by nucleoside triphosphates: generalized regulatory mechanism for cell wall biosynthesis. J. Bacteriol. **161** (1985): 1188—1194.
WOLLGIEHN, R., BRÄUTIGAM, E., SHEUHMANN, B., and ERGE, D.: Wirkung von Metalaxyl auf die Synthese von RNA, DNA und Protein in *Phytophthora nicotiana*. Z. Allgemeine Mikrobiol. **24** (1984): 269—279.

LYR, H. (Ed.): Modern Selective Fungicides — Properties, Applications, Mechanisms of Action. Longman Group UK Ltd., London, and VEB Gustav Fischer Verlag, Jena, 1987.

Chapter 19

2-Aminopyrimidine fungicides

D. W. Hollomon*) and H.-H. Schmidt**)

*) Crop Protection Division, Long Ashton Research Station, Long Ashton, University of Bristol, UK

**) Institute of Plant Protection Research Kleinmachnow of the Academy of Agricultural Sciences of the GDR

Introduction

Many important crop plants are attacked by powdery mildews, and potential markets for fungicides which control these diseases are large. Because they are easily handled in the greenhouse, mildews are invariably included by the agrochemical industry in fungicide screening programmes. Some 20 years ago, routine screening at Jealott's Hill Research Station, identified ceitain phosphorylated analogues of the insecticide diazinon (Fig. 19.1a) as having protectant activity against powdery mildews. Additional structure-activity studies revealed a series of 2-amino-4-hydroxy pyrimidines (Snell *et al.* 1966) which were not only specific against powdery mildews, but also systemic. Dimethirimol (Fig. 19.1b) was the first hydroxypyrimidine to be discovered and developed (Elias *et al.* 1968; Geoghegan and de Graaff 1969). A second related fungicide ethirimol (Fig. 19.1c) soon followed (Bebbington *et al.* 1969) because of its better activity, especially against barley powdery mildew. Advances in formulation and seed treatment technology enabled ethirimol to be marketed as a seed treatment, with considerable agronomic benefits to farmers. Its widespread use helped confirm that fungicides could be used economically to control cereal leaf diseases, and that

Table 19.1 Some chemical, physical an toxicological properties of ethirimol, dimethirimol and bupirimate (derived from Worthing, C. R., and Walker, S. B., 1983)

	Ethirimol	Dimethirimol	Bupirimate
code name	PP 149	PP 675	PP 588
trade marks	Milgo, Milgo E, Milstem, Milcurb Super	Milcurb	Nimrod
melting point	159—160 °C (with a phase change at about 140 °C)	102 °C	50—51 °C
vapour pressure	267 μPa at 25 °C	1.46 mPa at 30 °C	67 μPa at 20 °C
solubility at room temperature	200 mg/l water, sparingly soluble in acetone, slightly soluble in diaceton, alcohol, chloroform, trichlorethylene and aqueous solutions of strong acids and bases	1.2 g/l water 45 g/l acetone 1.2 kg/l chloroform 65 g/l ethanol 360 g/l xylene	22 mg/l water soluble in most organic solvents except paraffin hydrocarbons
acute oral LD_{50} for rats	6 340 mg/kg (female rat)	2 350 mg/kg	approx. 4 000 (female rat)
no effect level in 2-year feeding tests	200 mg/kg diet	300 mg/kg diet	100 mg/kg diet

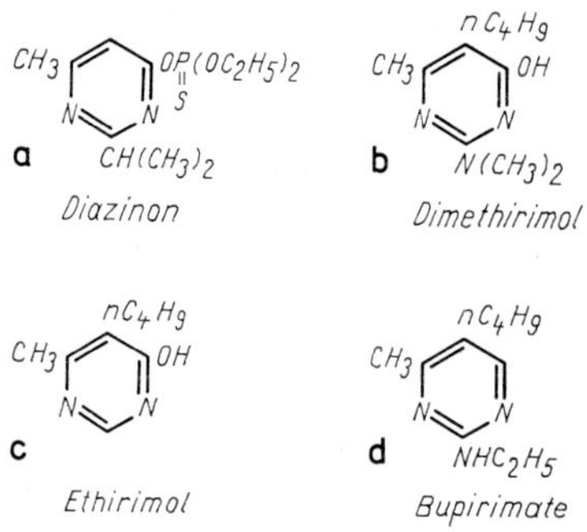

Fig. 19.1 Chemical structures of pyrimidine insecticides and fungicides. a) Diazinon, b) Dimethirimol, c) Ethirimol, d) Bupirimate. (corrections: add —NHC_2H_5 group in Ethirimol in 2-position, add —$SO_2N(CH_3)_2$ group in 4-position in Bupirimate)

chemicals offered a viable alternative to breeding for disease resistance. Dimethirimo and ethirimol were effective largely against powdery mildews of herbaceous plants; bupirimate (Fig. 19.1d) was developed later to control these diseases on woody plants and ornamentals (FINNEY *et al.* 1975).

Practical application

Ethirimol

Although used primarily to control powdery mildews of cereals *(Erysiphe graminis)*, ethirimol has also shown good effects against cucumber mildew (*E. cichoracearum* and *S. fuliginea*, BENT 1970; HARTMAN and SIEGEL 1973). Ethirimol also reduced damage caused by *E. betae*, and increased sugar content of beet, but its performance was inferior to that of sulphur and benomyl (BYFORD 1978). Yield increases in cereals following either protective or eradicative treatments to control mildew, have been obtained in many European countries (Tab. 19.2). Differences between trials may reflect variation in disease levels, and so responses frequently depend on the cultivars used, and their susceptibility to mildew (CLIFFORD *et al.* 1971; KRÜGER 1969; PETKOVA 1979). WOLFE (1969) obtained greatest control with spring sown barley and wheat; autumn sown crops were less responsive. Indeed, yield increases in wheat are often difficult to achieve (BAUERS 1972; HANSEN 1978), and may require two applications during each season, preferably as mixtures with broader spectrum fungicides.

Yield increases reflect either increased numbers of ears, numbers of grain per ear, or increases in grain size (D'ARBIGNY and DAWSON 1973; BAUERS 1972; JENKYN 1974, 1978), and effects on these components are influenced by application time (BETHUNE 1973; CARVER and GRIFFITHS 1981). Slight yield reductions on some mildew resistant spring barley cultivars were recorded by CHANNON and BOYD (1973) and GILMOUR (1971) after ethirimol seed dressing at above normal recommended rates. Significant reductions in root growth associated with a decrease in chromosome volume in the spring barley cultivar Julia, were observed after treatment with ethirimol (16 ppm) in a culture solution (BENNETT 1971). There was also a small reduction in chiasma frequency of pollen mother cells of cultivars Julia and Sultan, but yield losses due to infertility appeared unlikely. However, deleterious effects of ethirimol on plants were not demonstrated if recommended dose rates were used.

Although primarily a prophylactic measure against early infection, seed treatment has proved especially economical against barley powdery mildew (BEBBINGTON *et al.*

Table 19.2 Some examples of yield responses after application of ethirimol to control powdery mildew on barley and winter wheat

Country	Years	Number of tests	Dose rate (ai) seed dressing (g/100 kg seed)	Dose rate (ai) spray (g/ha)	Yield response (%)	Remarks	Reference
Spring barley							
Austria	1973	many	360	—	18 (10—24)		1
	1973	many	420	—	18 (10—31)		1
Belgium	1970—1974		—	280 (1 application)	10—20		2
Czecho-	1971	many	438	—	9		3
slowakia	1971		—	700	8		3
Denmark	1969—1976	72	240	480	7		4
FRG	1969—1971	15	600	—	111		5
	1971	6	360	—	6	mean results with 10 cultivars	6
GDR	1976	5	—	280	6 (4—8)		7
UK	1969—1970		1,200	—	4 (3—6)	Mildew susceptible cultivar 'Golden Promise'	8
	1971		1,200		26	Severe mildew attack in 1971 only	8
	1971	5	—	900	16		9
	1969—1971	28	573	—	8 (3—9)	Response dependent on cultivar used with yield losses on some cultivars	10
	1968—1974		400	—	12 (10—17)		11
	1970—1975		400	—	13 (7—30)	Yield loss recorded in 1975	12
Winter barley							
UK	1969—1971	18	700	—	6		13
Winter wheat							
Czecho-slowakia	1970	1	—	1,600 (2 applications)	26	Mildew susceptible cultivar 'Consul'	14
GDR	1975—1981	29	—	280 (2 applications)	3	Only 3 tests showed significant yield increases	15
Poland	1971—1975	223	—	750 (2 applications)	10		16

1 = Zwatz 1974; 2 = Meeus *et al.*, 1975; 3 = Benada 1972; 4 = Hansen 1978; 5 = Bauers 1972; 6 = Baumer and Ulonska 1971; 7 = Official registration tests; 8 = Channon and Boyd 1973; 9 = Gilmour 1971; 10 = Jenkyn and Moffat 1975; 11 = Shephard *et al.*, 1975; 12 = Wolfe 1975; 13 = Hall 1971; 14 = Benada 1971; 15 = Müller 1984; 16 = Jaczewska 1983.

1971; Hall 1971; Küthe 1972). Combinations with organic mercurials (Koch and Stark 1972), insecticides (Kunovski *et al.* 1973), and triazole fungicides (Northwood *et al.* 1984) have also been developed as seed treatments. Soil moisture levels must be adequate (Collier *et al.* 1979), otherwise insufficient ethirimol may be taken up by barley plants, so that seed dressings are not always effective under dry weather conditions (King 1977).

Spray applications enable farmers to respond with more flexibility to mildew attacks in different cereal species. Treatment of barley in the early phases of a mildew epidemic (3—5 % mildew infection on the first three leaves) are recommended (Carver and Griffiths 1981; Neuhaus and Reich 1975), whereas in wheat flag leaves and ears need to be protected. Müller (1984), who examined the efficacy of "Milgo E" in winter wheat, found good curative action against mildew when it was applied within the latent period of infection (up to 5 days after inoculation of cultivar Alcedo). Milgo E also showed stability against rainfall and in the GDR it has proved suitable for aerial application (spray rate: 50 litres/ha).

To enlarge the spectrum of activity tank mixtures of ethirimol with other fungicides may be used, and in France mixtures with captafol gave additional control of rust fungi, *Rhynchosporium* (Lescar 1977), and *Septoria* (Millou and D'Arbigny 1974).

Dimethirimol

Dimethirimol is limited mainly to treatment against *E. cichoracearum* and *S. fuliginea* on cucurbits. There are also records of positive results against ornamentals like cineraria, sweet pea (Geoghegan 1969) and chrysanthemum (Kobachidze and Toskina 1971). Dimethirimol can be applied either as a soil drench to the stem base of plants (e.g. 0.25 g a.i. per cucumber plant) or as a spray treatment. In glasshouses soil drenching is preferred. As with ethirimol uptake by plant roots is positively correlated with soil moisture, and dimethirimol is loosely adsorbed onto soil particles and released slowly (Geoghegan 1969). Its availability to plant roots is best in light alkaline soils.

Soil drenches have kept cucumber plants free from mildew for more than 6 weeks (Brooks 1970; Kobachidze and Toskina 1971). In some crops, especially irrigated melons (Brooks 1970), a granule formulation was very effective. Dimethirimol may also have some curative action against *E. pisi*, the cause of powdery mildew of peas (Gorska-Poczopko 1971).

Bupirimate

Bupirimate, the sulphamate ester of ethirimol, moves in woody plants (Shephard 1981; Teal and Cavell 1975), and its spectrum of activity includes powdery mildews of various fruit trees (Tab. 19.3). The effective spray rate depends on plant, mildew species and method of application, but ranges mostly between 50 to 500 ppm bupirimate. Finney *et al.* (1975) reported that bupirimate, at a concentration of 75 to 150 mg/l suppressed sporulation of *Podosphaera leucotricha* more effectively than binapacryl (500 mg/l), dinocap (250—300 mg/l), sulphur (3,200 mg/l) or triforine (250 mg per l), but showed activity similar to thiophanate-methyl (500 mg/l), benomyl (250 mg/l) and ditalimfos (375 mg/l). In potted apple plants under glass, both Hunter *et al.* (1981) and Jahn *et al.* (1986) found that bupirimate had significant curative effects against *P. leucotricha*, if applied within 4 days after inoculation. Applications 12 days before inoculation still gave 90 per cent disease control (Jahn *et al.* 1986). One reason for the remarkable protective and eradicative efficacy of bupirimate may

Table 19.3 Spectrum of activity of bupirimate

host plants	mildew species	good efficacy* reported by
woody plants		
apple	*Podosphaera leucotricha*	1, 2, 3
apricot	*Podosphaera tridactyla*	1
mango	*Oidium mangifera*	1
vine	*Uncinula necator*	1
peach	*Sphaerotheca pannosa*	1
rose	*Sphaerotheca pannosa*	1, 2
currants	*Sphaerotheca mors-uvae*	1
gooseberry	*Sphaerotheca mors-uvae*	4
herbaceous plants		
hop	*Sphaerotheca humuli*	2
strawberry	*Sphaerotheca humuli*	1
cucurbits	*Sphaerotheca fuliginea*	1, 2 (only cucumber)
	Erysiphe cichoracearum	1
comfreys	*Erysiphe cichoracearum*	2
peas	*Erysiphe pisi*	1
sugar beet	*Erysiphe betae*	1
chrysanthemum	*Oidium chrysanthemum*	1, 2
gerbera	*Oidium* sp.	2
pepper	*Leveillula taurica*	1

*1 = Finney 1975
1 = Finney *et al.* 1975
2 = registrations tests in the GDR 1978—1981 (unpublished data)
3 = Kundert 1977
4 = O'Riordain and Kennedy 1981

be its good vapour phase activity (Finney *et al.* 1975; Petsikol-Panyotarou 1980; Teal and Cavell 1975). Simulated rainfall (30 mm) did not alter the effectiveness of the EC-formulation of bupirimate (Nimrod 25 EC) against *P. leucotricha* on potted apple plants (Rathke and Jahn 1984), and the fungicide remained equally effective over the temperature range from 15 to 40 °C (Rathke 1984).

Phytotoxicity caused by bupirimate is not normally a serious problem if recommended application rates are used (Kundert 1977). However, Mantinger (cited by Kundert) recorded some detrimental effects (chlorosis, violet colour change and abscission of leaves) with the cultivars 'Morgenduft' and 'Jonathan', but this did not seem to affect yield. Investigating different cultivar × fungicide interactions (including bupirimate) with apples Jeger *et al.* (1983) found that spraying (irrespective of fungicide) was most effective on cultivars with a high level of mildew susceptibility (e.g. 'Golden Delicious'). Contrary to experience with ethirimol and dimethirimol, bupirimate did not effectively control foliar mildew infections when applied to young apple and cucumber plants as a soil drench (Finney *et al.* 1975).

The most effective way to apply bupirimate in orchards is by high volume spray (Cooke *et al.* 1977). In the GDR an application of 375 ml bupirimate/ha by helicopter (Spray volume 100 l/ha) only protects slight or moderately susceptible apple cultivars, when disease pressure is low. Surfactants and stickers may improve the fungicidal effectiveness of bupirimate (Shabi 1975). Equally, the effects of surfactants against overwintering mildew in apple buds were improved by addition of bupirimate (Bent *et al.* 1977). But it was necessary to reduce the amount of the surfactant PP 222 (a nonylphenolethoxylate) in this mixture, to avoid phytotoxicity. Bupirimate has a low toxicity against the predacious phytoselid mite *Typhlodromus pyri*, insects (EASTERBROOK *et al.* 1979), and fungal pathogens of aphids (HALL 1981). This, and its specificity as a powerful antisporulant, make bupirimate suitable for integrated pest management programmes in both orchard and glasshouse crops (BUTT *et al.* 1983).

Side-effects

2-aminopyrimidines generally have little effect on the soil fauna, including earthworms. Ethirimol seed dressings influenced neither the phylloplane microflora of barley leaves (DICKINSON 1973) nor *Sporobolomyces* spp. and *Cladosporium* spp. on flag leaves of winter wheat (JENKYN and PREW 1973a, b). However, other pathogenic fungi may colonize healthy leaves protected by treatment with mildew specific fungicides, and this may limit yield responses (JENKYN and MOFFATT 1975). Barley cultivars susceptible to brown rust were more severely infected by *P. hordei* after treatments with ethirimol, than were untreated mildewed plants (LITTLE and DOODSON 1971). In some instances, spray treatments seemed especially to favour rust development (JENKYN 1974), but not in others (BETHUNE 1973). ROUND and WHEELER (1978) examined the competition for space on the leaf surface of barley seedlings (cultivar 'Zephyr') and showed that an ethirimol soil drench applied at the time of rust inoculation, but six days after inoculation with *E. graminis* increased significantly both the numbers of rust pustules and their size.

In one of three trials carried out by KAMPE (1974), infection of spring barley with *R. secalis* following ethirimol seed treatment (350 g/dt) was twice that observed in untreated plots, and no yield responses were recorded. Ethirimol seed treatments also increased the severity of leaf stripe *(Helminthosporium sativum)* of winter barley (cultivar 'Dura') following artificial inoculation in the greenhouse (SAUR and SCHÖNBECK 1975). On agar, ethirimol (1 and 10 ppm) stimulated growth of *H. sativum*, and on barley leaves stimulated germination of conidia, penetration and colonization. Sporulation on treated plants occurred five days earlier than on untreated ones. Helminthosporal, a toxin of *H. sativum*, inhibited root growth, reduced IAA content, but increased activity of IAA oxidase, peroxidase, and ion efflux in ethirimol treated barley plants (SAUR 1976a, b; SAUR and SCHÖNBECK 1976). These changes are all likely to influence disease development, and it seems that there is some interaction with the host, perhaps involving damage to cell membranes.

In vitro effects

In vitro 2-aminopyrimidines inhibited spore germination, formation of appressoria and haustoria and, to a lesser extent, hyphal growth (BENT 1970; HOLLOMON 1979a). However, the concentration needed to affect germination were well above those normally found in plants (CALDERBANK 1971). Sporulation may also be reduced. Dimethirimol (50 ppm) failed to inhibit germination of conidia of *Botrytis fabae*, *Venturia inaequalis*, uredospores of *P. recondita*, but marginally (26%) inhibited germination of *Phytophthora infestans* zoosporangia. 2-aminopyrimidines also had little effect on the growth of *B. cinerea* (GRINDLE 1981). Susceptibility of different mildew species to 2-aminopyrimidines varied considerably, with the minimum inhibitory concentration ranging from 0.005 ppm to 10 ppm for *E. communis* on *Brassica juncea* (GORTER and NEL 1974). A similar range of variation was encountered in laboratory tests against *E. graminis* f. sp. *hordei*, reflecting a broad spectrum of genetic variability in response to ethirimol (HOLLOMON 1981). The intrinsic toxicity of ethirimol compared very favourably with eleven other mildew fungicides (KLUGE and LYR 1983). Unlike ethirimol, bupirimate has excellent vapour phase activity, and a deposit of 25 μg inhibited growth of *S. fuliginea* some 5 mm away. 20 mm away inhibiton of growth was about 50 per cent (PETSIKOL-PANAOTAROU 1980).

Uptake, metabolism and degradation

Availability to plants of 2-aminopyrimidines in soils depends very much on the organic matter present, and acidity. Once adsorbed onto soil particles these fungicides are released slowly into soil moisture so providing a continuous supply of fungicide for uptake by the crop. But the strong adsorption of ethirimol to acid peat prevents its use in these soils (COLLIER *et al.* 1979; GRAHAM-BRYCE and COUTTS 1971). Within herbaceous plants 2-aminopyrimidines move in the transpiration stream, accumulating around leaf margins. Some may even be exuded on the leaf surface. In woody plants 2-aminopyrimidines are much less mobile (SHEPHARD 1973). Bupirimate remains within veins when applied as a root drench (TEAL and CAVELL 1975), but is sufficiently volatile to be redistributed as vapour across leaf surfaces (FINNEY *et al.* 1975).

Degradation of 2-aminopyrimidines in both cucumber and barley leaves is rapid, with a half-life of no more than 4 days (CALDERBANK 1971). Polar metabolites are generated principally through N-dealkylation, conjugation to form glucosides, and hydroxylation of the 5 *n* butyl group (CAVELL *et al.* 1971). Some of these metabolites are also active against powdery mildews. Bupirimate is readily hydrolysed to ethirimol in acid solution on the leaf surface.

Isolated haustoria of *E. pisi* accumulated ethirimol to at least 60-fold the external concentration, which was in excess of its aqueous solubility (MANNERS and GAY 1980). Lipophilicity may play some role in this entry, but accumulation was similar regardless of whether ethirimol was protonated or not. It is not clear how relevant this uptake into haustoria is, since accumulation of a series 2-aminopyrimidines was not related to their fungitoxicity (HOLLOMON 1984). Furthermore, the main effect on powdery mildew fungi is seen at appressoria formation, which occurs before haustoria form.

2-aminopyrimidines readily polymerize on exposure to u/v light of concentrated solutions in the laboratory (WELLS *et al.* 1979). Similar polymers have not been detected after exposure to sunlight on leaf surfaces, although other photochemical degradation products of ethirimol have been indentified (CAVELL 1979).

Mechanism of action

Powdery mildews enter their hosts through the cuticle. This requires formation of an appressorium, and this step in development is inhibited by 2-aminopyrimidine fungicides. The critical biochemical events occur well before the appressoria are first seen, for only when applied within the first eight hours after inoculation did ethirimol prevent appressoria formation (HOLLOMON 1977). Exchange of cations and fluorescent dye between *E. graminis* and its host occurred soon after germination began (KUNOH *et al.* 1982), so fungicides may enter germinating conidia at an early stage. Other steps in mildew development may be inhibited, although this may be of limited practical significance.

Some indication of the likely mode of action of 2-aminopyrimidines followed attempts to reduce their toxicity with various metabolites (see HOLLOMON 1984 for refs.). Several, powdery mildews were examined, and the purine base adenine, and its ribonucleoside adenosine, were always good reversal agents. Kinetin (6-furfuryl adenine) and isopentenyl adenine both prevented appressoria formation in barley, and mildew strains resistant to ethirimol were cross-resistant to these purines. 2-aminopyrimidines seemed, therefore, to interfere with purine metabolism, although after the purine ring formed since *E. graminis* appears to be a purine auxotroph, obtaining purines it needs from its host (HOLLOMON 1979a).

Further studies showed that germinating *E. graminis* conidia incorporated adenine and adenosine into nucleic acids, but in the presence of ethirimol formation of inosine and adenosine nucleotides was inhibited and nucleic acid synthesis halted (HOLLOMON and CHAMBERLAIN 1981). Several enzymes involved in the reutilization of purines were examined in cell-free extracts from *E. graminis* conidia, but of these only adenosine deaminase (ADA-ase) was inhibited to any significant extent by ethirimol. ADA-ase catalyses the largely irreversible hydrolytic deamination of adenosine to inosine. Plants apparently lack ADA-ase, and instead deaminate 5′AMP at the nucleotide rather than nucleoside level. ADA-ase is present in many fungi, but only the enzyme from powdery mildew fungi was sensitive to ethirimol (HOLLOMON 1979b). This may account for the extreme specificity of 2-aminopyrimidines towards powdery mildews. Structure activity studies provided further evidence that ADA-ase was a site of action of these pyrimidine fungicides, for analogues that were poor inhibitors of this enzyme, were also poor fungicides. Exceptions occurred, but these analogues may have been activated by conversion to ethirimol within barley during bioassay (HOLLOMON and CHAMBERLAIN 1981).

Inhibition of ADA-ase by 2-aminopyrimidine fungicides was pH dependent and non-competitive. Indeed, ethirimol hardly resembles adenosine in shape, and it does not bind tightly to ADA-ase. The 5 *n* butyl group not only provides adequate lipophilicity for membrane permeability, but is essential for correct binding to the enzyme. Loss of both fungicide activity and ADA-ase inhibition follows hydroxylation of the butyl group, suggesting that it is buried within a hydrophobic region of ADA-ase, with the 2-substituted amino group towards the enzyme surface (HOLLOMON and CHAMBERLAIN 1981). In both mildew infected plants and *E. graminis* conidia, ADA-ase exists as a single, large molecular weight (MW 300,000) protein, with kinetic properties similar to those of ADA-ase from other organisms. In many of these organisms the catalytic subunit (MW 36,000) of ADA-ase is associated with a large accessory protein whose function is unknown (KELLEY *et al.* 1977). It is not known if mildew ADA-ase has a similar structure. Why inhibition of ADA-ase should have such drastic consequences on events leading to appressoria formation is not clear. *E. graminis* does not synthesise purines *de novo* so, apart from some limited reutilization of the purines already in conidia, infection requires a continuous supply of additional

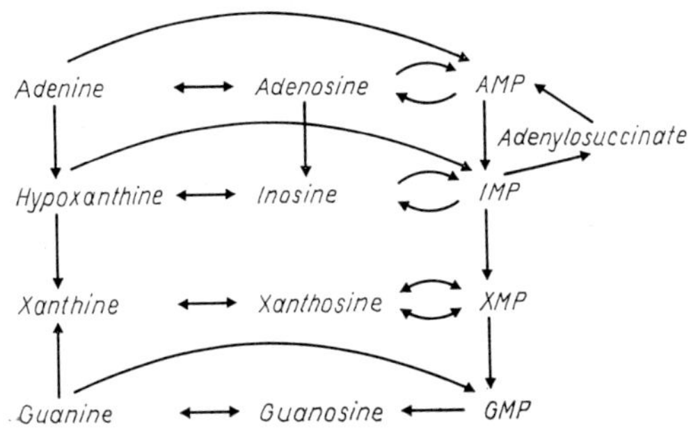

Fig. 19.2 Salvage pathway in purine metabolism.

purines from the host. ADA-ase may play some part in linking the purine metabolism of the host with that of the powdery mildew fungus (BUTTERS *et al.* 1985). Reutilization of purines involves a matrix of reactions (Fig. 19.2), in which ADA-ase would channel purines derived from the larger 5′ AMP pool to 5′GMP. However, neither inosine nor any subsequent metabolite involved in the pathway to 5-GMP synthesis reversed the toxicity of ethirimol (HOLLOMON 1979a), so clearly inhibition of ADA-ase must have other effects. ADA-ase is one of several enzymes that control adenosine levels and is likely, therefore, to alter the activity of enzymes such as S-adenosyl homocysteine hydrolase, which is regulated by adenosine and is involved in C-1 metabolism. Finally, ADA-ase ensures a "sink" for 5′AMP breakdown, so by increasing the energy charge (ATKINSON 1968) within conidia, many enzymes would be activated.

Resistance situation

Use of dimethirimol in 1970 to control cucumber powdery mildew *(Sphaerotheca fuliginea)* soon encountered difficulties in a number of glasshouses throughout N.W. Europe. Control failures were associated with a decline in sensitivity of the mildew (BENT *et al.* 1971) and withdrawal of dimethirimol was recommended. As a result, sensitivity has now increased, but attempts to reintroduce dimethirimol have only been partially successful as resistant strains can still be isolated from diseased cucumbers (SCHEPERS 1984). Control of the same disease on field grown cucurbits has not, so far, encountered problems. A more gradual decrease in sensitivy to ethirimol, followed its introduction as a seed treatment to control barley powdery mildew. Although continued use may not have led to any further decline in sensitivity, recommendation for use for ethirimol on autumn sown barley was withdrawn in 1973, in an attempt to reduce carry over of resistant strains which might otherwise infect spring sown crops. This may have reduced the frequency of resistant strains (SHEPHARD *et al.* 1975), which generally are less fit than wild-type strains (HOLLOMON 1978). The success of this policy in maintaining the effectiveness of ethirimol is difficult to judge, since broad spectrum triazole fungicides, introduced in 1978, soon replaced ethirimol. Population changes were further complicated by an association between ethirimol resistant strains and their virulence in barley cultivars with the Mla_{12} host-plant resistance gene (WOLFE and DINOOR 1973). No genetic link could be established (HOLLOMON 1981), suggesting that ethirimol resistance probably arose by chance in virulent strains on these Mla_{12} cultivars (Hassan, Sultan), which were widely grown at the time. These strains were then selected in the largely asexually reproducing mildew population. So far, the performance of bupirimate has not been eroded in this way by the appearance of resistant strains. When the triazole seed treatment, triadimenol, was first introduced, it was far superior to ethirimol for mildew control on many barley varieties (MARTIN *et al.* 1981). Since 1981, sensitivity of barley mildew to triadimenol has changed significantly (FLETCHER and WOLFE 1981; WOLFE *et al.* 1983), BUTTERS *et al.* 1984; HEANEY *et al.* 1984; and although control generally remains acceptable, some erosion of its performance has occurred, especially on mildew susceptible varieties. Triazole resistant mildew strains often showed wild-type sensitivity to ethirimol (BUTTERS *et al.* 1984), and in some situations ethirimol now provides better mildew control than triadimenol (NORTHWOOD *et al.* 1984; HOLLOMON *et al.* 1985).

Evidence of negatively correlated cross-resistance between triadimenol and ethirimol (BUTTERS *et al.* 1984), may simply reflect changes in field use of the two fungicides. In fact, a survey in 1984 revealed no significant correlations between ethirimol sensitivity and triadimenol sensitivity, but the mildew sampled did not include any populations that were either ethirimol resistant or triadimenol sensitive (Heaney *et al.* 1984). In field experiments, intensive selection can be applied to achieve populations with a wider range of sensitivities, especially to ethirimol. Selection for ethirimol resistance indeed increased sensitivity to triadimenol, but the reverse could not be shown, possibly because mildew at that site was already sensitive to ethirimol, and could not be shifted further (Hollomon *et al.* 1985). In a separate experiment use of each fungicide tended to induce greater sensitivity towards the other (Hunter *et al.* 1984), and mixtures of triadimenol and ethirimol limited development of resistance to each fungicide, gave better disease control, and increased yields. Alternating sequences may also give some of these benefits. A mixed seed treatment containing ethirimol and the triazole flutriafol (and thiabendazole to control *Fusarium* diseases) is now available (Northwood *et al.* 1984), and if used widely, will provide an opportunity to assess how readily strains resistant to both these fungicides will be selected.

Future prospects

The introduction of effective broad spectrum systemic fungicides in the late seventies greatly reduced the demand for 2-aminopyrimidines. But just as resistance to both dimethirimol and ethirimol had earlier prompted concern, so evidence of erosion in the level of mildew control achieved by fungicides that inhibit the C-14 demethylation step in sterol biosynthesis, has re-emphasised the need to diversify fungicide use. In powdery mildews shifts in fungicide sensitivity are generally gradual (Brent 1982), and overlap often remains between treated and untreated populations. This offers considerable scope for strategies using fungicides with different modes of action, either in mixtures, or as alternating programmes. Together with imaginative use of varietal resistance (Wolfe 1984), mildew populations may be stabilized, and loss of any one fungicide group through resistance prevented. Management of mildew populations in this way has generated renewed interest in ethirimol, and it is to be hoped that the mixed seed treatment with flutriafol will not herald a return to the resistance problems of some ten years ago. Sensible use of ethirimol in strategies with azole, morpholine and pyrazophos fungicides should maintain the usefulness of 2-aminopyrimidines for many years ahead.

References

Atkinson, D. E.: The energy charge of the adenylate pool as a regulatory parameter: interaction with feedback modifiers. Biochemistry **7** (1968): 4030—4034.

Bauers, C.: Erfahrungen bei der Mehltaubekämpfung im Getreide. Mitt. der DLG, Frankfurt (M.) **87** (1972): 347—350.

Baumer, M., and Ulonska, E.: Chemische Mehltaubekämpfung. Bayer. landwirtsch. Jahrb., München **48** (1971), Sonderheft 4: 107—113.

Bebbington, R. M., Brooks, D. H., Geoghegan, M. J., and Snell, B. K.: Ethirimol: a new systemic fungicide for the control of cereal powdery mildews. Chemistry and Industry (1969): 1512.

— Chamier, O. D., and Albrecht, J.: Erfahrungen mit Milstem-Saatgutbehandlungen zur Bekämpfung von *Erysiphe graminis* DC in der Bundesrepublik. Mitt. Biol. Bundesanstalt, Berlin-Dahlem **146** (1971): 264.

Benada, J.: Pouziti Milstemu proti padli na psenici. Agrochemia, Bratislava **11** (1971): 89—91.

— Pouziti systemovych fungicidu proti *Erysiphe graminis* DC. na jarnim jecmeni. Ochrana rostlin, Praha **45** (1972): 83—88.

BENNETT, M. D.: Effects of ethirimol on cytological characters in barley. Nature **230** (1971): 406.

BENT, K. J.: Fungitoxic action of dimethirimol and ethirimol. Ann. appl. Biol. **66** (1970): 103—113.

— COLE, A. M., TURNER, J. A. W., and WOOLNER, M.: Resistance of cucumber powdery mildew to dimethirimol. Proc. 6th Brit. Insect. Fungic. Conf., (1971): 274—282.

— SCOTT, P. D., and TURNER, J. A. W.: Control of apple powdery mildew by dormant season sprays — prospects for practical use. Proc. Brit. Crop Protection Conf. — Pests and Diseases (1977): 331—339.

BETHUNE, J. C.: Trois annees d'essais de lutte chemique contre l'oidium sur orge de printemps. In: La lutte contre maladies des cereales. Report of papers, Versailles 28th February (1973): 311—322.

BRENT, K. J.: Case study 4: Powdery mildews of barley and cucumber. In: DEKKER, J., and GEORGOPOULOS, S. G. (Eds.): Fungicide resistance on crop protection. Centre for Agricultural Publishing and Documentation, Wageningen 1982, pp. 219—230.

BROOKS, D. H.: The control of powdery mildew diseases with pyrimidine fungicides. In: VII. Intern. Congr. Plant Protect., Paris (1970): 247.

BUTT, D. J., JEGER, M. J., and SWAIT, A. A. J.: Supervised control of apple orchard diseases in England. Proc. 10th Intern. Congr. Plant Protect., Brighton (1983): 1005.

BUTTERS, J. A., BURRELL, M. M., and HOLLOMON, D. W.: Purine metabolism in barley powdery mildew and its host. Physiol. Pl. Path. **27** (1985): 65—74.

— CLARK, J., and HOLLOMON, D. W.: Resistance to inhibitors of sterol biosynthesis in barley powdery mildew. Meded. Fac. Landbouww. Rijksuniv. Gent. **49/2 a** (1984): 143—151.

BYFORD, W. J.: Field experiments on sugar-beet powdery mildew, *Erysiphe betae*. Ann. appl. Biol. **88** (1978): 377—382.

CALDERBANK, A.: Metabolism and mode of action of dimethirimol and ethirimol. Acta Phytopathol. Acad. Sci. Hungaricae, Budapest **6** (1971): 355—363.

CARVER, T. L. W., and GRIFFITHS, E.: Relationship between powdery mildew infection, green leaf area and grain yield of barley. Ann. appl. Biol. **99** (1981): 255—266.

CAVELL, B. D.: Methods used in the study of the photochemical degradation of pesticides. Pesticide Sci. **10** (1979): 177—180.

— HEMMINGWAY, R. J., and TEAL, G.: Some aspects of the metabolism and translocation of the pyrimidine fungicides. Proc. 6th Brit. Insect. Fungic. Conf. (1971): 431—437.

CHANNON, A. G., and BOYD, A. G.: The effect of some fungicides on mildew of spring barley in the South-West of Scotland. Proc. 7th Brit. Insect. Fungic. Conf., (1973): 21—28.

CLIFFORD, B. C., JONES, I. T., and HAYES, J. D.: Interaction between systemic fungicides and barley genotypes: their implications in the control of mildew. Proc. 6th Brit. Insect. Fungic. Conf., (1971): 287—294.

COLLIER, G. F., GRAHAM-BRYCE, I. J., KNIGHT, A. G., and COUTTS, J.: Direct observation of the distribution of radiolabelled ethirimol in soil by resin impregnation and autoradiography. Pesticide Sci. **10** (1979): 50—56.

COOKE, B. K., HERRINGTON, P. J., JONES, K. G., and MORGAN, N. G.: Progress toward economical and precise top fruit spraying. Proc. Brit. Crop Protect. Conf. — Pests and Diseases (1977): 323—329.

D'ARBIGNY, P., and DAWSON, M. G.: Trois annees d'essai de lutte contre l'oidium des orges avec l'ethirimol. In: La lutte contre les maladiés des ceréales. Report of papers, Versailles 28th February (1973): 236—249.

DICKINSON, C. H.: Effects of ethirimol and zineb on phylloplane microflora of barley. Trans. Brit. Mycol. Soc. **60** (1973): 423—431.

EASTERBROOK, M. A., SOUTER, E. F., SOLOMON, M. G., and CRANHAM, J. E.: Integrated pest management in English apple orchards. Proc. Brit. Crop Protect. Conf. — Pests and Diseases (1979): 61—67.

ELIAS, R. S., SHEPHARD, M. C., SNELL, B. K., and STUBBS, J.: 5-n-butyl-2-dimethylamino-4-hydroxy-6-methylpyrimidine: a systemic fungicide. Nature **219** (1968): 1160.

FINNEY, J. R.: PP 588 (Bupirimate: A new fungicide for the specific control of powdery mildews on apples and other crops. Proc. 8th Intern. Congr. Plant Protect., Moscow, Sect. III, Part I (1975): 235—236.

FLETCHER, J. T., and WOLFE, M. S.: Insensitivity of *Erysiphe graminis* f. sp. *hordei* to triadimefon, triadimenol and other fungicides. Proc. Brit. Crop Protect. Conf. — Pests and Diseases (1981): 633—640.

GEOGHEGAN, M. J. A.: Pyrimidine fungicide. Proc. 5th Brit. Insect. Fungic. Conf. (1969): 333—339.

— and DE GRAAFF, A.: Systemische fungicidenter bestrijding van meeldauw. Netherl. J. Plant Path. **75** (1969): 277—278.

GILMOUR, J.: Fungicidal control of mildew on spring barley in South-East Scotland. Proc. 6th Brit. Insect. Fungic. Conf. (1971): 63—74.

— Performance of barley mildew fungicides in South-East Scotland. Proc. 10th Intern. Congr. Plant Protect., Brighton (1983): 643.

GORSKA-POCZOPKO, J.: Preliminary investigations on PP 675 activity against pea mildew. Acta Mycol., Warszawa **7** (1971): 159—167.

GORTER, G. J. M. A., and NEL, D. D.: Systemic fungicidal effects on powdery mildews of metabolic inhibitors and related compounds in laboratory tests. I. Pyrimidine and purine analogues. Phytophylactica **6** (1974): 209—212.

GRAHAM-BRYCE, I. J., and COUTTS, J.: Interactions of pyrimidine fungicides with soil and their influence on uptake by plants. Proc. 6th Brit. Insect. Conf. (1971): 419—426.

GRINDLE, M.: Variations among field isolates of *Botrytis cinerea* in their sensitivity to antifungal compounds. Pesticide Sci. **12** (1981): 305—312.

HALL, D. W.: Control of mildew in barley with ethirimol. Proc. 6th Insect. Fungic. Conf. (1971): 26—32.

HALL, R. A.: Laboratory studies on the effects of fungicide, acaricides and insecticides on the entomopathogenic fungus, *Verticillium lecanii.* Ent. expt. et appl. **29** (1981): 39—48.

HANSEN, K. E.: Forsog med kemisk bekaempelse af meldug *(Erysiphe graminis)* pa korn 1969—1976. Tidsskrift for Planteavl, Kobenhavn **82** (1978): 289—306.

HARTMAN, J. K., and SIEGEL, M. K.: Cucumber *(Cucumis sativus)*, cantaloupe *(Cucumis melo)*, powdery mildew *(Erysiphe cichoracearum)*. Fungicide and Nematicide Tests, Bradenton **29** (1973): 58.

HEANEY, S. P., HUMPHREYS, G. J., HUTT, R., MONTEIL, P., and JEGERINS, P. M. F. E.: Sensitivity of barley powdery mildew to fungicides in the U.K. Proc. Brit. Crop Protect. Conf. — Pests and Diseases (1984): 459—464.

HOLLOMON, D. W.: Laboratory evaluation of ethirimol activity. In: MCFARLANE, N. R. (Ed.): Crop Protect. Agents. Academic Press, London 1977, pp. 505—515.

— Evidence that ethirimol may interfere with adenine metabolism during primary infection of barley powdery mildew. Pesticide Biochem. and Physiol. **10** (1979a): 181—189.

— Specificity of ethirimol in relation to inhibition of the enzyme adenosine deaminase. Proc. Brit. Crop Protect. Conf. — Pests and Diseases (1979b): 251—256.

— Genetic control of ethirimol resistance in a natural population of *Erysiphe graminis* f. sp. *hordei.* Phytopathology **71** (1981): 536—540.

— Antifungal activity of substituted 2-aminopyrimidines. In: TRINCI, A. P. J., and RYLEY, J. F. (Eds.): Mode of Action of Antifungal Agents. Cambridge Univ. Press, Cambridge 1984, pp. 185—205.

— FARRELL, G. M., and BENT, K. J.: Bupirimate — a new fungicide for the control of powdery mildews on apples and on other crops. Proc. 8th Brit. Insect. Fungic. Conf. (1975): 667—673.

— and CHAMBERLAIN, K.: Hydroxypyrimidine fungicides inhibit adenosine deaminase in barley powdery mildew. Pesticide Biochem. Physiol. **16** (1981): 158—169.

— LOCKE, T., and PROVEN, M.: Sensitivity of barley powdery mildew to ethirimol in relation to field performance. E. P. P. O. Bull (1985): 467—471.

HUNTER, L. D., BLAKE, P. S., and SOUTER, R. D.: Protectant and post-inoculation activities of fungicides for apple mildew control. Proc. Brit. Crop Protect. Conf. — Pests and Diseases (1981): 537—544.

HUNTER, T., BRENT, K. J., and CARTER, G. A.: Effects of fungicide regimes on sensitivity and control of barley powdery mildew. Proc. Brit. Crop Protect. Conf. — Pests and Diseases (1984): 471—476.

JACZEWSKA, A.: Evaluation of powdery mildew damage on wheat and attempts to chemical control. Zeszyty problemowe postępów nauk rolnicznych, Warszawa **275** (1983): 51—57.

Jahn, M., Burth, U., and Rathke, S.: Ein Beitrag zur differenzierten Beurteilung von Fungiziden am Beispiel von *Podosphaera leucotricha* (Ell. et Ev.) Salm and *Venturia inaequalis* (Cooke) Aderh. Arch. Phytopathol. u. Pflanzenschutz, Berlin **22** (1986): 205—212.

Jeger, M. J., Butt, D. J., and Swait, A. A. J.: Combining fungicide and partial resistance to control apple powdery mildew. Proc. 10th Intern. Congr. Plant Protect., Brighton (1983): 1004.

Jenkyn, J. F.: Effects of mildew on the growth and yield of spring barley: 1969—1972. Ann. appl. Biol. **78** (1974): 281—288.

— Effects of chemical treatments for mildew control at different times on the growth and yield of spring barley. Ann. appl. Biol. **88** (1978): 369—376.

— and Moffatt, J. R.: The effect of ethirimol seed dressings on yield of spring barley grown with different amounts of nitrogen fertilizer 1969—1971. Plant Pathol. **24** (1975): 16—21.

— and Prew, R. D.: Activity of six fungicides against cereal foliage and root diseases. Ann. appl. Biol. **75** (1973a): 241—245.

— — The effect of fungicides on incidence of *Sporobolomyces* spp. and *Cladosporium* spp. on flag leaves of winter wheat. Ann. appl. Biol. **75** (1973b): 253—256.

Kampe, W.: Verstärkter Befall durch *Rhynchosporium secalis* Dav. nach chemischer Bekämpfung von *Erysiphe graminis* D.C. bei Sommergerste (vorläufige Mitteilung). Nachr. Bl. Dt. Pflanzenschutzdienst, Braunschweig **26** (1974): 148—150.

Kelley, W. N., Daddona, P. E., and van der Weyden, M. B.: Characterization of human adenosine deaminase. In: Purine and Pyrimidine Metabolism, CIBA Foundation Symposium **NS48** (1977): 277—293.

King, J. E.: Surveys of foliar diseases of spring barley in England and Wales, 1972—1975. Plant Pathol. **26** (1977): 21—29.

Kluge, E., and Lyr, H.: Vergleichende Untersuchungen über die Wirkung von Systemfungiziden gegen *Erysiphe graminis* DC. In: Lyr, H., and Polter, C. (Eds.): Systemische Fungizide und antifungale Verbindungen. Abhandlungen der Akademie der Wissenschaften der DDR 1982, Nr. 1N, Akademie-Verlag, Berlin 1983, pp. 379—382.

Kobachidze, N. I., and Toskina, V. A.: Effectivnost' rjada sistemnych fungicidov v bor'be s nekotorymi mucnistorosjannymi zabolevanijami. Bjulleten' vsesojuznogo naucnoissledovatel'skogo Instituta sascity rastenij, Leningrad **22** (1971): 31—32.

Koch, W., and Stark, H.: Zur Verteilung von Universalbeizmitteln auf Gerste in Großbeizanlagen bei gleichzeitiger Saatgutbehandlung gegen Getreidemehltau mit Ethirimol. Nachr. Bl. Dt. Pflanzenschutzdienst, Braunschweig **24** (1972): 25—27.

Krüger, W.: Auftreten und Bekämpfung des „Echten Mehltaus" *Erysiphe graminis*. Jahresber. Biol. Bundesanstalt, Braunschweig (1969): A 97.

Kundert, J.: Apfelmehltaubekämfpung mit neuen organischen Fungiziden. Schweiz. Z. Obst- u. Weinbau, Wädenswil **113** (1977): 3—8.

Kunoh, H., Yamamori, K., and Ishizaki, H.: Cytological studies of the early stages of powdery mildew in barley and wheat. VIII Autofluorescence at penetration sites of *Erysiphe graminis hordei* on living barley coleoptiles. Physiol. Plant Path. **21** (1982): 373—379.

Kunovski, Z., Donceva, I., Racinska, C., and Gocova, V.: Study of the systemic fungicide ethirimol as affecting powdery mildew attacking wheat. Rasteniev"dni nauki, Sofia **10** (1973): 119—126.

Küthe, K.: Erfahrungen bei der Getreidemehltaubekämpfung (*Erysiphe graminis* DC) in Mittelhessen 1971. Gesunde Pflanzen, Frankfurt (M.) **24** (1972): 72—76.

Lescar, L.: Current practice in integrated cereal pest and disease control in North-Western-Europe (excluding Gr. Britain). Proc. Brit. Crop Protect. Conf. — Pests and Diseases (1977): 763—772.

Little, R., and Doddson, J. K.: The comparison of yields of some spring barley varieties in the presence of mildew and when treated with a fungicide. Proc. 6th Brit. Insect. Fungic. Conf., (1971): 91—97.

Manners, J. M., and Gay, J.: Fluxes and accumulation of ethirimol in haustoria of *Erysiphe pisi* and protoplasts of *Pisum sativum*. Ann. appl. Biol. **96** (1980): 283—293.

Martin, T., Morris, D. B., and Chipper, M. E.: Triadimenol seed treatment on spring barley: results of a 60 site evaluation in the United Kingdom. Proc. Brit. Crop Protect. Conf. — Pests and Diseases (1981): 299—306.

Meeus, P., Maddens, K., and Dohet, J.: Cinq annees de lutte contre l'oidium de l'orge de printemps en Belgique. Rev. Agric., Bruxelles **28** (1975): 285—306.

Millou, J., and D'Arbigny, P.: Utilisation d'associations d'ethirimol et de captafol contre les maladies foliares. Meded. Fac. Landbouww. Rijksuniv. Gent **39** (1974): 1051—1078.

Müller, P.: Untersuchungen zur Biologie und Bekämpfung des Weizenmehltaus (*Erysiphe graminis* DC f. sp. *tritici* Marchal) als Grundlage für die Modellierung des Epidemieverlaufes sowie die gezielte Anwendung von Fungiziden. Diss., Akademie der Landwirtschaftswissenschaften zu Berlin der DDR, 1984: 123 p.

Neuhaus, W., and Reich, R.: Bedeutung des Getreidemehltaus beim Anbau von Intensivgetreide und Möglichkeiten zur Bekämpfung. Nachr. Bl. Pflanzenschutz DDR, Berlin **29** (1975): 161—164.

Northwood, P. J., Paul, J. A., Gibbard, M., and Noon, R. A.: FF 4050 seed treatment — a new approach to control barley diseases. Proc. Brit. Crop Protect. Conf. — Pests and Diseases, (1984): 47—52.

O'Riordain, F., and Kennedy, D.: A comparison of eleven fungicides for the control of powdery mildew *(Sphaerotheca morusvae)* of gooseberry. Proc. Brit. Crop Protect. Conf. — Pests and Diseases (1981): 555—561.

Petkova, M.: Vazmoznosti za chimicna borba s brasnjankata po psenicata pri uslovijata na severo-zapadna balgarija. Rast. Nauki. Sofija **16** (1979): 129—134.

Petsikol-Panayotarou, N.: Action par vapeurs de quelques fongicides systemiques vis-a-vis de l'oidium des courges *Sphaerotheca fuliginea.* Meded. Fac. Landbouww. Rijksuniv. Gent **45** (1980): 199—206.

Rathke, S.: Einfluß der Temperatur auf die Wirkung ausgewählter Obstbaufungizide. Nachr. Bl. Pflanzenschutz DDR **38** (1984): 241—260.

— and Jahn, M.: Influence of rain on the effectiveness of fungicides against apple powdery mildew and apple scab. In: Lyr, H., and Polter, C. (Eds.): Systemic Fungicides and Antifungal Compounds. Tagungsbericht der Akademie der Landwirtschaftswissenschaften der DDR Nr. **222,** Berlin (1984): 253—258.

Round, P. A., and Wheeler, B. E. J.: Interactions of *Puccinia hordei* and *Erysiphe graminis* on seedling barley. Ann. appl. Biol. **89** (1978): 21—35.

Saur, R.: Untersuchungen über den Einfluß systemischer Fungizide auf den Befall der Gerste mit *Helminthosporium sativum* P., K. & B. unter besonderer Berücksichtigung von Ethirimol. Diss., Univ. Bonn (1976a): p. 93.

— Untersuchungen über den Einfluß von Ethirimol auf die Pathogenese einer Helminthosporiose *(H. sativum)* an Gerste. Phytopathol. Z., Berlin (West) und Hamburg **87** (1976b): 304—313.

— and Schönbeck, F.: Untersuchungen über den Einfluß systemischer Fungizide auf den Befall der Gerste mit *Helminthosporium sativum.* Z. Pflanzenkrankh. u. Pflanzenschutz **82** (1975): 173—175.

— — Einfluß von Ethirimol auf die Empfindlichkeit von Gerste gegenüber einem Toxin von *Helminthosporium sativum.* Meded. Fac. Landbouww. Rijksuniv. Gent **41** (1976): 511—516.

Schepers, H. T. A. M.: Persistence of resistance to fungicides in *Sphaerotheca fuliginea.* Netherl. J. Plant Path. **90** (1984): 165—171.

Shabi, E.: Control of apple mildew by bupirimate and the influence of added spreader-sticker on its performance. Proc. 8th Brit. Insect. Fungic. Conf. (1975): 711—714.

Shephard, M. C.: Barriers to uptake and translocation in herbaceous and woody plants. Proc. 7th Brit. Insectic. Fungic. Conf. (1973): 841—850.

— Factors which influence the biological performance of pesticides. Proc. Brit. Crop Protect. Conf. — Pests and Diseases, (1981): 711—721.

— Bent, K. J., Woolner, M., and Cole, A. M.: Sensitivity to ethirimol of powdery mildew from UK barley crops. Proc. 8th Brith. Insectic. Fungic. Conf., (1975): 59—66.

Snell, B. K., Elias, R. S., and Freeman, P. F. H.: Pyrimidine derivatives and the use there of as fungicides. GBP 1,182,584 (1966).

Teal, G., and Cavell, B. D.: Degradation of bupirimate fungicide on apples and in water. Proc. 8th Brit. Insect. Fungic. Conf. (1975): 25—30.

Wells, C. H., Pollard, S. J., and Sen, D.: Phytochemistry of some systemic pyrimidine fungicides. Pesticide Sci. **10** (1979): 171—176.

Wolfe, M. S.: Pathological and physiological aspects of cereal mildew control using ethirimol. Proc. 5th Brit. Insect. Fungic. Conf. (1969): 8—15.

— Pathogen response to fungicide use. Proc. 8th Brit. Insect. Fungic. Conf., (1975): 813—822.
— Trying to understand and control powdery mildew. Plant Path. **33** (1984): 451—466.
— and DINOOR, A.: The problems of fungicide tolerance in the field. Proc. 7th Brit. Insect. Fungic. Conf., (1973): 11—19.
— SLATER, S. E., and MINCHIN, P. N.: Fungicide insensitivity and host pathogenicity in barley mildew. Proc. 10th Intern. Congr. Plant Protect., Brighton (1983): 645.
WORTHING, C. R., and WALKER, S. B.: The pesticide manual — a world compendium. 7th edit. The Brit. Crop Protect. Council, London (1983).
ZWATZ, B.: Getreidemehltau: Erfahrungen 1973. Der Pflanzenarzt, Wien **27** (1974): 5—6.

LYR, H. (Ed.): Modern Selective Fungicides — Properties, Applications, Mechanisms of Action. Longman Group UK Ltd., London, and VEB Gustav Fischer Verlag, Jena, 1987.

Chapter 20

Organophosphorus fungicides

B. Schreiber
Hoechst AG, Frankfurt (Main), FRG

Introduction

Chemicals with direct conjunction of the phosphorus-atom and carbon-atom and derivates of phosphoric acid without C-P-conjunction are all counted as organophosphorus compounds. The organophosphorus fungicides include compounds especially active against blast and other diseases on rice, e.g. phosphorothiolates,

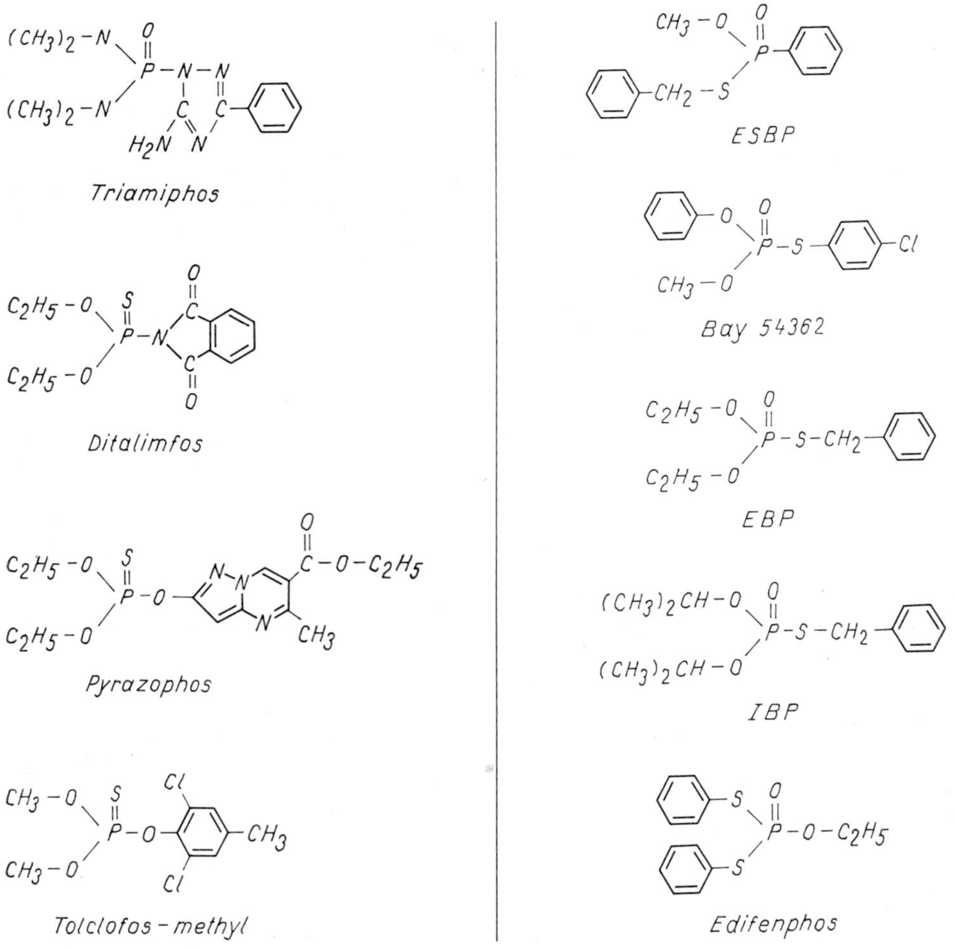

Fig. 20.1 Organophosphorus fungicides.

while others are primarily active against diseases caused by powdery mildew, e.g. phosphoric ester amides and phosphorothionates. This might be due to the more lipophilic properties of phosphorothionates and phosphorothiolates compared to the corresponding oxygene analogues, phosphates and phosphorothiolates. The uptake of phosphorothionates through the gelatinuous outer layer of the cell wall of powdery mildew fungi seems to be facilitated (McKeen *et al.* 1966 and 1967; Belal 1971). Thus the differences in lipophilic or hydrophilic character of the organophosphorus fungicides in relation to the cell wall composition of the fungi may account to some extent for their selectivity.

Practical application

Commercially accepted substances for control of rice blast are EBP, which was replaced by IBP, and edifenphos, while Bay 54362 could not be used in ordinary paddy fields because of the high toxicity to fish. The organophosphorus compounds most frequently used for control of powdery mildew fungi are ditalimfos and pyrazophos. Triamiphos has not found full acceptance in practice due to phytotoxicity problems in apples. Tolclofos-methyl was recently introduced for control of soil-borne diseases, especially *Rhizoctonia solani* (chapter 5 and 6).

Although the phosphorothiolate compounds such as IBP and edifenphos are mainly toxic to *Pyricularia oryzae*, but also effective against other fungi on rice such as *Corticium sasakii*, *Cochliobolus miyabeanus* and *Leptosphaeria salvanii*, they were also found to be slightly toxic to the powdery mildew fungi *Sphaerotheca fuliginea*

Table 20.1 Organophosphorus fungicides

Chemical group[1]	Common name	Trade mark	Inventor
Phosphoric acid esters	ESBP	Inezin®, Inejin®	Schrader (Bayer AG) 1957; Scheinpflug and Schrader (Bayer AG) 1965
Phosphorothiolates	Bay 54362	Cerezin®	Scheinpflug, Jung and Schrader (Bayer AG) 1963
	EBP	Kitazin®	Scheinpflug, Jung and Schrader (Bayer AG) 1963; Kado *et al.* (Ihara Noyaku Chem. Company) 1964
	IBP	Kitazin P®	(Kumiai Chemical Industry Co.) 1966
	Edifenphos	Hinosan®	Schrader, Mannes and Scheinpflug (Bayer AG) 1965
Phosphoric ester amides	Triamiphos	Wepsyn®	Koopmanns *et al.* (Philips-Duphar) 1957
	Ditalimfos, Dowco 199	Plondrel®, Laptran®, Frutogard®, Leucon®, Millie®	Tolkmith and Senkbeil (Dow Chem. Company) 1965
Phosphorothionates	Pyrazophos	Afugan®, Curamil®, Missile®	Scherer and Mildenberger (Hoechst AG) 1965
	Tolclofos-methyl	Rizolex®	(Sumitomo Chemical Company) 1979

[1]) According to C. Fest and K.-J. Schmidt (1973)

and *Erysiphe graminis* f. sp. *hordei* in spore germination tests and to a lesser extent in vivo (KADO and YOSHINAGA 1969; YOSHINAGA 1969; UESUGI 1970; UMEDA 1973; DE WAARD 1974). On the other hand, the phosphorothionate pyrazophos, primarily active against powdery mildew, also showed some fungitoxic action against *Pyricularia oryzae* on barley (SMIT 1969; DE WAARD 1974). Thus there seem to be only differences of degree between the antifungal spectra of phosphorothionate, phosphoric ester amide and phosphorothiolate compounds. But in this context differences in the acceptance of these compounds by various plant species have to be considered as well as the differences in the mode of action.

Moreover, it should be mentioned that ditalimfos is active against apple scab and some leaf spot diseases and pyrazophos against *Pyrenophora* spp. and *Colletotrichum* spp. (HUISMAN and PESKETT 1973; SCHREIBER et al. 1984). And both compounds showed no difference in sensitivity to *Sphaerotheca fuliginea* and *Erysiphe cichoracearum*. One exception among the fungicides referred to above is tolclofos-methyl, which differs remarkably in its fungicidal spectrum (OHTSUKI and FUJINAMI 1982; for details, see chapter 5).

Organophosphorus fungicides display protective as well as curative activity. This is due partly to their systemic properties.

The most systemic of these compounds is IBP (KOZAKA 1969). It is absorbed into the rice plant by foliar application and subsequently translocated within the plant. The absorption through the root system and rapid translocation in the transpiration stream to the site of pathogenic infection makes it easier to apply IBP to irrigation water in granular form. Since the active ingredient has only a limited degree of solubility in water (500 ppm), the granules appear to become active about three days after application to reach their maximum effectiveness after five to seven days and to persist for three to four weeks under field conditions (KOZAKA 1969; YOSHINAGA

Table 20.2 Organophosphorus fungicides — spectrum of antifungal activity

Plant pathogens	ESBP	Bay 54362	EBP IBP	Edifenphos	Triamiphos	Ditalimfos	Pyrazophos	Tolclofos-methyl
Pyricularia oryzae	+	+	+	+			0	
Corticium sasakii			0	0				
Cochliobolus miyabeanus			0	0				
Leptosphaeria spp.			0	0			0	
Rhynchosporium spp.				0			0	
Pyrenophora spp.							+	
Puccinia spp.				0				
Erysiphe spp.			0	0	+	+	+	
Sphaerotheca spp.			0	0	+	+	+	
Podosphaera leucotricha					+	+	+	
Uncinula necator					+	+	+	
Venturia spp.						0		
Monilinia spp.						0	0	
Colletotrichum spp.							+	
Diplocarpon rosae						0		
Rhizoctonia solani								+
Sclerotium rolfsii								+
Typhula spp.								+

+ = good activity, 0 = slight activity

Table 20.3 Organophosphorus fungicides — main field of practical application

Common name	Crop	Disease, Plant Pathogen	Formulation	Dosage (a.i.)	Application method	Remarks
IBP	Rice	*Pyricularia oryzae*	EC 48 Dust 2 % Granules 17 %	500—750 g a.i./ha 600—800 g a.i./ha 5,100—7,550 g a.i./ha	2—4 foliar sprays spread on leaves spread on water (1—2 times)	Side effects on *Corticium sasakii* + *Leptosphaeria salvanii* as well as on rice leaf and plant hoppers
Edifenphos	Rice	*Pyricularia oryzae*	EC, WP L 50 (ULV) Dust 2 %	350—500 g a.i./ha 500 g a.i./ha 600—800 g a.i./ha	2—4 foliar sprays foliar spray spread on leaves	Side effects on *Corticium sasakii*, *Leptosphaeria salvanii*, *Cochliobolus miyabeanus*, *Cercospora oryzae* and *Rhynchosporium oryzae* as well as on rice leaf and plant hoppers and caterpillars
Triamiphos	Apples, Pears, Vineyards, Ornamentals	Powdery mildew	WP 25 Flowable 100 g/l	0.025 % a.i.	protective sprays, every 10 days	Phytotoxic effects on Jonathan apples
Ditalimfos	Apples, Peaches, Cucurbits, Roses	Powdery mildew	WP 50	0.03—0.05 % a.i.	7—10 days spray interval	Side effects on *Venturia inaequalis*, *Monilinia* spp., *Coccomyces hiemalis* and *Diplocarpon rosae*. Russeting on Golden Delicious apples
Pyrazophos	Apples, Cucumbers, Ornamentals, Hops, Vineyards	Powdery mildew	EC 30	0.01—0.03 % a.i.	7—10 days spray interval	Russeting effects on some apple varieties, phytotoxic effects on some ornamental

	Cereals	*Erysiphe graminis* *Pyrenophora teres*	EC 30	450—600 g a.i./ha	1—2 applications at appearance of first symptoms	varieties. Good insecticidal effect on leaf miners. Side effect on *Rhynchosporium secalis*
Tolclofosmethyl	Cotton, Potatoes, Ornamentals	*Rhizoctonia* spp. *Sclerotium rolfsii*	WP 50, EC 20, Flowabel 25, Dust 5, 10, 20	200—400 g a.i./100 kg seeds 200 g a.i./tons of tuber 1,000 ppm a.i. tuber spray 10 kg a.i./ha soil incorpotation		

Table 20.4 Organophosphorus fungicides — oecotoxicological data[1])

	Bay 54362	IBP	Edifenphos	Triamiphos	Ditalimfos	Pyrazophos	Tolclofosmethyl
Toxicity acute oral LD 50 rat/mice	112—160 ppm	600 ppm	100—340 ppm	20—32 ppm	ca. 5,000 ppm	151—632 ppm	3,500—5,000 ppm
Fish toxicity LC 50 carp	0.06 ppm	5,1 ppm	1.3 ppm	—	19.4 ppm	4.5 ppm	non toxic
Toxicity to honey bees	—	—	—	not toxic up to 0.001 % a.i.	non toxic	not toxic up to 0.03 % a.i.	non toxic

[1]) According to Pesticide Manual (British Crop Protection Council) 1983

1969). The application of granules is obviously more effective than a three-spray-programme, even at the lowest rate (Yoshinaga 1969). An additional advantage is a shortening of the stalks which reduces the tendency for lodging without effecting length or number of ears. — Edifenphos has fewer systemic properties than IBP but a better residual activity, and is used therefore only by foliar application (Ou 1980).

In contrast to the systemic triamiphos, an absence of systemic activity was reported for ditalimfos, combined with effective persistence and high vapour phase activity (Tempel *et al.* 1968; Ellal and Dinoor 1973).

Pyrazophos was shown to penetrate the leaf surface after foliar spraying and to move laterally. After absorption through roots or stems, the fungitoxic activity of pyrazophos against powdery mildew is restricted to the vein (xylem) system. Possibly, due to the lipophilic character of the compound, the distribution from the veins to the mesophyll of leaves is reduced (Belal 1971; de Waard 1974; Köcher and Lötzsch 1976).

The organophosphorus fungicides are mainly applied protectively. Curative activity is most effective when applied not later than two to four days after infection. Eradicative effects were not described. So the first application in most annual crops should take place, when first symptoms appear. In top fruits protective applications are necessary.

In addition to the fungicidal activity with most of the organophosphorus fungicides insecticidal side-effects have also been observed. In rice, for example, IBP and edifenphos are effective against rice caterpillars as well as against rice leaf and plant hoppers acting as virus vectors. Pyrazophos shows some efficacy against aphids and leaf miners in cereals, cucumbers or ornamentals. Usually the insecticidal activity of organophosphorus fungicides is weaker than that of common insecticides, and due to the short residual insecticidal effect application times for optimal fungus and insect control rarely coincide. In general, effects on honey bees and fish are also possible with some compounds.

Negative effects on plants are to be expected with combined application of propanil in rice or delayed application of hormonal herbicides in cereals.

Mode of action and resistance situation

Infection structures of plant pathogens are influenced in different ways by various organophosphorus fungicides. IBP and edifenphos inhibit spore germination and appressorial formation of *Pyricularia oryzae* to a lesser extent than mycelium growth and sporulation (Ou 1980; Kodama *et al.* 1980). So curative activity is predominant compared to the protective activity. On the other hand, triamiphos stopped the development of *Erysiphe graminis* f. sp. *hordei* mainly because of malformation of haustoria (Magendans and Dekker 1966). Pyrazophos primarily inhibits spore germination and the appressorial formation of *Erysiphe graminis* and other powdery mildew fungi. However, abnormal mycelium growth, haustoria anomalies and reduced sporulation were also reported (Belal 1971; de Waard 1974; Sztejnberg *et al.* 1975; Blake *et al.* 1982; Smolka and Wolf 1984; Kohts 1985). With regard to *Pyrenophora* spp. the germination hyphal appressorial formation and sporulation are influenced by pyrazophos predominantly (Kohts 1985). So this fungicide has a protective action against *Pyrenophora* spp. and better protective than curative action against powdery mildew fungi.

The metabolic conversion of IBP and edifenphos by *Pyricularia oryzae* was demonstrated by Uesugi and Tomizawa 1971, Tomizawa and Uesugi 1972 and Kurogochi *et al.* 1985. In both fungicides a cleavage at P-S linkages was more remarkably ob-

Table 20.5 Organophosphorus fungicides — mode of action[1])

	IBP	Edifenphos	Pyrazophos	
Influenced by fungicides	*Pyricularia oryzae*	*Pyricularia oryzae*	*Erysiphe graminis*	*Pyrenophora teres*
Spore germination	++	++	++++	++
Appressorial formation	++	++	+++	++++
Haustorial formation			++	
Mycelium growth	+++	+++	++	++
Sporulation	+++	+++	+++	+++
Permeability of membranes	+++	+++		
Chitinsynthesis	+++	+++		
Synthesis of phospholipids	+++	+++	+++	+++
Accumulation of free fatty acids			+++	+++
Uptake of O_2 (respiration)			+++	+++
Carboxylesterase			+++	+++
Synthesis of melanines				+++
General Properties				
Systemic activity	++++	+	++	++
Residual activity	+	+++	+++	+++
Protective activity	++	++	+++	+++
Curative activity	+++	+++	++	+

[1]) Adapted from DE WAARD 1974; OU 1980; KODAMA *et al.* 1980; KOHTS 1985
[2]) Activity ranging from no (—) to slight (+) to strong (+++++)

served in sensitive strains than in resistant ones, additionally a very fast cleavage at S-C linkages of IBP and, less important, hydroxylation at p-position of the phenylring. Due to the neglegible fungitoxicity of the metabolites the biological meaning of P-S bond cleavage, possibly responsible for classification of sensitivity of *P. oryzae*, may be presumed as a metabolic activation of the fungicides.

The fungicidal activity of ditalimfos is associated with the N-phosphorylated dicarboximide moiety. Substitution of the thiono-group by oxygen and the introduction of methylene groups or of an oxygen atom between the nitrogen and phosphorus atom lead to a marked decrease in fungicidal activity (TOLKMITH and MUSSELL 1967; TOLKMITH and SENKBEIL 1967).

Pyrazophos was demonstrated to be metabolized in mycelial suspensions of *Pyricularia oryzae* into the oxygen-analogue PO-pyrazophos, followed by the pyrazolopyrimidin PP-pyrazophos (DE WAARD 1974). It was suggested that pyrazophos is largely absorbed to the mycelium and afterwards converted by microsomal mixed function oxidases to PP-pyrazophos, the actual fungitoxic principle of pyrazophos.

With regard to the biochemical mode of action IBP and edifenphos interfere with the synthesis of phospholipids by inhibiting the specific conversion of phosphatidylethanolamine to phosphatidylcholine by transmethylation of S-adenosylmethionine (AKATSUKA *et al.* 1977; KODAMA *et al.* 1980). Phosphatidylcholine is an important constituent of the membranes; about 50 % of the phospholipids of *Pyricularia oryzae* are represented by phosphatidylcholine (KODAMA *et al.* 1980). Due to the inhibition of the synthesis of phosphatidylcholine the structure and functions of the fungal membranes are influenced dramatically. So an increase in membrane permeability followed by an efflux of cellular components may explain the fungitoxicity of the compounds (DE WAARD 1974). In former reports indirect influences on the chitin synthesis due to membrane permeability were also mentioned (MAEDA *et al.* 1970; DE WAARD 1972).

In contrast, increased permeability of membranes and a change of efflux of cellular substances following contamination of fungal material with pyrazophos was not observed. Nevertheless, the membrane functions seem to be disturbed by an accumulation of free fatty acids and a reduction of synthesis of phospholipids (DE WAARD 1974; KOHTS 1985). However, effects on sterolbiosynthesis are neglegible. DE WAARD (1974) also described an inhibitory action of pyrazophos and PP-pyrazophos on oxygen uptake or respiration of the fungus as well as influences of PP-pyrazophos on protein synthesis. However, no correlation could be found between inhibition of carboxylesterase and fungitoxicity. In addition, a remarkable influence on the synthesis of melanines was observed with *Pyrenophora* spp. (KOHTS 1985).

In summary, the various modes of action indicate that the mechanism of fungicidal action of organophosphorus compounds is not necessarily the same in all cases. Nevertheless, there are also some important connections between the compounds in regard to the fungicidal spectrum, the influences on phospholipidsynthesis and cross-resistance with *Pyricularia oryzae* (KODAMA *et al.* 1980; DE WAARD 1980; KOHTS 1985).

Because of the numerous sites of action of the organophosphorus fungicides, only a few reports on the development of resistance are known. Mutants of *Pyricularia oryzae* resistant to IBP and edifenphos have been easily obtained under laboratory conditions, but first signs of development of resistance to the fungicides in the paddy field have been recognized only after more than ten years of intensive use (KATAGIRI *et al.* 1980). Among the few identified strains, different levels of resistance were described as well as cross-resistance between IBP and edifenphos.

Decreased sensitivity to pyrazophos was reported only from glasshouses in the Netherlands, where the fungicide was used intensively for control of cucumber and gherk in powdery mildew. Because of the lower fitness of resistant strains in comparison to more sensitive ones, the resistance level was reduced to a certain extent after reduction of selection pressure (DEKKER and GIELINK 1979; SCHEPERS 1985). Resistance development in fieldgrown crops is not known. — It is suggested that one mechanism of resistance to pyrazophos is related with the inability of a resistant strain to convert pyrazophos into PP-pyrazophos (DE WAARD 1980).

References

AKATSUKA, T., KODAMA, O., and YAMADA, H.: A novel mode of action of Kitazin P in *Pyricularia oryzae*. Agric. Biol. Chem. **41** (1977): 2111—2112.

BELAL, M. H.: Studien zur chemischen Bekämpfung des Echten Mehltaus. Thesis, Göttingen 1971.

BLAKE, P. S., HUNTER, L. D., and SOUTER, R. D.: Glasshouse tests of fungicides for apple powdery mildew control. 2. Eradicant and antisporulant activity. J. Horticult. Sci. **57** (1982): 407—412.

DEKKER, J., and GIELINK, A. J.: Decreased sensitivity to pyrazophos of cucumber and gherkin powdery mildew. Neth. J. Plant Pathol. **85** (1979): 137—142.

DE WAARD, M. A.: On the mode of action of the organosphosphorus fungicide Hinosan. Neth. J. Plant Path. **78** (1972): 186—188.

— Mechanism of action of the organophosphorous fungicide pyrazophos. Thesis, Wageningen 1974.

ELLAL, G., and DINOOR, A.: Systemic activity and/or vapour phase activity of fungicides against powdery mildew in cucurbits. Abstr. 2nd Int. Congr. Plant Pathol. Minneapolis, USA 1973.

FEST, C., and SCHMIDT, K.J.: The chemistry of organophosphorus pesticides. Reactivity — Synthesis — Mode of action — Toxicology. Springer-Verlag, Berlin 1973.

HUISMAN, A. H., and PESKETT, F. J.: 0,0-diethyl phthalimido-phosphonothionate, a new fungicide. Proc. 7th Br. Insect. Fungic. Conf. Brighton (1973): 687—693.

KADO, M., and YOSHINAGA, E.: Fungicidal action of organophosphorus compounds. Residue Rev. **25** (1969): 133—138.

KATAGIRI, M., UESUGI, Y., and UMEHARA, Y.: Development of resistance to organophosphorus fungicides in *Pyricularia oryzae* in the field. J. Pesticide Sci. **5** (1980): 417—421.

KODAMA, O., YAMASHITA, K., and AKATSUKA, T.: Edifenphos, inhibitor of phosphatidylcholine biosynthesis. Biol. Chem. **44** (1980): 1015—1021.

KÖCHER, H., and LÖTZSCH: Zur systemischen Wirkung von Afugan (Pyrazophos). Med. Fac. Landbouww. Rijksuniv. Gent **41/2** (1976): 635—644.

KOHTS, T.: Untersuchungen zu einigen neuen Aspekten der Wirkungsweise von Pyrazophos (Afugan) in *Erysiphe graminis* und *Helminthosporium*-Arten. Thesis, Bonn 1985.

KOZAKA, T.: Chemical control of rice blast in Japan. Rev. Pl. Protec. Res. **2** (1969): 53.

KUROGOCHI, S., KATAGIRI, M., TAGASE, I., and UESUGI, Y.: Metabolism of edifenphos by strains of *Pyricularia oryzae* with varied sensitivity to phosphorothiolate fungicides. J. Pesticide Sci. **10** (1985): 41—46.

MAEDA, T. H., ABE, H., KAKIKI, K., and MISATO, T.: Studies on the mode of action of organophosphorus fungicide, Kitazin. Part 2. Accumulation of an aminosugar derivative on Kitazin-treated mycelia of *Pyricularia oryzae*. Agric. biol. Chem. **34** (1970): 700—709.

MAGENDANS, J. F. C., and DEKKER, J.: A microscopic study of powdery mildew on barley after application of the systemic compound Wepsyn. Netherl. J. Plant Pathol. **72** (1966): 274—278.

MCKEEN, W. E., MITCHELL, N., JARVIE, W., and SMITH, R.: Electromicroscopy studies of conidial walls of *Sphaerotheca macularis, Penicillium levetium* and *Aspergillus niger*. Can. J. Microbiol. **12** (1966): 427—428.

— and SMITH, R.: The *Erysiphe cichoracearum conidium*. Can. J. Bot. **45** (1967): 1489—1496.

OHTSUKI, S., and FUJINAMI, A.: Rizolex (tolclofos-methyl). Japan Pestic. Inf. **41** (1982): 21—25.

OU, S. H.: A look at worldwide rice blast diesease control. Plant Disease **64** (1980): 439—445.

SCHEPERS, H. T. A. M.: Consequences of resistance to fungicides for the control of cucumber powdery mildew. Med. Fac. Landbouww. Rijskuniv. Gent **50** (1985).

SCHREIBER, B., KÖTTER, U., and WAGNER, H.-J.: Afugan- an alternative for the control of cereals diseases. Mitt. Biol. Bundesanst. **223** (1984): 221—222.

SMIT, M.: Diaethyl-methyl-ethoxycarbomyl-pyrazolo-pyrimidine-yl fosforothioate een systemisch werkzaam meeldauwbestrijdingsmiddel. Med. Fac. Landbouww. Rijksuniv. Gent **34** (1969): 763—771.

SMOLKA, S., and WOLF, G.: Cytological studies of the mode of action of systemic fungicides on the hostparasite relationship barley — powdery mildew *(Erysiphe graminis* f. sp. *hordei)*. Mitt. Biol. Bundesanst. **223** (1984): 249.

SZTEIJNBERG, A., BYRDE, J. W., and WOODCOCK, D.: Antisporulant action of fungicides against *Podosphaera leucotricha* on apple seedlings. Pesticide Sci. **6** (1975): 107—111.

TEMPEL, A., MELTZER, J., and VAN DEN BOS, B. G.: Systemic fungicidal and insecticidal activities of 1- and 2-[bis(dimethylamido)phosphoryl]-3-alkyl-5-anilino-1,2,4-triazoles. Neth. J. Plant Pathol. **74** (1968): 133.

TOLKMITH, H., and MUSSELL, D. R.: Novel N-heterocyclic fungicides. World Rev. Pest Control **6** (1967): 74—79.

— and SENKBEIE, O. H.: Fungicidal phthalimido-phosphonothioates. Science **155** (1967): 85—86.

TOMIZAWA, G., and UESUGI, Y.: Metabolism of S-benzyl 0,0-diisopropyl phosphorothioate (Kitazin P) by mycelial cells of *Pyricularia oryzae*. Agric. Biol. Chem. **36** (1972): 294—300.

UESUGI, Y.: Development of organosphosphorus fungicides. Japan Pestic. Inf. **2** (1970): 11—14.

— and TOMIZAWA, G.: Metabolism of 0-ethyl S,S-diphenyl phosphorodithioate (Hinosan) by mycelial cells of *Pyricularia oryzae*. Agric. Biol. Chem. **35** (1971): 941—949.

UMEDA, Y.: Hinosan, a fungicide for control of rice blast. Japan Pestic. Inf. **18** (1973): 25—34.

YOSHINAGA, E.: A systemic fungicide for rice blast control. Proc. 3th Br. Insect. Fungic. Conf. Brighton (1969): 593—599.

LYR, H. (Ed.): Modern Selective Fungicides — Properties, Applications, Mechanisms of Action. Longman Group UK Ltd., London, and VEB Gustav Fischer Verlag, Jena, 1987.

20*

Chapter 21

Other Fungicides

Maya Gasztonyi*) and H. Lyr**)

*) Institute for Plant Protection Research of the Hungarian Academy of Science, Budapest, Hungary
and
**) Institute for Plant Protection Research Kleinmachnow of the Academy of Agricultural Sciences of the GDR

Introduction

Some well known fungicides are not dealt with or were only briefly mentioned in other chapters. Therefore, we collected some information on fungicides which are partly still of practical importance or have contributed to some theoretical progress. The selection is of course very difficult, because some of these are still on the market, others had been there for some time and others again seem to have a come back. Some contributed to a better understanding of fungal biochemistry.

Dodine

1-dodecylguanidinium acetate (IUPAC nomenclature) (Fig. 21.1a) was introduced in 1956 by the American Cyanamid Co. as a protective foliar fungicide (USP 2867562). The w. p., liquid or dust formulations of dodine are known with the following trade names: "Cyprex", "Melprex", "Questuran", "Carpene", "Apadodine", "Venturol", "Vondodine", "Syllit".

The pure compound exists as colourless crystals, with m. p. 136 °C, v. p. 10^{-7} mbar at 20 °C. Its solubility in water at 25 °C: 0.063 %. It is soluble in hot water and alcohol, readily soluble in mineral acids, but insoluble in most organic solvents. The acute oral LD_{50} for male rats 1,000—2,000 mg/kg.

The formulations can be used for protective control of scab (*Venturia* spp.), leaf spot (*Mycosphaerella* spp.), blossom brown rot and leaf blight in fruit cultivation. Foliar application is recommended especially against *Venturia* spp. and cherry leaf spot against which dodine has some eradicant action (Worthing 1979). As a scab fungicide dodine has marked residual effect. Beside this, an excellent curative effect has been reported 30 hours after infection (Byrde 1969). Local systemic effect has been demonstrated, but there is only a minor translocation to new growth (1 mg/kg) on seedling apple trees sprayed with ^{14}C-dodine (2 kg a.i./ha). Although in a series of n-alkyl guanidine acetate homologues maximum phytotoxicity was found at C_{10} (Brown and Sisler 1960), damage caused by dodine on apples, pears, currants and gooseberries was also report ed.Green-skinned varieties of apples are particularly susceptible. Therefore in Europe the application of dodine on apples is recommended only before flowering (Grewe 1965). Dodine has also surface active properties. It is incompatible with anionic wetting agents, with lime and chlorobenzilate (Byrde 1969).

The free base of dodine is moderately strong, so that the acetate is largely ionised at physiological pH-values, but a varying degree of hydrolysis will occur depending on pH, causing twenty times higher fungitoxicity effect at pH 7.8 than at pH 5.1 (Hassall 1982). Accumulation of large quantities is required for adverse effects on germination of spores. ED_{50} values from 2,000 to 2,500 μg/g conidial weight were

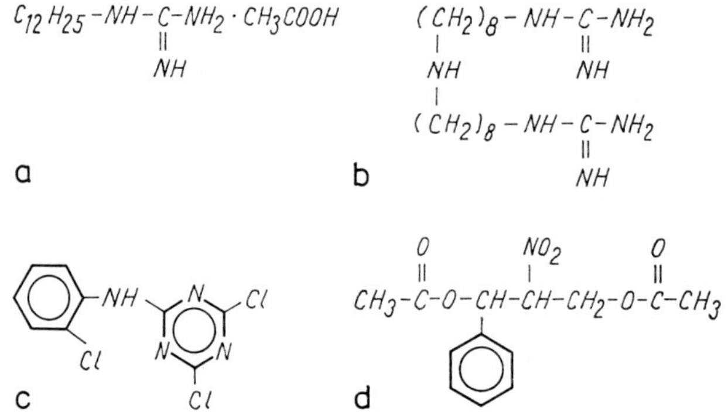

Fig. 21.1 Miscellaneous fungicides with broad spectrum activity. a) Dodine, b) Guazatine, c) Anilazine, d) Fenitropan.

obtained (Miller 1969). It appears to be rather firmly attached to immobile anionic cell constituents such as phosphate and carboxyl groups (Hassall 1982).

The site of action is considered to be the cell membranes. Its inclusion on membranes is presumably related to the ability of the lipophilic chain to dissolve in the lipid portion of the membrane, while the guanidine residue will tend to remain in the adjacent aqueous phase (Corbett 1974). The interaction with cell membranes leads to alterations in cellular permeability. The inhibition of acetate and glucose oxidation has been also observed, but there is no evidence of a direct inhibitory action on these metabolic pathways (Solel and Siegel 1984). 50% inhibition of electron transport in isolated mitochondria at a concentration of approximately 80 μM may be explained by interference with mitochondrial membranes as a primary site of action.

In plants dodine is converted to creatine via the action of a methyltransferase and a simultaneous oxidative cleavage of the dodecyl moiety. As intermediates guanidine as well as guanidine, substituted with short alkyl-groups, have been found (Vonk 1983).

Dodine is one of the few multi-site action fungicides, against which fungal resistance developed in the field (Szkolnik and Gilpatrick 1969). In case of laboratory-induced resistance the existence of at least three unlinked genes has been demonstrated, each conferring a different resistance level (Georgopoulos 1969, 1982). The mechanism of dodine resistance is not yet clear, but it is of interest, that negatively correlated cross-resistance between dodine and fenarimol (inhibitor of ergosterol biosynthesis) (chapter 13, 14, 24) was detected in laboratory isolates of *Aspergillus nidulans*, *Cladosporium cucumerinum*, *Penicillium expansum*, *P. italicum* and *Ustilago maydis* (de Waard and van Nistelrooy 1983).

Among the side effects of dodine may be mentioned the bactericidal action, as well as its effect at high concentration against most soil fungi except *Beauveria bassiana*. The latter property of dodine might be used in isolation techniques of certain fungi (Beilharz *et al.* 1982).

Guazatine

Bis(8-guanidino-octyl)amine (IUPAC nomenclature) (Fig. 21.1b) was introduced as its triacetate in 1968 under the code numbers "EM 379" (Evans Medical Ltd.) and "MC 25" (Murphy Chemical Ltd.) and protected by BP 1114155. Trade names are "Panoctine" and "Panolil". Its formulations include liquids and powders.

Physical properties of the triacetate salt: colourless crystals, m.p. 140 °C, readily soluble in water, but insoluble in organic solvents. Stable in neutral and aqueous acidic media. Acute oral LD_{50} for rats is 227—667 mg/kg (WORTHING 1979).

Guazatine salts are contact fungicides. In contrast to the structurally related dodine, the main field of guazatine application are seed dressings of cereals at 0.6—0.8 g a.i./kg seed and post-harvest dips of pineapple, citrus and potato (WORTHING 1979; SCHACHNAI and BARASH 1982). It is also used by foliar application against *Piricularia oryzae* in rice. Mixed formulations with imazalil and/or fenfuram are available.

Guazatine like dodine causes alterations in cellular permeability as well as in the acetate and glycose oxidation. Since experiments *in vitro* showed some enzymes involved in the oxidative processes to be not inhibited directly, the primary mechanism of action should involve a rapid effect on cellular permeability. The apparent loss in oxidative capacity might be explained by the inhibition of the uptake of certain substrates or by the loss of potassium from the cell (SOLEL and SIEGEL 1984).

The recently observed negative cross-resistance between dodine and fenarimol has been demonstrated to some extent also for guazatine and fenarimol DE WAARD and VAN NISTELROOY 1983 (chapter 24).

Anilazine

2,4-dichloro-6-(2-chloroanilino)-1,3,5-triazine (IUPAC nomenclature Fig. 21.1c) was chosen by WOLF *et al.* (1955) among 80 analogues of s-triazines for development as an agricultural fungicide. It was introduced between 1966—1968 by Bayer AG (Brit. 1120338), under the name "Dyrene".

The trade names of w. p. formulations are "Dyrene", "Kemate", "Zinochlor", "Triazine", "Botrysan", "Direx", "Triasyn".

The pure compound exists as colourless crystals, m. p. 159—160 °C; v. p. is very low. (Stable in neutral and weakly acid media, practically insoluble in water, soluble in most organic solvents.) Its acute oral LD_{50} for female rats is 2,700 mg/kg.

Anilazine (Dyrene) is a broad-spectrum protective leaf-fungicide. It was commercially used on turf grasses to control *Helminthosporium* blights, *Fusarium* snow mold and *Rhizoctonia* brown patch (WOLF *et al.* 1955). Its application was extended to the protection of cereals, coffee, vegetables and other crops against diseases caused by *Alternaria, Helminthosporium* and *Cercospora* spp. In Europe it is used preferably against *Botrytis cinerea* on ornamental plants (PERKOW 1983). On tomato plants it is highly effective against early blight *(Alternaria solani)* and grey leaf spot *(Botrytis cinerea)*, but less effective against late blight *(Phytophthora infestans)*. Blooming plants should not be treated. Anilazine is incompatible with oils and alkaline materials. The recently observed synergistic effect of Zn^{2+} and Cu^{2+} ions applied in combinations with anilazine (GOSS and MARSHALL 1985), makes it possible to reduce the concentration of a.i. It may promote the commercial application of anilazine.

Anilazine is quickly and strongly absorbed by fungal spores. ED_{50} value of spore germination in *Neurospora sitophila* is 1,530 μg/g. The ultimate site of action is not known, however the distribution of electrons in the triazine ring encourages nucleophilic substitution reactions (LUKENS 1969, 1971). Cellular amino and sulfhydryl groups are particularly reactive with anilazine (CORBETT 1974), so that it seems likely that it causes inhibition of a variety of cell processes by non-specific interactions with vital cell components. The fungicide character of this type of s-triazines depends on the phenylamino substituent, while the alkylamino derivatives possess herbicidal activity.

Non-essential thiol and amino groups of fungi can destroy the fungicide. One or two chlorine atoms of the triazine ring can take part in this reaction. The degradation in soil is very quick, with a half-life time of about 12 hours. Its photochemical degradation on silica gel is rather intensive, with a half-life time of 2 days (Hulpke *et al.* 1983).

Development of resistance in *Sclerotinia homoeocarpa* on turf grasses to anilazine has been reported from the USA (Nicholson *et al.* 1971).

Fenitropan

(1RS, 2RS)-2-nitro-1-phenyltrimethylene di(acetate) (IUPAC nomenclature, Fig. 21.1d) was introduced and patented by the EGYT Pharmacochemical Works (present name: EGIS) under the code number "EGYT 2248" (GBP 1 561 422, USP 4 160 035). The trade name is "Volparox".

Fenitropan forms colourless crystals, m.p. 70—72 °C. Its solubility in water is 30 mg/l, soluble in organic solvents (Worthing 1984). It is relatively stable in aprotic organic solvents, but it is gradually decomposed in water solution, producing the 1-desacetyl unsaturated derivative. This first product of transformation has a similar fungitoxic activity (Josepovits *et al.* 1984). Acute oral LD_{50} for rats is 3,240—3,850 mg/kg.

It is a contact fungicide with good preventive action against several diseases. At present it is recommended for seed dressing as a liquid formulation. The main field of application is cereals (0.2 g a.i./kg seed). It provides also a good control of seed-borne and soil-borne diseases on maize, rice and sugar beet (Kis-Tamás *et al.* 1981). Its application as a foliar fungicide against powdery mildew and scab on apple trees is under investigation.

Fenitropan is a fungicide with multi-site action. The molecular basis of its primary action seems to be the reaction with essential SH-groups. The above mentioned unsaturated desacetyl-derivative of fenitropan reacts more readily than fenitropan itself (Josepovits *et al.* 1982). Thus, the possibility of the formation of a double bond between carbon atoms 1 and 2 is an important requirement for the fungitoxicity (Mikite *et al.* 1982; Lopata *et al.* 1983). The inhibition of some essential SH-enzymes *in vitro*, as a consequence of blocking SH-groups, was demonstrated (Gullner and Mikite 1984). Addition of glutathion diminished both the inhibition of enzyme and the fungitoxic action.

Large amounts of free thiol-compounds (glutathion, cystein) in fungi can inactivate the fungitoxic agent and probably play a role in the resistance of some fungi (Josepovits *et al.* 1982). The acetyl-group at carbon atom 1 delays this inactivation.

The inhibition of the fixation of aromatic amino-acids on tRNAs in *Fusarium oxysporum* treated with fenitropan, was also observed (Király *et al.* 1985).

The first step of microbiological degradation of fenitropan does not differ greatly from that in sterile media, producing the 1-desacetyl derivative. A further continuous decrease of this compound leads to the formation of several saturated and unsaturated products, which is performed more intensively in living organisms. Addition reaction with glutathion also contributes to the transformation of fenitropan (Josepovits *et al.* 1984).

Dinocap

2(or 4)-(1-methylheptyl)-4,6(or 2,6)-dinitrophenyl crotonate (IUPAC nomenclature, Fig. 21.2a, b) was introduced in 1946 by Rohm and Haas Co., under the code number "CR-1693", and protected by USP 2526660; 2810767.

The trade names of w.p. or e.c. formulations are "Karathane", "Esenosan", "Crothotane". The mixed formulations of dinocap with folpet or dodine + monocrotophos are available. It is incompatible with oil-containing and alkaline preparations.

Technical dinocap is a mixture of about 65—70 % of 2,6-dinitro-4-octylphenyl crotonate (dinocap-4) (Fig. 21.2b) and 30—35 % of 2,4-dinitro-6-octylphenyl crotonate (dinocap-6) (Fig. 21.2a) (GREEN *et al.* 1979). Its physical form: brown liquid, b.p. 138—140 °C at 0.07 mbar. Dinocap is almost insoluble in water and soluble in common organic solvents. Hydrolysis of the ester group occurs in alkaline media. Acute oral LD_{50} for rats 980—1,190 mg/kg.

Dinocap was first developed as an acaricide, but it quickly became an important fungicide against powdery mildews, being often superior to sulphur because of its eradicant as well as protective action (GODFREY 1952; BRANDES 1964). It was the first organic fungicide specifically active against powdery mildews (SCHLÖR 1970) At present it is recommended for the control of powdery mildews on various fruits

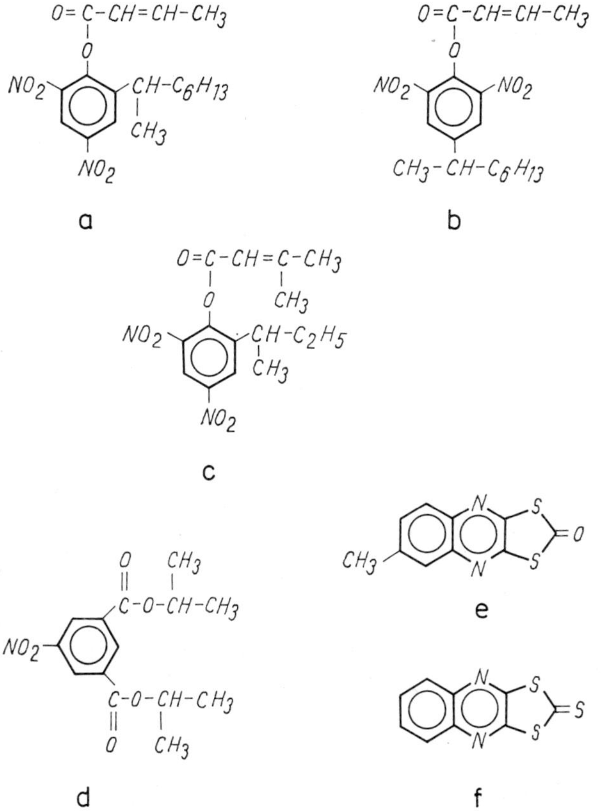

Fig. 21.2 Miscellaneous fungicides acting mainly against powdery mildew. a) Dinocap-6, b) Dinocap-4, c) Binapacryl, d) Nitrothal-isopropyl, e) Quinomethionate, f) Thioquinox.

grape vines and ornamentals at 70—1,120 g a. i./ha (WORTHING 1979). Red spider mites are also controlled by regular spraying. Dinocap-4 is responsible for the antifungal activity, while dinocap-6 for the acaricidal activity (GREEN *et al.* 1979). Crop tolerance is good for the recommended areas of use. Apple pollen was more sensible to dinocap used in a large volume spray than in ULV and LV applications (CHURCH *et al.* 1983).

By the analogy with the mode of action of dinitrophenols, it is a logical assumption that dinocap is hydrolyzed by fungal enzymes and the liberated dinitrophenol acts as uncoupler of mitochondrial oxidative phosphorylation (HASSALL 1982). The other parts of molecule presumably act as carriers for reaching the site of action within the fungus without damaging the higher plants.

Dinocap has a relatively short residual effect. In plants and other organisms both the enzymic reduction of nitro-groups and the hydrolysis of the ester bond are possible. The ortho-position of NO_2-group is more favourable for the enzyme reduction (JOSEPOVITS 1966). In soils dinocap is first transformed to dinitro-octyl-phenol (DNOP) and further to more polar metabolites. After 30 days about 11 % of the applied radioactivity was characterized as unchanged dinocap, while DNOP represented 5.5 %. The total mineralization was slower (MITTELSTAEDT and FÜHR 1984).

Binapacryl

2-sec-butyl-4,6-dinitrophenyl-3-methylcrotonate (IUPAC nomenclature, Fig. 21.2c) was introduced in 1960 by HOECHST AG (HOE 2784). Other name: dinoseb-methacrylate. Trade marks: "Acricid", "Endosan", "Morocide", as w.p., e.c. and dust formulations.

The pure compound exists as colourless crystals, m.p. 68—69 °C. The technical grade represented by yellow to brownish crystals, m.p. 65—69 °C, v.p. 0.421×10^{-6} mbar at 20 °C. Practically insoluble in water, soluble in organic solvents. It is unstable in alkalis. Acute oral LD_{50} for rats varies between 150 and 420 mg/kg (WORTHING 1979).

Chemically binapacryl is closely related to dinocap. It has a non-systemic acaricidal activity together with a fungitoxic effect against powdery mildews of top fruits at 0.025—0.05 % (a.i.) (WORTHING 1979), or 0.05—0.1 % a.i. (PERKOW 1983). As a fungicide binapacryl is less active than dinocap (GREWE 1965). It can be mentioned that both compounds differ not only in the chain-length of substituents but also in the position of nitro-groups. Binapacryl contains the NO_2-groups at 4 and 6 position, while dinocap consists of two isomers, one of which (2,6-dinitro) is more fungitoxic.

When applied to Cox apple trees, binapacryl had a favourable effect on the physiological state of plants (BYRDE *et al.* 1984). There is some risk of damage to young tomatoes, grapes and roses. It may be phytotoxic when mixed with organophosphorus compounds. Dinocap is also incompatible with alkaline preparations. The physical quality of suspension concentrates, a newly developed formulation of binapacryl, has been shown to be dependent on the presence of DNBP, a main impurity contained in technical grade binapacryl (FUJIMOTO *et al.* 1982).

As in case of dinocap, the mode of fungitoxic action of binapacryl is attributed to the free dinitrophenol acting as uncoupler or, according to ILIVICKY and CASIDA (1969), as an inhibitor of oxidative phosphorylation. The ester is less harmful to the plants than the parent phenol due to its higher lipophility, which allows its concentration in the external fatty layers of the leaves, where there is no hydrolysis. But it is probably hydrolyzed by enzymes in fungi providing an example of lethal metabolism (HASSALL 1982).

The main steps of metabolic degradation of binapacryl are the reduction of the nitro-groups to amino compounds and the hydrolysis of the ester bond. In animals the produced phenol is eliminated as the glucuronic acid conjugate.

Nitrothal-Isopropyl

Diisopropyl-5-nitroisophtalate (IUPAC nomenclature, Fig. 21.2d) was developed during 1969—1973 by BASF and introduced in 1973 under the code number "BAS 30000 F". It is used mainly as mixtures with sulphur ("Kumulan") or with metiram ("Pallinal"), formulated as w.p. or e.c.

Its physical form is pale yellow powder, m.p. 65 °C, v.p. 10^{-7} mbar at 20 °C. Solubility in water is 0.39 mg/l, better soluble in most organic solvents. It is hydrolyzed by strong acids and alkalis. Its acute oral LD_{50} for rats is about 10,000 mg/kg.

Nitrothal-isopropyl is a contact fungicide with specific action against powdery mildews (Phillips *et al.* 1973). The combined preparations are used mainly against powdery mildew and scab on apple trees at 0.05 % (a.i.) concentration.

A recently developed combination of nitrothal-isopropyl with tridemorph (BAS 38203 F) proved to be more effective against *Erysiphe cichoracearum* on melon, than fenarimol or triadimefon (Jennrich 1984). The prophylactic contact effect of nitrothal-isopropyl is an effective supplement to the systemic activity of tridemorph. There are no data on the mode of action.

Quinomethionate

6-methyl-1,3-dithiolo[4,5-b] quinoxalin-2-one (IUPAC nomenclature, Fig. 21.2e) was introduced in 1962 by Bayer AG under the code numbers "Bayer Ss2074" and "Bayer 36205". Other names: "Chinomethionate", "Oxythioquinox". It belongs to the quinoxaline derivatives. The trade name of w.p. or dust formulations is "Morestan".

It is a yellow crystalline compound, m.p. 172 °C, v.p. 2.7×10^{-7} mbar at 20 °C. Practically insoluble in water, slightly soluble in organic solvents. Under normal conditions it is relatively stable, but undergoes hydrolysis in alkaline media. Its acute oral LD_{50} for rats is 3,000 mg/kg.

Quinomethionate is a non-systemic fungicide specific to powdery mildews on fruits, vegetables and ornamentals at 0.0075—0.0125 % (a.i.) concentrations (Worthing 1979). The activity against *Podosphaera leucotricha* is superior to that of dinocap at the same concentrations (Grewe and Kaspers 1965). It is used also as acaricide. The thermal stability of quinomethionate makes it possible to use it as fumigant against powdery mildew and red spider mites in greenhouses (Schlör 1970). To enlarge the fungitoxic spectrum for other fungal diseases, a combination with propineb is used, but it was found by Carter *et al.* (1983), that quinomethionate itself has also some activity against several fungi *in vitro* except *Botrytis cinerea.*

The molecular basis of the mechanism of action might be the reactions with biologically important amino-groups (Schlör 1970) and the inhibition of enzymes possessing thiol groups. Inhibition of tricarboxylic acid cycle reactions was experimentally proven in rat liver (Carlson and du Bois 1970). Metabolically produced dithiol can also couple with essential metal ions. The metabolic conversion of quinomethionate by the cucumber plants leads to the disruption of the thiocarbonate linkage. The sulphur liberated is incorporated into sulphates and sulphur containing amino acids, while the quinoxaline nucleus is catabolized. Microorganisms can share in fungicide biodegradation (Metche and Piffaut 1983). The terrestrial green algae *Ankistrodesmus falcatus* vigorously degrade the quinomethionate, producing an aminochlorinated compound identified as 6-methyl-2-amino-3-chlorquinoxaline. The latter compound possess herbicidal activity towards the cucumber seedlings (Piffaut and Metche 1983).

In the environment quinomethionate is relatively quickly destroyed by photodegradation, too. The rate of its photodegradation measured on silica gel plates was similar to that of fuberidazole and anilazine, but much more intensive than in the case of other fungicides (Hulpke *et al.* 1983). Two products of photochemical degradation in benzene were identified as 6-methyl-1,2,3,4-tetrahydroquinoxaline-2,3-dione and 6-(or 7)-methyl-3-phenyl-1,2-dihydroquinoxalin-2-one. The latter compound proved to be fungitoxic against *Podosphaera leucotricha*, but the dione showed no activity (Clark and Loeffler 1980).

Thioquinox

1,3-dithiolo 4,5-b quinoxaline-2-thione (IUPAC nomenclature, Fig. 21.2f) was introduced in the sixties by Bayer AG. Other names: quinothionate, chinothionate. It closely resembles quinomethionate in structure. The trade names of w.p. formulations are "Eradex", "Eradition", "Erazidon".

The technical compound is a brownish powder, m.p. 165 °C, m.p. of the pure compound is 180 °C, v.p. 1.3×10^{-7} mbar at 20 °C. It is stable against hydrolytic effects and heating, but sensitive to light, especially in damp conditions. Practically insoluble in water, slightly soluble in aceton, ethanol, chloroform and n-hexane. Acute oral LD_{50} for rats is 3,400 mg/kg.

Thioquinox is mainly used as acaricide in fruit and vine cultivations, but it controls also powdery mildews.

Its mode of action is considered to be similar to that of quinomethionate (Schlör 1970).

Fenaminosulf

This compound (Fig. 21.3a) with interesting features was introduced in 1955 by Bayer AG (Bayer 22 555) (Patent DAS 1 028 828), later known (US Patent 24 960) as "Dexon" or "Lesan" developed by CHEMAGRO Corp. Because of its instability against daylight (Hills and Leach 1962), it was used as a soil and seed fungicide, which exhibited an astonishing high selectivity against Oomycetes such, as *Pythium*, *Aphanomyces*, *Phytophthora* spp. (Tab. 21.1). For seed treatment 45—110 g a.i./100 kg seed have been recommended. The acute oral LD_{50} dose for rats is 60 mg a.i./kg. It can be stabilized by sulphite or in alkaline media. Hills (1962) stated that dexon is taken up by roots and is transported to the hypocotyl in sugar beet seedlings. Even in the dark it was rapidly converted to 4 other products within the plantlets. Exposure to light of treated hypocotyls alleviated the fungistatic effect. *Pythium ultimum*, *Aphanomyces cochlioides* and *P. aphanidermatum* revealed highly sensitive against this compound, and sporulation was inhibited. It has been recommended for control of *Pythium* (damping-off) in seedlings of corn, cotton, sorghum, sugar beets, beans, spinach, cucumber, flax, peanut, vegetables and ornamentals, and against root rot in sugar cane, avocado, pineapple, scrubs and turf. In field plots, treated with dexon high yield increase in peas was obtained by Mitchell and Hagedorn (1971). Low doses of the compound persisted in the soil, high enough for suppression of zoospore

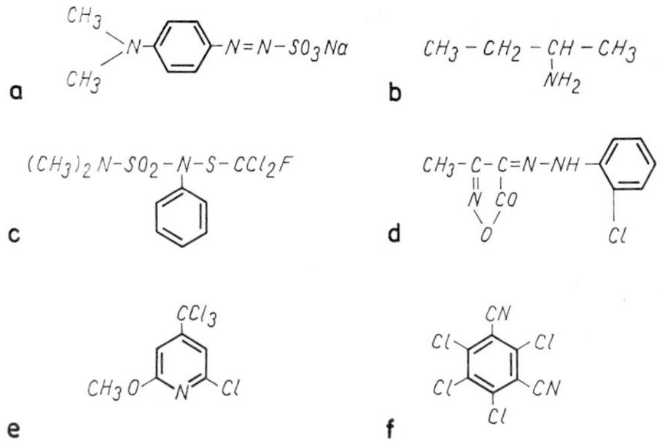

Fig. 21.3 Miscellaneous fungicides with more or less selective spectrum of activity. a) Fenaminosulf, b) Sec-butylamine, c) Dichlofluanide, d) Drazoxolon, e) Pyroxychlor, f) Chlorothalonile.

formation, therefore, a lasting effect was obtained. Application of dexon suppressed the attack of *Aphanomyces cochlioides* in infested soil and in combination with aldicarb produced normal yields in sugar beets even in beet-monoculture (STEUDEL 1972). Dexon controlled root rot in avocado seedlings and in larger trees after repeated applications. It proved to be fungistatic, but at 5 ppm reduced the formation of sporangia and chlamydospores of *Phytophthora cinnamomi* (ZENTMYER 1973).

Comparing dexon and terrazole, WHEELER *et al.* (1970) found similar properties in the control in *Phytophthora* and *Pythium* spp., although a certain variance in the effect against different species exists. Because of its narrow spectrum of activity, its relatively high acute toxicity and its instability the use of fenaminosulf decreased in practical application.

The mechanism of action of fenaminosulf was investigated by TOLMSOFF (1962) and more recently by MÜLLER and SCHEWE (1977), SCHEWE and MÜLLER (1979). TOLMSOFF demonstrated, that dexon inhibits site I (NADH cytochrome c reductase) in the respiratory chain of *Pythium* but nearly does not inhibit site II (succinate dehydrogenase). Mitochondria of Rhizoctonia and sugar beet decomposed fenaminosulf in the presence of NADH, which might be a reason for their insensivity.

HALANGK and SCHEWE (1975) confirmed these effects and demonstrated that electron-transporting particles (ETP) from bovine heart mitochondria are highly

Table 21.1 ED_{50} values for some fungicides on malt agar dishes (inhibition of mycelial radial growth in percent of controls)

	Fenamino-sulf	Dichlo-fluanid	Pyroxy-chlor	Chloro-thalonil	Draz-oxolon
Phytophthora cactorum	7	4	5	10	20
Pythium ultimum	30	2	0.1	15	15
Botrytis cinerea	120	0.5	100	1	0.5
Fusarium oxysporum	250	5	100	6	100
Colletotr. lindemuth.	250	10	100	—	0.5
Cochliobolus carboneum	80	30	100	—	1
Rhizoctonia solani	250	2	100	26	200
Verticillium albo-atrum	20	9	—	65	0.5

sensitive, and stated that dexon inhibits the FMN of the NADH dehydrogenase in contrast to rotenone which interacts at the Fe-S-region of site I (Fig. 10.1). They found several sensitive and insensitive NADH-dependent flavin enzyme systems from various sources.

In sensitive flavin enzymes probably an interaction of dexon with both NAD(P)H and FMN occurs. The concentration in half inhibition by dexon in the NADPH-cytochrome c reductase from yeast corresponds to its FMN concentration. Dexon must be activated and could produce its enzyme inhibition by coplanar complexation of the reduced pyridinnucleotide, the isoalloxazine ring of FMN or both in a sandwich-like manner (MÜLLER and SCHEWE 1977). The selectivity of this compound seems to be mainly caused by differences in degradation or penetration processes in the fungal cells.

Sec-Butylamine (SBA)

This compound (Fig. 21.3b) was introduced in 1962 by the University of California (Riverside) and described by ECKERT and KOLBEZEN (1964). Trade names are "Tutane" (Eli Lilly & Co.) and "Butafume" (BASF AG). The acute oral LD_{50} for rats is 380 mg/kg.

Its main use is for the control of fruit rotting fungi. In potatoes gangrene and skin spot can be controlled. It is used as dips, sprays or as fumigant and is not phytotoxic in 10-fold concentration necessary for fungal control.

The compound has a narrow spectrum of activity. It inhibits fungistatically spore germination and hyphal growth of *Penicillium digitatum*, *P. italicum*, *P. expansum*, *Phomopsis citri* and *Monilinia fructicola*. Growth of other species of *Penicillium* is not inhibited as well as that of other genera even in high concentrations (ECKERT *et al.* 1975). The authors found no evidence for metabolism of SBA in inhibited hyphae, therefore sec-butylamine should be the active fungicide. The (+) isomer is less fungitoxic than the (—) isomer. Both were actively accumulated to the same level, but were not firmly bound within the cell. Sensitivity or resistance of species are not correlated to differences in accumulation or metabolism. The target of SBA in mitochondria is the pyruvic acid dehydrogenase. Cell permeability is not disturbed, but active amino acid transport is inhibited (BARTZ and ECKERT 1972). Pyruvic acid accumulates 7 times over the control under the influence of SBA in *P. digitatum*. This means that the decarboxylase is the primary target of this compound which inhibits respiration and secondarily other metabolic processes (YOSHIKAWA and ECKERT 1976). This was confirmed with isolated mitochondria. The inhibition proved to be a competitive one (YOSHIKAWA *et al.* 1976).

SBA is practically applied in aqueous solutions to oranges and lemons after harvest in order to control *Penicillium rots*. Appearance of resistance within several years after the introduction of the fungicide was observed as could be expected by its selective action. ECKERT (1984) discussed the possibilities to avoid a resistance break through in citrus fruit packing houses by various strategies of application of fungicides. The cause of resistance is not a difference in accumulation, but more probably a mutation of the structure of the target enzyme.

Dichlofluanid (DCF)

DCF was introduced in 1965 by Bayer AG (Bayer 47 531) (Fig. 21.3c) and protected by DAS 1 193 498, trade names are "Euparen" and "Elvaron". The acute oral ED_{50} for rats is 2,500 mg/kg. It has a relative broad spectrum of activity (Tab. 21.1) and is active already at low doses in sensitive fungi.

It is a protective fungicide and has been used against *Venturia* spp. in apples and pear cultures. It controls *Botrytis cinerea* and downy mildews and has some side effects against powdery mildews and red spider mites (GREWE 1968).

Its development was stimulated by the valuable properties of the related compounds captan and folpet. Decisive was an exchange of one chlorine atom in the perchloromethylmercapto moiety by a fluorine atom. Although the reaction of this group within a target organism should be rather unspecific, DCF has a certain selectivity.

A higher activity against *Botrytis cinerea, Plasmopara viticola, Venturia inaequalis, Phytophthora infestans* and *Alternaria solani* in comparison to TMTD, zineb, captan and folpet could be stated. Important for its action seems to be an additional activity over the vapour phase.

Its practical application was stimulated by its protective and curative control at lower doses than captan of apple and pear scab. Additionally favoured its practical application good effects against *Gloeosporium* fruit decay as preharvest spray (KASPERS 1968).

A side effect against *Podosphaera leucotricha* in apples could be stated which allowed to lower the concentration of morestan (chinomethionate) in mixtures. At concentrations of 0.1 % side effects against spider mites in apples and hop were observed which allowed to reduce acaricidal sprays.

A good effectivity against *Plasmopara viticola* and *Botrytis cinerea* in vineyards and against *Pseudoperonospora humuli* and partly *Sphaerotheca humuli* as well as against *Botrytis cinerea* recommended its practical application (KOLBE and KASPERS 1968).

Besides the application in crop protection its properties allow use as an antifouling agent or as an additive for paints for wood preservation especially against blue staining fungi.

The mechanism of action could be similar to that of captan or folpet, but a certain selectivity similar to chloroneb (Tab. 2.5 and Tab. 2.1) can be stated which is rather surprising if the mechanism of action is a multiside one as described for captan or folpet (LUKENS and SISLER 1958; SIEGEL 1971). Resistance or cross resistance to AHF or other fungicides was not observed (LYR and CASPERSON 1982).

Drazoxolon

Its IUPAC name is 4-(2-chlorophenylhydrazono)-3-methyl-5-isoxazolone (Fig. 21.3d). It was synthesized by ICI Ltd. in 1960, produced under the code number "PP781" and protected by BP 999097. It is almost insoluble in water, but can be dissolved in aqueous alkali under salt formation (WORTHING 1979). It should not be combined with lime sulphur or dodine because of instability and phytotoxicity.

Drazoxolon has been used in the control of powdery mildews on roses, blackcurrants and other crops and is active against other foliar diseases. As seed treatment it controls *Pythium* and *Fusarium* spp. in peas, maize, beans, grass and can be used as a soil treatment to control damping off in ornamental seedlings. The acute oral LD_{50} is for rats 126 mg/kg. It was commercialized as "MIL-COL" ((330 g a.i./l), Saisan (300 g/l) or "Ganocide" (for control of *Ganoderma* spp. in rubber trees *(Hevea)*).

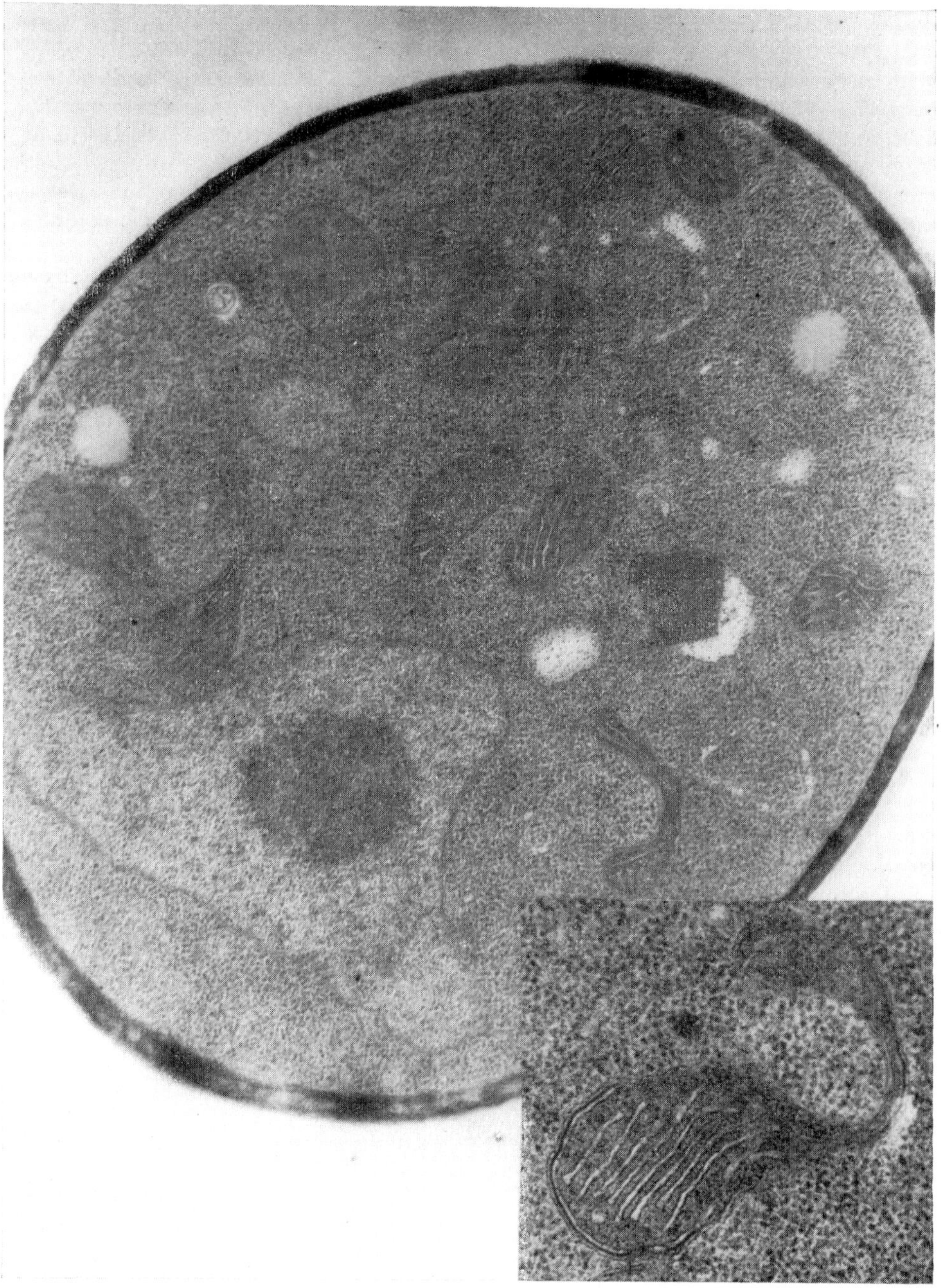

Fig. 21.4 Ultrastructural changes in *Mucor mucedo*, 2 hours after application of 20 ppm Drazoxolon. The matrix of the mitochondria is very dense, the cell wall thickness increased. 36,000 : 1 and 60,000 : 1 (phot. Dr. CASPERSON).

The compound is of medium activity and found no wide distribution. Regarding its mechanism of action there was described an uncoupling activity in rat liver mitochondria by PARKER and SUMMERS (1970). This effect needs to be reinvestigated, but electron microscopic pictures (Fig. 21.4) seem to confirm this effect. Mitochondria in *Mucor* are well preserved regarding their inner mitochondrial membranes, but somewhat distorted just like the nucleus, which resembles to the effect of trichlorophenol. In contrast to true uncoupling compounds the cell wall gets thickened in *Mucor* under the influence of drazoxolone (Fig. 6.7). Recent results revealed that drazoxolon induces a lipid peroxidation (LYR, unpubl.).

Pyroxychlor

This compound (2-chloro-6-methoxy-4-trichloromethylpyridine) (Fig. 21.3e) was developed by the DOW Chemical Corp. and tested as "DOWCO 269", "Nurelle", "Lorvek". It is worth mentioning by its selectivity against Oomycetes such as *Peronosporaceae*, *Pythium* and *Phytophthora*. Other fungi are hardly affected (Tab. 21.1) (KLUGE 1978).

An unique and surprising feature is its ambimobility, that means it can be transported in the xylem as well as in the phloem. In its mechanism of action it should be related to the group of Aromatic Hydrocarbon Fungicides, but detailed investigations were not presented.

Chlorothalonile

Chlorothalonile (tetrachlorisophthalonitrile) (Fig. 21.3f) was already introduced in 1963 by Diamond Alkali Co (Diamond Shamrock Corp.) under trade names as "Bravo" "Daconil" 2787, "Exotherm Termil" and protected by USP 3 280 353, 3 331 735. It is a stable and non corrosive compound with a relatively broad spectrum of activity (Tab. 21.1). It can be used for the protection of many vegetable and agricultural crops at rates of 0.63—2.52 kg a.i./ha. It controls *Botrytis cinerea* and other pathogens, but has a certain selectivity (Tab. 21.1). The acute LD_{50} for albino rats is extremely high (more than 10,000 mg/kg) (WORTHING 1979). Because of its simple structure, good efficiency, low toxicity and valuable other properties it is still in practical use on a broad scale. Beside its fungicidal properties it is used as algicide or as a preservative in paints and adhesives.

The mechanism of action was thoroughly investigated by VINCENT and SISLER (1968), TURNER and BATTERSHELL (1970) and TILLMAN *et al.* (1973).

Chlorothalonile reacts *in vitro* with glutathione, coenzyme A, 2-mercaptoethanol and other compounds forming several S-derivatives (VINCENT and SISLER 1968). By this the SH-content in cells is significantly reduced which results in a lethal inhibition of a number of thiol dependent reactions (enzymes) in fungal cells.

The glutathione level is an important regulator of normal cell metabolism which is readily affected by chlorothalonile. In addition a direct interaction with SH-groups of enzymes can occur, leading to irreversible enzyme inactivations.

Because of this unspecific attack within the cell a certain selectivity is surprising.

References

BARTZ, J. A., and ECKERT, J. W.: Studies on the mechanism of action of 2-aminobutane. Phytopathology 62 (1972): 239—245.

BEILHARZ, V. C., PARBERY, D. G., and SWART, H. J.: Dodine: A selective agent for certain soil fungi. Trans. Brit. Mycol. Soc. 79 (1982): 507—511.

BRANDES, G. A.: The dinitros, "Elder Statesman" of the pesticides. Farm Chem. 172 (1964): 46—48.

BROWN, J. F., and SISLER, H. D.: Mechanisms of fungitoxic action of n-dodecyl-guanidine acetate. Phytopathology 50 (1960): 830—839.

BYRDE, R. J. W.: Non aromatic Organics. In: TORGESON, D. C. (Ed.): Fungicides. Vol. 2. Academic Press, New York and London 1969, pp. 531—578.

— HUTCHEON, J. A., COYLE, J., and HOLGATE, M. E.: Cumulative effects of apple mildew fungicides 1982—1983. Brit. Crop Protect. Conf. 1984. Proc. Vol. 3. 1984: 1143—1148.

CARLSON, G. P., and DU BOIS, K. P.: Studies on the toxicity and biochemical mechanism of action of 6-methyl-2,3-quinoxalinedithiol cyclic carbonate (Morestan). J. Pharmacol. Exp. Ther. 173 (1970): 60—70.

CARTER, G. A., CLARK, T., and JAMES, C. S.: Fungicidal activity of substituted quinoxalines. Pesticide Sci. 14 (1983): 135—141.

CHURCH, R. M., MORGAN, N. G., and WILLIAMS, R. R.: The effect of spray volume on the toxicity of captan and dinocap to apple pollen in the orchard. J. Horticult. Sci. 58 (1983): 165—168.

CLARK, T., and LOEFFLER, R. S. T.: The photolysis of quinomethonate in benzene solution. Pesticide Sci. 11 (1980): 451—457.

CORBETT, J. R.: The Biochemical Mode of Action of Pesticides. Academic Press, London-New York 1974.

DE WAARD, M. A., and VAN NISTELROOY, J. G. M.: Negatively correlated cross-resistance to dodine in fenarimol-resistant isolates of various fungi. Netherl. J. Plant Pathol. 89 (1983): 67—73.

ECKERT, J. W.: Etiology and control of *Penicillium* races resistant to postharvest fungicides. In: LYR, H., and POLTER, C. (Eds.): Systemic Fungicides and Antifungal Compounds. Tagungsber. Akad. Landwirtschaftswiss. DDR 222 (1984): 165—177.

— BRETSCHNEIDER, B. F., and RAHM, M.: Studies on the selective fungitoxicity of sec-butylamine. Proc. VIII Intern. Plant Protect. Congr. Moscow (1975) Sect. III, pp. 215—227.

— and KOLBEZEN, M. J.: 2-Aminobutane salts for control of postharvest decay of citrus, apple, pear, peach, and banana fruits. Phytopathology 54 (1964): 978—986.

FUJIMOTO, M., NAKAMURO, T., and MURAOKO, E.: Influence of the impurity in technical grade binapacryl on the particle growth of binapacryl in suspension concentrates. J. Pesticide Sci. 7 (1982): 507—512.

GEORGOPOULOS, S. G.: The problem of fungicide resistance. Bio Science 19 (1969): 971—973.

— Genetical and biochemical background of fungicide resistance. In: DEKKER, J., and GEORGOPOULOS, S. G. (Eds.): Fungicide Resistance in Crop Protection. Pudoc, Wageningen 1982, pp. 46—62.

GODFREY, G. H.: Cantaloup powdery mildew control with dinitro capryl phenyl crotonate. Phytopathol. 42 (1952): 335—337.

GOSS, V., and MARSHALL, W. D.: Synergistic antifungal interactions of zinc or copper with anilazine. Pesticide Sci. 16 (1985): 163—171.

GREEN, M. B., HARTLEY, G. S., and WEST, T. F.: Chemicals for Crop Protection and Pest Control. Pergamon Press, Oxford-New York-Toronto-Sydney-Paris-Frankfurt 1979.

GREWE, F.: Rückblick auf 25 Jahre Fungizidforschung. Pflanzenschutz-Nachrichten BAYER 18, Sonderheft (1965): 45—74.

— Euparen (Dichlofluanid), ein neues polyvalentes Fungizid mit besonderer Wirkung gegen Grauschimmel (*Botrytis cinerea* Pers.). Pflanzenschutz-Nachrichten BAYER 21 (1968): 147—170.

— and KASPERS, H.: Morestan, ein neues Fungizid aus der Gruppe der 2,3-disubstituierte Chinoxaline zur Bekämpfung echter Mehltaupilze. Pflanzenschutz-Nachr. Bayer 18 (1965): 1—23.

GULLNER, G., and MIKITE, G.: Mutual effect of nitroolefine fungicides with thiol-containing dehydrogenase enzymes. Növényvédelem 20 (1984): 261.

HALANGK, W., and SCHEWE, T.: Untersuchungen zum biochemischen Wirkungsmechanismus von Dexon. In: LYR, H., and POLTER, C. (Eds.): Systemfungizide (Symposium). Akademie-Verlag, Berlin 1975, S. 177—182.

HASSALL, K. A.: The Chemistry of Pesticides. The Macmillan Press Ltd, London and Basingstoke 1982.

HILLS, F. J.: Uptake, translocation, and chemotherapeutic effect of p-dimethylamino-benzene-diazo-sodium-sulfonate (Dexon) in sugar beet seedlings. Phytopathology **52** (1962): 389—392.

— and LEACH, L. D.: Photochemical decomposition and biological activity of p-dimethyl-amino-benzene-diazo-sodium-sulfonate (dexon). Phytopathology **52** (1962): 51—56.

HULPKE, H., STEIGH, R., and WILMES, R.: Light-induced transformations of pesticides on silica gel as a model system for photodegradation on soil. In: MATSUNAKA, S., HUTSON, D. H., and MURPHY, S. D. (Eds.): Pesticide Chemistry. Human Welfare and the Environment. Vol. 3. Pergamon Press. Oxford-New York-Toronto-Sydney-Paris-Frankfurt 1983, pp. 323—332.

ILIVICKY, J., and CASSIDA, J. E.: Uncoupling action of 2,4-dinitrophenols and certain other pesticide chemicals upon mitochondria and its relation to toxicity. Biochem. Pharmacol. **18** (1969): 1389—1401.

JENNRICH, H.: Control of powdery mildew *(Erysiphe cichoracearum)* with BAS 38203 on melons in Spain. Proc. Brit. Crop Protect. Conf. 1984, Vol. 3 (1984): 1167—1168.

JOSEPOVITS, G.: Biologische Reduktion von nitroaromatischen Fungiziden. In: M. GIRBARDT (Ed.): Mechanisms of Action of Fungicides and Antibiotics. Akademie-Verlag, Berlin 1966, pp. 341—348.

— GASZTONYI, M., KIS-TAMÁS, A., and MIKITE, G.: Fungal metabolism and fungitoxicity of the diacetyl derivative of 1-phenyl-2-nitro-1,3-propandiol. V. IUPAC Conf., Kyoto 1982. Abstr. IVd—5.

— — — — and GULLNER, G.: Chemical and microbiological transformations of the fungicide fenitropan. Tagungsber. Akad. Landwirtschaftswiss. DDR, Berlin **222** (1984): 225—228.

KASPERS, H.: Über Anwendungsmöglichkeiten von Euparen zur Bekämpfung von Kelch- und Lagerfäulen. Pflanzenschutz-Nachrichten BAYER **21** (1968): 243—256.

KIRÁLY, I., JAKUCS, E., RÁCZ, I., TAMÁS, L., and LÁSZTITY, D.: Effect of the fungicide fenitropan on the metabolism of some higher plants and *Fusarium oxysporum.* Pesticide Sci. **16** (1985): 1—9.

KIS-TAMÁS, A., JAKUCS, E., and MIKITE, G.: (1RS, 2RS)-2-nitro-1-phenyltrimethylene-di-acetate a novel fungicide. Proc. Brit. Crop Protect. Conf., Vol. **1** (1981): 29—33.

KLUGE, E.: Vergleichende Untersuchungen über die Wirksamkeit von Systemfungiziden gegen Oomyzeten. Arch. Phytopath. u. Pflanzenschutz, Berlin, **14** (1978): 115—122.

KOLBE, W., and KASPERS, H.: Die Bekämpfung pilzlicher Krankheiten im Hopfenbau mit organischen Fungiziden unter Berücksichtigung der akariziden Nebenwirkung. Pflanzenschutz-Nachrichten BAYER **21** (1968): 278—303.

LOPATA, A., DARVAS, F., VALKÓ, K., MIKITE, G., JAKUCS, E., and KIS-TAMÁS, A.: Structure-activity relationships in a series of new antifungal nitroalcohol derivatives. Pesticide Sci. **14** (1983): 513—520.

LUKENS, R. J.: Heterocyclic nitrogen compounds. In: TORGESON, D. C. (Ed.): Fungicides. Vol. II. Academic Press, New York and London 1969, pp. 396—446.

— Chemistry of Fungicidal Action. CHAPMAN and HALL, London 1971.

— and SISLER, H. D.: Chemical reactions involved in the fungitoxicity of Captan. Phytopathology **48** (1958): 235—244.

METCHE, M., and PIFFAUT, B.: Study of the toxicity of quinomethionate (6-methyl-2,3-dithioquinoxaline cyclocarbonate). I. Penetration and metabolism of quinomethionate in cucumber *(Cucumis sativus).* Agric. Biol. Chem. **47** (1983): 1725—1732.

MIKITE, G., JAKUCS, E., KIS-TAMÁS, A., DARVAS, F., and LOPATA, A.: Synthesis and antifungal activity of new nitro-alcohol derivatives. Pesticide Sci. **13** (1982): 557—562.

MILLER, P.: Mechanisms for reaching the site of action. In: TORGESON, D. C. (Ed.): Fungicides. Vol. II. Academic Press, New York-London 1969, pp. 2—60.

MITCHELL, J. E., and HAGEDORN, D. J.: Residual dexon and the persistent effect of soil treatments for control of pea root rot caused by *Aphanomyces euteiches.* Phytopathology **61** (1971): 978 to 983.

MITTELSTAEDT, W., and FÜHR, F.: Degradation of dinocap in three german soils. J. Agr. Food Chem. **32** (1984): 1151—1155.

Müller, W., and Schewe, T.: Zum Wirkungsmechanismus der Hemmung von Pyridinnukleotid abhängigen Flavinenzymen durch das Systemfungizid Dexon. Acta biol. med. Germ. **36** (1977): 320—327.

Nicholson, J. F., Meyer, W. A., Sinclair, J. B., and Buller, J. D.: Turf isolates of *Sclerotinia homeocarpa* tolerant to dyrene. Phytopathol. Z. **72** (1971): 169—172.

Parker, V. H., and Summers, L. A.: Uncoupling of oxidative phosphorylation by arylhydrazono-isoazolone fungicides. Biochem. Pharmacol. **19** (1970): 315—317.

Perkow, W.: Wirksubstanzen der Pflanzenschutz- und Schädlingsbekämpfungsmittel. Paul Parey, Berlin-Hamburg 1983.

Phillips, W. H., Pommer, E. H., and Löcher, F.: Field evaluation of 5-nitro-isophtalic diisopropyl ester (BAS 3000 F) for the control of apple mildew, *Podosphaera leucotricha*, in the European community. Proc. Brit. Insect. Fungicid. Conf. 7th, Vol. 2 (1973): 673—680.

Piffaut, B., and Metche, M.: Study of the toxicity of quinomethionate (6-methyl-2,3-dithioquinoxaline cyclocarbonate). II. Biotransformation of quinomethionate into toxic derivatives by a green alga *(Ankistrodesmus falcatus)*. Agric. Biol. Chem. **47** (1983): 1733—1740.

Schachnai, A., and Barash, I.: Evaluation of the fungicides CGA 64251, guazatine, sodium o-phenylphenate and imazalil for control of sour rot on lemon fruit. Plant Dis. **66** (1982): 733—735.

Schewe, T., and Müller, W.: Zum molekularen Wirkungsmechanismus von Dexon. In: H. Lyr and C. Polter (Eds.) Systemic Fungicides. Akademie Verl. Berlin, 1979, S. 317—324.

Schlör, H.: Chemie der Fungizide. In: Wegler, R. (Hrsg.): Chemie der Pflanzenschutz- und Schädlingsbekämpfungsmittel. Band 2. Springer-Verlag, Berlin-Heidelberg-New York 1970, S. 45—171.

Siegel, M.: Reactions of the fungicide folpet (N-[trichloromethylthio]phthalimide) with a non-thiol protein. Pesticide Biochem. and Physiol. **1** (1971): 234—240.

Solel, Z., and Siegel, M. R.: Effect of the fungicide guazatine and dodine on growth and metabolism of *Ustilago maydis*. Z. Pflanzenkrankh. Pflanzenschutz. **91** (1984): 273—285.

Steudel, W.: Neue Ergebnisse zur Frage der Schädigung von Zuckerrüben durch pilzliche Wurzelparasiten. Phytopathol. Z. **75** (1972): 202—214.

Szkolnik, M., and Gilpatrick, J. D.: Apparent resistance of Venturia inaequalis to dodine in New York apple orchards. Plant Dis. Rep. **53** (1969): 861—865.

Tolmsoff, W. J.: Biochemical basis for biological specificity of Dexon (p-dimethylamino-benzene-diazosodium-sulfonate) as a fungistat. Phytopathology **52** (1962): 755.

Vonk, J. W.: Metabolism of Fungicides in Plants. In: Hutson, D. H., and Roberts, T. R. (Eds.): Pesticide Biochemistry and Toxicology. Vol. **3**. John Wiley and Sons, Chichester-New York-Brisbane-Toronto-Singapore 1983, pp. 111—162.

Wheeler, J. E., Hine, R. B., and Boyle, A. M.: Comparative activity of dexon and terrazole against *Phytophthora* and *Pythium*. Phytopathology **60** (1970): 561—562.

Wolf, C. N., Schuldt, P. H., and Baldwin, M. M.: s-triasine derivatives — a new class of fungicides. Science **121** (1955): 61—62.

Worthing, C. R.: The Pesticide Manual. 6th edit. Brit. Crop Protect. Council. Croydon 1979.

— The Pesticide Manual. 7th edition. Brit. Crop Protect. Council, Croydon 1984.

Yoshikawa, M., and Eckert, J. W.: The mechanism of fungistatic action of sec-butylamine. I. Effects of sec-butylamine on the metabolism of hyphae of *Penicillium digitatum*. Pesticide Biochem. Physiol. **6** (1976): 471—481.

— — and Keen, N. T.: The mechanism of fungistatic action of sec-butylamine. II. The effect of sec-butylamine on pyruvate oxidation by mitochondria of *Penicillium digitatum* and on the pyruvate dehydrogenase complex. Pesticide Biochem. Physiol. **6** (1976): 482—490.

Zentmyer, G. A.: Control of *Phytophthora* root rot of avocado with p-dimethylaminobenzene-diazo-sodium-sulfonate (dexon). Phytopathology **63** (1973): 267—272.

Lyr, H. (Ed.): Modern Selective Fungicides — Properties. Applications, Mechanisms of Action. Longman Group UK Ltd., London, and VEB Gustav Fischer Verlag, Jena, 1987.

Chapter 22

Computer design of fungicides

A. F. Marchington and Sandra A. Lambros
ICI Plant Protection Division, UK

Introduction

The fundamental aim of research departments in the agrochemical and pharmaceutical industries is the search for biologically active molecules of commercial worth. The search for such a compound is usually started through the discovery of a lead compound, found from either random screening, other companies patients, natural substrates or, in theory, more rational biochemical design. Synthetic chemists then try to enhance the biological signal by synthesising modifications of this lead compound. The problem has always been in deciding which are the most advantageous modifications, and even by careful consideration of scientific principles the success rate is very low — only one molecule in approximately 15—20,000 becomes a development compound. Many companies, being aware of the large cost which this entails, have recently concentrated in trying to increase their chances of success through the use of theoretical chemistry and recent advances in computer graphics (Richards 1984; Marchington, Robins and Richards 1982).

Usually, but not always, a molecule possesses biological activity primarily because it binds to a biological macromolecule such as a protein. Although a couple of hundred X-ray structures have now been determined for proteins (Taylor and Bernstein) nearly all of the target sites of interest to drug and, especially, agrochemical companies are unknown. Many times more primary sequences of proteins have been determined but until such techniques as protein folding (Ghelis and Yon 1982) and subsequent energy minimisation programmes (Weiner and Kollman 1981; Brooks *et al.* 1983) have been perfected, such information is of limited value unless the structure of a homologous protein is known. Much current work is at present being devoted to these areas (Capaldi, Marshall and Staples 1983).

Even when the X-ray structure of the protein is known, several questions still remain unanswered. For instance, what changes do:

(a) the substrate; and
(b) the receptor

undergo as one approaches the other? How can we understand the complicated environment of solvent, lipid, counter-ions, and proteins which surround these molecules in real life? Will the molecule ever reach the site of interest in the first place? Theoretical methods may seem inappropriate under these conditions since they are primarily concerned with the isolated molecule. However, the recent successes of such techniques are due to judicious application in conjunction with other experimental disciplines which provide physical and spectral information to supplement calculations on the small molecule protein interactions. In addition, where possible, a series of similar molecules is often studied and environmental factors assumed constant.

In molecular terms, how one molecule appears to another can be divided into 2 parts:

(a) **Where are the Nuclei?**

This is not just a question of equilibrium shape as measured by n.m.r., X-ray or neutron spectroscopy, but also concerns what possible shapes the molecule can assume as it interacts, namely what flexibility it possesses. This applies to both the protein as well as to the substrate.

(b) **Where are the Electrons?**

In this area, theoretical chemistry methods, based upon the solution of the Schrodinger equation (RICHARDS and HORSLEY 1970; POPLE and BEVERIDGE 1970), offer advantages since, unlike the available experimental methods, non-equilibrium as well as equilibrium states can be studied. The question which needs to be answered is how the distribution of these electrons around the nuclei determine the likelihood of effective collision and consequent interaction. Molecules interact most strongly at their surfaces and many workers through computer graphics techniques are beginning to look at the appearance of these surfaces.

Both these questions can be answered through the use of theoretical chemistry, programmes for which are readily available through the Quantum Chemistry Programme Exchange (QCPE) at Indiana University, and indeed the most important advance in the last few years has not been in these methods themselves but in the computer technology which is necessary to apply them.
The two most important technological advances are:

(i) Increased CPU and storage has meant that these techniques can be applied to larger and larger molecules which are of biological interest.
(ii) The advent of computer graphics technology at an affordable price, which has enabled the rapid visualisation of such results and hence enabled non-specialist, such as synthetic chemists, and specialists alike, to readily apply theoretical techniques to pesticide and drug design. In particular molecular graphics provides the link between a chemist's intuition and a vast array of chemical, physical and biological information (VINTER 1985; HASSALL 1985).

In trying to design new and better inhibitors which have a similar mode of action, it is first necessary to collate all the known information available both from the literature and from experiment. This includes all the biochemical, physical and biological data which can then be used to construct a crude, two-dimensional picture of the site of action.

Through the use of computer graphics in conjunction with the available crystal data, theoretical calculations, including the use of both molecular orbital and molecular mechanics calculations, infra-red and n.m.r. spectroscopic studies, it is then possible to construct a three dimensional model of the target enzyme active site which whilst exhibiting high selectivity for the natural substrate, is also capable of binding a range of known lead antagonists.

The design triazole fungicides

As an example of the use of theoretical methods in conjunction with computer graphics we describe the design of triazole fungicides such as flutriafol (Fig. 22.1). This is the latest molecule from a general class of compound which is now attracting wide commercial interest both as crop protection fungicides and as pharmaceutical

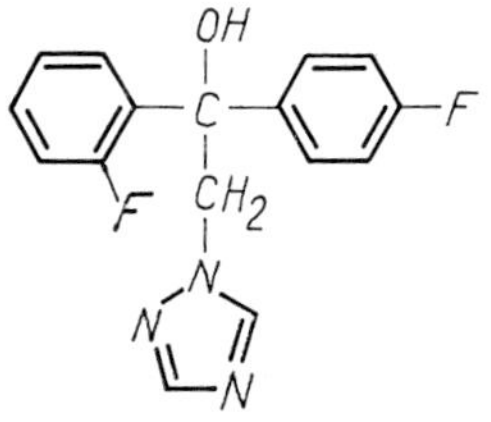

Fig. 22.1 The chemical structure of flutriafol, a new broad spectrum triazole fungicide.

anti-mycotics, and include such materials as triadimefon, triadimenol, prochloraz, propiconazole, penconazole, imazalil and diclobutrazol.

The model therefore was designed specifically to accommodate both the natural substrate (24-methylene-24,25-dihydrolanosterol) as obtained from its X-ray structure and also the known antagonists in conformations which were either their global minimum energy conformer or a relatively low energy form.

It was then possible to use this model to suggest new structure activity relationships and contribute towards the design of new novel fungicides.

The triazole fungicides have been shown to inhibit the 14-alpha-demethylation of 24-methylene-25,25-dihydrolanosterol, the ergosterol precursor (GADHER *et al.* 1983). This is known to be an important step in ergosterol biosynthesis since once it is completed a number of other steps can begin more or less in tandem. When one of these inhibitors is added to a rat liver cytochrome P-450 preparation, for example, an unmistakable Type II Soret difference spectrum is observed. This indicates that the triazole 4-nitrogen has co-ordinated to the heme ferric ion (which remains in the ferric not ferrous state) in its low spin resting state. Obviously this antagonist will have to displace the natural sixth ligand of the heme which may be a water molecule (GRIFFIN and PETERSON 1975) or less likely an imidazole group available from a protein histidine. The other axial ligand, below the heme plane, is believed to be a cysteine sulphur (MURAKAMI and MASON 1967).

In computer modelling the antagonism of these heme binding fungicides, it is necessary to consider only the first oxidation of the parent lanosterol to the 14-methyl alcohol. This is because the 14-alpha-demethylation of dihydrolanosterol proceeds in three main stages with the two intermediates — the alcohol, 5-alpha-lanost-8-ene-3-beta,32-diol, and the aldehyde, 3-beta-hydroxy-5-alphalanost-8-en-32-al both tightly protein bound. The oxidation of the 14-alpha-methyl group is initiated by the cytochrome P-450 component of the enzyme system but this is not required for the subsequent oxidation steps, which ultimately lead to the elimination of formic acid (GIBBONS, PULLINGER and MITROPOULOS 1979). In addition the alcohol and aldehyde metabolites seem to act as inhibitors of this first oxidation process. Thus only the oxidation of lanosterol to the 14-methyl alcohol component was considered.

The suggested mechanism for this oxidation is as follows (SLIGAR, KENNEDY and PEERSON 1980):

— On binding the substrate the ferric porphyrin is converted from low to high spin due to the displacement of the high field sixth ligand. The bound substrate molecule has been shown, using spin labelled compounds, to lie very close to the iron (PIRRWITZ *et al.* 1982).
— This complex is then reduced to Fe2+ which can then bind molecular oxygen.
— Further, one electron reduction yields a species which is less well defined but corresponds to the hypothetical state [Fe(3+) 02(2—)] which has all the electron equivalents required for methyl hydroxylation, water production and regeneration of the ferric resting state.

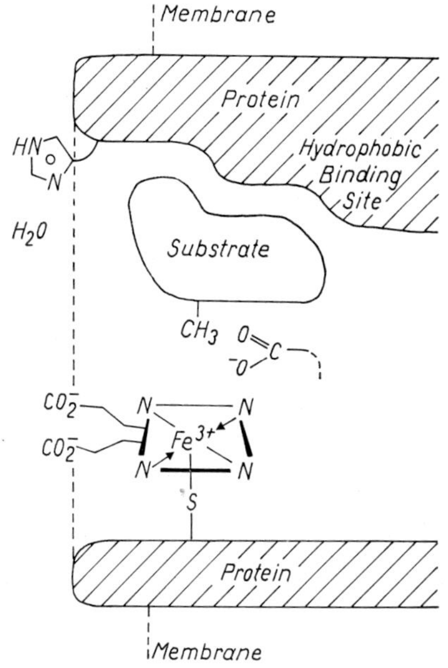

Fig. 22.2 The two dimensional model of the Cytochrome P-450 active site.

— The final step requires an effector molecule — a free acylating group, provided in bacterial hydroxylase by the carboxy terminal tryptophan, or the penultimate glutamine of putidaredoxin.
— This acyl group is responsible through a peracyl group of generating the final iron-oxene intermediate.

Figure 22.2 shows the crude two dimensional model of the P-450 active site.

In designing an inhibitor for the process there are three central features to consider:

(i) The heme prostetic group available for complexation.
(ii) The hydrophobic substrate binding site able to accommodate lanosterol. Indeed a recent paper (Dus 1980) implicates two binding sites for various cytochrome P-450s — one for substrate and the other for nascent product, and both with activated thiol groups.
(iii) The occurrence of the various hydrophilic groups in an otherwise grossly hydrophobic environment:
  — the porphyrin propionate side chains (Peterson *et al.* 1978);
  — the catalytic acyl-effector group which could intervene between the bound substrate and the plane of the heme;
  — there is also the possibility of hydrogen bonding with the displaced histidine (if present);
  — a general polar interaction which is a consequence of the interface which exists by virtue of the enzyme sitting in a membrane.

From these facts it was then necessary to locate the natural substrate and the flexible inhibitors in a three dimensional computer model of the enzyme active site.

Computer graphics model of the enzyme site

All the initial work was done on a DEC PDP 11/60 minicomputer with a GT40 Decgraphics terminal and a software package available through the Video Vector Dynamics Chemical Graphics System, VVD Ltd., in Glasgow. This equipment has now been upgraded to a VAX 11/750 and Evans and Sutherland Multi-Picture System. The software used has been written in-house.

(a) **The natural substrate**

The first task was to build up a three dimensional model for the binding of the natural substrate lanosterol to the protoporphyrin ring. Crystal structures for both the porphyrin ring and the lanosterol nucleus were obtained from the Crystal Structure Search and Retrieval (ČSSR) library provided by the SERC, which was mounted on the digital computer at Edinburgh and available through a computer link. The database can now be purchased, and hence available in-house, by application to Dr. Olga Kennard at Cambridge.

Although ideally it might be thought that it would be preferable to study the transition state of the reaction of the oxene with the 14-methyl group, the intermediary alcohol could well be an inhibitor for the enzyme and hence it was felt better to model the ground state geometry. An iron oxygen distance of 19A and a carbon-oxygen-iron angle of 130° were chosen, as these were the values obtained theoretically by Loew for an iron-carbene system (Loew 1980). There are now three single bonds about which the bound lanosterol can exercise rotations, namely iron-oxygen, oxygen-carbon and carbon-carbon. The computer graphics can now be used to investigate the possible orientations of the lanosterol relative to the porphyrin ring and calculate simultaneously, by molecular mechanics, the total internal energy of interaction.

In Figure 22.3 a computer drawn line drawing shows the lanosterol placed so that the beta-hydroxyl polar group lies over the propionate side chains.

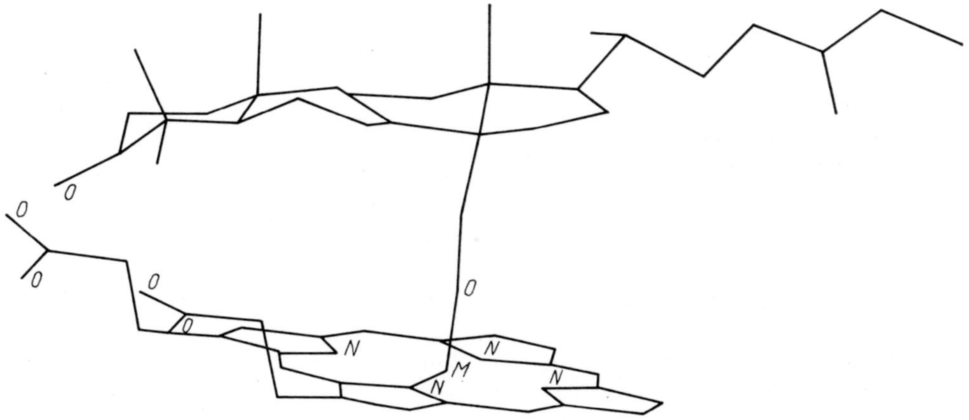

Fig. 22.3 A computer drawn stick model of the lanosterol natural substrate on the porphyrin prosthetic group.

(b) The antagonists

At the start of the project the crystal structures of the antagonists which were to be studied had not been determined. Theoretically the task of calculating all the low energy shapes for just one molecule of interest is considerable. For example, a complete study of the potential surface for a typical triazole fungicide e.g. diclobutrazol with five axes of rotation, sampled at 30° intervals, involves about a quarter of a million individual calculations. Obviously it was necessary to reduce this number to a manageable proportion. Therefore a very crude method based upon van der Waals contacts was used to eliminate from a full conformational search all those shapes which are sterically too high in energy to be considered for further analysis. All the remaining steric minima were then analysed using semi-empirical molecular-orbital methods and subject to a single full *ab initio* calculation to obtain the absolute minimum energy conformation. The crystal structure of RR-diclobutrazol was later determined (Branch *et al.* 1983) and the calculated structure was found to be in good agreement with this. In addition the calculated structure for the less fungicidally active isomer (RR-) of triadimenol was found to agree with the X-ray structure (Spitzer, Kopf and Nickless 1982).

In setting up these calculations the hydroxyl proton was placed so as to be unavailable for possible hydrogen bonding with the 2-nitrogen of the triazole. Theoretically it is well known that extraordinary lengths have to be undertaken to account for this phenomenon properly, even for simple molecules. It seemed more sensible to calculate the other energy contribution theoretically but to look for the formation of internal hydrogen bonding in a dilution experiment in the infra-red.

In the related triazole fungicides intra-molecular hydrogen bonds are not observed in the available crystal structures. Infra-red dilution studies show internal hydrogen bonds in both diclobutrazol diastereoisomers but in neither the active RS-, SR- or the less active RR-, SS-triadimenol. To summarise, the presence of an intra-molecular hydrogen bond between the hydroxyl group and the triazole 2-nitrogen does not in itself seem essential to fungicide activity.

Results

As an example of the above, Figure 22.4 shows a comparison of one of the fungicidally active triazole compounds, namely RR-diclobutrazol, with the natural substrate lanosterol as positioned above the iron porphyrin during initial oxidation through the C-32 position.

The main central features of this model are the hydrophobic binding site, the polar region between this hydrophobic region and the heme plane, and a common complexation to the porphyrin iron.

In more detail these features consist of:

(a) The hydrophobic substrate binding site consisting of three distinct volumes:
  (i) A region corresponding to the lanosterol A ring which terminates in a polar group, the 3-beta-hydroxyl. The inhibitor makes no use of this space in the enzyme cleft.
  (ii) A bulky volume occupied by the inhibitor tertiary butyl group and in part by the sterol 6-alpha-methyl. It might be expected therefore that extension of the t-butyl group other than onto the A ring would reduce activity. In fact *in vitro* activity has been shown to be highly sensitive to the size of this lipophilic moiety.

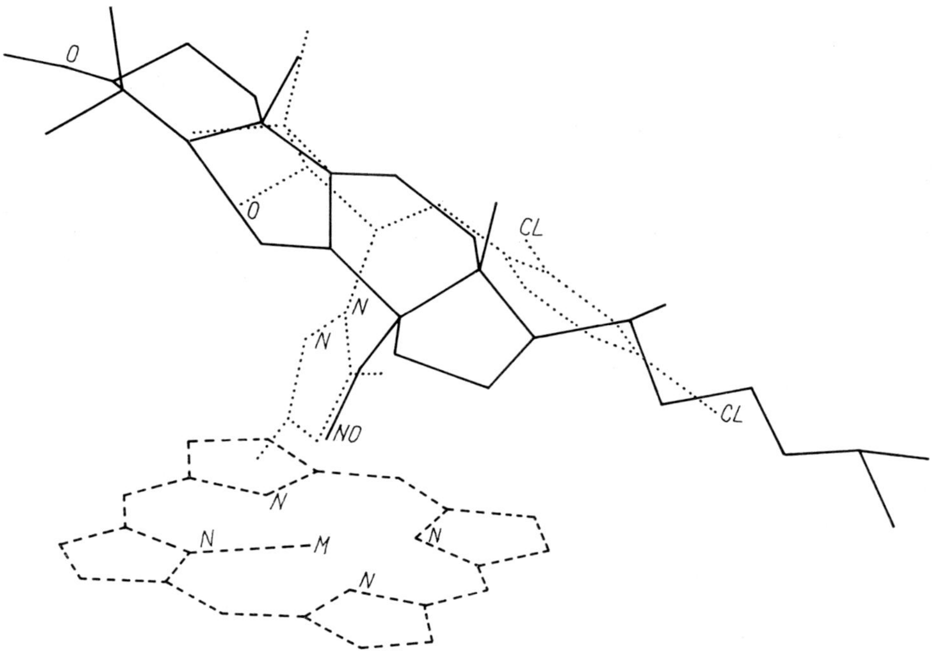

Fig. 22.4 A comparison of lanosterol and RR-diclobutrazol in the Cytochrome P-450 active site

(iii) A deep cavity into which the lanosterol molecule protrudes its side chain and the triazole fungicide projects the benzyl group. This suggests that the benzyl group could be greatly extended, which again agrees with in vitro data. The para-phenyl benzyl compound for example, shows good activity.

(b) The antagonist hydroxyl function lies at a distance relative to the heme group which would make it a candidate for hydrogen bonding to either a heme propionate side chain or an effector acyl group. More generally the polar hydroxyl and the triazole 2-nitrogen could make the interface with polar protein, membrane phospholipid head groups, or solution.
This agrees very much with the model proposed (by PETERSON *et al.* 1978) for the 5-exo hydroxylation of d and l camphor in mammalian cytochrome P-450, and is also consistent with the relationship they noted from steroid metabolism by cytochrome P-450, between the position hydroxylated and its relation to a polar functional group.

(c) The triazole group binds perpendicularly to the heme group and gauche to the iron-nitrogen bonds in the porphyrin plane.

Conclusions

Figure 22.5 now summarises the model requirements for in vitro anti 14-demethylase activity.

A gauche conformation is required between the polar function and the iron chelating group leading to restrictions on the substitution pattern at A, B, C and D. Obviously these substituents must be compatible in size and shape with the lipophilic recogni-

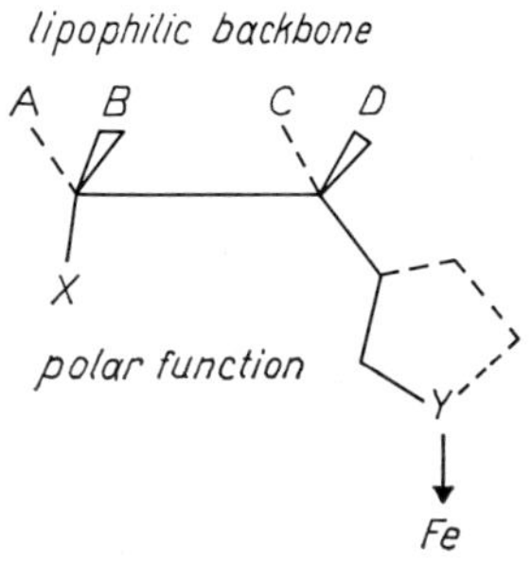

Fig. 22.5 A conceptional Cytochrome P-450 inhibitor. A, B: rigid, limited length; C, D: articulated, extended length.

tion site which accommodates lanosterol. Logically this leads to the possibility of other substitutions patterns which will achieve this gauche conformational requirement with groups of the right kind. Substitution at A and B, for example, leaving C and D as hydrogen yields a series of compounds which have all the correct requirements for activity and yet are different in overall appearance. The new ICI fungicide flutriafol (PP 450) has orthofluorophenyl and parafluorophenyl in these two positions. Theoretical calculations on flutriafol (PP 450) gives excellent agreement with a recent crystal structure determination by KENDRICK and OWSTEN at the Polytechnic of North London, which again shows no intra-molecular hydrogen bonding, but a gauche relationship between the hydroxyl function and the triazole ring.

All the graphical representations so far have consisted of only showing the nuclei positions — through the use of stick diagrams. Obviously a better representation of the spatial requirements of the enzyme active site is provided by the use of space filling diagrams which represent the van der Waals surface of the molecule. Figures 22.6 and 22.7 respectively compare the space-filling and stick representations for flutriafol in the P-450 active site.

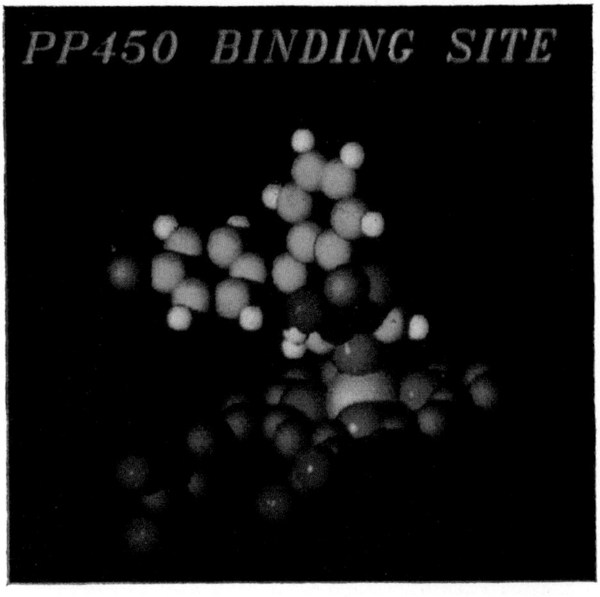

Fig. 22.6 A ball-and-stick model corresponding to Figure 22.7.

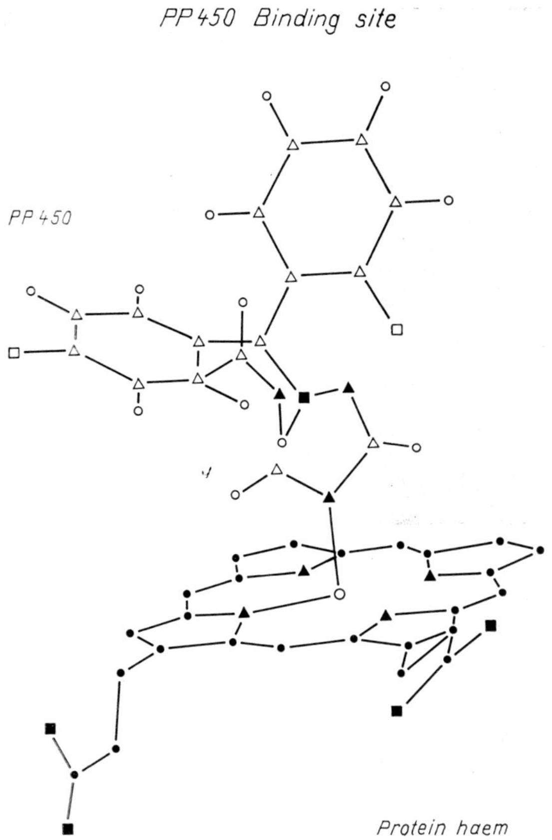

Fig. 22.7 A computer drawn space filling model of flutriafol in the Cytochrome P-450 active site.

A technique in common use today is to perform logical operations on these surfaces to produce other types of surfaces (BASH *et al.* 1983) such as:

— union surfaces which represent the space occupied by any of the designated molecules. This gives an indication of the minimum spatial requirement of the active site.
— intersection surfaces which represent the region of space occupied by all the designated molecules.
— exclusion surfaces which represent the region of space which is occupied by one molecule but not by another. This is useful in the design process as it is easy to visualise where the compound being designed protrudes into unexplored or unfavourable space.

Other groups instead of using the van der Waals surface make use of the surface accessible for interaction, which is obtained by rolling a water molecule over the van der Waals surface of the molecule and noting its path of closest approach (CONNOLLY 1983).

So far we have dealt only with nuclei positions which is how chemists have traditionally thought of molecules. In reality it is the electrons which are of importance as chemical reactions involve interactions between the electronic clouds and not between the nuclei. The electronic properties are generally much harder to deal with and have to be calculated using a quantum-mechanical approach. Many methods have been developed for the display of electron density (both experimental [COPPENS 1975]

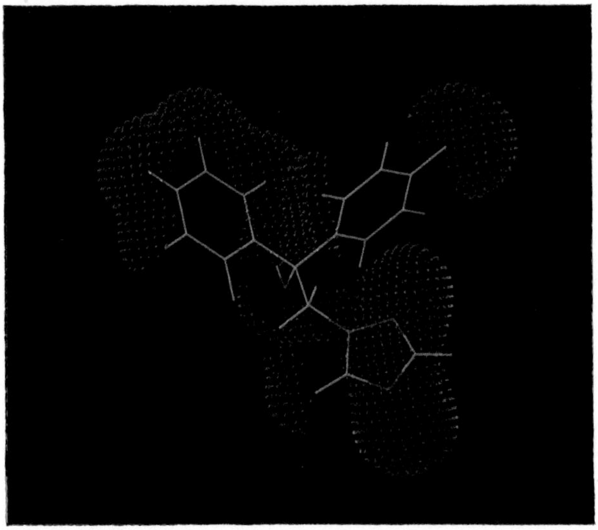

Fig. 22.8 Electrostatic potential energy contours (red-positive, blue-negative) surrounding the fungicide flutriafol.

and calculated) or the electrostatic potential. The major difficulty has been in defining a meaningful contour level which gives an unbiased faithful representation of the molecule (SMITH *et al.* 1977).

Much of the work upon electron density plots has concluded that the "best" contour level to portray the molecule is that which lies between 0.002 and 0.004 atomic units, which coincides with the approximate van der Waals radii of the atoms (BADER *et al.* 1967; KOLOS *et al.* 1980). Therefore, it is becoming increasingly common to look at this problem in reverse, i.e. to define the shape of the molecule as given by its van der Waals surface and to display upon this surface the values of the electron density or electrostatic potential through the use of colour (COHEN 1979; WEINER *et al.* 1981).

This has the additional advantage that the molecule is recognisable. The electrostatic potential is the most commonly displayed, and consequently probably the most useful electronic property considered. It is defined as the energy of interaction between a free proton and the molecule under study and is hence an indication of electrophilic attack. Ideally the use of such methods enable the electronic as well as spatial requirements for binding to be identified (QUARENDON 1984).

Work in this area is proceeding for the triazole fungicides and an example of this kind of approach is shown in Figure 22.8 showing the principal electrostatic contour surrounding flutriafol. Thus, it is possible to design inhibitors which not only comply with the spatial requirements of the receptor but that also possess the necessary electronic features as well.

The future

A criticism of this types of theoretical methods so far, and one which is generally true of quantum pharmacology, is that it only deals with the isolated molecule. In reality a substrate will normally undergo reaction by being transferred from one type of environment, such as solution, to another, such as protein, and in most instances the nature of this protein binding site is unknown. As more X-ray structures are determined and protein folding with energy minimisation through the use of parameterised programmes improves, our knowledge of the binding site will be such as to allow us to design molecules more precisely to block the receptor.

The energy of the molecule may be considerably altered by the perturbing effect of the environment. It is possible to understand some of these effects by making assumptions about the nature of the binding and using empirical intermolecular potential functions to account for electrostatic, polarisation, hydrogen bonding and other contributions — the so called molecular mechanics method (WEINER and KOLLMAN 1981; BROOKS *et al.* 1983). This, however, only gives information regarding the conformations of the molecule and not their electronic properties. Inclusion of the complete biological environment into direct quantum-mechanical calculations is not possible, since the computational time and disk space required, would be astronomical. One method to overcome this is to incorporate only a few atoms of the enzyme- binding site, although this is obviously a drastic approximation (OSMAN and WEINSTEIN 1980). An alternative method, which can incorporate all the enzyme binding site for little extra computational cost, consists of representing the atoms of the protein by a series of point charges determined from calculations on model peptides (HAYES and KOLLMAN 1976). Using this method it is possible to evaluate a binding enthalpy for a small molecule to an enzyme (LAMBROS *et al.* 1984; RICHARDS and CUTHBERTSON 1984) but more importantly it allows the study of changes in electronic distribution of the substrate as it approaches the enzyme.

References

BADER, R. F. W., HENNECKER, W. H., and CADE, P. E.: Molecular charge distributions and chemical binding. J. Chem. Phys. **46** (1967): 3,341—3,363.

— KEAVENY, I., and CADE, P. E.: Molecular charge distributions and chemical binding II. First row diatomic hydrides AH. J. Chem. Phys. **47** (1967): 3,381—3,402.

BASH, P., HUANG, C., PATTABIRAMAN, N., FERRIN, T., and LANGRIDGE, R.: van der Waals surfaces in molecular modelling and implementations using real-time computer graphics. Science **222** (1983): 1,325—1,327.

BRANCH, S. K., ANDERSON, N. H., LOEFFLER, R. S. T., MARCHINGTON, A. F., and NOWELL, I. N.: Ergosterol Biosynthesis Inhibitors. S. C. I. Symposium, Reading, UK, 1983.

BROOKS, B. R., BRUCCOLERI, R. E., OLAFSON, B. D., STATES, D. J., SWAMINATHAN, S., and KARPLUS, M.: CHARMM: A programme for macromolecular energy, minimisation and dynamics calculations. J. Comput. Chem. **4** (1983): 187—217.

CAPALDI, R. A., MARSHALL, F. A., and STAPLES, S. J.: Structure of intrinsic membrane proteins and their amino acid sequences. Comments Mol. Cell. Biophys. **1** (1983): 365—382.

COHEN, N. C.: Beyond the 2-D chemical structure. Computer Assisted Drug Design. ACS Symposium. **112** (1979): 371—381.

CONNOLLY, M.: Solvent-accessible surfaces of proteins and nucleic acids. Science **221** (1983): 709—713.

COPPENS, P.: Experimental charge densities in solids and their significance for theoretical calculations. NATO Adv. Study Inst. Ser., Ser. B9 (1975): 227—257.

DUS, K. M.: Insights into the active site of the cytochrome P-450 haemoprotein family — a unifying concept based on structural consideration. Xenobiotica **12** (1980): 745—722.

GADHER, P., MERCER, E. I., BALDWIN, B. C., and WIGGINS, T. E.: A comparison of the potency of some fungicides as inhibitors of sterol 14-demethylation. Pesticide Biochem. Physiol. **19** (1983): 1—10.

GHELIS, C., and YON, J.: Protein Folding. Academic Press Inc., New York 1982.

GIBBONS, C. F., PULLINGER, C. R., and MITROPOULOS, K. A.: Studies on the mechanism of lanosterol 14-alpha-demethylation. Biochem. J. **183** (1979): 309—315.

GRIFFIN, B. W., and PETERSON, J. A.: *Pseudomonas putida* cytochrome P-450. J. Biol. Chem. **250** (1975): 6,445—6,451.

HASSALL, C. H.: Computer graphics as an aid to drug design. Chem. in Britain **21** (1985): 39—46.

HAYES, D. M., and KOLLMAN, P. A.: Electrostatic potential of proteins. 1. Carboxypeptidase A. J. Am. Chem. Soc. **98** (1976): 3,335—3,345.

KOLOS, W., RANGHINO, G., CLEMENTI, F., and NOVARO, D.: Interaction of methane molecules. Int. J. Quantum Chem. **17** (1980): 429—448.

LAMBROS, S. A., MARCHINGTON, A. F., and RICHARDS, W. G.: Theoretical calculations on enzyme-substrate interactions: The binding of n-alkylboronic acids to alpha-chymotrypsin A. J. Molec. Strut. Theochem. **109** (1984): 61—71.

LOEW, G. H.: Electronic spectra of model oxy, carboxy P-450 and carboxy heme complexes. J. Am. Chem. Soc. **102** (1980): 3,655—3,657.

MARCHINGTON, A. F., ROBINS, S. A., and RICHARDS, W. G.: Chemistry computers and commerce. T.I.P.S. **3** (1982): 425—428.

MURAKAMI, K., and MASON, H. S.: An electron spin resonance study of microsomal Fe. J. Biol. Chem. **242** (1967): 1,102—1,110.

OSMAN, R., and WEINSTEIN, H.: Models for active sites of metalloenzymes: Comparison of zinc and beryllium containing complexes. Isr. J. Chem. **19** (1980): 149—153.

PETERSON, J. A., O'KEEFFE, D. H., WERRINGLOER, J., EBEL, R. E., and ESTABROOK, R. W.: Micro-environments and Metabolic Compartmentation. In: SRERE, P. A., and ESTABROOK, R. W. (Eds.): Academic Press, New York 1978, pp. 433—446.

PIRRWITZ, J., SCHWARZ, D., REIN, J., RISTAU, O., JANIG, G. R., and RUCKPAUL, K.: Studies on the active center of cytochrome P-450 using a spin-labelled type I substrate analogue. Biochem. et Biophysica Acta **708** (1982): 42—48.

POPLE, J. A., and BEVERIDGE, D. L.: Approximate Molecular Orbital Theory. McGraw-Hill, New York 1970.

QUARENDON, P., NAYLOR, C. B., and RICHARDS, W. G.: Display of quantum mechanical properties on van der Waals surfaces. J. Molec. Graphics **2** (1984): 4—7.

RICHARDS, W. G.: Quantum Pharmacology. Endeavour **8** (1984): 172—178.

— and CUTHERBERTSON, A. F.: Binding of methotrexate to dihydrofolate reductase by quantum chemical calculation. J. Chem. Soc. Chem. Commun. **3** (1984): 167—168.

— and HORSLEY, J. A.: Ab initio Molecular Orbital Calculations for Chemists. Clarendon Press, Oxford 1970.

SLIGAR, S. C., KENNEDY, K. A., and PEERSON, D. C.: Chemical mechanisms for cytochrome P-450 hydroxylation. Evidence for acylation of heme bound dioxygen. Proc. Natl. Acad. Sci. USA **77** (1980): 1,240—1,244.

SMITH, V. H., PRICE, P. F., and ABSAR, I.: Representations of the electron density and its topographical features. Isr. J. Chem. **16** (1977): 187—197.

SPITZER, T., KOPF, J., and NICKLESS, G.: 1-(4-Chlorophenoxy)-1-(1,2,4-triazolyl)-3,3-dimethyl-2-butanol,$C_{14}H_{18}O_2N_3Cl$. Cryst. Struct. Comm. **11** (1982): 319—325.

TAYLOR, K. K., and BERNSTEIN, H.: Brookhaven National Laboratory, Upton, New York, USA.

VINTER, J. G.: Molecular graphics for the medicinal chemist. Chem. in Britain **21** (1985): 32—38.

WEINER, P. K., LANGRIDGE, R., BLANEY, J., SHAEFER, R., and KOLLMAN, P.: Electrostatic potential molecular surfaces. Proc. Natl. Acad. Sci. USA **79** (1981): 3,754—3,758.

— and KOLLMAN, P. A.: AMBER: Associated model building with energy refinement. A general programme for modelling molecules and their interactions. J. Comput. Chem. **2** (1981): 287 303.

LYR, H. (Ed.): Modern Selective Fungicides — Properties, Applications, Mechanisms of Action. Longman Group UK Ltd., London, and VEB Gustav Fischer Verlag, Jena, 1987.

Chapter 23

Disease control by nonfungitoxic compounds

Hugh D. SISLER* and Nancy N. RAGSDALE**
* Department of Botany, University of Maryland, USA
and
** Cooperative State Research Service, USDA, Washington, USA

Introduction

Nonfungitoxic disease control compounds are chemicals which have little or no effect on the rate of growth of the pathogen in vitro at concentrations which control pathogenic activity in vivo. The antifungal activity of these compounds, therefore, is based on actions which lead to inhibition of growth and reproduction of the pathogen in its parasitic phases. The compound may act directly on the pathogen to prevent it from becoming established in plant tissue or from causing disease once it has become established. On the other hand, it may affect the host-parasite interaction in such a way that host defense mechanisms kill or halt encroachment of the fungal pathogen.

Compounds which are converted to a derivative in vivo that is directly toxic to the pathogen do not belong in the group of compounds under consideration here. However, with systemic fungicidal compounds, disease control may involve both direct fungitoxicity and enhanced host resistance mechanisms. These combination effects are discussed later in connection with the fungicide metalaxyl.

Nonfungitoxic disease control chemicals offer several advantages over conventional fungitoxic compounds. First, these compounds are more likely to affect target sites specific to fungi or higher plants than conventional fungicides and would therefore constitute less of an environmental hazard. Second, nonfungitoxic compounds which regulate host resistance are likely, in many cases, to be active at extremely low levels. Third, compounds that enhance host defense activity are less likely to encounter fungal resistance than are many conventional systemic fungicides.

The various aspects of plant disease control by nonfungitoxic chemicals have been discussed previously. The reader is referred to the following references for information or points of view that may differ in some respect from those presented here (DIMOND and RICH 1977; SISLER 1977; LANGCAKE 1981; WADE 1984; DEKKER 1983).

Action on host defense systems

Higher plants have evolved passive as well as active defense systems to ward off potential pathogenic fungi and other microorganisms in the environment. Only those fungi which have developed mechanisms to render these defense systems ineffective are able to cause disease. Gaps in host defense systems which allow these fungi to become established as pathogens have created the need to develop antifungal compounds for use in disease control. Thus far, the chemicals used for this purpose have been almost exclusively fungitoxic compounds that act directly on the pathogen to prevent or eradicate infections. While this simple and straightforward approach will almost certainly remain in use, an attractive alternative is to chemically modify the host-parasite interaction so that the host defense mechanisms provide the antifungal activity needed to control the pathogen.

Recent concepts of host-parasite interactions as they relate to host susceptibility and resistance responses have been discussed by HEATH (1981), BUSHNELL and ROWELL (1981) and TEPPER and ANDERSON (1984). It is believed by some investigators that resistance genes are regulator or sensor genes that recognize elicitors produced by challenging microorganisms. When recognition does not occur or if the elicitor is not produced, the plant does not respond in a resistant manner. A model has been proposed wherein regulatory genes can trigger the expression of many structural genes which leads to a multitude of events that constitute a hypersensitive or resistance reaction (TEPPER and ANDERSON 1984). These events may include phytoalexin production, phenol production, lignification and plasma membrane changes, all of which may be detrimental to the pathogen as well as to the higher plant cells in the local vicinity. It seems reasonable that resistance in some cases might result from eliciting only one of the aforementioned types of events.

In a chemical regulation of resistance that is consistent with this model system, the compound applied could induce pathogen (or elicitor) recognition, but in the absence of the pathogen, should not trigger expression of the structural genes concerned with resistance. The compound (nonfungitoxic disease control agent) might act by changing the conformation of the elicitor binding site (receptor), by preventing the pathogen from producing a suppressor, by blocking binding of this suppressor to the receptor or to the elicitors or by inducing elicitor production by the pathogen in cases where production is weak or absent. KUĆ (1984) has discussed the aforementioned type of regulation of resistance by substances which systemically immunize the plant so that it responds only in the presence of the pathogen.

Compounds such as probenazole (SEKIZAWA and MASE 1981) and 2,2-dichloro-3,3- dimethylcyclopropane carboxylic acid (LANGCAKE 1981), which will be discussed later, are reputed to increase host resistance possibly by one of the mechanisms discussed above.

Action on pathogenic mechanisms

In addition to the effects on the pathogen that accentuate host resistance responses, nonfungitoxic compounds may act on the pathogen in a variety of ways to block pathogenicity. They could block the induction or action of enzymes involved in penetration of the host or spread of the pathogen within the host tissue. Ruffianic acid, for example, is a compound reported to inhibit pectolytic and cellulolytic enzymes of *Fusarium* and *Verticillium* species (GROSSMANN 1968). The action of cutinase inhibitors as antipenetrants is described in a later section of this article.

Another example of antipathogenic action is interference with appressorial development or function as has been observed for tricyclazole, pyroquilon and other compounds described in the section on antipenetrants.

Phytotoxins produced by fungal pathogens in some cases play a critical role in pathogenesis or severity of a disease. The role of toxins in disease development is firmly established in the case of the "host specific toxins" (SCHEFFER 1976), but is less clear in the case of many nonspecific fungal toxins. Nevertheless, some of the latter toxins seem necessary for infection of the host while others play a role in determining disease severity (RUDOLPH 1976). Blocking production or counteracting effects of a phytotoxin, therefore, can be a disease control mechanism. As pointed out by DEKKER (1983) there are few cases where this has been done. On the other hand, there are studies which indicate that it might not be advisable to interfere with the production or action of some phytotoxins because they may play a role in eliciting host resistance

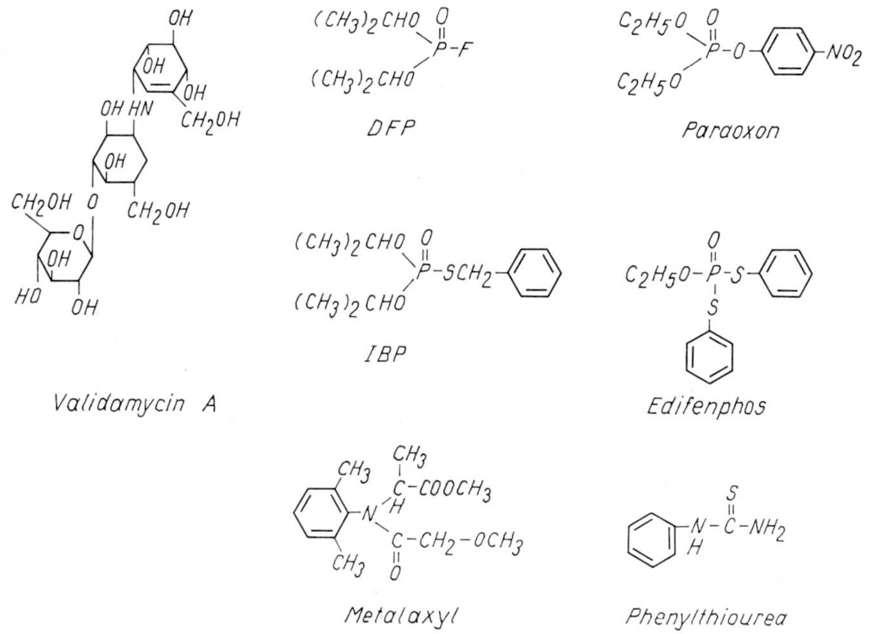

Fig. 23.1 Structures of several plant disease control chemicals discussed in the text. DFP represents diisopropylfluorophosphate, IBP represents S-benzyl-0,0-diisopropylphosphorothiolate.

responses. For example, LANGCAKE *et al.* (1983) showed that picolinic acid, which is a nonspecific phytotoxin produced by *Pyricularia oryzae*, causes a hypersensitive response much like that produced by *P. oryzae* in rice plants treated with the sensitizing agent 2,2-dichloro-3,3-dimethylcyclopropane carboxylic acid. Moreover, tenuazonic acid, another toxin produced by *P. oryzae*, is reported to elicit defense reactions in rice leaves (LEBRUN *et al.* 1984). The reactions were more intense in those varieties with a high level of general resistance to *P. oryzae* than in those with a low level.

One possible mechanism for fungal pathogens to overcome plant resistance is through metabolic detoxication of phytoalexins (VAN ETTEN 1982). A number of cases of phytoalexin degradation by plant pathogenic fungi have, in fact, been reported (SISLER 1977). The extent to which this mechanism is used by fungi to overcome plant resistance is unclear; however, there are indications that detoxication of the phytoalexin pisatin is required for pathogenicity by *Nectria haematococca* MP VI on peas (VAN ETTEN 1982). Blocking induction or action of enzymes involved in phytoalexin degradation may prove to be a useful method of plant disease control by nonfungitoxic compounds.

In the following section, the action of several chemicals on fungal pathogenicity or host resistance will be examined in some detail.

Mode of action of various chemicals

Cutinase inhibitors

The plant epidermis and the cuticle in particular, are formidable barriers which many fungal pathogens penetrate in order to gain access to plant tissue. Penetration is believed to be accomplished by mechanical forces, enzymatic action or a combination of both. There is now appreciable evidence that infection can be prevented by compounds or agents which act specifically on the penetration process but are nonfungitoxic to growth of the pathogens in vitro.

Experimental control of certain diseases has been obtained by compounds which block fungal cutinase activity. Specific cutinase antiserum or nonfungitoxic concentrations of the potent cutinase inhibitor diisopropylfluorophosphate (Fig. 23.1) protect pea epicotyls from infection by *Fusarium solani* (Maiti and Kolattukudy 1979) and papaya fruit from infection by *Colletotrichum gloeosporioides* (Dickman *et al.* 1982).

Several organic phosphorus pesticides (insecticides and fungicides) inhibit cutin esterase of *F. solani* (Köller *et al.* 1982a) and *C. gloeosporioides* (Dickman *et al.* 1983) and protect plant tissue from infection at concentrations which are not fungitoxic in vitro. Infection is prevented only when the tissue is unwounded, which suggests that protection results from an antipenetrant action. There is, in fact, a general degree of correspondence between effectiveness of various compounds as inhibitors of cutinase of *F. solani* and protection from infection by this pathogen. Among the highly effective compounds in both types of activity are the insecticides paraoxon (Fig. 23.1), 0,0-dimethyl-0-(2,4,5-trichlorophenyl)phosphate and 0,0-diethyl-0-(3,5,6-trichloro-2-pyridyl) phosphate. The I_{50} for each of these 3 compounds as cutinase inhibitors is well below 1 μM while a high degree of protection against infection is afforded by a concentration of 50 nM or less. The latter two compounds are potent inhibitors of cutinase of *C. gloeosporioides*, and effectively inhibit infection of papaya fruit by this pathogen at nM concentrations.

The organic phosphorus fungicides Kitazin (IBP) and Hinosan (edifenphos) (Fig. 23.1) are appreciably less effective than the aforementioned insecticides both as cutinase inhibitors and as protectants of pea tissue from infection by *F. solani*. They do, however, give good protection at concentrations which show little or no toxicity to growth of *F. solani* in vitro. IBP (S-benzyl-0,0-diisopropylphosphorothiolate), and edifenphos (0-ethyl-S,S-diphenylphosphorodithiolate) are used primarily to control rice blast disease caused by *P. oryzae*. In this organism they inhibit phospholipid N-methyltransferase (Akatsuka *et al.* 1977; Kodama *et al.* 1980), a fungitoxic mechanism which is presumed to be the basis for their fungitoxicity and plant protective action. The role that cutinase inhibition and antipenetrant action plays in their protective activity against *P. oryzae* has apparently not been determined. In antipenetrant tests made with rice sheaths and Cellophane film by Araki and Miyagi (1977), 10 μg/ml of IBP proved to be quite inhibitory to spore germination and appressorial formation by *P. oryzae;* however, appressorial penetration of both rice sheaths and Cellophane film was strongly inhibited by the compound at 2 μg/ml. Since the antipenetrant action in the case of Cellophane is not dependent on the inhibition of cutinase, it is doubtful that control of rice blast disease by IBP is due specifically to cutinase inhibition. It would seem that highly potent cutinase inhibitors such as paraoxon would be more effective than the fungicides edifenphos or IBP for rice blast control if antipenetrant activity based on cutinase inhibition were the primary mode of protection.

In other studies, it has been shown that the fungicide benomyl but not its fungitoxic degradation product carbendazim (MBC) protects pea stems from infection by *F. solani* (Köller *et al.* 1982b). Benomyl breaks down to yield the transient toxicant butylisocyanate in addition to carbendazim, and since the toxicity of carbendazim to growth of *F. solani* in vitro is essentially the same as that of benomyl (Köller *et al.* 1982b), it was suggested that inhibition of cutinase by butylisocyanate is responsible for preventing infection by *F. solani*. The authors also suggest that inhibition of cutinase by butylisocyanate may explain why benomyl is superior to MBC for the control of certain other fungal diseases.

Melanin biosynthesis inhibitors

Among the most interesting and successful nonfungitoxic disease control chemicals are the melanin biosynthesis inhibitors (Fig. 23.2). While members of this group block melanin biosynthesis in a variety of Ascomycetes and imperfect fungi (Toukousbalides and Sisler 1978; Wheeler 1983) they give practical control only of rice blast disease caused by *Pyricularia oryzae* and experimental control of certain diseases caused by *Colletotrichum* species.

Melanin biosynthesis inhibitors (MBI) are antipenetrants that act on the pathogen to prevent it from piercing the plant epidermis. The antipenetrant action of pentachlorobenzyl alcohol (PCBA) (Ishida *et al.* 1969; Araki and Miyagi 1977) and of fthalide (Araki and Miyagi 1977) was known before the compounds were recognized as melanin biosynthesis inhibitors (Woloshuk *et al.* 1982; Yamaguchi *et al.* 1982). MBI compounds at concentrations which control diseases, do not inhibit spore germination or appressorial formation on epidermal surfaces (Woloshuk and Sisler 1982; Yamaguchi *et al.* 1982; Inoue *et al.* 1984) or on the surface of barriers such as Cellophane (Araki and Miyaki 1977; Yamaguchi *et al.* 1982), nitrocellulose (Kubo *et al.* 1982b) or Formvar plastic (Woloshuk *et al.* 1983; Wolkow *et al.* 1983). These compounds do, however, prevent penetration of epidermal or other barriers by appressoria of *P. oryzae* (Ishida *et al.* 1969; Araki and Miyagi 1977; Woloshuk and Sisler 1982; Chida *et al.* 1982; Okuno *et al.* 1983; Inoue *et al.* 1984) or by appressoria of *C. lagenarium* (Kubo *et al.* 1982b) and *C. lindemuthianum* (Wolkow 1982). The fact that these compounds do not control disease if the epidermal wall is punctured (Woloshuk *et al.* 1983; Wolkow 1982; Inoue *et al.* 1984) or if their application is delayed for more than 8 hr after inoculation (Inoue *et al.* 1984) is consistent with their specific mechanism of antipenetrant action.

The antipathogenic activity of tricyclazole parallels the ability of the compound to block the polyketide pathway (Fig. 23.3) leading to melanin biosynthesis (Tokousbalides and Sisler 1978; Woloshuk *et al.* 1980; Yamaguchi *et al.* 1982). This tricyclazole sensitive pathway of melanin biosynthesis is widely distributed among Ascomycetes and imperfect fungi, but apparently does not occur in Basidiomycetes (Wheeler 1983). At concentrations which control rice blast disease, these compounds appear to affect only melanin biosynthesis in the plant pathogen, *P. oryzae*. Tricyclazole, for example, completely blocks melanin biosynthesis at 0.1 μg/ml but growth is not inhibited by concentrations as high as 20 μg/ml (Tokousbalides and Sisler 1978). Tricyclazole blocks the polyketide pathway to melanin (Fig. 23.3) between 1,3,6,8-THN and scytalone and between 1,3,8-THN and vermelone (Tokous-

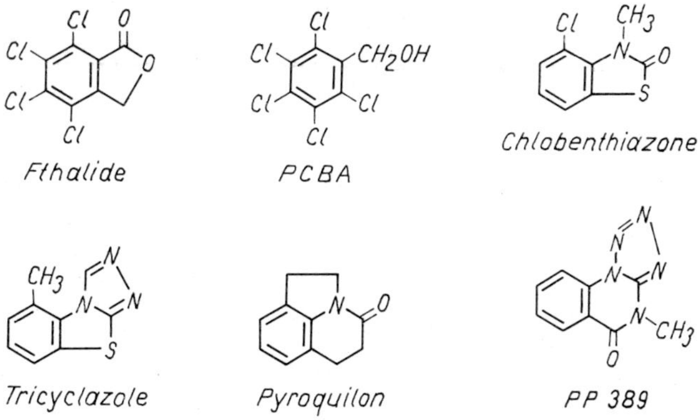

Fig. 23.2 Structures of several melanin biosynthesis inhibitors.

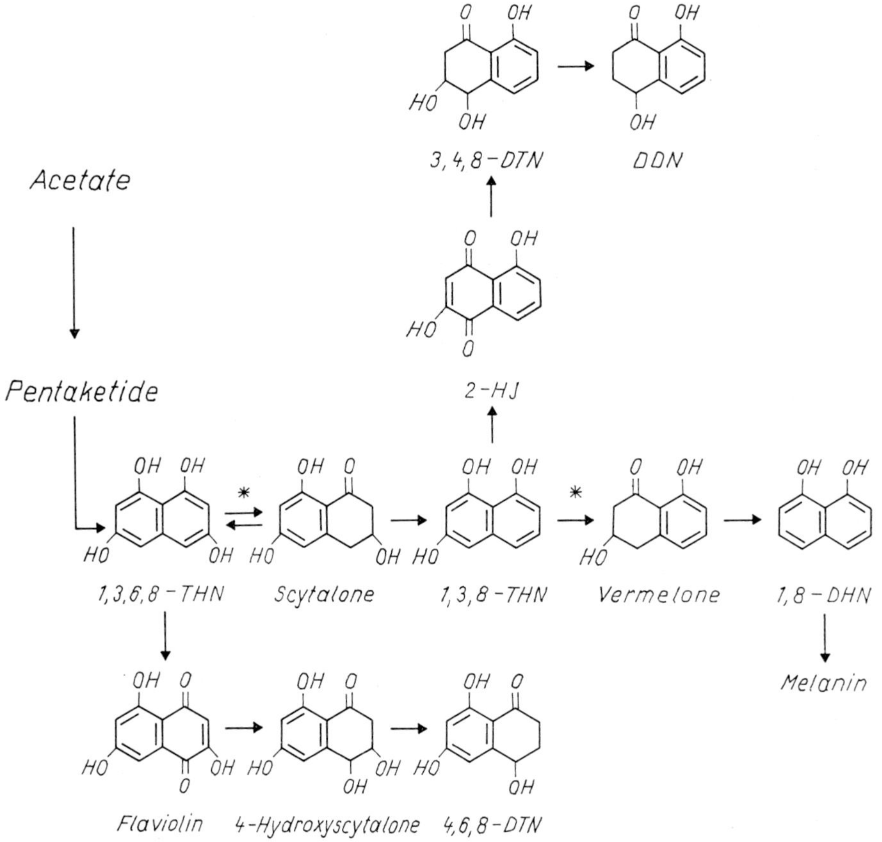

Fig. 23.3 Polyketide pathway to melanin in *Pyricularia oryzae* with branches at **1,3,6,8-THN** (1,3,6,8-tetrahydroxynaphthalene) and **1,3,8-THN** (1,3,8-trihydroxynaphthalene). Other abbreviations indicate the following compounds: 2HJ (2-hydroxyjuglone); **3,4,8-DTN** (3,4-dihydro-3,4,8-trihydroxy-1-(2H) naphthalenone); DDN (3,4-dihydro-4,8-dihydroxy-1-(2H) naphthalenone); **4,6,8-DTN** (3,4-dihydro-4,6,8-trihydroxy-1-(2H) naphthalenone; **1,8-DHN** (1,8-dihydroxy-naphthalene). The asterisks indicate main sites of action of melanin biosynthesis inhibitors such as tricyclazole.

BALIDIS and SISLER 1979; WHEELER and STIPANOVIC 1979; WOLOSHUK *et al.* 1980). The inhibited steps are NADPH dependent reduction reactions (WHEELER 1982). Since 2HJ accumulates more than flaviolin at low tricyclazole concentrations and flaviolin accumulation is accentuated as the tricyclazole concentration is increased, it is believed that the 1,3,8-THN to vermelone step is more sensitive to the inhibitor than the 1,3,6,8-THN to scytalone step in *V. dahliae* and *P. oryzae* (TOKOUSBALIDES and SISLER 1979; WOLOSHUK *et al.* 1980). However, WHEELER and STIPANOVIC (1985) have demonstrated an alternate pathway to 2HJ through flaviolin in the human pathogenic yeast, *Wangiella dermatitidis*. If this alternate pathway also exists in *V. dahliae* and *P. oryzae* then the basis for deducting a greater tricyclazole sensitivity for the 1,3,8-THN reduction to vermelone than for the 1,3,6,8-THN reduction to scytalone (TOKOUSBALIDES and SISLER 1979; WOLOSHUK *et al.* 1980) might be questioned. Additional studies in cell-free systems are needed to clarify this point. There is no doubt, however, that the conversion of 1,3,8-THN to vermelone is quite sensitive to tricyclazole in *P. oryzae* and *V. dahliae*. This conversion in *C. lagenarium* is also sensitive to tricyclazole (KUBO *et al.* 1985).

One may raise the question of how the blocking of melanin biosynthesis is related to antipenetrant action. *P. oryzae* as well as *C. lindemuthianum* and *C. lagenarium* develop an appressorial structure for penetrating host epidermal cell walls. The appressorial walls of these fungi become darkly melanized prior to penetration. In the presence of an MBI such as tricyclazole, or if the fungus is genetically deficient in melanin biosynthesis, penetration of epidermal or other barriers does not occur. For example, buff mutants of *P. oryzae* which are essentially identical with the tricyclazole treated wild type, do not form melanized appressoria and are nonpathogenic (Woloshuk *et al.* 1980) because they cannot penetrate the host epidermis. Treatment of *C. lindemuthianum* (Wolkow 1982) or *C. lagenarium* (Kubo *et al.* 1982b) with tricyclazole or pyroquilon blocks appressorial melanization and also appressorial penetration of barriers such as epidermal walls or nitrocellulose membranes. These appressoria germinate to produce hyphae which grow laterally along the barrier surface. Albino mutants of *C. lagenarium* form appressoria without melanized walls that behave essentially the same as those developed from the tricyclazole treated wild type fungus (Kubo *et al.* 1982a; 1982b). These albino mutants are not pathogenic.

Melanization of the appressorial wall appears to be necessary for the architecture and rigidity needed to support and focus the mechanical forces involved in the penetration process (Woloshuk *et al.* 1983; Wolkow *et al.* 1983). Increased wall rigidity and focus might result from simple deposition of melanin between wall fibrils or through cross-linking of wall polymers by an oxidation product of an immediate melanin precursor (Sisler *et al.* 1984) such as that which occurs in insect cuticle tanning (Jacobs 1980). If the latter is the critical mechanism, then black melanin deposition may be unimportant or even competitive with this cross-linking process. There is need for additional studies on the role of terminal metabolites of the melanin biosynthetic pathway as they relate to the melanization or cross-linking in the appressorial walls of *P. oryzae* and *Colletotrichum* species.

While present evidence suggests that the ultimate effects of MBI which result in antipenetrant action are on the appressorial wall structure, there may be other explanations of the antipenetrant action such as a cytotoxic effect of pentaketide metabolites which accumulate when melanin biosynthesis is blocked by these compounds (Yamaguchi *et al.* 1983b). This explanation, however, almost certainly cannot account for the failure of albino mutants of *C. lagenarium* (Kubo *et al.* 1982a) to penetrate epidermal or nitrocellulose membrane barriers. Moreover, it also appears unlikely because the antipenetrant action of MBI can be partially reversed by adding vermelone or 1,8-dihydroxynaphthalene, which are metabolites on the melanin side of the tricyclazole inhibition sites in the biosynthetic pathway (Woloshuk *et al.* 1983; Wolkow *et al.* 1983; Yamaguchi *et al.* 1983a; Okuno *et al.* 1983; Kubo *et al.* 1985). Even DOPA (3,4-dihydroxyphenyalanine) has been shown to restore capacity of tricyclazole treated appressoria of *P. oryzae* to penetrate nitrocellulose membranes (Okuno *et al.* 1983).

Validamycin

Validamycin (Fig. 23.1) is a water soluble, weakly basic compound classified as an aminoglucoside antibiotic (Suami *et al.* 1980; Trinci, 1985) which is active primarily for control of diseases caused by *Rhizoctonia* type fungi (Wakae and Matsuura 1975). It is widely used in Japan to control sheath blight of rice caused by *Rhizoctonia solani* = *(Pellicularia sasakii)*. Under poor nutritional conditions, the compound inhibits radial growth of *R. solani* on agar media. Nioh and Mizushima (1974) observed, however, that the antibiotic does not reduce the mycelial mass when the fungus is grown in liquid culture, even though it alters the morphology. Trinci

(1985) made similar observations for the action of validamycin A on growth of *Rhizoctonia cerealis*. When grown on Vogel's solid medium containing 50 mM glucose, the radial growth rate is markedly inhibited by 10 μM validamycin, but the doubling time in Vogel's liquid medium is not affected by this concentration of the antibiotic. The antibiotic induces excessive branching which results in denser colonies that expand in diameter more slowly than control colonies (NIOH and MIZUSHIMA 1974; TRINCI 1985). The pathogenicity of *P. sasakii* is markedly reduced on rice plants treated with validamycin. The addition of meso-inositol partially restores pathogenicity and also prevents the morphological effect produced in *R. solani* in vitro. It was suggested that the antibiotic interferes with the biosynthesis of meso-inositol which is necessary for pathogenicity of *P. sasakii* (WAKAE and MATSUURA 1975). Meso-inositol, however, did not counteract the inhibitory effect of validamycin on radial growth of *R. cerealis* in vitro (TRINCI 1985). On the other hand, the onset of inhibition of radial growth of both *R. solani* and *R. cerealis* by 0.2 μM validamycin is delayed in direct proportion to the glucose concentration present between 1.5 and 10 mM, but glucose concentration does not affect the ultimate growth rate attained in antibiotic treated cultures. The mechanism underlying this antagonism by glucose is not understood. There are indications that validamycin inhibits the synthesis of a hyphal extension factor in certain fungi. Inhibition of the formation of such a factor by validamycin A in *R. solani* has been reported by SHIBATA *et al.* (1982). Validamycin may control diseases by preventing the penetration of the host by the pathogen or by reducing the rate of spread of the pathogen in the host so enough time is allowed for the host to mobilize its defense system (TRINCI 1985). While it appears that disease control results from the action of validamycin on the pathogen, the primary biochemical mechanism of this action remains obscure.

Probenazole

Probenazole (Oryzemate), a relative of saccharin, is a systemic compound that controls rice blast disease caused by *P. oryzae* and bacterial leaf blight caused by *Xanthomonas oryzae*. Probenazole (Fig. 23.4) is ordinarily applied as a submerged treatment to the roots of rice p'ants. The compound does not show appreciable toxicity to hyphal growth of *P. oryzae* on agar media or to conidial germination on glass slides; however, conidial germination on rice leaf sheaths as well as appressorial formation and penetration are quite sensitive to the compound (WATANABE 1977). While these observations suggest that probenazole may act prior to penetration, the compound is known to suppress lesion spread after the fungus has penetrated into the plant tissue (WATANABE 1977). Evidence indicates that root application of probenazole leads to blast control through enhancement of the resistance response of the rice plant rather than through direct fungitoxicity of the applied chemical (WATANABE *et al.* 1979; SEKIZAWA and MASE 1981). Analysis of probenazole treated plants inoculated with *P.*

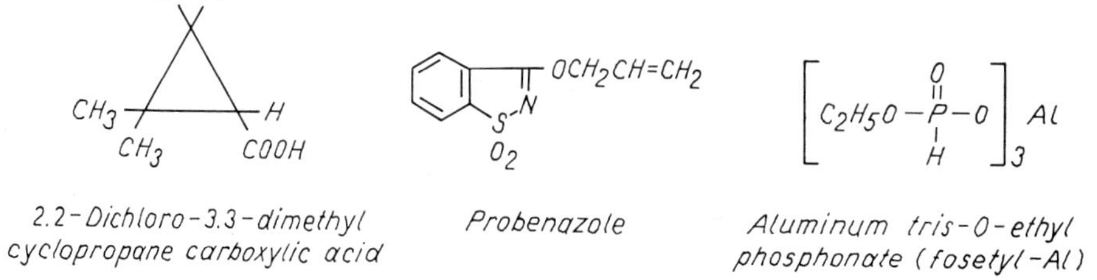

Fig. 23.4 Structures of three compounds reported to increase host resistance responses.

oryzae reveals an accumulation of α-linolenic acid and three other fungitoxicants with similar properties. These toxicants are believed to form a chemical barrier to the invading pathogen. There is also an augmentation of peroxidase, phenylalanine ammonia lyase and catechol-O-methyltransferase in treated plants inoculated with *P. oryzae* which is believed to facilitate the formation of a lignoid barrier around invaded cells (SEKIZAWA and MASE 1981). Although the protective action of probenazole is apparently not due to direct fungitoxicity, it is conceivable that the fungus is the primary target of probenazole action which leads to the enhanced resistance response.

DDCC

The compound 2,2-dichloro-3,3-dimethylcyclopropane carboxylic acid (DDCC) at nonfungitoxic concentrations, specifically controls rice blast disease caused by *P. oryzae* (LANGCAKE and WICKINS 1975). While it is an interesting experimental chemical, DDCC (Fig. 23.4) is not used for the practical control of rice blast disease. In contrast to tricyclazole, DDCC does not prevent epidermal penetration by *P. oryzae,* but promotes a markedly enhanced host response once penetration has occurred (CARTWRIGHT *et al.* 1980). Rice plants treated with DDCC respond in a rapid hypersensitive fashion to penetration by *P. oryzae* whereas untreated plants respond in a delayed and mild fashion. There is a rapid development of intracellular hyphae in cells of untreated plants, but hyphal development is halted soon after penetration in cells of treated plants (CARTWRIGHT *et al.* 1980). The suppression of hyphal development in DDCC treated plants apparently results from the accumulation of the fungitoxic substances, momilactones A and B, in the tissue surrounding the invasion site. The accumulation of these phytoalexins is more rapid and far greater in magnitude in leaves of DDCC treated plants than in leaves of untreated plants.

The data concerning control of rice blast disease by DDCC clearly indicate that host defense systems are the source of antifungal substances which halt the invasion of *P. oryzae*. However, the mechanism leading to the DDCC enhancement of the resistance response is unknown. There are several lines of evidence to suggest that the primary target of DDCC action is in the host plant rather than in the pathogen (LANGCAKE *et al.* 1983). First, picolinic acid applied to wounds on DDCC treated plants causes an intense hypersensitive type reaction but a much less intensive response when applied to wounds on untreated plants. Picolinic acid is reputed to be a phytotoxin produced by *P. oryzae*. Second, greater momilactone accumulation occurs in response to mycelial extracts of *P. oryzae* in treated plants than in untreated plants. Finally, DDCC is most effective when applied via the roots and before infection. This suggests that prior interaction with the plant is necessary for DDCC activity (LANGCAKE *et al.* 1983).

The DDCC-*P. oryzae*-rice plant system is an interesting experimental model which can contribute further to an understanding of the chemical regulation of host/parasite interactions. Perhaps the most critical bit of information needed in this system is the identity of the primary target of DDCC which triggers the enhanced resistance response.

Phenylthiourea

Phenylthiourea (PTU) prevents disease development in certain plants at concentrations exhibiting little or no fungitoxicity in vitro. The compound (Fig. 23.1), however, is primarily of research interest since it has not been adopted for practical use.

When cucumber seedlings are allowed to take up PTU by root absorption from solutions containing 50 μg/ml, they are protected from infection by *Cladosporium cucumerinum* (Kaars Sijpesteijn 1969). The sap of these seedlings contains 10 to 20 μg/ml of PTU, but is not toxic to growth of the pathogen in vitro. Lignification is enhanced in PTU treated plants around sites of penetration of *C. cucumerinum*, and this is believed to be the basis of protection (Kaars Sijpesteijn and Sisler 1968; Kaars Sijpesteijn 1969). PTU is a potent inhibitor of polyphenol oxidase (tyrosinase) activity. According to the hypothesis of Kaars Sijpesteijn (1969), PTU inhibits polyphenol oxidase in the plant tissue as well as in the invading pathogen. As a consequence, phenolic precursors of lignin accumulate and are rapidly converted to lignin by the elevated levels of peroxidase found in the PTU treated tissue.

PTU controls rice blast disease when used at 100 to 200 μg/ml. The ED_{95} values of PTU for inhibition of mycelial growth, spore germination and appressorial formation by the pathogen, *P. oryzae*, are greater than 1,000 μg/ml. PTU acts as an antipenetrant toward *P. oryzae* on both Cellophane film and rice sheaths (Araki and Miyagi 1977). The compound inhibits penetration of Cellophane film by both appressoria and hyphae at 25 μg/ml. This is in contrast to the action of fthalide (Rabcide) which results only in the inhibition of appressorial penetration. These results indicate that PTU can act directly on the pathogen as an antipenetrant.

Fosetyl-Al

Aluminum tris-0-ethylphosphonate (fosetyl-Al) known as Aliette is a systemic, organic phosphite compound used to control certain diseases caused by Peronosporales. The compound (Fig. 23.4) is degraded in buffer or plant tissue to phosphorous acid (Bompeix *et al.* 1980), a product on which protective activity is apparently based. Fosetyl-Al controls plant diseases at concentrations which have little or no effect on growth of the pathogen in vitro under certain conditions. The low in vitro toxicity has been interpreted as an indication that antifungal activity of the compound is indirect and mediated through host defense systems (Bompeix *et al.* 1980; Fettouche *et al.* 1981).

Fosetyl-Al treatment enhances phenol accumulation and produces necrotic blocking (defense) reactions around infection sites on tomato leaves inoculated with *Phytophthora capsici* (Bompeix *et al.* 1980; Durand and Sallé 1981). Induction of these defense reactions is blocked by phosphate ions. Fosetyl-Al also enhances production of antifungal stilbenes and flavanoids and reduces disease symptoms in grape leaves inoculated with *Plasmopara viticola* (Langcake 1981). These observations in addition to its low toxicity to mycelial growth in vitro, support the idea that fosetyl-Al acts by stimulating the natural defense system of the host plant.

Farih *et al.* (1981) have observed, however, that fosetyl-Al at a concentration of 10 μg/ml strongly inhibits sporangial formation or zoospore release in some *Phytophthora* species. Moreover, a recent study by Fenn and Coffey (1984) indicates that a direct fungitoxic effect of fosetyl-Al (or of H_3PO_3) in vivo is involved in plant protective activity. The fungitoxicity of fosetyl-Al is affected by the amount of phosphate present (Bompeix *et al.* 1980; Fenn and Coffey 1984) and probably also by the rate of release of H_3PO_3. The high level of phosphate present in culture medium used for *in vitro* tests and possibly a more rapid conversion of the compound to H_3PO_3 *in vivo* than *in vitro* may have led some investigators to underestimate the potential direct fungitoxicity of fosetyl-Al in vivo.

Fenn and Coffey (1984) found that H_3PO_3 is quite fungitoxic in vitro to *Phytophthora cinnamomi*, being appreciably more so than fosetyl-Al. The EC_{50} values of these compounds for this fungus growing on low phosphate medium were 0.05 PO_3

milliequivalents of H_3PO_3 (4 μg/ml) and 0.45 PO_3 milliequivalents of fosetyl-Al (54 μg/ml). FENN and COFFEY (1984) found similar in vitro and in vivo activity for H_3PO_3, but lower activity for fosetyl-Al in vitro than in vivo. However, efficacy of the two compounds in vivo was similar for control of seedling root rot of *Persea indica* caused by *P. cinnamomi*. Although the data of FENN and COFFEY (1984) clearly indicate that direct fungitoxicity is involved in plant protection by fosetyl-Al and H_3PO_3, there is need to clarify what role, if any, plant defense systems play in the effectiveness of these compounds. A phenomenon which needs explanation is the suppression of fosetyl-Al "induced resistance" in tomato to *P. capsici* by the phenolic pathway inhibitors, glyphosate and α-aminooxyacetic acid (FETTOUCHE *et al.* 1981). The suppression of "triphenylphosphite induced resistance" to rust in *Phaseolus vulgaris* by α-amino-oxyacetic acid observed by RUSUKU *et al.* (1984), may be analogous to the aforementioned suppression of "fosetyl-Al induced resistance" in tomato.

Whatever the relationships may be between the host, the pathogen and fosetyl-Al or H_3PO_3, there is essentially nothing known about the primary biochemical action of these compounds or of the subsequent biochemical events which lead to disease control. An interesting and valuable property of fosetyl-Al and H_3PO_3 is their capacity to control root diseases when applied to above-ground parts of plants (DARVAS *et al.* 1984; ROHRBACH and SCHENCK 1985). This implies a downward translocation of these compounds or of other factors which leads to root protection from pathogen attack.

Metalaxyl

Metalaxyl, an acylalanine fungicide (Fig. 23.1) (chapter 17, 18) is an important control agent for pathogenic fungi in the order Peronosporales, such as soil-borne *Pythium* and *Phytophthora* spp., downy mildews and potato late blight (DAVIDSE and DE WAARD 1984). Metalaxyl interferes with a template-bound RNA polymerase (DAVIDSE 1984) and is quite toxic in vitro to the aforementioned types of fungi. Nevertheless, the antifungal activity of metalaxyl in plants sometimes appears accentuated, promoting the thought that the fungicide also affects pathogenicity or resistance mechanisms (WADE 1984). This idea is supported by studies showing that potato tubers treated with metalaxyl are resistant to decay caused by *Fusarium sambucinum* and *Alternaria solani*, fungi to which metalaxyl shows no direct toxicity (BARAK and EDGINGTON 1983).

Metalaxyl promoted glyceollin production in soybean seedlings infected with a compatible race of *Phytophthora megasperma*. This phenomenon was believed to contribute to disease control (WARD *et al.* 1980). Further study (STÖSSEL *et al.* 1982) indicated that metalaxyl acted directly on the fungus and not on the host and that necrosis and enhanced glyceollin production in plant tissue are secondary effects, possibly due to release of elicitors from metalaxyl damaged hyphae, failure to produce a suppressor or a reduction in growth rate thus allowing host cells enough time to develop a resistance response. Another investigation (LAZAROVITS and WARD 1982) led to the conclusion that metalaxyl alone and not glyceollin restricted spread of the pathogen in tissue, even though appreciable accumulation of the latter had occurred. More recently BÖRNER *et al.* (1983), working also with a compatible race of *P. megasperma*, found that while metalaxyl did not increase glyceollin production per treated hypocotyl, it did affect glyceollin distribution in the tissue so that very high levels accumulated around infection sites. BÖRNER *et al.* (1983) suggested that an increased release of elicitor from fungal cell walls together with growth inhibition of hyphae can explain high and localized glyceollin accumulation. Thus, it seems that metalaxyl may increase host resistance through its toxic action on a compatible fungal pathogen within plant tissue. WARD (1984) provided further evidence for this. Using glyphosate to block the shikimic acid pathway in tissue infected with a compatible race of *P. megasperma*, the effectiveness of metalaxyl was appreciably reduced when it was present in the tissue at concentrations marginally inhibitory to the fungus. Whether this was due to inhibitory effects of glyphosate on glyceollin production alone, however, is uncertain.

Miscellaneous compounds

There is an extensive literature concerning the effect of plant hormones, growth regulator chemicals and various metabolites such as amino acids on disease development, but space permits only a few comments to be made about such compounds. This subject has recently been discussed by DEKKER (1983).

The use of plant hormones and plant growth regulators has been studied extensively, particularly in respect to the control or reduction of severity of vascular wilt diseases caused by *Fusarium*, *Verticillium* and *Ceratocystis* species. The literature on this subject has been reviewed by ERWIN (1977). Among the mechanisms proposed to explain vascular wilt control by plant hormones or growth regulators are restriction of propagule spread and increased levels of antifungal compounds in the infected tissue. In no case is the molecular basis of the disease control mechanism known. Some of the most promising results have been obtained with growth retardants exhibiting antigibberellin activity. For example, a marked reduction of *Verticillium* propagules occurred in cotton plants treated with 2-chloroethyl trimethylammonium chloride (CCC). Yield of cotton seed and lint was increased 16 and 12 percent by 10 and 25 g/ha doses of CCC respectively. Yields were unaffected by 50 g/ha and were reduced 6 percent by 75 g/ha (ERWIN 1977). In another case, promising results were obtained by using the non-bactericidal growth retardant 3,5-dichlorophenoxy acetic acid for control of potato scab caused by *Streptomyces scabies* (MCINTOSH *et al.* 1968).

When growth regulators affect plant growth but do not reduce crop yield, there may be promise for their use in disease control; however, as DEKKER (1983) has pointed out, limited disease control and adverse side effects produced by chemicals of this type greatly restrict their value as disease control agents.

Because of highly specific associations with the host, a modification of host or pathogen physiology might be expected to succeed more often for control of obligate parasites than for other types of fungal pathogens. A number of cases of experimental control of powdery mildews and rusts have, in fact, been described. The growth regulator kinetin, for example, has proven to be active against powdery mildews of cucumber (DEKKER 1963) and tobacco (COLE and FERNANDES 1970). Other natural metabolites such as methionine control powdery mildew on cucumber (DEKKER 1969) and barley (AKUTSU 1977). Since folic acid reversed methionine activity toward powdery mildew of cucumber, DEKKER (1969) suggested that the amino acid interfered with folic acid metabolism. In another case, soybean lecithin controlled strawberry and cucumber powdery mildew but was not toxic to the fungi in vitro (MISATO *et al.* 1977). Disease control was believed to result from an action of lecithin on the host plants.

Synthetic chemicals with no apparent fungitoxicity also control powdery mildews and rusts. One of these is procaine hydrochloride which controls powdery mildews, apparently through effects on host metabolism (DEKKER 1983). Another is the compound Indar (4-n-butyl-1,2,4-triazole) which is specific for control only of brown rust of wheat caused by *Puccinia recondita* (VON MEYER *et al.* 1970). The mechanism of action of this compound remains obscure, although WATKINS *et al.* (1977) suggest from ultrastructural studies that the compound acts on the pathogen rather than through defense mechanisms of the host.

Treatment of tomato or eggplant seedlings with dinitroaniline herbicides results in markedly increased resistance to vascular wilts caused by *Fusarium* and *Verticillium* species (GRINSTEIN *et al.* 1984). Tomato plants susceptible to *Fusarium oxysporum* f. sp. *lycopersici* when pretreated with the dinitroaniline herbicide, trifluralin, and inoculated with the aforementioned pathogen, accumulate fungitoxic compounds. These toxins do not accumulate in comparable plants not pretreated with trifluralin. However, toxins do accumulate in inoculated, monogenic resistant plants without treatment with trifluralin. This herbicide is regarded as a sensitizer which conditions the plant to produce toxins, which are probably phytoalexins, upon challenge by the pathogen (GRINSTEIN *et al. et al.* 1984). It thus appears to correct some deficiency in the host/parasite interaction necessary for a resistance response by the host. In this respect, the action of trifluralin resembles that of a dichlorocyclopropane carboxylic acid in the rice/*P. oryzae* system (LANGCAKE 1981).

Summary and conclusions

Although practical control of fungal diseases by nonfungitoxic compounds has been achieved in several instances, the area still remains one with few successes. Among the very successful compounds are the melanin biosynthesis inhibitors which directly block pathogenicity of *P. oryzae.* If the mechanism of appressorial penetration used by *P. oryzae* were utilized widely by plant pathogenic fungi, this group of compounds alone, would have had a major impact on the control of fungal diseases.

Validamycin, like the melanin biosynthesis inhibitors, also acts on the pathogen, but its success may depend in part on host resistance responses.

Disease control by probenazole and DDCC is mediated through host defense systems which actually produce the antifungal action. However, much of the biochemical basis for this action is unclear; it is uncertain, for example, whether the primary target of the compounds is in the host or in the pathogen.

In the case of fosetyl-Al and H_3PO_3, a question remains concerning the role of direct fungitoxicity and host resistance responses in the ultimate antifungal action. Disease control obtained with these compounds may be due to an enhanced host resistance response resulting from weak fungitoxic action of the chemicals on the pathogen.

The operation of a disease control mechanism of this type appears to occur in metalaxyl treated soybean tissue infected with *P. megasperma.* In this case, the contribution of the host resistance system is probably important only at marginally toxic concentrations of the fungicide. The extent to which the fungitoxic action of various systemic fungicides is supported by host resistance mechanisms remains an interesting question.

There are good prospects that chemical regulation of host resistance or of fungal pathogenicity by nonfungitoxic compounds will increase in significance as mechanisms of disease control. Progress in this area is presently hindered by lack of fundamental knowledge concerning factors involved in fungal pathogenicity and hostparasite interactions which lead to resistance or susceptibility.

References

Akatsuka, T., Kodama, O., and Yamada, H.: A novel mode of action of Kitazin P in *Pyricularia oryzae*. Agric. Biol. Chem. **41** (1977): 2111—2112.

Akutsu, K., Amano, K., and Ogasawara, N.: Inhibitory action of methionine upon barley powdery mildew *(Erysiphe graminis* f. sp. *hordei)*. I. microscopic observation of development of the fungus on barley leaves treated with methionine. Ann. Phytopath. Soc. Japan. **43** (1977): 33—39.

Araki, F., and Miyagi, Y.: Effects of fungicides on penetration by *Pyricularia oryzae* as evaluated by an improved cellophane method. J. Pesticide Sci. **2** (1977): 457—461.

Barak, E., and Edgington, L. V.: Bioactivity of the fungicide metalaxyl in potato tubers against *Phytophthora infestans* and other fungi. Can. J. Plant Pathol. **5** (1983): 200.

Börner, H., Schatz, G., and Grisebach, H.: Influence of the systemic fungicide metalaxyl on glyceollin accumulation in soybean infected with *Phytophthora megasperma* f. sp. *glycinea*. Physiol. Plant Pathol. **23** (1983): 145—152.

Bompeix, G., Ravisé, A., Raynal, G., Fettouche, F., and Durand, M. C.: Modalités de l'obtention des nécroses bloquantes sur feuilles détachees de tomate par l'action du tris-0-éthyl phosphonate d'aluminium (Phoséthyl d'aluminium), hypothèses sur son mode d'action in vivo. Ann. Phytopathol. **12** (1980): 337—351.

Bushnell, W. R., and Rowell, J. B.: Suppressors of defense reactions; a model for roles in specificity. Phytopathology **71** (1981): 1012—1014.

Cartwright, D. W., Langcake, P., and Ride, J. P.: Phytoalexin production in rice and its enhancement by a dichlorocyclopropane fungicide. Physiol. Plant Pathol. **17** (1980): 259 to 267.

Chida, T., Uekita, T., Satake, K., Hirano, K., Aoki, K., and Noguchi, T.: Effect of fthalide on infection process of *Pyricularia oryzae* with special observation of penetration site of appressoria. Ann. Phytopath. Soc. Japan. **48** (1982): 58—62.

Cole, J. S., and Fernandes, D. L.: Changes in the resistance of tobacco leaf to *Erysiphe cichoracearum* D.C. induced by topping, cytokinins and antibiotics. Ann. appl. Biol. **66** (1970): 239 to 243.

Darvas, J. M., Toerien, J. C., and Milne, D. L.: Control of avocado root rot by trunk injection with phosethyl-Al. Plant Dis. **68** (1984): 691—693.

Davidse, L. C.: Antifungal activity of acylalanine fungicides and related chloroacetanilide herbicides. In: Trinci, A. P. J., and Riley, J. F. (Eds.): Mode of Action of Antifungal Agents. Cambridge Univ. Press, London 1984, pp. 239—255.

— and de Waard, M. A.: Systemic fungicides. Adv. Plant Pathology **2** (1984): 191—257.

Dekker, J.: Effect of kinetin on powdery mildew. Nature (London) **197** (1963): 1027—1028.

— L-Methionine induced inhibition of powdery mildew and its reversal by folic acid. Netherlands J. Plant Pathol. **75** (1969): 182—185.

— Non-fungicidal compounds which prevent disease development. In: Plant Protection for Human Welfare. Proc. 10th Int. Congr. Plant Protect. **1** (1983): 237—248.

Dickman, M. B., Patil, S. S., and Kolattukudy, P. E.: Purification, characterization and role in infection of an extracellular cutinolytic enzyme from *Colletotrichum gloeosporioides*. Penz. on *Carica papaya* L. Physiol. Plant Pathol. **20** (1982): 333—347.

— — Effects of organophosphorous pesticides on cutinase activity and infection of papayas by *Colletotrichum gloeosporioides*. Phytopathology **73** (1983): 1209—1214.

Dimond, A. E., and Rich, S.: Effects on physiology of the host and on host/pathogen interactions. In: Marsh, R. W. (Ed.): Systemic Fungicides, 2nd edition. Longman, London 1977, pp. 115—130.

Durand, M. C., and Sallé, G.: Effet du tris-0-éthyl phosphonate d'aluminium sur le couple *Lycopersicum esculentum* Mill. — *Phytophthora capsici* Leon. Etude cytologique et cytochimique Agronomie **9** (1981): 723—732.

Erwin, D. C.: Control of vascular pathogens. In: Siegel, M. R., and Sisler, H. D. (Eds.): Antifungal Compounds. V. 1. Marcel Dekker, New York 1977, pp. 163—224.

Farih, A., Tsao, P. H., and Menge, J. A.: Fungitoxic activity of efosite aluminum on growth, sporulation and germination of *Phytophthora parasitica* and *P. citrophthora*. Phytopathology **71** (1981): 934—936.

FENN, M. E., and COFFEY, M. D.: Studies on the in vitro and in vivo antifungal activity of Fosetyl-Al and phosphorous acid. Phytopathology **74** (1984): 606—611.

FETTOUCHE, F., RAVISÉ, A., and BOMPEIX, G.: Suppression de la résistance induite-phosétyl-Al chez la tomate à *Phytophthora capsici* avec deux inhibiteurs-glyphosate et acide α-amino-oxyacétique. Agronomie **9** (1981): 826.

GRINSTEIN, A., LISKER, N., KATAN, J., and ESHEL, Y.: Herbicide induced resistance to plant wilt diseases. Physiol. Plant Pathol. **24** (1984): 347—356.

GROSSMANN, F.: Studies on the therapeutic effects of pectolytic enzyme inhibitors. Netherland J. Plant Pathol. **74**, Supplement 1 (1968): 91—103.

HEATH, M. C.: A generalized concept of host-parasite specificity. Phytopathology **71** (1981): 1121—1123.

INOUE, S., UEMATSU, T., and KATO, T.: Effects of chlobenthiazone on the infection process by *Pyricularia oryzae*. J. Pesticide Sci. **9** (1984): 689—695.

ISHIDA, M., SUMI, H., and OKU, H.: Pentachlorobenzyl alcohol, a rice blast control agent. Residue Rev. **25** (1969): 139—148.

JACOBS, M. E.: Influence of β-alanine on ultrastructure, tanning, and melanization of *Drosophila melanogaster* cuticles. Biochem. Genet. **18** (1980): 65—76.

KAARS SIJPESTEIJN, A.: Mode of action of phenylthiourea, a therapeutic agent for cucumber scab. J. Sci. Fd. Agric. **20** (1969): 403—405.

— and SISLER, H. D.: Studies on the mode of action of phenylthiourea, a chemotherapeutant for cucumber scab. Neth. J. Plant Pathol. **74** (1968), Supplement 1: 121—126.

KODAMA, O., YAMASHITA, K., and AKATSUKA, T.: Edifenphos, inhibitor of phosphatidylcholine biosynthesis in *Pyricularia oryzae*. Agric. Biol. Chem. **44** (1980): 1015—1021.

KÖLLER, W., ALLAN, C. R., and KOLATTUKUDY, P. E.: Protection of *Pisum sativum* from *Fusarium solani* f. sp. *Pisi* by inhibition of cutinase with organophosphorus pesticides. Phytopathology **72** (1982a): 1425—1430.

— — — Inhibition of cutinase and prevention of fungal penetration into plants by benomyl-A possible protective mode of action. Pesticide Biochem. Physiol. **18** (1982b): 15—25.

KUBO, Y., SUZUKI, K., FURUSAWA, I., ISHIDA, N., and YAMAMOTO, M.: Relation of appressorium pigmentation and penetration of nitrocellulose membranes by *Colletotrichum lagenarium*. Phytopathology **72** (1982a): 498—501.

— — — and YAMAMOTO, M.: Effect of tricyclazole on appressorial pigmentation and penetration from appressoria of *Colletotrichum lagenarium*. Phytopathology **72** (1982b): 1198—1200.

— — — — Melanin biosynthesis as a prerequisite for penetration by appressoria of *Colletotrichum lagenarium:* Site of inhibition by melanin-inhibiting fungicides and their action on appressoria. Pesticide Biochem. Physiol. **23** (1985): 47—55.

KUĆ, J.: Systemic plant immunization. In: LYR, H., and POLTER, C. (Eds.): Systemic Fungicides and Antifungal Compounds. Tagungsber. Akad. Landwirtschaftswiss. **222**. DDR, Berlin (1984): 189—198.

LANGCAKE, P.: Alternative chemical agents for controlling plant disease. Phil. Trans. Royal Soc. London B **295** (1981): 83—101.

— CARTWRIGHT, D. W., and RIDE, J. P.: The dichlorocyclopropanes and other fungicides with indirect mode of action. In: LYR, H., and POLTER, C. (Eds.): Systemische Fungizide und antifungale Verbindungen. Akademie-Verlag, Berlin 1983, S. 199—210.

— and WICKINS, S. G. A.: Studies on the action of the dichlorocyclopropanes on the host-parasite relationship in the rice blast disease. Physiol. Plant Pathol. **7** (1975): 113—126.

LAZAROVITS, G., and WARD, E. W. B.: Relationship between localized glyceollin accumulation and metalaxyl treatment in the control of *Phytophthora* rot in soybean hypocotyls. Phytopathology **72** (1982): 1217—1221.

LEBRUN, M. H., ORCIVAL, J., and DUCHARTRE, C.: Resistance of rice to tenuazonic acid, a toxin from *Pyricularia oryzae*. Rev. Cytol. Biol. Végét-Bot. **7** (1984): 249—259.

MAITI, I. B., and KOLATTUKUDY, P. E.: Prevention of fungal infection of plants by specific inhibition of cutinase. Science **205** (1979): 507—508.

MCINTOSH, A. H., BATEMAN, G. L., CHAMBERLIN, K., DAWSON, G. W., and BURRELL, M. M.: Decreased severity of potato common scale after foliar sprays of 3,5-dichlorophenoxy acetic acid, a possible antipathogenic agent. Ann. appl. Biol. **99** (1968): 275—281.

MISATO, T., HOMMA, Y., and KO, K.: The development of a natural fungicide, soybean lecithin. Netherl. J. Plant Pathol. **83**, Supplement 1 (1977): 395—402.

Nioh, T., and Mizushima, S.: Effect of validamycin on the growth and morphology of *Pellicularia sasakii*. J. Gen. appl. Microbiol. **20** (1974): 373—383.

Okuno, T., Matsuura, K., and Furusawa, I.: Recovery of appressorial penetration by some melanin precursors in *Pyricularia oryzae* treated with tricyclazole and in a melanin deficient mutant. J. Pesticide Sci. **8** (1983): 357—360.

Rohrbach, K. G., and Schenck, S.: Control of pineapple heart rot caused by *Phytophthora parasitica* and *P. cinnamomi* with metalaxyl, fosetyl-Al and phosphorous acid. Plant Dis. **69** (1985): 320—323.

Rudolph, K.: Nonspecific toxins. In: Heitefuss, R., and Williams, P. H. (Eds.): Physiological Plant Pathology. Springer-Verlag, Berlin 1976, pp. 270—315.

Rusuku, G., Lepoivre, P., Meulemans, M., and Semal, J.: Effects of triphenylphosphite on bean rust development. Plant Disease **68** (1984): 154—155.

Scheffer, R. P.: Host-specific toxins in relation to pathogenesis and disease resistance. In: Heitefuss, R., and Williams, P. H. (Eds.): Physiological Plant Pathology. Springer-Verlag, Berlin 1976, pp. 247—269.

Sekizawa, Y., and Mase, S.: Mode of controlling action of probenazole against rice blast disease with reference to the induced resistance mechanism in rice plant. J. Pesticide Sci. **6** (1981): 91—94.

Shibata, M., Mori, K., and Hamashima, M.: Inhibition of hyphal extension factor formation by validamycin in *Rhizoctonia solani*. J. Antibiot. **35** (1982): 1422—1423.

Sisler, H. D.: Fungicides: Problems and prospects. In: Siegel, M. R., and Sisler, H. D. (Eds.): Antifungal Compounds. V. 1. Marcel Dekker, New York 1977, pp. 531—547.

— Woloshuk, C. P., and Wolkow, P. M.: Specific regulation of appressorial function. In: Lyr, H., and Polter, C. (Eds.): Systemic Fungicides and Antifungal Compounds. Tagungsber. Akad. Landwirtschaftswiss. **222**. DDR, Berlin (1984): 17—28.

Stössel, P., Lazarovits, G., and Ward, E. W. B.: Light and electron microscopy of *Phytophthora* rot in soybeans treated with metalaxyl. Phytopathology **72** (1982): 106—111.

Suami, T., Ogawa, S., and Chida, N.: The revised structure of validamycin A. J. Antibiot. **33** (1980): 98—99.

Tepper, C. S., and Anderson, A. J.: The genetic basis of plant-pathogen interaction. Phytopathology **74** (1984): 1143—1145.

Tokousbalides, M. C., and Sisler, H. D.: Effect of tricyclazole on growth and secondary metabolism in *Pyricularia oryzae*. Pesticide Biochem. Physiol. **8** (1978): 26—32.

— — Site of inhibition by tricyclazole in the melanin biosynthetic pathway of *Verticillium dahliae*. Pesticide Biochem. Physiol. **11** (1979): 64—73.

Trinci, A. P. J.: Effect of validamycin A and L-sorbose on the growth and morphology of *Rhizoctonia cerealis* and *Rhizoctonia solani*. Exp. Mycol. **9** (1985): 20—27.

Van Etten, H. D.: Phytoalexin detoxification by monooxygenases and its importance for pathogenicity. In: Asada, Y., Bushnell, W. R., Ouchi, S., and Vance, C. P. (Eds.): Plant Infection — The Physiological and Biochemical. Basis. Japan Sci. Soc. Press, Tokyo 1982, pp. 315—327.

Von Meyer, W. C., Greenfield, S. A., and Seidel, M. C.: Wheat leaf rust control by 4n-butyl-1,2,4-triazole, a systemic fungicide. Science **169** (1970): 997—998.

Wade, M.: Antifungal agents with an indirect mode of action. In: Trinci, A. P. J., and Riley, J. F. (Eds.): Mode of Action of Antifungal Agents. Cambridge Univ. Press, Cambridge 1984, pp. 283—298.

Wakae, O., and Matsuura, K.: Characteristics of validamycin as a fungicide for *Rhizoctonia* disease control. Rev. Plant Protec. Res. **8** (1975): 81—92.

Ward, E. W. B.: Suppression of metalaxyl activity by glyphosate: evidence that host defense mechanisms contribute to metalaxyl inhibition of *Phytophthora megasperma* f. sp. *glycinea* in soybeans. Physiol. Plant Pathol. **25** (1984): 381—386.

— Lazarovits, G., Stössel, P., Barrie, S. D., and Unwin, C. H.: Glyceollin production associated with control of *Phytophthora* rot of soybeans by the systemic fungicide, metalaxyl. Phytopathology **70** (1980): 738—740.

Watanabe, T.: Effects of probenazole (Oryzemate) on each stage of rice blast fungus (*Pyricularia oryzae* Cavara) in its life cycle. J. Pesticide Sci. **2** (1977): 395—404.

— Sekizawa, Y., Shimura, M., Suzuki, Y., Matsumoto, K., Iwata, M., and Mase, S.: Effects of probenazole (Oryzemate) on rice plants with reference to controlling rice blast. J. Pesticide Sci. **4** (1979): 53—59.

WATKINS, J. E., LITTLEFIELD, L. J., and STATLER, G. D.: The effect of the systemic fungicide 4-n-butyl-1,2,4-triazole on the development of *Puccinia recondita* f. sp. *tritici* in wheat. Phytopathology **67** (1977): 985—989.

WHEELER, M. H.: Melanin biosynthesis in *Verticillium dahliae:* Dehydration and reduction reactions in cell-free homogenates. Exp. Mycol. **6** (1982): 171—179.

— Comparisons of fungal melanin biosynthesis in Ascomycetous, imperfect and Basidiomycetous fungi. Trans. Br. Mycol. Soc. **81** (1983): 29—36.

— and STIPANOVIC, R. D.: Melanin biosynthesis in *Thielaviopsis basicola*. Exp. Mycol. **3** (1979): 340—350.

— — Melanin biosynthesis and the metabolism of flaviolin and 2-hydroxyjuglone in *Wangiella dermatitidis*. Arch. Microbiol. **142** (1985): 234—241.

WOLKOW, P. M.: Compounds which specifically block epidermal penetration by *Colletotrichum lindemuthianum*. M. S. Thesis, Univ. of Maryland, 1982.

— SISLER, H. D., and VIGIL, E. L.: Effect of inhibitors of melanin biosynthesis on structure and function of appressoria of *Colletotrichum lindemuthianum*. Physiol. Plant Pathol. **23** (1983): 55—71.

WOLOSHUK, C. P., and SISLER, H. D.: Tricyclazole, pyroquilon, tetrachlorophthalide, PCBA, coumarin, and related compounds inhibit melanization and epidermal penetration by *Pyricularia oryzae*. J. Pesticide Sci. **7** (1982): 161—166.

— — TOKOUSBALIDES, M. C., and DUTKY, S. R.: Melanin biosynthesis in *Pyricularia oryzae:* Site of tricyclazole inhibition and pathogenicity of melanin-deficient mutants. Pesticide Biochem. Physiol. **14** (1980): 256—264.

— — and VIGIL, E. L.: Action of the antipenetrant, tricyclazole, on appressoria of *Pyricularia oryzae*. Physiol. Plant Pathol. **22** (1983): 245—259.

YAMAGUCHI, I., SEKIDO, S., and MISATO, T.: The effect of non-fungicidal anti-blast chemicals on the melanin biosynthesis and infection by *Pyricularia oryzae*. J. Pesticide Sci. **7** (1982): 523 to 529.

— — — Inhibition of appressorial melanization in *Pyricularia oryzae* by non-fungicidal anti-blast chemicals. J. Pesticide Sci. **8** (1983 a): 229—232.

— — SETO, H., and MISATO, T.: Cytotoxic effect of 2-hydroxyjuglone, a metabolite in the branched pathway of melanin biosynthesis in *Pyricularia oryzae*. J. Pesticide Sci. **8** (1983 b): 545—550.

LYR, H. (Ed.): Modern Selective Fungicides — Properties, Applications, Mechanisms of Action. Longman Group UK Ltd., London, and VEB Gustav Fischer Verlag, Jena, 1987.

Chapter 24

Synergism and antagonism in fungicides

M. A. DE WAARD
Laboratory of Phytopathology, Agricultural University, Wageningen, The Netherlands

Introduction

Fungicides are often combined in mixtures to extend the spectrum of antifungal activity and to counteract development of fungicide resistance. Additional advantages of mixtures may be due to synergistic interactions by which the efficiency of the individual components can be increased or the amount of expensive active ingredients can be reduced. The latter consideration stimulated DIMOND and HORSFALL (1944) to develop synergistic mixtures of cuprous oxides and sulphur in order to conserve the war-short material copper. Up till now, the intentional use of fungicidal mixtures with a synergistic interaction has remained in general rather limited. This contrasts to the commercial development of synergistic mixtures of insecticides (O'BRIEN 1967; WILKINSON 1976) and drugs (KERRIDGE and WHELAN 1984). Due to recent developments the prospects for use of fungicide synergism may improve. The purpose of this chapter is to review the concepts of joint action in mixtures and the (bio)-chemical mechanisms which may underlie the synergistic action.

Definitions and test methods

The expected response of a mixture of two chemicals is the sum of the effects of the components separately. This is usually referred to as additive action. It may occur between chemicals with an identical or a different mode of action and has been designated by BLISS (1939) and FINNEY (1947) as similar and independent joint action, respectively. In case of similar joint action, dosage-response curves of the chemicals are often parallel, so that one component can be substituted at a constant proportion for the other, without altering the toxicity of the mixture. The toxicity of the mixture is directly predictable from the relative proportions of the constituents. With independent joint action dosage-response curves are often not parallel and the additive toxicity of a mixture can only be predicted from the dosage-response curve for each component applied alone and the correlation in sensitivity to the two toxicants.

In case of synergism the effectiveness of the mixture can not be computed from that of the individual ingredients. Synergism is defined as the simultaneous action of two or more compounds in which the total response of an organism to the pesticide combination is greater than the sum of the individual components (NASH 1981). Antagonism is the opposite.

In order to test whether an observed response can be considered as synergistic or antagonistic, the expected response in the absence of synergism or antagonism should be well defined. It requires a null hypothesis or reference model (MORSE 1978). This presents no great difficulty if only one component of the mixture affects the test

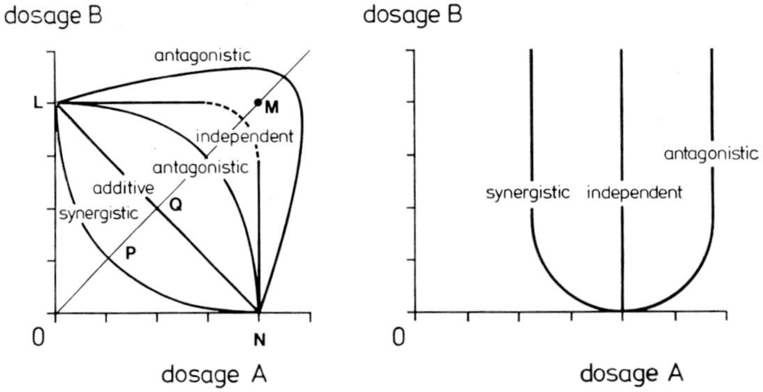

Fig. 24.1 Growth isoboles for combinations of fungicides A and B at a defined growth response e.g. the EC_{50}. Left: both chemicals A and B fungitoxic; right: only chemical A fungitoxic. Explanation of letters in text.

organism when applied on its own, but if more components are active the reference model is sometimes hard to define. In studies on joint action of herbicides (MORSE 1978) and NASH (1981) refer to two types: the additive dose model (ADM) and the multiplicative survival model (MSM). These models are also valid for other pesticide combinations.

A common ADM is the isobole method in which curves or isoboles represent lines of equal response, e.g. the EC_{50} (TAMMES 1964). With mixtures of two toxic components the line of equalresponse representing additive action is a straight line LN (Fig. 24.1; left). This reference isobole is valid for most cases of similar joint action. Isoboles drawn from observed responses to mixtures indicate synergism when they are located below this reference and antagonism or independent interaction when above. The interaction can be quantified by the ratio OQ/OP. An independent interaction represents the exceptional situation that the response is that of the most potent ingredient of the mixture. Interactions for mixtures with only one toxicant are much more simple (Fig. 24.1; right).

The MSM holds for mixtures of which the constituents produce their effect in entirely different ways. It is, therefore, a reference model for chemicals displaying independent joint action. The reference isobole for additive action of this model does not coincide with the straight line LN but lies within triangle LMN for most response levels; for high response levels it can even lie within triangle LON (MORSE 1978). The recognition that the location of the reference isobole may vary, often fails. Methods to assess accurately the location of the reference isobole in the MSM are, however, not conclusive. Improved analytic procedures to do so are described by MORSE (1978) and NASH (1981).

A simple mathematical method to test the additivity of a pesticide combination is based on the equation used by GOWING (1960)

$$E = X + \frac{Y(100 - X)}{100}$$

in which X and Y represent the percentage-of-inhibition of growth by toxicant A (dosage p) and B (dosage q), respectively. E is the expected percentage-of-inhibition of growth by the mixture A and B (dosage p + q). The equation has been modified by COLBY (1967) in

$$E_1 = \frac{X_1 Y_1}{100}$$

in which X_1 and Y_1 represent growth as a percentage-of-control with toxicant A (dosage p) and toxicant B (dosage q), respectively. E_1 is the expected growth as a percentage-of-control with the mixture A and B (dosage p + q). The observed response is obtained with the mixture containing the same rate of the constituents as applied singly. A deviation of the expected response from the observed response would indicate synergism or antagonism. The method can be considered as a MSM. The method is subject to serious criticism (MORSE 1978; NASH 1981) and should, therefore, be used with care.

Another, less common method has been described by WADLEY (1945, 1967) as cited by GISI *et al.* (1983). In this case, the theoretical response $EC_{90(th)}$ to a two component mixture is calculated according to the formula

$$EC_{90(th)} = \frac{a + b}{\dfrac{a}{EC_{90(A)}} + \dfrac{b}{EC_{90(B)}}}$$

A and B in the formula represent the two components and a and b the ratio of the components in the mixture. The level of interaction R can be calculated as $R = EC_{90(th)}/EC_{90(obs)}$ in which $EC_{90(obs)}$ is the observed EC_{90} value of the specific mixture.

All methods described above can be used for a wide variety of tests for various types of pesticides. The methods described below only apply to in vitro tests for fungicides or bactericides. A rapid, qualitative method is the crossed-paper strip bioassay, originally described by BONIFAS (1952) and modified by others (CORBAZ 1963; KATAGIRI and UESUGI 1977; DE WAARD and VAN NISTELROOY 1982). Filter paper strips impregnated with test chemicals are placed cross-wise on agar seeded with a fungus. Measurement and visual assessment of the inhibition zones at the crossing of the strips indicate the type of the interaction. A disadvantage of the method is sometimes to differentiate with certainty between synergistic, additive or independent action. Since visualization of the interaction depends on the formation of concentration gradients in the agar, the method is not appropriate for water-insoluble chemicals or chemicals with a relatively high water solubility.

A semi-quantitative method is the paper disk bioassay, in which disks impregnated with a chemical are placed on agar seeded with the test organism. Interaction with another chemical can be assessed by measuring the inhibition zone around disks on agar amended or not with a second chemical at a sublethal concentration. The method can be modified in various ways (e.g. GOSS and MARSHALL 1985).

Another semi-quantitative method is again an adoption of the poisoned food technique (LEROUX and GREDT 1978). Agar amended with test compounds, singly and in mixture, is inoculated with mycelial agar disks of a test fungus. The interaction can be assessed with the formula

$$A_{a,f} = 100\left(\frac{V_{a,f}}{V_{a,o}} - \frac{V_{o,f}}{V_{o,o}}\right)$$

$V_{o,o}$ stands for the mycelial growth rate in the absence of agonist and fungicide, $V_{o,f}$ in the absence of agonist and presence of fungicide, $V_{a,o}$ in the presence of agonist and absence of fungicide and $V_{a,f}$ in the presence of agonist and fungicide. Negative values of $A_{a,f}$ indicate synergism and positive values antagonism.

Mechanism of synergism and antagonism

Synergism and antagonism may be operative via different mechanisms. A most obvious one is based on chemical interactions between components of mixtures before or after fungal uptake. Interactions of this type have already been described by

HORSFALL (1945) under the name "potentiated" synergism or antagonism. Mixtures of sulphur and toxic metallic oxides like those of cuprous, cupric, zinc and lead are, for instance, synergistic due to the formation of metallic sulphides. Antagonism based on chemical detoxication has been described for several conventional fungicides (TORGESON 1967). Thiols, for instance, reduce the toxicity of phthalimides and of chlorothalonil (LUKENS and SISLER 1958; VINCENT and SISLER 1968). Another classical example of interaction is the complex formation of dithiocarbamates with copper ions, resulting in a bimodal dosage response curve (GOKSØYR 1955). Complex formation between polyene macrolides and certain sterols is also the reason why sterols antagonize the toxicity of filippin and related antibiotics (DEKKER 1969).

Other mechanisms may relate to interaction with physiological processes of the target organism. From a theoretical point of view interaction with one of the following processes can be involved (DE WAARD 1985). 1. Non-mediated diffusion across the plasma membrane. 2. Carrier-mediated transport across the plasma membrane. 3. Energy-dependent efflux from the fungal cell. 4. Transport to the target site. 5. Activation. 6. Detoxication. 7. Affinity for the target-site. 8. Circumvention of the target site. 9. Compensation of the target-site. The companion chemical may affect these processes in such a way that the efficacy or toxicity of the fungicide is enhanced. Such specific physiological interactions may particularly explain synergism or antagonism with modern, site-specific fungices. Examples are described elsewhere in this chapter.

Effects of companion chemicals on disease control are not necessarily based on chemical or physiological interactions with the fungicide. It may also be caused by a better field performance of the mixture or by the composition of the pathogen population. Improved field performance is usually achieved with formulation agents. They improve, for example, the distribution of the active ingredient over the plant surface, prevent run-off from leaves, increase persistance and enhance penetration into plants and thereby enhance their efficacy. The composition of a pathogen population is important in the case of disease complexes and of fungicide resistance. Pathogens of a disease complex may have a differential sensitivity to fungicides, by which the action of single fungicides may be ineffective but effective when combined (BERGER and WOLF 1974; EISA and BARAKAT 1978; FUSHTEY 1975; PAPAVIZAS and LEWIS 1977). Similarly, a disease may be simultaneously incited by fungicide-sensitive and fungicide-resistant strains of the same pathogen. Again, the single fungicide will fail to control the disease but a mixture with a companion fungicide to which all resistant strains display negatively-correlated cross-resistance can be successful. The only example known so far is the control of sensitive and benzimidazole-resistant *Botrytis* in French vineyards by a mixture of carbendazim and isopropyl N-(3,4-diethoxyphenyl)-carbamate (LEROUX *et al.* 1985).

Increased field performance of mixtures with formulation agents or in case of disease complexes or fungicide resistance is sometimes described as synergism. This is questionable since the intrinsic toxicity of the active ingredients to the individuals separately is not changed. Therefore, it is suggested that these mechanisms should be indicated by the expression pseudo-synergism. In the case of formulation agents it should be realized that they may not only act in such a pseudo-synergistic way but can also behave as true synergists. Detergents, for instance, enhance the in vitro toxicity of sulphur to *B. cinerea*, possibly by changing its membrane permeability to sulphur (PEZET and PONT 1984).

Another exceptional type of synergistic interaction has been described for the fungicide tridemorph which enhances the inhibitory action of its formulating product Nekanil LN on oxygen consumption by several fungi (KERKENAAR and KAARS SIJPESTEIJN 1979). The interaction between tridemorph and several other respiratory inhibitors (Bergmann 1979) seems to have a similar character. An explanation for the synergism may be that tridemorph enhances the transport of the companion chemicals to their site of action.

Examples of antagonism

The action of antagonistic compounds is often based on circumvention or compensation of the target site of the fungicide in a biosynthetic pathway. In consequence, search for such compounds often forms a part of mode of action studies. Examples cited in this paragraph only relate to site-specific fungicides.

A classical example of compensation has been demonstrated for the experimental fungicide 6-azauracil (AzU). AzU is metabolized to 6-azauridine-5′-phosphate which inhibits the decarboxylation of orotidine-5′-phosphate to form uridine-5′-phosphate. Uracil and other precursors of uridine-5′-phosphate reverse AzU toxicity because of compensation (Dekker and Oort 1964).

Toxicity of 2-aminopyrimidines towards powdery mildews can be reduced by adenine, guanine, folic acid, pyridoxal-5-phosphate and adenosine. A correlation between the activity of these chemicals and the mode of action of 2-aminopyrimidines, inhibition of adenosine deaminase activity, has not been established (c.f. Hollomon 1984). The reversal by riboflavin was caused by direct photo-inactivation of the fungicides (Bent 1970).

The toxicity of carboxamide fungicides depends of the carbon source in the nutrient medium. Toxicity with an acetate substrate is higher than with glucose. This relates to the inhibitory action of carboxamides on succinate dehydrogenase activity in the Krebs cycle. Inhibition of energy generation is, to a certain extent, circumvented in the presence of glucose (Ragsdale and Sisler 1970; Georgopoulos *et al.* 1972; Ritter *et al.* 1973).

Validamycin A causes in *Rhizoctonia* spp. various morphological effects on mycelial growth like a decrease in the maximum rate of hyphal extension and an increase in hyphal branching. Growth yield is not inhibited. Validamycin A is, therefore, regarded as an antifungal agent with an indirect mode of action (Trinci 1984). Its effectiveness in preventing infection of rice by *Rhizoctonia solani* is reduced by meso-inositol. Meso-inositol formation in liquid cultures is inhibited by validamycin (Wakae and Matsura 1975). The antagonism may, therefore, be based on compensation for a reduced meso-inositol content of the mycelium.

The toxicity of fungicides which inhibit sterol biosynthesis (SBIs) can be alleviated in certain fungi by a variety of lipophylic compounds like ergosterol. This is an example of compensation. Other lipophilic chemicals with antagonistic activity are fatty acids, phospholipids, triglycerides and non-ionic detergents (c.f. Sisler and Ragsdale 1984). They may act as a substitute for ergosterol in fungal membranes. The antagonism may also result from complex formation or partitioning of the SBIs into undissolved residues of the chemicals. However, these physico-chemical mechanisms do not explain the specificity in antagonism towards different fungi. Other mechanisms, which, therefore, might be involved are decreased affinity for SBIs to the membrane-bound target enzyme or a reduction in membrane permeability for the fungicide (Sherald *et al.* 1973; de Waard and van Nistelrooy 1982). The mechanism of antagonistic action of calcium chloride, tetracaine, carboxin and dialkyldithiocarbamates on fungitoxicity of fenarimol is unknown (de Waard and van Nistelrooy 1982).

The mode of action of etridiazole is probably based on stimulation of phospholipases and induced lipid peroxidation (chapter 6). Procaine is an inhibitor of phospholipase activity. This explains why this local anesthetic antagonizes the toxicity of the fungicide to *Mucor mucedo*. Calcium ions have a similar effect. The antagonistic activity of tocopherol may be due to its property to protect cell membranes from lipid peroxidation by oxygen radicals (Lyr *et al.* 1977; Radzuhn and Lyr 1984).

Examples of synergism

Numerous examples of synergism among fungicides are mentioned in reviews and papers by Dimond and Horsfall (1944), Corbaz (1963) and Scardavi (1966). Additional reports are listed in Tab. 24.1. It is not the goal of this paper to discuss these cases since virtually no information on the mechanisms involved is available. The same applies to the numerous patents on synergistic fungicidal compositions. The object of this paragraph is to discuss the well-documented synergism among organophosphorus compounds, SBIs and phenylamides.

Crossed-paper strip bioassays have revealed synergistic interactions between organophosphorus thiolate fungicides (PTLs) and phosphoramidates like dibutyl **N**-methyl **N**-phenylphosphoramidate (BPA). The toxic action of PTLs in *Pyricularia oryzae* is caused by inhibition of phosphatidylcholine biosynthesis. The toxic action of BPA is probably reduced by rapid fungal hydroxylation and **N**-demethylation resulting in the formation of non-toxic products. PTLs like diisopropyl **S**-benzyl phosphorothiolate (IBP) inhibit the metabolism of BPA, thereby enhancing BPA toxicity. BPA detoxication by certain PTL-resistant mutants was much slower than by the wild type. This probably explains the higher sensitivity of these mutants to BPA and the absence of synergism (Uesugi and Sisler 1978). Isoprothiolane has a similar mode of action as PTLs and also displays synergism with BPA (Katagiri and Uesugi 1977).

Crossed-paper strip bioassays revealed also synergistic interactions in wild-type and SBI-resistant strains of *Aspergillus nidulans* between the SBI fenarimol and a variety of other chemicals (Tab. 24.2; de Waard and van Nistelrooy 1982, 1984a). Fenarimol and sodium orthovanadate showed in a similar test synergism to wild-type and SBI-resistant strains of *Penicillium italicum* (de Waard and van Nistelrooy 1984). The synergism with the inorganic chemicals and the detergents might be due to increased solubility of fenarimol in the medium. However, for the respiratory

Table 24.1 Cases of fungicide synergism

Components in mixture	Pathogen	Reference
Anilazine/zinc or copper	*Botrytis cinerea* *Colletotrichum coccodes*	Goss and Marshall 1985
Zineb/polyram	*Plasmopara vizicola*	Flieg and Pommer 1965
Carboxin/mancozeb	various	Baicu and Nägler 1974
Chloroneb/thiram	*Pythium ultimum*	Richardson 1973
Copper/zineb	*Plasmopara viticola*	Thiolliere 1985
Dimethyldithiocarbamates/complex-forming agents	*Botrytis cinerea*	Matolcsy *et al.* 1971
Dodine/captan	*Venturia inaequalis* *Xanthomonas prunei*	Powell 1960 Diener and Carlton 1960
Elemental sulfur/surfactants	*Botrytis cinerea*	Pezet and Pont 1984
Ethazole/pentachloronitrobenzene	*Pythium aphanidermatum*	Rahimian and Banihashemi 1982

Table 24.2 Chemicals which show synergism with fenarimol to wild-type and EBI-resistant strains of *Aspergillus nidulans*

Anionic agents	sodium lauryl sulphate sodium tetradecyl sulphate
Cationic agents	dodecylamine dodine
Inorganic chemicals	hydrochloric acid sodium hydroxide
Phthalimide fungicides	captafol captan folpet
Respiratory inhibitors	carbonyl cyanide 3-chlorophenylhydrazone (CCCP) dicyclohexylcarbodi-imide (DCCD)

inhibitors, phthalimides and sodium orthovanadate a mechanism based on interference with accumulation of fenarimol in mycelium has been proposed. Accumulation by both fungi mentioned above is the sum of passive influx by non-mediated diffusion and energy-dependent efflux. The sensitive and resistant strains differ from each other in efflux activity, being inducible and constitutive, respectively. The constitutive efflux activity in resistant strains results in a permanent operation of a permeability barrier by which fenarimol does not accumulate in mycelium, as is the case with the sensitive strain. This is probably the mechanism of resistance to fenarimol (DE WAARD and VAN NISTELROOY 1979, 1980, 1981, 1984b). Phthalimide fungicides, respiratory inhibitors and sodium orthovanadate do not interfere with passive influx of fenarimol but do inhibit its energy-dependent efflux. In consequence, the permeability barrier for fenarimol is annihilated and the fungicide accumulates to similar levels in sensitive and resistant strains. The anihilation of fenarimol efflux activity probably explains the synergistic interaction, being highest in resistant isolates. It is suggested that the fenarimol efflux is a membrane process, related with maintenance of an electrochemical proton gradient over membranes by plasma membrane ATPase (DE WAARD and VAN NISTELROOY 1984b). In this concept, the synergism is due to interference with this enzyme, either directly by inhibition of its activity (orthovanadate) or indirectly by depletion of its substrate (mitochondrial respiratory inhibitors, phthalimides).

Fenarimol accumulation by various other pathogens is also energy-dependent. This also applies to accumulation of imidazole and triazole EBIs by *P. italicum* (DE WAARD unpublished results). Therefore, the phenomenon seems to be of general relevance and useful for the design of synergistic mixtures with practical value. Many EBIs are already marketed in mixtures with conventional fungicides which inhibit respiration. It may, therefore, be possible that synergism as described above, already operates in practice. It has for instance been observed for a mixture of the EBI fenpropimorph and chlorothalonil in control of *Pyrenophora teres* on barley (HAMPEL and LARTAUD 1983).

Recently, synergistic interactions in mixtures of phenylamides with several other fungicides have been described. Synergism of oxadixyl activity against *Phytophthora infestans* on potato and *Plasmopara viticola* on grape vines was observed when it was combined with mancozeb, cymoxanil or phosetyl-Al (GISI *et al.* 1983, 1985). Synergistic activity of the same mixtures was also observed with phenylamide-resistant isolates of *P. infestans* (GRABSKI and GISI 1985). A mixture of metalaxyl and mancozeb was synergistic in control of downy mildew of cucumber (SAMOUCHA and COHEN 1984). Synergism with phenylamides seems therefore of general significance and might be useful for practice. The synergistic mechanism of action is still speculative (GISI *et al.* 1985).

Concluding remarks

Optimization of the field performance of fungicide compositions implies minimization of antagonistic and exploitation of synergistic interactions. Both types of interactions are described in this chapter. Formulating agents also improve field performance but are not regarded as synergists. In fact, true fungicide synergists are not yet commercially available, in contrast to insecticide synergists. The action of the latter chemicals is often based on inhibition of detoxication reactions which cause in resistance to insecticides (Wilkinson 1976). Detoxication only plays a minor role in fungicide toxicity and resistance. This may reflect the differential importance of fungicide and insecticide synergists. This situation may change in future since synergistic interactions with fungicides have recently been demonstrated for various experimental compounds and existing commercial fungicides. This is often limited to in vitro bioassays or tests under well-controlled conditions, which is of course no guarantee for in vivo activity. Fungicide and synergist may, for instance, have a different residual activity or may become spatially separated in the plant by systemic activity. These factors might explain the lack of synergism between fenarimol and sodium orthovanadate against *P. italicum* on oranges (de Waard and van Nistelrooy 1984b). The ideal combination would be a systemic fungicide and a systemic synergist, a goal probably difficult to realize. Other problems which may be involved in the development of fungicide synergists are: (1) difficulty in finding safe compounds, (2) the costs of potential synergists in relation to the price of the fungicide, (3) the ratio of synergist and fungicide necessary in the combination, (4) the simultaneous formulation of synergist and fungicide and (5) variation in the spectrum of antifungal activity. Despite these difficulties it is believed that synergistic mixtures of fungicides may be developed in future. Fundamental studies on fungicide action and resistance as described for PLTs and SBIs may provide rational leads. First of all synergism in mixtures of already labelled fungicides should be explored. In the case of SBIs, such mixtures might extend their antifungal spectrum to pathogens like *B. cinerea* or *P. oryzae* and counteract development of resistance as has been observed in *Sphaerotheca fuliginea* (Schepers 1983, 1985; Huggenberger *et al.* 1984). In other words, synergistic mixtures might be a tool to cope with natural and acquired resistance to SBI fungicides (de Waard 1984). A similar suggestion was made for phenylamide fungicides (Grabski and Gisi 1985). Development of mixtures with newly developed synergists may not always be feasible but the possibility should not be overlooked.

References

Baicu, T., and Nägler, M.: Die Wirkung der Fungizidmischung Carboxin und Mancozeb bei der Bekämpfung einiger Weizen- und Gerstenkrankheiten. Arch. Phytopathol. und Pflanzensch., Berlin **10** (1974): 395—404.

Bent, K. J.: Fungitoxin action of dimethirimol and ethirimol. Ann. appl. Biol. **66** (1970): 103 to 113.

Berger, R. D., and Wolf, E. A.: Control of seedborne and soilborne mycoses of 'Florida Sweet' corn by seed treatment. Plant Dis. Rep. **58** (1974): 922—923.

Bergmann, H.: Wirkung von Tridemorph auf *Torulopsis candida* Berlese. Z. Allg. Mikrobiol. **19** (1979): 155—162.

Bliss, C. I.: The toxicity of poisons applied jointly. Ann. appl. Biol. **26** (1939): 585—615.

Bonifas, V.: Détermination de l'association synergique binaire d'antibiotes et de sulfamides. Experienta **8** (1952): 234—235.

Colby, S. R.: Calculating synergistic and antagonistic responses of herbicide combinations. Weeds **15** (1967): 20—22.

CORBAZ, R.: Recherches en laboratoire concernant l'association binaire de fongicides: synergie et antagonisme. Phytopathol. Z. **48** (1963): 337—347.

DEKKER, J.: Antibiotics. In: TORGESON, D. C. (Ed.): Fungicides, an Advanced Treatise. Academic Press, New York 1969.

— and OORT, A. J. P.: Mode of action of 6-azauracil against powdery mildew. Phytopathology **54** (1964): 815—818.

DE WAARD, M. A.: Negatively correlated cross-resistance and synergism as strategies in coping with fungicide resistance. 1984 Brit. Crop Protect. Conf. — Pests and Diseases **2** (1984): 573—584.

— Fungicide synergism and antagonism. 1985 Fungicides for Crop Protection. BCPC monograph **31** (1985): 89—95.

— and VAN NISTELROOY, J. G. M.: Mechanism of resistance to fenarimol in *Aspergillus nidulans*. Pesticide Biochem. Physiol. **10** (1979): 219—229.

— — An energy-dependent efflux mechanism for fenarimol in a wild-type strain and fenarimol-resistant mutants of *Aspergillus nidulans*. Pesticide Biochem. Physiol. **13** (1980): 255—266.

— — Induction of fenarimol-efflux activity in *Aspergillus nidulans* by fungicides inhibiting sterol biosynthesis. J. gen. Microbiol. **126** (1981): 483—489.

— — Antagonistic and synergistic activities of various chemicals on the toxicity of fenarimol to *Aspergillus nidulans*. Pesticide Sci. **13** (1982): 279—286.

— — Effects of phthalimide fungicides on the accumulation of fenarimol by *Aspergillus nidulans*. Pesticide Sci. **15** (1984a): 56—62.

— — Differential accumulation of fenarimol by a wild-type isolate and fenarimol-resistant isolates of *Penicillium italicum*. Neth. J. Pl. Path. **90** (1984b): 143—153.

DIENER, U. L., and CARLTON, C. C.: Dodine-captan combination controls bacterial spot of peach. Plant Dis. Rep. **44** (1960): 136—138.

DIMOND, A. E., and HORSFALL, G.: Synergism as a tool in the conservation of fungicides. Phytopathology **34** (1944): 136—139.

EISA, N. A., and BARAKAT, F. M.: Relative efficiency of fungicides in the control of damping-off and *Stemphilium* leafspot of broad bean ó*Vicia fabaá* **62** (1978): 114—118.

FINNEY, D. J.: Probit Analysis. A statistical treatment of the sigmoid reponse curve. Cambridge Univ. Press, Cambridge 1947.

FLIEG, O., and POMMER, E.-H.: Polyram Combi. Die Landwirtschaftliche Versuchsstation Limburgerhof 1914—1964, BASF AG, Ludwigshafen 1965, S. 319—328.

FUSHTEY, S. G.: The nature and control of snow mold of fine turfgrass in Southern Ontario. Can. Plant Dis. Survey **55** (1975): 87—96.

GEORGOPOULOS, S. G., ALEXANDRI, E., and CHRYSAYI, M.: Genetic evidence for the action of oxathiin and thiazole derivatives on the succinic dehydrogenase system of *Ustilago maydis* mitochondria. J. Bacteriol. **110** (1972): 809—817.

GISI, U., BINDER, H., and RIMBACH, E.: Synergistic interactions of fungicides with different modes of action. Trans. Br. mycol. Soc. **85** (1985): 299—306.

— HARR, J., SANDMEYER, R., and WIEDMER, H.: A new systemic oxazolidone fungicide (SAN 371 F) against diseases caused by *Peronosporales*. Meded. Fac. Landbouww. Rijksuniv. Gent **48** (1983): 541—549.

GOKSØYR, J.: The effect of some dithiocarbamyl compounds on the metabolism of fungi. Physiologia Plantarum **8** (1955): 719—827.

GOSS, V., and MARSHALL, W. D.: Synergistic interactions of zinc or copper with anilazine. Pesticide Sci. **16** (1985): 163—171.

GOWING, D. P.: Comments on tests of herbicide mixtures. Weeds **8** (1960): 379—391.

GRABSKI, C., and GISI, U.: Mixtures of fungicides with synergistic interactions for protection against phenylamide resistance in *Phytophthora*. 1985 Fungicides for Crop Protection BCPC monograph **31** (1985): 315—318.

HAMPEL, H., and LARTAUD, G.: The control of important cereal diseases using fenpropimorph mixtures. Proc. 10th Intern. Congr. Plant Protect., Brighton, England **3** (1983): 926.

HOLLOMON, D. W.: Antifungal activity of substituted 2-aminopyrimidines. In: TRINCI, A. P. J., and RYLEY, J. F. (Eds.): Mode of Action of Antifungal Agents. Cambridge Univ. Press, Cambridge 1984.

HORSFALL, J. G.: Synergism and antagonism. Plant Disease Rep. **157** (1945): 162—166.

HUGGENBERGER, F., COLLINS, M. A., and SKYLAKAKIS, G.: Decreased sensitivity of *Sphaerotheca fuliginea* to fenarimol and other ergosterol-biosynthesis inhibitors. Crop Protect. **3** (1984): 137—149.

KATAGIRI, M., and UESUGI, Y.: Similarities between the fungicidal action of isoprothiolane and organophosphorus thiolate fungicides. Phytopathology **67** (1977): 1415—1417.

KERKENAAR, A., and KAARS SIJPESTEIJN, A.: On a difference in the antifungal activity of tridemorph and its formulated product Calixin. Pesticide Biochem. Physiol. **12** (1979): 124—129.

KERRIDGE, D., and WHELAN, W. L.: The polyene macrolide antibiotics and 5-fluorocytosine: molecular action and interaction. In: TRINCI, A. P. J., and RYLEY, J. F. (Eds.): Mode of Action of Antifungal Agents. Cambridge Univ. Press, Cambridge 1984.

LEROUX, P., and GREDT, M.: Etude de l'action antagoniste d'acides gras, de stérols et de divers dérivés isopréniques vis à vis de quelques fongicides. Phytopath. Z. **91** (1978): 177—181.

— — MASSENOT, F., and KATO, T.: Activite du phenylcarbamate: S32165 sur *Botrytis cinerea*, agent de pourriture grise de la vigne. 1985 Fungicides for Crop Protection BCPC monograph **31** (1985): 443—446.

LUKENS, R. J., and SISLER, H. D.: Chemical reactions involved in the fungitoxicity of captan. Phytopathology **48** (1958): 235—244.

LYR, H., CASPERSON, G., and LAUSSMANN, B.: Wirkungsmechanismus von Terrazol bei *Mucor mucedo*. Z. Allg. Mikrobiol. **17** (1977): 117—129.

MATOLCSY, G., HAMRÁN, M., and BORDÁS, B.: Increased antifungal action of zinc dimethyldithiocarbamate in the presence of complex forming compounds. Pesticide Sci. **2** (1971): 229—231.

MORSE, P. M.: Some comments on the assessment of joint action in herbicide mixtures. Weed Sci. **26** (1978): 58—71.

NASH, R. G.: Phytotoxic interaction studies — Techniques for evaluation and presentation of results. Weed Sci. **29** (1981): 147—155.

O'BRIEN, R. D.: Insecticides, Action and Metabolism. Academic Press, New York 1967.

PAPAVIZAS, G. S., and LEWIS, J. A.: Effects of cottonseed treatment with systemic fungicides on seedling disease. Plant Dis. Rep. **61** (1977): 538—542.

PEZET, P., and PONT, V.: Sensitivity of *Botrytis cinerea* Pers. to elemental sulfur in the presence of surfactants. Experienta **40** (1984): 354—356.

POWELL, D.: The inhibitory effects of certain fungicide formulations to apple scab conidia. Plant Dis. Rep. **44** (1960): 176—178.

RADZUHN, B., and LYR, H.: On the mode of action of the fungicide etridiazole. Pesticide Biochem. Physiol. **22** (1984): 14—23.

RAGSDALE, N. N., and SISLER, H. D.: Metabolic effects related to fungitoxicity of carboxin. Phytopathology **60** (1970): 1422—1427.

RAHIMIAN, M. K., and BANIHASHEMI, Z.: Synergistic effects of ethazole and pentachloronitrobenzene on inhibition of growth and reproduction of *Pythium aphanidermatum*. Plant Disease **66** (1982): 26—27.

RICHARDSON, L. T.: Synergism between chloroneb and thiram applied to peas to control seed rot and damping-off by *Pythium ultimum*. Plant Dis. Rep. **57** (1973): 3—6.

RITTER, G., KLUGE, E., and LYR, H.: Beziehungen zwischen Carboxin-Resistenz und glykolytischer Potenz bei Pilzen. Z. Allg. Mikrobiol. **13** (1973): 243—250.

SAMOUCHA, Y., and COHEN, Y.: Synergy between metalaxyl and mancozeb in controlling downy mildew in cucumbers. Phytopathology **74** (1984): 1434—1439.

SCARDAVI, A.: Synergism among fungicides. Ann. Review Phytopathol. **4** (1966): 335—348.

SCHEPERS, H. T. A. M.: Decreased sensitivity of *Sphaerotheca fuliginea* to fungicides which inhibit ergosterol biosynthesis. Neth. J. Pl. Path. **89** (1983): 185—187.

— Changes during a three-year period in the sensitivity to ergosterol biosynthesis inhibitors of *Sphaerotheca fuliginea* in the Netherlands. Neth. J. Plant Path. **91** (1985): 36—49.

SHERALD, J. L., RAGSDALE, N. N., and SISLER, H. D.: Similarities between the systemic fungicides triforine and triarimol. Pesticide Sci. **4** (1973): 719—727.

SISLER, H. D., and RAGSDALE, N. N.: Biochemical and cellular aspects of the antifungal action of ergosterol biosynthesis inhibitors. In: TRINCI, A. P. J., and RYLEY, J. F. (Eds.): Mode of Action of Antifungal Agents. Cambridge Univ. Press, Cambridge 1984.

TAMMES, P. M. L.: Isoboles, a graphic representation of synergism in pesticides. Neth. J. Plant Path. **70** (1974): 73—80.

THIOLLIERE: Progres apportes par cuprosan — Première association synergique organo — cuprique. 1985 Fungicides for Crop Protection. BCPC monograph **31** (1985): 227—230.

TORGESON, D. C. (Ed.): Fungicides, an Advanced Treatise, vol. 2. Academic Press, New York 1967.
TRINCI, A. P. J.: Antifungal agents which effect hyphal extension and hyphal branching. In: TRINCI, A. P. J., and RYLEY, J. F. (Eds.): Mode of Action of Antifungal Agents. Cambridge Univ. Press, Cambridge 1984.
UESUGI, Y., and SISLER, H. D.: Metabolism of a phosphoramidate by *Pyricularia oryzae* in relation to tolerance and synergism by a phosphorothiolate and isoprothiolane. Pesticide Biochem. Physiol. **9** (1978): 247—254.
VINCENT, P. G., and SISLER, H. D.: Mechanism of antifungal action of 2,4,5,6-tetrachloroisophthalonitril. Physiologia Plantarum **21** (1968): 1249—1264.
WADLEY, F. M.: The evidence required to show synergistic action of insecticides and a short cut in analyses. ET-223, U.S. Department of Agriculture, 1945.
— Experimental Statistics in Entomology. Graduate School Press, U.S.D.A., Washington 1967.
WAKAE, O., and MATSUURA, K.: Characteristics of validamycin as a fungicide for *Rhizoctonia* disease control. Review of Plant Protect. Res. **8** (1975): 81—92.
WILKINSON, C. F.: Insecticide synergism. In: METCALF, R. L., and MCKELVEY, J. J. (Eds.): The Future for Insecticides. John Wiley & Sons, New York 1976.

LYR, H. (Ed.): Modern Selective Fungicides — Properties, Applications, Mechanisms of Action. Longman Group UK Ltd., London, and VEB Gustav Fischer Verlag, Jena, 1987.

Chapter 25

Outlook

H. Lyr
Institute for Plant Protection Research Kleinmachnow of the Academy of Agricultural Sciences of the GDR

Introduction

During the past two decades the introduction of a number of new fungicides has led to marked improvements in the control of fungal diseases. This has contributed to a world wide increase of crop yields. Many groups of fungicides described in this book are systemic in the plant, and thus exhibit internal therapeutic effects against invading fungi. Of course the question arises as to whether the needs of agriculture, horticulture and silviculture have been adequately fulfilled. Surely, this is not yet the case and further progress is needed.

New toxophores

Although numerous compounds already exist with fungicidal or antifungal activity, compounds with new mechanisms of action must be developed. A driving force is the appearance of resistance in fungal populations which requires a rotation of fungicides that are not subject to the cross resistance patterns of the fungal pathogen.

Additionally, there exist several problems (e.g. vascular mycoses, storage diseases, certain soil pathogens) which are not yet adequately controlled.

It is true, that most fungicides have been found by a more or less empirical screening of chemicals. But costs for advanced screening and new syntheses are increasing rapidly and the probability of finding desirable new toxophores is decreasing (Ryley and Rathmell 1984). On the other hand we know relatively little about the biochemical or molecular targets of the new compounds or about the differential biochemistry of fungi and plants. However, as new groups of herbicides have demonstrated (e.g. chlorosulfuron, glyphosate) new targets still exist in the plant which are very different from the classic targets in the photosynthetic apparatus. These permit the development of highly effective compounds with new and surprising properties. The same is almost certainly true for fungi. The present known points of attack of modern fungicides are very few and therefore new ones can certainly be detected. Analogous syntheses including those of natural compounds, can optimize or modify properties of new fungicides, but very often such compounds share undesirable properties, for example, susceptibility to the same types of fungal resistance. The relative success of screening programs in the past did not stimulate alternative programs (Geissbühler, 1984).

A biorational design of new compounds is an attractive alternative to conventional screening, although it has not contributed much thus far to the detection of new toxophores. These limitations of the biorational approach are due mainly to superficial and poor biochemical analyses of target organisms and the lack of detailed

descriptions of essential target structures at the molecular level (SCHWINN and GEISSBÜHLER, 1986).
Although this book includes chapters on the mechanism of action of bioactive compounds that demonstrate much progress of our understanding in this areas, many important or decisive details, even about well known fungicides, are still too obscure to permit construction of new molecules. The exact description of target enzymes (or structures) and their interaction with antifungal compounds needs to be clarified. As chapter 22 demonstrates, computer application can be very useful, but the prerequisites for such models are still very limited. The rapid progress in biochemistry in combination with genetic engineering and the broader use of computer programs may produce a new situation. Directed biochemical investigations on relevant targets in fungi (LYR 1979; BALDWIN 1984) and a deeper understanding of the interaction of host and parasites are needed. In conclusion, at present no fundamental factor is apparent which would prohibit the development of new toxophores with valuable properties.

New or better quality of fungicides

Although we usually speak of "systemic fungicides", it should be kept in mind that only a xylem transport of fungicides with the water stream is realized in most cases (chapter 1). Of course even this type of systemic activity was a great advance in the development of fungicides. But this property is only rarely taken advantage of in practice where leaf application dominates. This property is of importance only in seed dressing or root drenches. The most significant is the tolerance of systemic fungicides by the plant cells, a factor which permits intratherapeutic action against pathogens. A highly desirable feature would be a phloem transport that would allow protection of roots, bulbs, new sprouts and leaves by leaf application as well as a redistribution of active compounds in the leaf system of a plant, or protection of the bark of a tree. It is not by chance that this property is rare among active compounds because it has as prerequisites high antifungal activity, special chemical structures and low phytotoxicity. The mass flow through the phloem is quantitatively much inferior to that of the water flow through the xylem and, moreover, entrance into the phloem and exit therefrom is usually an active process that is quite specific. Phloem transport is well documented among fungicides only for fosethyl aluminium (Aliette) and Dowco 296 (pyroxychlor). However, many herbicides and some insecticides demonstrate, that this principle can be achieved by various chemical structures. These compounds have new, valuable features for practical applications.

At present there is need for new compounds to control several important diseases which are not controlled by fungicides presently available (e.g. vascular mycoses, some soil borne diseases, persistent structures, tree and storage diseases).

Of course a minimum risk for resistance development is a highly desired feature of new compounds which have selective systemic properties and low phytotoxicity. Low mammalian toxicity and moderate prices are especially important. New and better formulations and advanced application techniques can contribute to higher quality of control but the impact of these should not be overestimated. For seed dressings, compounds with improved performance are very welcome.

Indirectly acting fungicides

As described in chapter 23, several compounds that are not directly fungitoxic prevent fungal diseases of plants by specific, although different mechanisms. Such substances can be detected only in host parasite combinations in a screening program, but not *in vitro* tests. Whether fungicides which act indirectly have advantages compared with normal fungicides, remains on open question. Compounds in this category have proven to be highly effective for the control of rice blast disease caused by *Pyricularia*. With these compounds it is not yet known whether resistance is unlikely to occur, or whether fungi are able to adapt to the new situation as in resistance to breeding. The specificity for fungi or host/parasite combinations is rather high. It may be that this principle is already in operation with some well known fungicides (phenylamides, benzimidazoles). Of course, such combined properties can contribute to a multisite mechanism of action that possibly can lower the risk of resistance development. Therefore new substances acting by this mechanism alone or combined with a fungitoxic mechanism are welcome.

The practical significance of induced resistance by a systemically transportable (still unknown) principle (Kuć 1985; Salt and Kuć 1980; Kuć and Tuzun 1983) remains an open question. According to Rathmell (1983) the resistance seems to be a rather unspecific response to a sublethal effect on a plant. The practical value of such responses for fungal disease control is still uncertain and the phenomenon may be limited to certain host/parasite combinations.

If special physiological conditions are required to express such an effect, it would be hard to manage. It seems that this principle is working in nature (at subpathological infections), although it does not appear to be strong enough to prevent the outbreak of epidemic diseases where applications of fungicides are needed.

Synergism and antagonism in fungicides

These phenomena (chapter 24) need further attention from a theoretical as well as from a practical point of view. The directed use of synergists can be very favourable for increasing antifungal spectrum or decreasing the danger of resistance formation in the fungal population. Synergistic interactions with simultaneously applied herbicides and insecticides should not be neglected (Yegen and Heitefuss 1970; Heitefuss 1970; Heitefuss 1972), as well as possible antagonistic interactions, which diminish the fungicidal effect (Kataria and Dodan 1983).

Biological control of plant diseases

Due to some drawbacks of chemical control of plant diseases and pests, the possibilities for biological control have been widely discussed and tested. Biological control mentioned here means the inoculation or application of antagonistic fungi or bacteria. Fungi of the genus *Trichoderma* have long been used for biological control experiments, more or less successfully. Of course other fungal organism such as *Peniophora gigantea* (against *Fomes annosus*) or bacteria such as *Pseudomonas* spp. or sporulating genera *(Bacillus)* have also been used. It seems that there are possibilities for successful applications especially against soil borne pathogens (dressing, inoculation of seeds, combinations with soil disinfectants). Whether vascular diseases such as Dutch elm disease can be controlled by *Pseudomonas* spp. (Strobel and Lanier 1981) or other bacteria remains to be observed. A disadvantage of working with living organisms is their dependence on the actual ecological conditions which are hard to predict or to control. Therefore, both positive and negative results have been described. The directed genetic manipulation of certain isolates can perhaps increase the effect of biological control measures, but due to their inherent limitations they can solve only a very small part of the overall problem.

Breeding of resistant plant varieties

We can look back on several decades of intensive efforts to select cultivars resistant to fungal diseases. In spite of many successes in the past, a breakthrough of new aggressive strains or new pathogens could not be avoided. The successes did not lead to such a stabilization of plant production that the application of chemicals could be eliminated. Recently there is an increased optimism for solving the problem of fungal disease resistance with the help of genetic engineering methods and new selection methods using cell cultures. But at present, the genetic base of resistance in plants is very obscure and even biochemically only scarcely understood. If one considers that each pathogen-host-interaction probably has different causes of resistance and eventually also various plant species or cultivars differ in this respect, the difficulties for the gene technique to construct R-genes and to insert them with high stability into a productive plant variety are evident. Therefore very quick progress can not be expected (Chilton 1984). Regarding horizontal resistance, breeders can be confronted with toxicological problems (accumulation of defense substances such a alkaloids, glycosides, phytoalexins and other secondary plant products which may even have carcinogenic or estrogenic properties) that require residue analysis and toxicological evaluations equal to those of pesticides.

Because many other problems such as high productivity, stress tolerance, product qualities, special demand from a technological point of view must be realized by breeding programs (Spaar and Lyr 1984) application of fungicides (and other pesticides and growth regulating substances) will be required to stabilize plant production for a long time. The possibility of breeding varieties resistant to toxigenic, common molds such as *Fusarium*, *Penicillium*, *Aspergillus* and other generea which are known as producers of highly toxic mycotoxins is nearly on unsolvable problem (Lyr 1985). The long life span of trees make breeding for resistance very ineffective because of the high adaptibility of fungal species. The main line of progress will probably be the selection of more tolerant varieties with a broad genetic variability (Bremermann 1983).

Conclusions

The system of fungal diseases control in the future will be more sophisticated and contain more options than those presently available. In some cases, epidemiological information may be used to guide biological/chemical control strategies in good agricultural practise, or integrated systems of plant protection.

There will still be a requirement for new compounds with new mechanisms of action, low succeptibility to resistance, low mammalian toxicity and some new features (for example phloem mobility). These compounds will contribute to a still higher efficiency of fungal disease control.

New toxophores will be developed either by recent screening methods or by an advanced biorational design. In any case, advances will depend very much on a deeper understanding of the biology and biochemistry of at least the main diseases and the main crop plants and their varieties. Efforts of plant protection science in cooperation with those of other disciplines of science all over the world should find solutions to many of the unsolved problems of plant protection in the coming decades.

References

Baldwin, B. C.: Potential targets for the selective inhibition of fungal growth. In: Trinci, A. P. J., and Ryley, J. F. (Eds.): Mode of Action of Antifungal Agents. Brit. Mykol. Soc. (1984): 43—62.

Bremermann, H. J.: Theory of catastrophic diseases of cultivated plants. J. theor. Biol. **100** (1983): 255—274.

Chilton, Mary-Dell: Genetic engineering-prospects for use in crop management. Brit. Crop Protect. Conf. (1984): 1 A, 3—9.

Geissbühler, H.: Biorational reflections in agricultural chemical research. Chimia **38** (1984): 307—316.

Heitefuss, R.: Nebenwirkungen von Herbiziden auf Pflanzenkrankheiten und deren Erreger. Z. Pflanzenkrankh. u. Pflanzenschutz, Sonderheft V (1970): 117—127.

— Ursachen der Nebenwirkungen von Herbiziden auf Pflanzenkrankheiten. Z. Pflanzenkrankh. u. Pflanzenschutz, Sonderheft VI (1972): 80—87.

Kataria, H. R., and Dodan, D. S.: Impact of two soil-applied herbicides on damping-off of cowpea caused by *Rhizoctonia solani*. Plant and Soil **73** (1983): 275—283.

Kuć, J., and Tuzun, S.: Immunization for disease resistance in tobacco. Recent Adv. in Tobacco Sci. **9** (1983): 174—213.

— Induced systemic resistance to plant disease and phytointerferons — are they compatible? Fitopathologia Brasilia (1985): 17—40.

Lyr, H.: Differentialmerkmale zwischen Pilzen und höheren Pflanzen als Basis für eine selektive Wirkung systemischer Fungizide. In: Lyr, H., and Polter, C. (Eds.): Systemic Fungicides. Abh. Akad. Wiss. DDR, 2 N. Akademie-Verlag, Berlin 1979, pp. 7—13.

— Mykotoxine — eine vermeidbare Gefahr. Biol. Rdsch. **23** (1985): 285—293.

Rathmell, W. G.: The discovery of new methods of chemical disease control: current developments, future prospects and the role of biochemical and physiological research. Advanc. in Plant Pathol. **2** (1983): 259—288.

Ryley, J. F., and Rathmell, W. G.: Discovery of antifungal agents: *in vitro* and *in vivo* testing. In: Trinci, A. P. J., and Ryley, J. F. (Eds.): Mode of Action of Antifungal Agents. Brit. Mykol. Soc. (1984): 63—87.

Salt, St. D., and Kuć, J.: Elicitation of disease resistance in plants by the expression of latent genetic information. In: Hedin, P. A. (Ed.): Bioregulators for Pest Control. ACS Symposium, Series **276** (1985): 47—68.

Schwinn, F., and H. Geissbühler: Towards a more rational approach to fungicide design. Crop protection (in press) (1986).

Spaar, D., and Lyr, H.: The use of fungicides and resistant crop varieties as basis of modern plant protection strategies. In: Lyr, H., and Polter, C. (Eds.): Systemic Fungicides and Antifungal Compounds. Tagungsber. Akad. Landwirtschaftswiss. DDR, **222** (1984): 7—16.

Strobel, G. A., and Lanier, G. N.: Dutch elm disease. Scientific American **245** (1981): 56—66.

Yegen, O., and Heitefuss, R.: Nebenwirkungen von Natriumtrichloracetat (NaTA) auf den Wurzelbrand der Rüben und das antiphytopathogene Potential des Bodens. Zucker **23** (1970): 694—700; 723—729.

Lyr, H. (Ed.): Modern. Selective Fungicides — Properties, Applications, Mechanisms of Action. Longman Group UK Ltd., London, and VEB Gustav Fischer Verlag, Jena, 1987.

Subject index